Inspiring Environmental Stewardship Through Sustainable Leadership

Qaisar Iqbal
Al Yamamah University, Saudia Arabia

IGI Global
Scientific Publishing
Publishing Tomorrow's Research Today

Published in the United States of America by
IGI Global Scientific Publishing
701 East Chocolate Avenue
Hershey, PA, 17033, USA
Tel: 717-533-8845 | Fax: 717-533-7115
Website: https://www.igi-global.com E-mail: cust@igi-global.com

Library of Congress Cataloging-in-Publication Data

Names: Iqbal, Qaisar, 1986- editor
Title: Inspiring environmental stewardship through sustainable leadership / Edited by Qaisar Iqbal.
Description: Hershey, PA : IGI Global Scientific Publishing, [2027] | Includes bibliographical references and index. | Summary: "This book bridges the conceptual gap between leadership theory and practical environmental action. While much has been written on both sustainability and leadership, this book provides a cohesive framework that specifically links the qualities and behaviors of leaders to the cultivation of stewardship in others"-- Provided by publisher.
Identifiers: LCCN 2026003342 (print) | LCCN 2026003343 (ebook) | ISBN 9798337378770 hardcover | ISBN 9798337378787 paperback | ISBN 9798337378794 ebook
Subjects: LCSH: Environmental management | Environmental protection--Citizen participation | Sustainable development--Management | Leadership--Philosophy
Classification: LCC GE300 .I565 2026 (print) | LCC GE300 (ebook)
LC record available at https://lccn.loc.gov/2026003342
LC ebook record available at https://lccn.loc.gov/2026003343

British Cataloguing in Publication Data
A Cataloguing in Publication record for this book is available from the British Library.

Table of Contents

Detailed Table of Contents

Sustainable leadership incorporates ethical decision making, environmentalism and long-term value generation in the company strategy. Since sustainability is an international requirement, leaders should take their solutions beyond the greenwashing to quantifiable and transparent initiatives. This chapter looks at sustainable leadership-its dimensions, the comparison with the traditional models as well as how it is played out in actual organizational change. Based on the environmental management, organizational behavior, and ethics, it emphasizes the need to balance between the economic performance, social equity, and ecological preservation. Sustainable leaders facilitate responsibility, creativity, and involvement of stakeholders, which entails the aspect of sustainability in decision making. These major features are systems thinking, ethical stewardship, vision and emotional intelligence. It also gives approaches to evaluate the authenticity of leadership and states that the real sustainability is based on the morality and strategic vision rather than on marketing.

In response to the challenge of corporate greenwashing, this chapter advances a governance-centered redefinition of sustainable leadership. Sustainable leadership scholars emphasize ethical behavior, visionary thinking, and stakeholder engagement; however, such perspectives inadequately address the structural conditions that may enable deceptive environmental claims. By drawing on agency theory, sustainability governance research, and global reporting standards, this chapter proposes a novel framework for sustainable leadership. The novel framework integrates rigorous governance to eliminate corporate greenwashing and approaches greenwashing as a governance failure, rather than a communication deficit. This chapter introduces the concept of a sustainable leader as a governance architect, whose core responsibility is the design and implementation of institutional systems that can ensure transparency, accountability, and verifiable sustainability.

The phenomenon of greenwashing indicates that sustainability initiatives are often reduced to image-building efforts without genuine moral commitment. Contemporary environmental ethics—both anthropocentric and ecocentric—have proven insufficient in preventing information manipulation, as they fail to provide leaders with a profound ethical foundation. This chapter proposes theocentric sustainable leadership as a conceptual alternative rooted in spiritual values, particularly the principles of amanah (trust) and the human role as khalifah (steward) within Islamic tradition. A theocentric approach frames sustainability as a moral–transcendental responsibility, thereby strengthening integrity, transparency, and accountability in decision-making. By integrating spiritual values into organizational vision, governance, and culture, this model offers a more robust ethical framework to counter greenwashing and foster authentic sustainability for future generations.

As businesses deal with escalating environmental issues and sustainability demands, sustainable leadership has grown in significance. There is still little empirical data connecting sustainable leadership to environmental stewardship, despite the fact that leadership and sustainability have been discussed extensively. The purpose of this study is to investigate how sustainable leadership contributes to the advancement of ecologically conscious business practices. Utilising survey information gathered from managerial staff, a quantitative research methodology was used. The results

introduces the concept of representational readiness, emphasizing that leadership is central to ensuring employees embody organizational sustainability authentically. The findings reveal that booths serve as visible markers of environmental commitment, yet many representatives appeared uncertain, hesitant, or unprepared when discussing their company's practices. The findings also exposed that employees' participation in events appeared more driven by organizational requirements or stakeholder expectations than by direct involvement in sustainability initiatives.

Climate change has emerged as a systemic source of economic and financial risk, yet existing climate risk assessments remain fragmented, often relying on descriptive indices or opaque modelling approaches that limit interpretability and policy relevance. This study develops a transparent and explainable artificial intelligence (AI) enabled framework to classify cross-country climate risk exposure while explicitly incorporating economic capacity, demographic scale, and systemic climate–financial spillover potential. Using publicly available Climate Risk Index data for 2022, alongside population, income, and policy indicators for a diverse sample of countries, the analysis constructs interpretable AI-based climate risk regimes through rule-based classification.

The chapter explores leadership for sustainability transformations in U.S. higher education with a focus on senior international officers (SIOs) and environmental sustainability in education abroad. Student mobility for cross-border learning has been increasing, often without accounting of its environmental impacts. This research surveyed U.S. SIOs and found that many had the personal conviction and competencies for leading sustainable internationalization, but few had promoted institution-wide strategies addressing environmental impacts of education abroad. Many believed their institutions had the will to advance environmental sustainability, but few saw institutional action to reconcile the international agenda with sustainability goals. The chapter concludes by suggesting that U.S. SIOs seek training, communicate with sustainability experts, and assess their institutions' readiness for sustainable practices in education abroad. Furthermore, by engaging in sustainability-oriented

the importance of sustainability competencies and environmental stewardship in leadership development.

Whether stewardship leaders are products of educational design or biographical accident has never been straightforward, though literature increasingly favours the former. This chapter takes that view seriously. Drawing on peer-reviewed scholarship from 2021 to 2026, it examines how formal schooling, transformative pedagogy, outdoor and nature-based learning, service learning, indigenous knowledge integration, and digital modalities contribute—independently or in combination—to the knowledge, values, and competencies required for environmental stewardship leadership. The analysis is framed by Education for Sustainable Development and transformative learning theory, proposing a five-stage model mapping development from ecological awareness to sustained leadership. A comparative table and effectiveness graph support the discussion. The chapter concludes with recommendations for educators, curriculum designers, and researchers.

In traditional learning environments, there is less interaction between learners and the surrounding environment. Considering employers' needs and graduates' adaptive profile that meets the labor market, transformation of the learning environment should be done. The shift in pedagogy from the traditional learning approach to project-based learning (PrBL) is more suitable for new human capital demands of industries and workplaces, as well as sustainable development specialists. By participation in PrBL, both educator and student can gain information, skills, and attitudes for a sustainable environment in the community; beside developing other soft skills such as critical thinking, problem-solving skills, teamwork, adaptability, promoting opportunities to practice and solve real contextual problems. The aim of this chapter is to provide the leader with a better understanding on PrBL, integration of PrBL on sustainability and its impacts.

Preface

INTRODUCTION

The accelerating pace of environmental change has made sustainability one of the defining challenges of the twenty-first century. Climate change, ecosystem degradation, biodiversity loss, and increasing pressure on natural resources have reshaped the way societies, governments, industries, and communities must think about development and progress. While advances in technology, scientific research, and environmental policy continue to provide valuable solutions, meaningful and lasting transformation ultimately depends upon leadership. It is through visionary, ethical, and purpose-driven leadership that organizations and individuals are inspired to move beyond compliance toward genuine environmental stewardship.

Inspiring Environmental Stewardship Through Sustainable Leadership has been developed to explore this critical intersection between leadership and sustainability. The chapters assembled in this edited volume demonstrate that environmental stewardship is not merely a collection of sustainable practices but a philosophy that must be cultivated through values, organizational culture, education, collaboration, and long-term strategic thinking. Sustainable leadership extends beyond short-term objectives or isolated environmental initiatives; it requires leaders who recognize the interconnectedness of economic prosperity, environmental protection, and social well-being while inspiring others to embrace shared responsibility for the future of our planet.

One of the distinguishing features of this volume is its multidisciplinary perspective. Drawing upon contributions from scholars and practitioners across diverse fields, the book bridges environmental science, leadership studies, organizational behavior, public policy, education, ethics, and sustainability management. By integrating these perspectives, the volume presents readers with a comprehensive understanding of how sustainable leadership can be translated into practical action across a wide range of organizational and societal contexts.

The chapters collectively examine both theoretical foundations and real-world applications of sustainable leadership. Readers will encounter discussions of leadership models that foster environmental responsibility, governance frameworks that support sustainable decision-making, organizational strategies for embedding sustainability into institutional culture, and case studies that illustrate successful stewardship initiatives across corporate, governmental, educational, and community settings. Throughout the volume, contributors emphasize that effective environmental stewardship emerges not solely from regulations or technological innovation but from the ability of leaders to cultivate intrinsic motivation, encourage collaboration, and empower stakeholders to become active participants in sustainable development.

Another important contribution of this book lies in its emphasis on practical implementation. While sustainability has become an increasingly prominent area of research, many organizations continue to struggle with translating conceptual frameworks into meaningful action. The authors provide actionable insights, evidence-based strategies, and transferable leadership principles that can assist decision-makers in navigating complex environmental challenges. These practical perspectives offer valuable guidance for leaders seeking to build resilient organizations capable of balancing environmental responsibility with organizational performance and long-term societal impact.

As editors, we believe this volume will serve a diverse readership. Researchers and graduate students will find a rich collection of contemporary scholarship that advances understanding of sustainable leadership and environmental stewardship. Corporate executives, sustainability professionals, and organizational leaders will discover practical approaches for integrating environmental responsibility into strategic planning and organizational culture. Policymakers and public administrators will gain insights into leadership approaches that strengthen environmental governance and public engagement, while educators, consultants, and nonprofit leaders will benefit from frameworks that support teaching, advocacy, and community-based sustainability initiatives.

The contributions presented in this volume collectively reinforce an important message: environmental stewardship is ultimately a human endeavor. Technologies may enable progress, policies may establish direction, and scientific discoveries may deepen our understanding, but sustainable transformation depends upon leaders who inspire vision, cultivate responsibility, and empower others to act. By fostering cultures of stewardship rather than mere compliance, sustainable leadership has the capacity to create enduring environmental, social, and economic value for present and future generations.

It is our sincere hope that *Inspiring Environmental Stewardship Through Sustainable Leadership* will encourage thoughtful dialogue, stimulate innovative research, and support meaningful action among scholars, practitioners, educators, policymakers, and organizational leaders alike. More importantly, we hope this volume inspires readers to recognize that leadership is not simply the management of organizations but the stewardship of our shared future. Through informed, ethical, and sustainable leadership, we each possess the opportunity—and the responsibility—to contribute to a healthier, more resilient, and more sustainable world.

ORGANIZATION OF THE BOOK

The opening chapter establishes the conceptual foundation for the volume by distinguishing authentic sustainable leadership from superficial environmental messaging. It examines how ethical decision-making, systems thinking, stakeholder engagement, and long-term strategic vision enable leaders to embed sustainability into organizational culture rather than treating it as a marketing exercise. By emphasizing measurable environmental responsibility, social equity, and economic resilience, the chapter provides readers with a comprehensive framework for understanding sustainable leadership as an enduring commitment to organizational transformation.

Building upon this foundation, Chapter 2 advances the discussion by introducing an innovative governance-centered perspective that positions sustainable leaders as architects of institutional integrity. Rather than viewing greenwashing as a communication failure, the chapter reframes it as a governance challenge requiring robust systems of accountability, transparency, and independent oversight. Through its Governance-Architect Model, the chapter demonstrates how carefully designed governance structures can ensure that sustainability claims are supported by verifiable evidence and embedded within organizational decision-making.

Chapter 3 contributes a distinctive ethical perspective by exploring sustainable leadership through a faith-based framework. Drawing upon the principles of stewardship, trust, and moral accountability, the chapter illustrates how spiritual values can strengthen organizational integrity and reinforce authentic environmental responsibility. By integrating ethical convictions with organizational governance and culture, the authors demonstrate how leadership grounded in deeper moral purpose offers a compelling alternative to symbolic sustainability initiatives.

Chapter 4 shifts attention toward empirical evidence, presenting a research-based framework that links sustainable leadership behaviors with measurable improvements in environmental stewardship. The chapter demonstrates how leaders who prioritize resource efficiency, environmental innovation, and responsible governance can positively influence organizational sustainability performance. Its findings provide valuable guidance for practitioners seeking to strengthen environmental, social, and governance (ESG) initiatives through effective leadership practices.

Expanding the organizational perspective, Chapter 5 introduces Sustainable Stewardship Leadership as a behavioral and systems-oriented framework for institutional transformation. The chapter illustrates how leadership extends beyond individual influence by shaping organizational culture, aligning institutional systems, and empowering employees and communities to embrace environmental responsibility. Through its integration of behavioral science and systems thinking, the chapter highlights pathways for creating lasting organizational resilience and ecological sustainability.

Chapter 6 explores the increasingly important relationship between sustainable leadership and intergenerational trust. Recognizing that different generations possess distinct values, ethical expectations, and consumption behaviors, the chapter examines how leaders can cultivate authentic organizational cultures that resonate across diverse stakeholder groups. The discussion emphasizes that environmental stewardship requires not only responsible leadership but also transparent communication, accessible sustainability practices, and genuine organizational commitment capable of fostering long-term trust and loyalty.

Chapter 7 broadens the conversation by examining environmental leadership through both international perspectives and the Indonesian experience. Drawing upon examples from government, industry, education, and community organizations, the chapter demonstrates how transformational and adaptive leadership enable organizations to navigate environmental uncertainty while advancing sustainability goals. The discussion highlights both the progress achieved through national sustainability commitments and the practical challenges associated with implementation, equity, and long-term environmental governance.

Chapter 8 offers a timely examination of sustainability communication by investigating how organizations represent their environmental commitments during corporate events. The authors introduce the concept of representational readiness, illustrating that authentic sustainability depends not only upon organizational messaging but also upon employees' understanding and embodiment of environmental values. This chapter provides valuable insight into the leadership responsibilities required to ensure that sustainability is consistently reflected in both organizational culture and stakeholder interactions.

Moving into the realm of emerging technologies, Chapter 9 demonstrates how explainable artificial intelligence can strengthen climate risk assessment and environmental decision-making. By developing transparent, interpretable AI frameworks for evaluating cross-country climate risks and climate-financial interactions, the chapter illustrates how technological innovation can enhance policy development, governance, and strategic planning. This contribution underscores the importance of combining advanced analytical tools with transparent leadership practices to address complex global environmental challenges.

Chapter 10 examines sustainable leadership within higher education by focusing on the role of senior international officers in advancing environmentally responsible internationalization. As global academic mobility continues to expand, the chapter highlights the need for institutional leaders to balance international engagement with environmental sustainability. The discussion emphasizes leadership development, organizational readiness, and cross-sector collaboration as essential components of creating more sustainable approaches to global education.

Chapter 11 presents a unique socio-structural analysis of leadership within Malaysia's construction industry through the lens of Bourdieusian field theory. By examining leadership as a dynamic interaction of social capital, institutional influence, and competitive positioning, the chapter provides valuable insights into how leaders can strengthen organizational resilience while responding to globalization and sustainability challenges. The discussion broadens the understanding of sustainable leadership by considering the structural forces that shape organizational success.

Chapter 12 explores workforce localization through the lens of sustainable leadership development. Using the context of Saudization, the chapter demonstrates how national human capital initiatives can simultaneously strengthen leadership capacity, organizational performance, and environmental stewardship. By aligning leadership development with broader national sustainability goals, the chapter illustrates how strategic workforce policies can contribute to both economic transformation and environmental responsibility.

Recognizing education as the foundation for future leadership, Chapter 13 investigates how formal learning experiences cultivate environmental stewardship competencies. Integrating contemporary scholarship on transformative learning, experiential education, indigenous knowledge, and digital learning environments, the chapter presents a developmental model that traces the evolution of learners into stewardship-oriented leaders. Its recommendations offer valuable guidance for educators and institutions seeking to prepare future generations to address increasingly complex sustainability challenges.

The volume concludes with Chapter 14, which highlights the transformative role of project-based learning in advancing sustainability education. By emphasizing experiential learning, critical thinking, collaboration, and real-world problem solving, the chapter demonstrates how innovative pedagogical approaches prepare learners to become adaptable leaders capable of addressing environmental challenges. This final contribution reinforces one of the central messages of the book: sustainable leadership is cultivated through education that empowers individuals to translate knowledge into meaningful action for the benefit of society and the environment.

CONCLUDING REMARKS

The contributions presented in this volume collectively affirm that sustainable leadership is not simply an emerging management philosophy—it is an essential imperative for addressing the environmental challenges of our time. Across diverse disciplines, organizational settings, cultural perspectives, and geographic contexts, the chapters demonstrate that meaningful environmental stewardship is achieved when leadership is grounded in ethical responsibility, transparency, innovation, collaboration, and a genuine commitment to long-term societal well-being. Together, these contributions move the conversation beyond aspirational sustainability toward practical strategies that empower leaders to inspire measurable and lasting environmental change.

One of the greatest strengths of this collection lies in its diversity of perspectives. The chapters examine sustainable leadership through governance, organizational behavior, education, spirituality, public policy, artificial intelligence, workforce development, higher education, and industry-specific applications, illustrating that environmental stewardship is not confined to any single sector or profession. Rather, it is a shared responsibility that requires cooperation among scholars, practitioners, policymakers, educators, and community leaders alike. This multidisciplinary approach reflects the complexity of today's environmental challenges while offering readers a rich collection of ideas that can be adapted across varied institutional and cultural contexts.

As editors, we hope this volume serves not only as a scholarly resource but also as a catalyst for continued dialogue, research, and action. The insights presented throughout these chapters underscore the importance of cultivating leaders who possess both the strategic vision and the moral courage to place sustainability at the heart of organizational and societal decision-making. By embracing authentic stewardship rather than symbolic commitment, leaders have the capacity to influence institutions, inspire communities, and create resilient systems capable of meeting the needs of present and future generations.

It is our sincere hope that *Inspiring Environmental Stewardship Through Sustainable Leadership* will encourage readers to rethink the role of leadership in shaping a more sustainable future. Whether advancing new research, informing policy, transforming organizations, or educating future leaders, the ideas presented in this volume demonstrate that sustainable leadership begins with individual responsibility but achieves its greatest impact through collective action. We trust that this book will serve as both a valuable reference and a source of inspiration for all those committed to protecting our shared environment and fostering a more equitable, resilient, and sustainable world.

Qaisar Iqbal
Al Yamamah University, Saudia Arabia

Chapter 1
Beyond the Green Façade:
The Imperative of Authentic Sustainable Leadership

Richa Singh
http://orcid.org/0009-0007-0196-4919
Lovely Professional University, India

Lovleen Marwaha
http://orcid.org/0000-0002-5175-2590
Lovely Professional University, India

ABSTRACT

Sustainable leadership incorporates ethical decision making, environmentalism and long-term value generation in the company strategy. Since sustainability is an international requirement, leaders should take their solutions beyond the green-washing to quantifiable and transparent initiatives. This chapter looks at sustainable leadership- its dimensions, the comparison with the traditional models as well as how it is played out in actual organizational change. Based on the environmental management, organizational behavior, and ethics, it emphasizes the need to balance between the economic performance, social equity, and ecological preservation. Sustainable leaders facilitate responsibility, creativity, and involvement of stake-holders, which entails the aspect of sustainability in decision making. These major features are systems thinking, ethical stewardship, vision and emotional intelligence. It also gives approaches to evaluate the authenticity of leadership and states that the real sustainability is based on the morality and strategic vision rather than on marketing.

DOI: 10.4018/979-8-3373-7877-0.ch001

1. INTRODUCTION

The concept of leadership in the twenty-first century has transformed to a new perspective of power, financial gain, and immediate development. The world problems that have transformed people on how leaders should carry their activities in organizations and the society include climate change, social inequality and economic instability. Here, the concept of sustainable leadership has become an important tool that considers environmental, social, and ethical issues as an in-built part of the decision-making process (Adams *et al.,* 2011). Sustainable leadership does not just ensure that business operates are financially successful but also act in a manner that they enhance the wellbeing of people and the planet. But with the rising popularity of the concept of sustainability in the corporate and institutional circles, the threat of greenwashing, or the act of creating a false image of environmental or ethical accountability, has grown. Sustainability is a language or a symbol many organizations use without incorporating real sustainable practices in its operations. Such a shallow strategy undermines the effectiveness of sustainability programs and makes actual progress impossible (Rosario & Boechat, 2025). As such, it is crucial to know what constitutes true sustainable leadership and greenwashing to create confidence, honesty, and lasting influence. Sustainable leadership entails a change in management thinking to long-term value generation as opposed to short-term performance. It focuses on transparency, inclusiveness, and accountability and leaders are seen as custodians of organizational prosperity and social welfare. This type of leadership is based on the systems thinking where the complex systems of economy, environment, and social systems are interrelated. Sustainable leaders create innovative cultures, promote ethical thinking, and focus strategic objectives on sustainability agendas in the world, like United Nations Sustainable Development Goals (SDGs) (Saxena *et al.,* 2021). The chapter is focused on determining the essence of sustainable leadership, discussing its peculiarities, and how it would go beyond greenwashing. It attempts to give a clear picture of how leaders can be able to drive sustainable change in their organizations and communities in an authentic manner by incorporating theoretical knowledge and practical example.

2. CONCEPTUAL FOUNDATIONS OF SUSTAINABLE LEADERSHIP

Sustainable leadership focuses on the long term organizational success based on ethics in stewardship, social responsibility and environmental awareness. It is a result of the intersection of leadership theory, sustainability science, and corporate ethics. Here the paper examines the development of sustainable leadership, its principles, moral and strategic implications of sustainable leadership in responsible corporate and institutional practices.

2.1 Evolution of Leadership Theories toward Sustainability

The leadership theory has changed to include the traditional hierarchical and performance-driven model to those that are inclusive, values-oriented and incorporate sustainability. The initial theories which included trait, transactional and transformational leadership theories were devoted primarily to influence, performance and motivation. Nevertheless, such global challenges as climate change, inequality, and environmental degradation made the scholars extend the paradigm of leadership to being responsible and ethical. The sustainable leadership model presented by highlights the organizational sustainability, stakeholder inclusiveness, and economic, social and environmental performance equilibrium. This paradigm is opposed to the idea of short-termism which supports the notion that resilience and social equality are the key factors of success. Similarly, (Anos, 2025) opine that the sustainability of leadership is based on people development, institutional memory and continuity beyond the life of individual leaders. As time progressed, sustainable leadership developed as a comprehensive approach to the combination of systems thinking and moral integrity. It goes beyond the rationality of economics by giving sustainability a place in the purpose and identification of organizations. Scholars like (Virapongse *et al.*, 2016) refer to it as leadership-by-values where moral consciousness, empathy and vision integrate business results with health of planet. Therefore, the development of leadership indicates the paradigm shift between resource management and stewardship of systems equalizing the resources of the current and future generations. Theoretical convergence within the various fields of study, namely ecology, ethics, and organizational behavior, places sustainable leadership as a dynamic activity to the global challenges in such a way that leadership is a transformative, enduring, and regenerative process, instead of a purely managerial role.

2.2 Core Principles and Dimensions of Sustainable Leadership

The principles of sustainable leadership are based on the premises that help leaders to strive towards organizational success without undermining the integrity of the environment and social justice. These values, including visionary orientation, ethical behavior, inclusiveness of stakeholders, flexibility, and resilience are what inculcate the sustainability in the strategic and cultural context of institutions. Based on (Voulvoulis *et al.*, 2022), sustainable leaders embrace the paradigm of systems thinking in order to foresee change and develop an adaptive capacity. They establish work cultures that promote innovation and cooperation in solving problems. All sustainable practices rely on ethical behavior, which makes the practices transparent, fair, and trust-based among the stakeholders. On the same note, inclusivity of the stakeholders focuses on participatory governance where different voices are regarded in establishing resilient decisions and social legitimacy. Visionary orientation is what sets the sustainable leaders and the traditional managers apart. They can act in the long-term perspective, compromising between the profitability, ecological and social demands. Adaptability enables the leaders to cope with uncertainty and changing sustainability issues and resilience enables individuals and organizations to continue moving forward with stress or crisis. The findings of (Boeske, 2023) support the idea that the leaders motivated by sustainability concepts combine personal ethics and organizational objectives and align the personal purpose with the common course of actions. Their behavior is driven by their integrity and moral vision as opposed to being regulated by outsiders. All these principles transform leadership effectiveness into the capacity to create shared value, which is economic, environmental, and societal value. Sustainable leadership is therefore a transformational practice which does not only provide results but also makes natural systems, communities and institutions viable over a long term.

Table 1. Evolution of leadership models toward sustainability

Leadership Model	Core Characteristics	Focus Dimension	Example	Contribution to Sustainable Leadership
Transactional Leadership	Based on rewards, performance targets, and corrective actions	Efficiency and productivity	Henry Ford's production management system emphasizing standardized processes and performance-based outcomes	Established accountability, structure, and operational discipline within organizations
Transformational Leadership	Inspires vision, innovation, and organizational change	Employee motivation and transformation	Satya Nadella's leadership at Microsoft, promoting innovation, learning culture, and organizational renewal	Introduced purpose-driven change and value-based leadership
Servant Leadership	Prioritizes service, empathy, and employee well-being	Human development and social responsibility	Herb Kelleher at Southwest Airlines, focusing on employee care and customer service excellence	Strengthened stakeholder orientation and ethical responsibility

continued on following page

Table 1. Continued

Leadership Model	Core Characteristics	Focus Dimension	Example	Contribution to Sustainable Leadership
Responsible Leadership	Integrates ethics, governance, and stakeholder engagement	Corporate citizenship and societal impact	Paul Polman at Unilever, embedding sustainability into corporate strategy through the Sustainable Living Plan	Connected leadership with corporate responsibility and sustainable business practices
Authentic Leadership	Emphasizes self-awareness, transparency, and integrity	Ethical behavior and trust building	Howard Schultz at Starbucks, promoting ethical sourcing, employee welfare, and transparent leadership	Enhanced organizational trust and moral credibility
Sustainable Leadership	Balances economic performance, social equity, and environmental stewardship	Long-term sustainability and resilience	Yvon Chouinard at Patagonia, integrating environmental activism and responsible business practices into organizational strategy	Represents a holistic leadership approach that aligns profit, people, and planet

Changes in leadership models reflect a gradual progression from a performance orientated leadership to a value orientated and sustainability orientated leadership. Transactional and transformational leadership added to the efficiency and change in the organization, but in today's world ethical responsibilities, stakeholder engagement, transparencies and environmental stewardship are key concerns of leaders. While some of the key concepts of the preceding forms of leadership are still relevant to the current issues of sustainability, social responsibility, and long-term resiliency within an organization, sustainable leadership combines these strengths with the needs of the day (Table 1).

2.3 Ethical and Moral Underpinnings of Sustainable Decision-Making

Ethical consciousness, as a moral system of making decisions, which influence people, the environment, and future generations, is a core of sustainable leadership. Ethical leadership incorporates the virtue ethics, deontology, and utilitarianism to strike a balance between conflicting priorities. Leaders are not only requested to act out of compliance but because it is a moral obligation and a purpose. As (Newstead *et al.,* 2020) comments, humility, empathy, and justice are some of the types of moral virtues that sustainable leaders portray and which help foster trust and cooperation. Sustainability ethics is not just a matter of individual morality but it is an organizational standard within the culture and strategy. Transparency and accountability making sure that sustainability claims can be checked which negates greenwashing risks and reputation damage. Empirical research indicates that ethical leadership is positively related to employee engagement, organizational commitment and innovation (Jia *et al.,* 2022). Employees would be more in line with the values of sustainability when leaders are more focused on the principles of fairness and respect, which would lead to collective responsibility. Ethical stewardship therefore makes sustainability an actual rather than a rhetorical concept. Ethics in sustainable decision-making requires that a balance between economic development and environmental constraints and social justice are established. This triadic balance indicates the third bottom line principle which guarantees the success of organizations is not only quantified by profit, but also by people and the planet. Finally, the ethical aspect of sustainable leadership makes leaders the custodians of intergenerational equity. Ethical thinking is what makes the leadership process not focused on short-term but rooted in a higher cause the common good man and nature.

2.4 Relationship between Leadership, Sustainability, and Corporate Responsibility

Corporate responsibility and sustainable leadership are closely connected to each other since they focus on accountability, transparency, and stewardship. Corporate responsibility brings the ethical duty of leadership to the external stakeholders (society and environment). The result of this alignment is organizations that are beneficial to social and ecological systems and are still economical. According to (Maak & Pless, 2006), responsible leadership is a relational process where the leaders become the citizens of the corporate ecosystems and be able to become moral relationships through dialogue, trust and common purpose. On the same note, (Alkhodary, 2023) assert that sustainable leadership is critical in integrating corporate responsibility into strategic management and the sustainability programs should have a measurable and long-term effect. Sustainable leadership in corporate responsibility goes beyond philanthropy to strategic sustainability where social innovation is incorporated into business models. This is through implementation of sustainable supply chains, ethical governance, and inclusion of labor practices. Leaders are role models who are able to shape organizational ethics and culture and hence sustainability is not a symbolic act, but more of a practice. Besides, the inclusion of environmental, social, and governance (ESG) concepts into the sphere of leadership is now one of the distinguishing features of responsible organizations. Sustainable leaders associate ESG performance and organizational reputation, investor confidence and stakeholder trust. Finally, both sustainable leadership and corporate responsibility create synergistic organizations that are agents of change, not only profit-making organizations. This intersection is what strengthens the idea that sustainability is a mutual moral agreement between businesses and society which is brought about by ethical, visionary, and accountable leadership.

3. GREENWASHING-UNDERSTANDING THE CHALLENGE

Greenwashing is a misleading statement where the organizations are depicted as environmentally friendly even when they hide the unsustainable behaviors. This section explores its definition, intentions, and implications, with special focus on the way in which false sustainability communication leads to lack of trust among stakeholders as well as genuine leadership and sustainability credibility of the organization in the long run. It ends by comparing real sustainability to the symbolic or superficial sustainability.

3.1 Definition and Characteristics of Greenwashing

Greenwashing is a situation where groups of organizations overstate or lie about their sustainability accomplishments in order to attain reputational or financial rewards. The term was initially popularized by the environmental activists in the 1980s, but now covers various forms of misleading actions, such as misleading labeling, selective disclosure, and exaggerated environmental advertising statements. It is an alleged strategic communication failure that misleads the citizenry on corporate sustainability. Such a study by (Crilly *et al.*, 2016) classifies greenwashing on two levels which include the dimension of communication and performance. Companies can either engage in high levels of environmental rhetoric and low levels of environmental performance (decoupling) or can engage in low environmental performance and high environmental performance (silent green). These are both harmful to the perception of the stakeholders, but the latter is more damaging because it alters trust and decreases transparency. Common examples of greenwashing are the use of vague terms (eco-friendly, natural), unverifiable information, heavy use of visuals as opposed to quantitative data, and the cherry-picking of insignificant successes. According to (Jini, 2024), these practices are the result of the competitive environment and poor environmental laws. These attributes are important to comprehend in order to achieve sustainable leadership characterized by honesty, accountability, and quantifiable improvement. By detecting the greenwashing indicators, the leaders will be able to develop the authentic reporting of sustainability based on the verified performance indicators and third-party auditing.

Over the last few years, there has been a significant boost in action to tackle greenwashing and enhance the validity of sustainability credentials by governments, regulatory bodies and international organizations. The Green Claims Directive of the European Union sets clear standards for companies to make environmental claims and will no longer allow the use of "green" statements that are vague or misleading. Likewise, the United States Federal Trade Commission (FTC) Green Guides offer guidance about environmental marketing claims to promoting an accurate and transparent way to communicate sustainability information. International standards, like ISO 14021, are agreed sets of principles for self-declared environmental claims, to ensure environmental communications are not misleading, verifiable and true. These frameworks help to create more accountability and incentivize organisations to provide substantiated evidence to back up claims of sustainability, rather than make symbolic or promotional claims. With growing regulatory attention all over the world, true sustainable leadership should go beyond mere compliance to encourage transparency and ethical accountability within the organization.

3.2 Organizational Motives and Drivers Behind Greenwashing

Companies are practicing greenwashing because of reputational, financial and regulatory reasons. Companies tend to capitalize on the existence of consumer-related demand toward eco-friendly products or investor concerns regarding Environmental, Social, and Governance (ESG) to pretend to be sustainable without altering anything. As Szabo & Webster (2021) state, greenwashing takes place when there are greater marketing advantages in the sustainability claims than the perceived risks of disclosure. According to the institutional theory, greenwashing may be due to the symbolic conformity, when organizations imitate the sustainability norms to sustain their legitimacy instead of improving the environment. It was found that the more companies are being scrutinized by the public, the more selective disclosure they practice, disclosing positive information and neglecting the negative effects (Marquis *et al.*, 2016). Greenwashing is also motivated by economic factors. Companies can save the cost of enacting some changes through advertising sustainable conduct rather than enacting real reforms. The weak or weakened regulatory systems particularly in the emerging markets contribute to the issue, as they do not implement the reporting standards. Leadership ethics and corporate culture also predetermine the predisposition: leaders whose profit orientations are short-term tend to put more emphasis on appearance than on accountability. On the other hand, business entities that have good ethical leaders and open governance systems will be less susceptible to this kind of manipulation. Real sustainability does not entail a shallow marketing program. Therefore, being aware of these drivers helps the leaders to develop governance mechanisms, including third-party checks, stakeholder consultation and whistleblower safeguards, that minimize the chances of deceptive practice and enhance the trust in the institutions.

3.3 Impacts of Greenwashing on Stakeholder Trust and Credibility

Greenwashing destroys the trust of the stakeholders, which forms the basis of the organizational legitimacy in the long-term. False pretensions leave a doubt on the minds of consumers, employees, investors and regulators. Reputational damage can be more than short-term marketing payoffs when it is disclosed. Zhao *et al.*, (2024) show that the perception of deception decreases the intention to purchase the product and brand loyalty by a significant margin. At the organizational level, greenwashing compromises psychological contracts the implicit trust of corporate communication by the stakeholders. Once the credibility is breached, it is hard to reestablish it even with the true improvements. The exposure to greenwashing is associated with a negative effect on investor confidence and engagement among employees (Shahid *et al.,* 2025). On a social level, greenwashing undermines the sustainability of the population, as it creates distrust in the corporate programs. It water-downs the definition of such concepts as green and ethical and it becomes more difficult to make responsible organizations stand out. Moreover, it is able to misinform policymakers and consumers, moving resources towards literally sustainable businesses. Ethical leadership is the answer to these impacts by ensuring that there is a clear communication line and quantifiable sustainability objectives. Restoration of credibility is achieved through independent audits, transparent measures and transparent reporting systems like the Global Reporting Initiative (GRI). Sustainable leaders should always report on their progress and restrictions truthfully and transparency should be presented as a responsibility and not vulnerability.

3.4 Distinguishing Authentic Sustainability from Symbolic Practices

True sustainability involves integrating the elements of environmental and social responsibility within the organization i.e. strategy, culture and operations as opposed to considering it as a branding tool. The difference is based on content as opposed to form. It is demonstrated that (Manning *et al.,* 2019) companies that govern with sustainability enable them to have excellent performance and stakeholder relationships over a long term. In contrast to this, symbolic sustainability is based on marketing campaigns, limited reporting or token CSR projects out of strategic intent. According to Christensen & Cornelissen, (2015), real companies report not only achievements but also show failures, whereas symbolic ones will not be transparent. The main indicators of authenticity are quantifiable objectives, third-party validation, involvement of stakeholders and compatibility with international models like the UN Sustainable Development Goals (SDGs). Sustainable leaders are also consistent in their words and actions, the so-called behavioral integrity, which contributes to trust in the long term. True sustainability takes heart to reveal flaws and readiness to educate. When it is built into governance systems it makes leadership an image management process but value creation. Therefore, it is not only an ethical requirement, but a strategic requirement towards sustainable competitive advantage to differentiate authenticity and the symbolic (Figure 1).

Figure 1. Greenwashing-Understanding the challenge

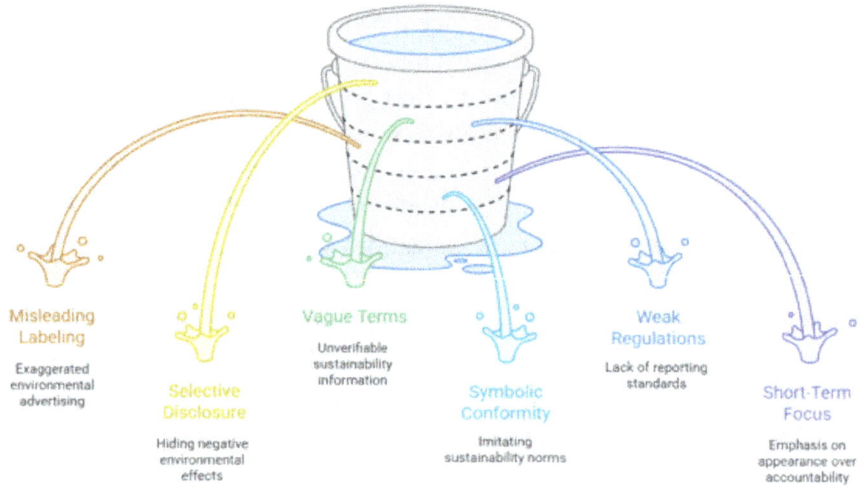

Figure 1 presents the major forms of greenwashing, including misleading labeling, selective disclosure, vague sustainability claims, symbolic conformity, weak regulations, and short-term organizational focus. Together, these practices create a false perception of environmental responsibility and undermine authentic sustainability efforts

4. BUILDING AUTHENTIC SUSTAINABLE LEADERSHIP

True sustainable leadership is the aspect that reconciles between ethical integrity and long-term sustainability. It has pointed out transparency, accountability, self-awareness and relational trust as pillars of responsible governance. In this section, the theory is discussed in terms of practical applications, organizational strategies, and cultural enablers that make sustainability values translate into consistent leadership behavior, and this way, ethical, resilient, and credible institutions will be achieved.

4.1 Authentic Leadership in Sustainable Contexts

True leadership is about being aware of oneself, having a moral outlook, and having open relationships qualities that build trust and responsibility in sustainable organizations. It also makes leaders engage internal values with external activities so that sustainability is exercised in a genuine and not a performative manner. Zehir *et al.*, (2014) assert that charismatic leaders establish ethical climates that contribute to health in the long run of the organization. This authenticity is brought about by a balanced processing, objective assessment of the information, and relational transparency, which is open communication with the stakeholders. Researchs affirms that authentic leadership positively affects employee commitment levels, engagement and ethical conducts that are the determinants to transitions of sustainability. Authenticity in the sustainable settings is a protection against greenwashing by inculcating a moral consistency in leadership behavior. According to Alavi (2024), authentic leaders qualify as morally anchored ones because they can balance profit motives and social and environmental responsibility. Equally, Franklin (2010) discovered that authenticity-based leadership development programs enhance self-awareness and empathy enabling leaders to make morally sound decisions. Authentic leadership is hence not a personal attribute, but a strategic skill that makes organizations remain credible as they strive to attain sustainability. It turns sustainability into a rhetorical declaration and makes it the daily operational reality, which strengthens the integrity as the pillar of corporate resilience.

4.2 Transparency and Accountability as Cornerstones of Sustainability

Authentic sustainable leadership has transparency and accountability as its ethical pillars. They include effective communication of objectives, activities and outcomes-enabling the stakeholders to assess integrity and performance of an organization. Matasick, (2017) discovered that open government increases trust and institutional credibility of stakeholders, more so in health and government sectors. Mol (2015) states that transparency is a moral and managerial tool: it lessens information asymmetry and makes sure that the sustainability practices are not but a show. Accountability is a complementary factor to transparency because it enables leaders and organizations to be responsible to the environmental and social consequences of their choices. According to research conducted by Iqbal & Piwowar-Sulej (2022), transparency and accountability are mediator variables in the relationship between leadership and organizational sustainability, especially in higher education. By making the sustainability issues of the institutions transparent, the institutions are able to not only increase trust but also create communal problem solving. Equally, Hoang (2018) pointed out that the corporate accountability tools, including ethical codes and reporting on the part of the company, are proven to enhance ESG (Environmental, Social, Governance) performance considerably. Real leaders carry these values to their work by setting examples and demonstrating their bias toward both victory and defeat, as well as by consulting the stakeholders in a dialogue. The transparency can be an organizational strength in lieu of a weakness through open communication and moral responsibility that bolster the long-term organizational legitimacy.

4.3 Building Organizational Culture for Authentic Leadership

A genuine sustainable leadership culture is built when the shared values, learning, and inclusivity are entrenched into the systems of the organisation. The culture also determines how employees perceive the leadership behavior and principles of sustainability which will affect the implementation effectiveness. The concept of culture according to Walumbwa & Schaubroeck, (2009) denoted a collection of shared assumptions that direct the behavior of organizations. True leaders ensure sustainable cultures by setting a good example of integrity, ensuring psychological safety, and encouraging ethical discussion. A study conducted by Malik & Mehmood (2022) established that authentic leadership promotes collective learning and innovation, essential in adjusting to the sustainability challenges. The article by Alemu (2025) offered the so-called Sustainable Leadership Pyramid, where the authors highlighted the importance of values-based culture and empowerment of staff members as the keys to successful performance in the long term. Equally, Joo *et al.,* (2016) showed that authenticity fosters desirable emotional climates that have a positive impact on employee engagement and commitment to the organization. Authenticity is institutionalized through organizational practices, which encompass ethical training, transparent decision-making and open feedback systems. Cultural resilience to unethical or short-term demands is empowered by leaders who promote reflective thinking and collaboration between shareholders. Thus the genuine sustainable leadership thrives whereby the organizational culture aligns purpose and practice. A trust based, learning culture, where there is a sense of shared accountability does not just maintain ethical conduct but turns sustainability into a permanent organizational identity (Table 2).

Table 2. Framework for implementing authentic sustainable leadership

Implementation Dimension	Strategic Objective	Practical Mechanisms	Expected Outcomes
1. Ethical Governance	Ensure integrity and fairness in decision-making.	Develop transparent reporting, ethical audits, and leadership codes.	Trust, accountability, and reduced greenwashing.
2. Organizational Culture Integration	Embed sustainability in values and behaviors.	Ethical training, internal sustainability champions, value-based incentives.	Employee engagement and ethical alignment.
3. Stakeholder Collaboration	Develop group sustainability ownership.	Social collaborations, inter-sector discussion, community planning.	Inclusive decision making and collective impact.
4. Transparency & Accountability	Maintain open communication of sustainability goals.	Public sustainability reports, disclosure of open-data and reviews by third parties.	Strengthened legitimacy and stakeholder confidence.
5. Continuous Learning & Innovation	Adapt leadership to evolving sustainability challenges.	Leadership development programs, SDG mapping, innovation incubators.	Resilient and future-ready organizations.

4.4 Leadership Strategies for Embedding Authentic Sustainability

Authentic sustainability entails strategic plans that are purposeful in aligning ethical vision to organizational systems. Good leaders will ensure that values of sustainability are brought to tangible policies so that the ethical responsibility is implemented instead of being symbolic. According to Boeske (2023), sustainable leadership approaches combine individual values and mission of the organization. To achieve coherence between intent and action, leaders should put in place feedback mechanisms, sustainability auditing and cross sector collaborations. On the same note Galpin & Whittington (2012) holds the view that sustainability leadership entails transformative intent, in other words, redefinition of business success as the development of social and ecological value.

Key strategies include:

1. **Ethical Decision Structures:** Designing some formal systems of evaluating social and environmental impact.
2. **Stakeholder Engagement:** The agendas on sustainability are co-created with employees, communities and partners.
3. **Grounded on global standards (e.g., GRI, SDGs) Transparent Reporting**: This involves validating performance using global standards.

4. **Leadership Development**: Incorporating ethics and systems thinking into executive schools.

According to Murphy (2022), responsible leaders are moral entrepreneurs who incorporate sustainability in all their decisions. This will demand the bravery to deal with ethical crisis, humility to learn through mistakes and the vision to implement long term change.

Finally, authentic sustainability implanted in its leadership makes it less of a management approach and more of a steward-like approach- building organizations that succeed by doing good in the society and the planet.

4.5 Case Studies of Authentic Sustainability and Greenwashing

The best way to understand what sustainable leadership is is to see it in action, and learn about some of the most successful sustainability practices and the results of greenwashing. An effective analysis of organizational experiences can give us some clues as to how leaders make decisions affect stakeholder trust, organizational reputation, and important outcomes for sustainability. Patagonia is an industry that is widely known for its sustained leadership. Yvon Chouinard made sustainability a business strategy and not a marketing ploy, as Patagonia became one of the first companies to make sustainability a core part of its business model (Shan, 2025). The company was working on responsible-sourcing, environmental activism, product repair programs, and transparency of the supply chain. The "Worn Wear" initiative was one of the many programs implemented to promote the reuse of products and waste reduction among customers. Patagonia's efforts to be good stewards of the environment are a testament to the fact that good leadership, delivered in a sustainable manner, can drive long-term value and keep stakeholders happy while maintaining the integrity of the business. In contrast, the Volkswagen's Dieselgate scandal is one of the biggest corporate greenwashing scandals in recent history (Boeske, 2023). The company was touting its diesel cars as being eco-friendly, while at the same time they were using software that they could use to unfairly manipulate emissions tests. Once the fraud was discovered, Volkswagen was hit with significant monetary fines, legal action and bad reputation. The incident was a reminder of the dangers of focusing upon image management instead of true sustainability performance. It also illustrated the potential dangers that can be created by false environmental signs of a threat that erode public confidence and the legitimacy of the organization. As these seemingly opposing cases show, there is a core principle of sustainable leadership. The principle of authenticity demands that values, values that are documented in corporate sustainability commitments, and sustainability practices be congruent. Patagonia built trust with stakeholders by being transparent about its environmental efforts and being responsible, while Volkswagen suffered much from the disconnect between what they claimed and what they did. Sustainable leadership is more than just about the communication and the branding, it's about taking ethical decisions, being accountable and acting in a measurable way that is connected to long-term environmental and social objectives.

4.6 Leadership for Environmental Stewardship

Environmental stewardship is a key aspect of Sustainable Leadership. But good leaders can help make sustainability policies and reporting systems a reality, as well as getting their staff, stakeholders and communities to take active steps to protect the environment and conserve resources. Sustainable leadership, in other words, means to establish an organization that promotes environmental responsibility as a value to be shared and not just a compliance law. Leaders help to establish the stewardship behaviours by leading their staffs to join in with sustainability efforts (Ims, 2019). More and more, companies enable employee-led initiatives for waste reduction, biodiversity protection, the use of renewable energies, and circular economy. These programs enable staff to provide innovative solutions, and reinforce their dedication to organisational sustainability targets. Research indicates that "the role modelling of the leaders and their environmental decisions and actions positively influences that of their employees, who are more likely to act pro-environmentally. Moreover, working together outside of an organization is the key to environmental stewardship. Sustainable leaders involve customers, suppliers, communities and policy makers in collaborative action to tackle environmental issues. By collaborating and engaging stakeholders, their sustainability efforts can reach out to a larger scale and help achieve greater ecological objectives. Community tree-planting initiatives, ecosystem restoration projects, sustainable supply-chain collaborations and climate action measures are just some examples. Some companies, like Patagonia, Interface, and Unilever, have proven that leadership is a key to engaging in sustainability, making it part of organizational culture and prompting employees to be part of the solution regarding environmental innovation (Kelvin, 2024). These are organizations that demonstrate how there can be both ecological and economic gains in being good stewards of the environment and building stakeholder relationships. Finally, environmental stewardship is an expression of compliance-based sustainability to values-based sustainability. Sustainable leaders bring about an organizational culture that encourages ongoing learning and innovation, and promotes long-term ecological responsibility by inspiring people to take responsibility for the environment.

5. FRAMEWORKS AND STRATEGIES FOR IMPLEMENTATION

Sustainable leadership implementation entails strategic alignment of the vision, values, and operational systems. This part examines frameworks and practices that can help the leader to actualize the sustainability principles into action. It underlines cultural integration, accountability measures and compliance with global sustainability in order to secure authentic organizational change in the long term.

5.1 Integrating Sustainability into Organizational Culture

Organizational culture practices sustainability to make the values of the environment and ethics a part of decision-making. Culture is a social fabric, which determines the behaviour of employees and social responsibility. (Berkes, 2009) argues that leadership modelling, joint learning and core values reinforcement are used to create a sustainable culture. Grooms & Reid-Martinez (2011) have suggested the model of sustainable leadership that focuses on long-term orientation, the development of staff, and the relations built on trust as the cultural anchors. Likewise, Bertels *et al.,* (2010) has found cultural transformation as the most essential and difficult part of the sustainability integration that needs profound changes in the mindset and management styles. According to a study by Walkiewicz *et al.,* (2021), sustainability needs to be ingrained within the mission statements, reward system, and internal communication channel to become a part of corporate DNA. Once staff members feel genuine dedication on the part of the leadership, they develop purpose driven actions and creativity. Moreover, Bennett *et al.,* (2018) also claimed that collective stewardship, the collective duty at all levels of the organization, is the basis of the sustainability of the long-term viability. Therefore, genuine leaders achieve cultures of inclusivity, ethical reflection and continuous improvement, which allow sustainable leadership to succeed as an organizational culture, rather than a management trend.

5.2 Measuring and Evaluating Leadership Sustainability

The ability to measure the performance of sustainability leadership offers accountability and stimulates continuous improvement. Good indicators not only measure qualitative but also quantitative results so that the aspect of sustainability is not just a form of pretense. Caraiani *et al.,* (2012) suggested multidimensional models of integrating economic, social, and environmental indicators to assess the performance of the leaders. Measuring leadership sustainability, Engelbrecht *et al.,* (2014) proposed measuring leadership in terms of employees engagement, stakeholder trust, ability of the company to innovate and ethical consistency. In the meantime, Thoradeniya *et al.,* (2015) have added that sustainability reporting should not be limited to financial reporting but should extend to cover cultural, behavioral and governance reporting. Nevertheless, there are tools like the Global Reporting Initiative (GRI), the ISO 26000 standards, which are useful in offering all-inclusive evaluation frameworks enabling organizations to measure ethical and environmental performance. Besides that, Nakamura et al., (2023) also insisted that stakeholder feedback and self-assessment loop of leadership that is of paramount importance in behavioral integrity and leadership authenticity checks. The definition of sustainability is not only about results, as measuring it, that is, is about the necessity of keeping the values, actions, and the long-term effect in balance with each other.

Table 3. Key performance indicators for evaluating sustainable leadership beyond greenwashing

Dimension	Key Performance Indicator (KPI)	Measurement Method
Climate Performance	Scope 1, Scope 2, and Scope 3 Emissions	Carbon accounting and emissions reporting
Energy Efficiency	Energy intensity	Energy consumption per unit of output
Resource Management	Water consumption and waste reduction	Sustainability audits
Social Responsibility	Employee engagement and retention rates	Employee surveys and HR records
Governance	ESG disclosure quality	Independent sustainability assessment
Transparency	Frequency and quality of sustainability reporting	Annual reports and external verification
Stakeholder Trust	Stakeholder satisfaction index	Stakeholder surveys
Ethical Leadership	Number of ethics violations or compliance incidents	Compliance monitoring systems
Innovation	Revenue generated from sustainable products or services	Financial reporting

Measurable indicators enable organizations to determine if their sustainability initiatives are making a difference. The indicators offer objective evidence of the environmental, social and governance performance, and help prevent greenwashing. When leadership decisions are connected to the measurable sustainability outcomes, organizations can boost accountability, transparency, and stakeholder trust (Table 3).

5.3 Aligning Leadership with the United Nations Sustainable Development Goals (SDGs)

The United Nations Sustainable Development Goals (SDGs) offer an international guideline to help reconcile the leadership vision with internationally accepted priorities in sustainability. Realistic sustainable leadership links these aims to organizational missions so that the success of the businesses leads to social and environmental prosperity. The SDGs are an integrated policy framework that interrelates the economic growth, environmental protection and social equity as explained by (Adanma & Ogunbiyi (2024). SDGs ensure the leaders support systemic change and global responsibility. Jimenez *et al.,* (2021) determined that alignment with the SDGs leads to improved corporate reputation and stakeholder engagement, particularly in its association with quantifiable goals. According to Tan and Sim (2023), SDGs are used by organizations that are more innovative and employee-driven. Localizing global ethics will be achieved by leaders using SDG principles to plan their corporations. Real sustainable leadership therefore becomes a way of turning its global responsibility into day to day action and closing the divide between the rhetoric of corporate sustainability and the real global change.

5.4 Overcoming Barriers to Authentic Implementation

Although there is increasing awareness, there is a systemic and behavioral barrier to the implementation of authentic sustainable leadership. These are the pressures related to profitability in the short term, a poor cooperation with stakeholders, and cultural opposition to change. According to Galpin & Whittington (2012), the lack of integration between sustainability and strategy and fragmentation of leadership models do not allow authentic implementation. Lamm *et al.,* (2015) emphasized that employees who are not psychologically empowered will be less engaged in the sustainability activities. They can be overcome through leadership education and capacity-building programs which would promote systems-thinking and ethical awareness. Dumitru, (2021) stressed the importance of moral courage and institutional support of leaders who bring into question the unsustainable norms. The cross-sector collaboration is also important; business-government-academia partnerships increase the impact and resources distribution. To have authenticity, the leaders need to move away with compliance-based sustainability and adopt transformational approaches based on transparency, inclusivity, and shared accountability. The solution to these blocks therefore does not only lie in structural change, but also redefining success as a leader not as a profit-maximizer, but as someone who creates holistic value (Figure 2).

Table 4. Major barriers to sustainable leadership implementation and mitigation strategies

Barrier	Impact Level	Mitigation Strategy
Short-term profit pressures	High	Align incentives with long-term sustainability goals
Organizational resistance to change	High	Leadership training and stakeholder engagement
Limited sustainability knowledge	Medium	Continuous education and capacity building
Resource constraints	Medium	Strategic partnerships and innovation investments
Regulatory uncertainty	Medium	Continuous monitoring of sustainability regulations
Leadership turnover	High	Succession planning and institutionalization of sustainability practices
Greenwashing culture	High	Transparency audits and independent verification

Knowing what challenges are likely to arise on the way to implementation helps organizations to plan ahead and make sustainability performance better. To overcome these challenges, commitment of leadership, participation of stakeholders and organization learning are required. Mitigation strategies – which are effective – not only lead to better sustainability results, but also make organizations more resilient and adaptable in increasingly dynamic business environments (Table 4).

Figure 2. Sustainable Leadership Implementation

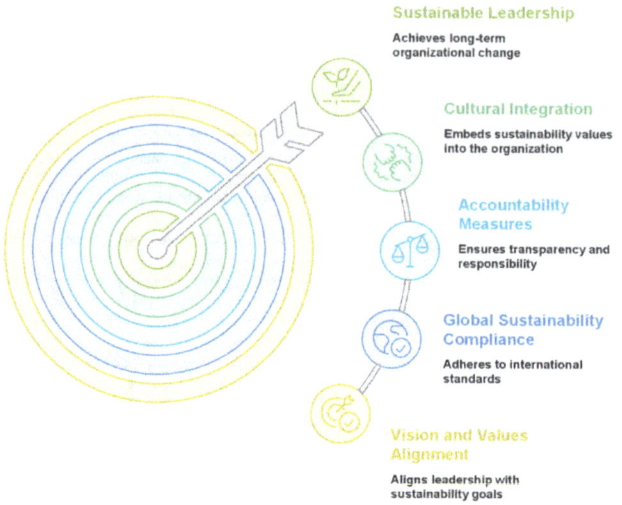

Figure 2 illustrates the key components of authentic sustainable leadership, including vision and values alignment, global sustainability compliance, accountability measures, cultural integration, and long-term organizational change. Together, these elements support credible sustainability practices and help organizations move beyond greenwashing

6. FUTURE DIRECTIONS AND EMERGING PERSPECTIVES

This is the way forward of sustainable leadership as it is capable of incorporating ethics, innovation and systemic adaptability in the face of global uncertainty. With the growing world business challenges regarding the environment, social and technological dynamics, organizational leadership should not be constrained to compliance and image management but adopt the component of authenticity and transformative sustainability as a long term global survival strategy.

6.1 Integrating Digital Transformation with Sustainability

This means that the digital revolution is transforming the way organizations have approached sustainability hence making digital literacy a critical skill that upcoming leaders should adopt. In a sustainable leadership, today it requires the implementation of data-driven transparency, AI-supported decision-making, and green digital infrastructure to keep track of environmental and social performance and to make it better. Digital tools also help leaders to quantify emissions, ethical supply chains and involve stakeholders in an even more inclusive fashion. Studies also emphasize that, with ethical principles leading digital transformation, corporate responsibility and efficiency of operations improve. Nevertheless, excessive dependence on technology without moral regulation may lead to supporting techno-greenwashing as the rhetoric of sustainability without actual action. Digital ethics should therefore be incorporated in sustainable leaders and technology should be used to promote human and ecological wellbeing. The new leadership model must afford a balance between innovation and responsibility, which focuses on reporting on data and governing digitally. As an example, the IoT and blockchain technologies have the potential to increase traceability and minimize unethical conduct in manufacturing systems. Collaborative ecosystems can also be created through digital empowerment to enable cross-sector partnerships to address common sustainability challenges (Pedersen *et al.,* 2021). Thus, the future of sustainable leadership will lie in the incorporation of the ethical use of digital stewardship, which is based on technology enhancing purpose, rather than profit. When leaders embrace digital change in a responsible manner, they will influence an organization that is transparent, efficient, and concerned with the well-being of the planet in the long term.

6.2 Cross-Cultural and Inclusive Leadership Models

The world is changing to be more globalized and demographically diverse and thus needs a leadership framework that meets the needs of cultural flexibility and inclusivity. The new generation of leaders in the field of sustainability has to operate within a complicated value system so that sustainability projects should not cause disrespect to the local traditions and at the same time should be oriented to the international ethical objectives. Studies show that empathetic-based culturally sensitive leadership approaches that are founded on justice and shared learning promote innovation and social justice (Brown, 2004). The cross-cultural cooperation establishes a trusting relationship, which leads to the joint development of the solution to the sustainability in accordance with the local demands instead of being imposed on the top-down, Western-centric, and so on. Besides, the future leadership should aim at equity-based sustainability in which different voices are used to establish ecological and ethical priorities. This practice makes leadership to surpass national, gender and disciplinary boundaries. True inclusivity boosts group intelligence that enables organizations to make ethical adjustments according to the intricacies of the global environment (Bai *et al.,* 2023). When coupled with cultural humility, sustainable leadership generates strong governance systems that embrace power relations and structural inequalities. The forthcoming leaders will be anticipated to develop a sense of psychological safety and respect towards each other, which are the staples of a prolonged cooperation and creativity. Therefore, intercultural sustainability leadership is required in the future, leaders who are both ethically conscious and globally competent. This is a balanced strategy that will not only entail environmental and economic change through sustainable leadership, but also human dignity and inclusiveness.

6.3 The Role of Education and Leadership Development

Today's global business landscape is rapidly evolving, and leadership is needed to effectively address environmental, social, technological and economic challenges. Hence, leadership development has increasingly become a key component of education for sustainable leadership. Today, leadership learning is not just about management skills and organisational performance, but also sustainability, ethics, digital transformation, and systems thinking are becoming essential competencies for long-term success. In today's day and age, companies are mandated to respond to climate change, biodiversity loss, resource constraints and social inequality as well as face new challenges like artificial intelligence, big data, automation, digital governance etc. These changes have changed the nature of leadership needs and the ability to be "sustainability literate" with the capacity to make "ethical choices" are now seen as essential skills that future leaders require. This means that leadership development programs need to develop people to cope with complexity, uncertainty and fast change and to keep organizational responsibility and accountability. HEIs are becoming more and more important by embedding principles of Environmental, Social and Governance (ESG), Sustainable Development Goals (SDGs), principles of the circular economy, climate literacy and responsible innovation in university courses. Students have the opportunity to learn through experiential methods, projects on sustainability, working on cases, interdisciplinary collaboration and community engagement projects, which enable them to acquire practical skills and problem-solving skills for problem solving in the real world. These pedagogical approaches promote that students are aware of the interdependence of the environmental, social and economic systems. There is also an acute focus on ongoing leadership development initiatives with a particular focus on stakeholder engagement, sustainable innovation, ethics in governance, and digital responsibility. The use of virtual learning platforms, simulations, AI-supported learning systems, online professional development, etc. has increased leadership learning more broadly and has provided more individualized learning experiences. Moreover, sustainable leadership development is seen as an on-going learning journey, not a single educational event. Leaders have to be continually inquisitive and knowledgeable, seeking new opportunities and challenges to sustainability and the needs of stakeholders. Hence, the importance of education as a key building block for future leaders who can help to advance genuine sustainability, inspire responsible organizational actions, drive innovation, and add value for the society, the economy and environment.

6.4 Limitations and Future Research Directions

While this chapter fully examines sustainable leadership and its contributions to combating greenwashing, there are a number of caveats worth noting. Firstly, the chapter takes a mainly conceptual and theoretical approach, extensively referenced from the literature to explain the development, traits and uses of sustainable leadership. This method yields a solid basis to comprehend the subject under investigation, but it does not provide much empirical evidence to inform researchers about the effectiveness of specific leadership practices in various organizational and cultural settings. Second, sustainable leadership remains a work in progress, constantly changing in face of new environmental, technological and social issues, challenges and opportunities. Therefore, some of the sustainability practises mentioned in this chapter could benefit from further qualification in the form of longitudinal studies that look at the sustainability practices' long-term effectiveness. Long-term studies on the effects of sustainable leadership on organizational performance, employee behavior, stakeholder trust, and environmental results are encouraged for future research. Thirdly, cultural and regional variations can impact the understanding and practice of sustainable leadership. What works with one approach might not work with another approach. Future research is thus needed to further investigate cross-cultural views of sustainable leadership and to understand what factors in the context helps it be implemented successfully. Last but not least, there are new opportunities and challenges for leadership research due to the fast-paced developments in AI, digital transformation, and sustainability reporting. Further research and studies should explore and highlight the potential of new technologies to aid transparency, accountability, and sustainability efforts, while mitigating potential risks like digital greenwashing or misinformation. The solutions to these research gaps will help increase the understanding of sustainable leadership and help pave the way for better strategies to achieve long-term organizational and societal sustainability.

7. CONCLUSION

Sustainable leadership is a revolutionary leadership style; it incorporates ethics, long-term vision, as well as environmental and social responsibility in the leadership practice itself. It goes beyond the gloss and slogans of greenness and photo-op management, and reinvents leadership as a moral and holistic process that aims at maintaining integrity in an organization and the wellbeing of the planet. It has been shown in this chapter that sustainable leadership developed as a development of traditional theories, i.e., not as a transactional and transformational theory but as an inclusive and ethically-based theory. Authenticity, transparency, and accountability, which are the conceptual basis, are needed to make sure that the principles of sustainability are not merely organizational rhetoric but commitment of behavior on a daily basis. Leaders also promote trust, innovation and resilience in their institutions by instilling these values into culture and decision making. The greenwashing debate has revealed how moral consistency and an open administration is of paramount importance. True leaders do not comply with performative sustainability as they align the intent and the quantifiable results, thereby safeguarding the organizational credibility. It was discovered through case studies and real-life examples that authenticity and sustainability flourish where leadership practices are based on truth, openness, and ethical reflections. Implementation must be done through structures that will incorporate sustainability in culture, strategy, and assessment. By matching the leadership practice to the global objectives, especially the United Nations Sustainable Development Goals (SDGs), the intent of an ethical act is rendered into measurable social and environmental outcomes. Systemic barriers that hinder change, like short-termism, change resistance, etc., cannot be conquered without moral courage, collective learning, and cross-sector collaboration. In the future, sustainable leadership has to change in adaptive, inclusive, and ethically aware forms. The next generation of leaders will have to juggle between digital change and environmental sustainability, international aspirations and national intelligence. After all, sustainable leadership is not a style as such, but it is a commitment to create a fair, open, and restorative future in which prosperity is not just a matter of profitability, but of long-lasting good health of individuals and the environment.

REFERENCES

Adams, C. A., Heijltjes, M. G., Jack, G., Marjoribanks, T., & Powell, M. (2011). The development of leaders able to respond to climate change and sustainability challenges: The role of business schools. *Sustainability Accounting. Management and Policy Journal*, *2*(1), 165–171. DOI: 10.1108/20408021111162191

Adanma, U. M., & Ogunbiyi, E. O.Uwaga Monica AdanmaEmmanuel Olurotimi Ogunbiyi. (2024). A comparative review of global environmental policies for promoting sustainable development and economic growth. *International Journal of Applied Research in Social Sciences*, *6*(5), 954–977. DOI: 10.51594/ijarss.v6i5.1147

Alavi, S. B. (2024). The Making of an Authentic Leader's Internalized Moral Perspective: The Role of Internalized Ethical Philosophies in the Development of Authentic Leaders' Moral Identity. *Journal of Business Ethics*, *190*(1), 77–92. DOI: 10.1007/s10551-023-05442-9

Alemu, B. A. (2025). Cultivating a Culture of Sustainability: The Role of Organizational Values and Leadership in Driving Sustainable Practices. *Business Ethics and Leadership*, *9*(1), 79–94. DOI: 10.61093/bel.9(1).79-94.2025

Alkhodary, D. (2023). Integrating sustainability into strategic management: A path towards long-term business success. *International Journal of Professional Business Review: Int. J. Prof. Bus. Rev.*, *8*(4), 39.

Anos, C. (2025). Ensuring leadership continuity: An integrative review of succession planning, leadership development and organisational memory. *Edelweiss Applied Science and Technology*, *9*(5), 2869–2877. DOI: 10.55214/25768484.v9i5.7607

Bai, A., Vahedian, M., Ghahreman, R., & Piri, H. (2023). Elevating women in the workplace: The dual influence of spiritual intelligence and ethical environments on job satisfaction. *arXiv preprint arXiv:2310.16341*.

Bennett, N. J., Whitty, T. S., Finkbeiner, E., Pittman, J., Bassett, H., Gelcich, S., & Allison, E. H. (2018). Environmental stewardship: A conceptual review and analytical framework. *Environmental Management*, *61*(4), 597–614. DOI: 10.1007/s00267-017-0993-2 PMID: 29387947

Berkes, F. (2009). Evolution of co-management: Role of knowledge generation, bridging organizations and social learning. *Journal of Environmental Management*, *90*(5), 1692–1702. DOI: 10.1016/j.jenvman.2008.12.001 PMID: 19110363

Bertels, S., Papania, L., & Papania, D. (2010). Embedding sustainability in organizational culture. *A systematic review of the body of knowledge*, 25.

Boeske, J. (2023). Leadership towards sustainability: A review of sustainable, sustainability, and environmental leadership. *Sustainability (Basel), 15*(16), 12626. DOI: 10.3390/su151612626

Boeske, J. (2023). Leadership towards sustainability: A review of sustainable, sustainability, and environmental leadership. *Sustainability (Basel), 15*(16), 12626. DOI: 10.3390/su151612626

Boeske, J. (2023). Leadership towards sustainability: A review of sustainable, sustainability, and environmental leadership. *Sustainability (Basel), 15*(16), 12626. DOI: 10.3390/su151612626

Brown, K. M. (2004). Leadership for social justice and equity: Weaving a transformative framework and pedagogy. *Educational Administration Quarterly, 40*(1), 77–108. DOI: 10.1177/0013161X03259147

Caraiani, C., Lungu, C. I., Dascalu, C., Cimpoeru, M. V., & Dinu, M. (2012). Social and environmental performance indicators: Dimensions of integrated reporting and benefits for responsible management and sustainability. *African Journal of Business Management, 6*(14), 4990.

Christensen, L. T., & Cornelissen, J. (2015). Organizational transparency as myth and metaphor. *European Journal of Social Theory, 18*(2), 132–149. DOI: 10.1177/1368431014555256

Crilly, D., Hansen, M., & Zollo, M. (2016). The grammar of decoupling: A cognitive-linguistic perspective on firms' sustainability claims and stakeholders' interpretation. *Academy of Management Journal, 59*(2), 705–729. DOI: 10.5465/amj.2015.0171

Dumitru, D. (2021). Critical thinking, moral courage, and sustainable development goals. Universities as a driving force for societal change. In *Universities, Sustainability and Society: Supporting the Implementation of the Sustainable Development Goals* (pp. 171–179). Springer International Publishing. DOI: 10.1007/978-3-030-63399-8_12

Engelbrecht, A. S., Heine, G., & Mahembe, B. (2014). The influence of ethical leadership on trust and work engagement: An exploratory study. *SA Journal of Industrial Psychology, 40*(1), 1–9. DOI: 10.4102/sajip.v40i1.1210

Franklin, R. S. (2010). *Exploring the moral development and moral outcomes of authentic leaders*. Regent University.

Galpin, T., & Lee Whittington, J. (2012). Sustainability leadership: From strategy to results. *The Journal of Business Strategy, 33*(4), 40–48. DOI: 10.1108/02756661211242690

Galpin, T., & Lee Whittington, J. (2012). Sustainability leadership: From strategy to results. *The Journal of Business Strategy, 33*(4), 40–48. DOI: 10.1108/02756661211242690

Grooms, L. D., & Reid-Martinez, K. (2011). Sustainable leadership development: A conceptual model of a cross-cultural blended learning program. *International Journal of Leadership Studies, 6*(3), 412–429.

Hoang, T. (2018). The role of the integrated reporting in raising awareness of environmental, social and corporate governance (ESG) performance. In *Stakeholders, governance and responsibility* (pp. 47-69). Emerald Publishing Limited.

Ims, K. J. (2019). Caring Entrepreneurship and Ecological Conscience—The Case of Patagonia Inc. In *Caring management in the new economy: socially responsible behaviour through spirituality* (pp. 197–220). Springer International Publishing. DOI: 10.1007/978-3-030-14199-8_11

Iqbal, Q., & Piwowar-Sulej, K. (2022). Sustainable leadership in higher education institutions: Social innovation as a mechanism. *International Journal of Sustainability in Higher Education, 23*(8), 1–20. DOI: 10.1108/IJSHE-04-2021-0162

Jia, K., Zhu, T., Zhang, W., Rasool, S. F., Asghar, A., & Chin, T. (2022). The linkage between ethical leadership, well-being, work engagement, and innovative work behavior: The empirical evidence from the higher education sector of China. *International Journal of Environmental Research and Public Health, 19*(9), 5414. DOI: 10.3390/ijerph19095414 PMID: 35564809

Jimenez, D., Franco, I. B., & Smith, T. (2021). A review of corporate purpose: An approach to actioning the sustainable development goals (SDGs). *Sustainability (Basel), 13*(7), 3899. DOI: 10.3390/su13073899

JINI. T. (2024). *A STUDY ON AWARENESS OF GREEN WASHING IN ERNAKULAM* (Doctoral dissertation, St. Teresa's College (autonomous) Ernakulam).

Joo, B. K., Lim, D. H., & Kim, S. (2016). Enhancing work engagement: The roles of psychological capital, authentic leadership, and work empowerment. *Leadership and Organization Development Journal, 37*(8), 1117–1134. DOI: 10.1108/LODJ-01-2015-0005

Kelvin, K. (2024). Global Leadership and Sustainability. [IJSRM]. *Int. J. Sci. Res. Manag., 12*, 7015–7026.

Lamm, E., Tosti-Kharas, J., & King, C. E. (2015). Empowering employee sustainability: Perceived organizational support toward the environment. *Journal of Business Ethics, 128*(1), 207–220. DOI: 10.1007/s10551-014-2093-z

Maak, T., & Pless, N. M. (2006). Responsible leadership in a stakeholder society–a relational perspective. *Journal of Business Ethics*, *66*(1), 99–115. DOI: 10.1007/s10551-006-9047-z

Malik, H., & Mehmood, M. (2022). Impact of authentic leadership, sustainable leadership on sustainable growth with mediating role of organizational learning. [AJSS]. *Academic Journal of Social Sciences*, *6*(1), 48–69. DOI: 10.54692/ajss.2022.06011681

Manning, B., Braam, G., & Reimsbach, D. (2019). Corporate governance and sustainable business conduct—E ffects of board monitoring effectiveness and stakeholder engagement on corporate sustainability performance and disclosure choices. *Corporate Social Responsibility and Environmental Management*, *26*(2), 351–366. DOI: 10.1002/csr.1687

Marquis, C., Toffel, M. W., & Zhou, Y. (2016). Scrutiny, norms, and selective disclosure: A global study of greenwashing. *Organization Science*, *27*(2), 483–504. DOI: 10.1287/orsc.2015.1039

Matasick, C. (2017). Open government: How transparency and inclusiveness can reshape public trust. *Trust and Public Policy*, 105.

Mol, A. P. (2015). Transparency and value chain sustainability. *Journal of Cleaner Production*, *107*, 154–161. DOI: 10.1016/j.jclepro.2013.11.012

Murphy, C. (2022). *Sustainable leadership: Lessons of vision, courage, and grit from the CEOs who dared to build a better world*. John Wiley & Sons.

Nakamura, Y. T., Williams, E. N., Pyle, L., & Grande, S. E. (2023). Cultivating authentic leaders: Toward conceptual coherence and sustainable practice. *The Journal of Values Based Leadership*, *16*(1), 19.

Newstead, T., Dawkins, S., Macklin, R., & Martin, A. (2020). The virtues project: An approach to developing good leaders. *Journal of Business Ethics*, *167*(4), 605–622. DOI: 10.1007/s10551-019-04163-2

Pedersen, E. R. G., Lüdeke-Freund, F., Henriques, I., & Seitanidi, M. M. (2021). Toward collaborative cross-sector business models for sustainability. *Business & Society*, *60*(5), 1039–1058. DOI: 10.1177/0007650320959027

Rosario, A. T., & Boechat, A. C. (2025). How Sustainable Leadership Can Leverage Sustainable Development. *Sustainability (Basel)*, *17*(8), 3499. DOI: 10.3390/su17083499

Saxena, A., Ramaswamy, M., Beale, J., Marciniuk, D., & Smith, P. (2021). Striving for the United Nations (UN) sustainable development goals (SDGs): What will it take? *Discover Sustainability*, *2*(1), 20. DOI: 10.1007/s43621-021-00029-8

Shahid, A. U., Nagarajan, C. D., Tufail, H. S., & Imran, M. (2025). Investors' pro-environmental attitude, spirituality and socially responsible investing: A mediating role of greenwashing practices. *Journal of Accounting & Organizational Change*, *21*(4), 711–735. DOI: 10.1108/JAOC-03-2024-0102

34. Shan, Y. (2025, July). Analysis of the Causes and Countermeasures of "Greenwashing"-Taking Volkswagen's "Dieselgate" as an Example. In *2025 International Conference on Financial Risk and Investment Management (ICFRIM 2025)* (pp. 666-675). Atlantis Press. DOI: 10.2991/978-94-6463-748-9_74

Szabo, S., & Webster, J. (2021). Perceived greenwashing: The effects of green marketing on environmental and product perceptions. *Journal of Business Ethics*, *171*(4), 719–739. DOI: 10.1007/s10551-020-04461-0

Tan, J., & Sim, J. (2023). Making Employee-Driven Innovation Achievable: Paving the Way. In *Making Employee-Driven Innovation Achievable* (pp. 1-10). Routledge.

Thoradeniya, P., Lee, J., Tan, R., & Ferreira, A. (2015). Sustainability reporting and the theory of planned behaviour. *Accounting, Auditing & Accountability Journal*, *28*(7), 1099–1137. DOI: 10.1108/AAAJ-08-2013-1449

Virapongse, A., Brooks, S., Metcalf, E. C., Zedalis, M., Gosz, J., Kliskey, A., & Alessa, L. (2016). A social-ecological systems approach for environmental management. *Journal of Environmental Management*, *178*, 83–91. DOI: 10.1016/j.jenvman.2016.02.028 PMID: 27131638

Voulvoulis, N., Giakoumis, T., Hunt, C., Kioupi, V., Petrou, N., Souliotis, I., Vaghela, C. J. G. E. C., & binti Wan Rosely, W. (2022). Systems thinking as a paradigm shift for sustainability transformation. *Global Environmental Change*, *75*, 102544. DOI: 10.1016/j.gloenvcha.2022.102544

Walkiewicz, J., Lay-Kumar, J., & Herzig, C. (2021). The integration of sustainability and externalities into the "corporate DNA": A practice-oriented approach. *Corporate Governance (Bradford)*, *21*(3), 479–496. DOI: 10.1108/CG-06-2020-0244

Walumbwa, F. O., & Schaubroeck, J. (2009). Leader personality traits and employee voice behavior: Mediating roles of ethical leadership and work group psychological safety. *The Journal of Applied Psychology*, *94*(5), 1275–1286. DOI: 10.1037/a0015848 PMID: 19702370

Zehir, C., Müceldili, B., Altindağ, E., Şehitoğlu, Y., & Zehir, S. (2014). Charismatic leadership and organizational citizenship behavior: The mediating role of ethical climate. *Social Behavior and Personality*, *42*(8), 1365–1375. DOI: 10.2224/sbp.2014.42.8.1365

Zhao, H., Xu, Y., & Wang, X. (2024). Greenwashing in workplace: The influencing mechanism of psychological contract fulfillment and the moderating role of ethical work climate. *Journal of Hospitality & Tourism Research (Washington, D.C.)*, *48*(3), 420–433. DOI: 10.1177/10963480221109551

Chapter 2
Beyond Greenwashing:
The Governance-Architect Model of Authentic Sustainable Leadership

Zahra Mohamed El Shlmani
http://orcid.org/0009-0004-8084-1455
Universiti Sains Malaysia, Malaysia

Hong Jin Liew
http://orcid.org/0009-0007-5860-5618
HELP University, Malaysia

Ibrahim O. A. Eriqat
http://orcid.org/0000-0002-6777-9472
Smart University College for Modern Education, Palestine

Mushtaq Yousif Alhasnawi
http://orcid.org/0009-0009-2346-9681
University of Thi-Qar, Iraq

Awaisu Adamu Salihi
http://orcid.org/0000-0002-5045-468X
Kano State Polytechnic, Nigeria

ABSTRACT

In response to the challenge of corporate greenwashing, this chapter advances a governance-centered redefinition of sustainable leadership. Sustainable leadership scholars emphasize ethical behavior, visionary thinking, and stakeholder engagement; however, such perspectives inadequately address the structural conditions that may enable deceptive environmental claims. By drawing on agency theory,

DOI: 10.4018/979-8-3373-7877-0.ch002

sustainability governance research, and global reporting standards, this chapter proposes a novel framework for sustainable leadership. The novel framework integrates rigorous governance to eliminate corporate greenwashing and approaches greenwashing as a governance failure, rather than a communication deficit. This chapter introduces the concept of a sustainable leader as a governance architect, whose core responsibility is the design and implementation of institutional systems that can ensure transparency, accountability, and verifiable sustainability.

1. INTRODUCTION

The concept of sustainable leadership has emerged as an essential component of corporate governance, environmental management, and organizational strategy. The concept has garnered considerable attention, as organizations face increasing pressure to address environmental and social challenges. Moreover, sustainable leadership has become a central theme in contemporary management discussions, especially as global environmental challenges increase and organizational legitimacy becomes tied with environmental, social, and governance (ESG) performance (Ahmed, 2024). Business leaders are facing growing pressure from investors, regulators, civil society, and consumers to authentically commit to climate action and sustainability. The prevalence of greenwashing in modern businesses has created a need to redefine sustainable leadership beyond superficial commitment. However, despite its prevalence and the surge of interest in sustainability, a wide gap remains between ambitious commitment and verifiable impact (Du Toit, 2024). This gap has contributed to the proliferation of greenwashing, which is reflected in misleading claims about environmental practices or advantages, or fabricated environmental performance. The phenomenon has defined governance challenges and cast a shadow over genuine efforts. Greenwashing can ultimately lessen consumer trust, invite regulatory scrutiny, and, most importantly, divert resources from authentic sustainability initiatives.

Regulators in the European Union, the United States, and Asia have tightened disclosure rules; however, greenwashing has persisted, which has revealed a fundamental conceptual flaw in our understanding of sustainable leadership. In addition, despite the updating of reporting norms of the International Sustainability Standards Board (ISSB) and Corporate Sustainability Reporting Directive (CSRD), practical challenges in implementation continue to undermine proven sustainability outcomes (Apostol et al., 2025). Current dominant definitions of sustainable leadership emphasize values such as ethics, long-termism, social responsibility, and stakeholder inclusion. Although important, such values are inadequate (Sajjad et al., 2024). This chapter argues that greenwashing is a problem rooted in accountability and integrity and typically inherited by an organizational structure that reflects deep structural deficiencies in governance and verification.

Therefore, this chapter advances a central thesis: A truly sustainable leader cannot exist without a governance framework that can essentially restrict greenwashing. The thesis proposes governance-reformulated concepts that move beyond declaration, ethical goals, and personal influence. To develop this theoretical contribution, this chapter extracts insights from different contemporary sources and proposes an innovative organizational mechanism, that is, a stewardship and integrity committee. Although the argument is primarily conceptual, this chapter provides illustrative explanations to show how governance dysfunction has enabled greenwashing and interprets the conceptual claims without drifting into the case study mode. Moreover, this chapter provides a critical theoretical redefinition and a practical governance roadmap to environmental leadership, climate governance, and organizational resilience. Furthermore, this chapter offers a new model for organizational leaders, boards, and policymakers committed to strengthening sustainability integrity and making greenwashing structurally unattainable.

2. BACKGROUND OF THE PROBLEM

The issue of sustainability, accompanied by an increase in strategic misrepresentation, has been raised as a corporate priority. The practice of making unsubstantiated or misleading environmental claims is no longer a fringe behavior; it has become an organizational risk documented in up to 70% of industries (Kwon et al., 2024). Organizations are increasingly relying on sustainability narratives to attract impact-oriented consumers and meet stakeholders' expectations. However, the disconnect between public claims and meaningful action has become pronounced. Regulatory screenings indicate that around 40% of environmental claims examined in consumer markets lack adequate evidence or may be misleading (European Commission, 2021). Such distortion of market signals can erode public trust and divert resources from genuine innovation. Such activities are illustrated in the DWS Group's 2022 "ESG investing" fraud case, in which the Deutsche Bank asset management misled investors about its ESG integration, which resulted in a U.S. Securities and Exchange Commission (SEC) settlement and raids by German authorities (U.S. SEC, 2023). Despite such cases, the sustainable leadership discourse emphasizes communication strategies and outward-facing initiatives. The persistence of greenwashing suggests that the core problem lies in weak internal accountability mechanisms; thus, greenwashing must be reframed as a fundamental governance deficit rather than a public relations issue. Critics point to inconsistent reporting standards and assurance practices and the absence of enforceable penalties for sustainability misrepresentation

(Sneideriene & Legenzova, 2026). Organizations typically operate within voluntary frameworks, such as the Global Reporting Initiative (GRI), the Sustainability Accounting Standards Board (SASB), the Task Force on Climate-related Financial Disclosures, and the CDP (formerly the Carbon Disclosure Project). Such bodies permit interpretive flexibility and selective disclosure, which highlights favorable metrics while concealing negative performance. Empirical research showed that firms that engage in dummy ESG activities typically score high on sustainability ratings but perform poorly in objective environmental benchmarks. Thus, research has exposed a contradiction between reported performance and actual impact (Christensen et al., 2022).

Therefore, greenwashing has been recognized as misalignment rooted in incentives, information asymmetry, and weak controls, particularly under resource constraint conditions (Li & Liu, 2025). Organizations face competing pressures to deliver financial returns while maintaining their progress within finite budget, time, and expertise constraints. Substantial transformation is resource-intensive; however, symbolic disclosure through flexible reporting frameworks can serve as a low-cost and fast-return alternative. Agency theory explains the following rational choice: When governance systems fail to verify a major action, limited-resource firms will prioritize false disclosure over costly reform (Hao et al., 2025; Janik & Ryszko, 2026). Thus, greenwashing reflects not only ethical decay but also failed governance systems that cannot align resource allocation with sustainability claims.

The foundations of sustainable leadership have yet to manage such structural realities. The literature is mostly rooted in normative paradigms of ethics and stakeholder empathy (Visser & Courtice, 2011; Sajjad et al., 2024). Although important, such perspectives underestimate the systemic nature of sustainability deception. Traditional models highlight long-term orientation and social purpose but inadequately address governance weaknesses that can enable greenwashing. This systemic flaw has prompted scholars to call for the development of systematic approaches for sustainability governance (Forliano et al., 2025). Leadership commitment statements risk becoming disconnected from organizational behavior with the absence of mechanisms for data integrity, incentive alignment, and cross-functional accountability, which is known as the leadership accountability gap. This gap is evident in practice. For example, H&M's 2019 "Conscious Collection" relied on unverifiable data and misleading labeling and thus was flagged by European regulators. Moreover, several energy companies promote climate-neutral initiatives while expanding their fossil fuel investment. Board governance research showed that, though many boards engage in sustainability issues, few possess the expertise or authority to audit ESG information (Sahu et al., 2025). Instead, they rely on management-generated data, which may be biased or incomplete (Sánchez et al., 2026). The literature reveals that greenwashing persists not because leaders lack values, but because organizations lack systems. It is driven by an accountability gap and a verification gap. In addition, a shared operative definition of greenwashing is lacking, which can fragment research and weaken policy responses (Spaniol et al., 2024). This framing leads to a critical question: Can sustainable leadership that fails to emphasize accountability within the governance structure be immune to the contradiction between its stated values and actual performance?

3. THEORETICAL FOUNDATIONS

A strong new definition of sustainable leadership must be built on clear theories and be able to explain the potential value of sustainable leadership and the persistent problem of inauthentic claims. Three bodies of studies are particularly relevant to the research context: scholarly discourse on sustainability leadership, agency theory studies, and governance research. The three research streams can clarify why values-based approaches for sustainability have been inadequate and why structural or systematic approaches are necessary to eliminate greenwashing. This section synthesizes the perspectives and shows how they converge on a central insight. Organizations cannot rely solely on ethical aspirations but must embed sustainability into a robust governance structure.

Sustainability Leadership Literature

The field of sustainable leadership has expanded over the past two decades. The concept emerged from broad conversations about transformational leadership, ethical leadership, and corporate social responsibility. Early studies emphasized leaders' personal qualities, such as moral consciousness, environmental awareness, and long-term thinking, as the primary drivers of sustainability. For instance, Avery and Bergsteiner (2011), Pless and Maak (2011), and Ferdig (2007) framed sustainable leaders as individuals who integrate social and ecological commitments into organizational strategy. Moreover, the authors argued that responsible leaders cultivate a culture of care, embrace ecological limits, and engage stakeholders with empathy and transparency. This research changed corporate discourse from shareholder primacy to broad societal obligations and established sustainability as a central component of leadership ethics.

However, a close examination of the idea reveals an important limitation: overreliance on normative qualities (Taylor, 2024). Leadership is typically presented as a matter of personal character, rather than institutional design. The presumption in much of the literature is that leaders who possess the correct ethical outlook can smoothly steer their organization toward sustainable outcomes. Although this assumption has obvious appeal, it does not fully align with real-world evidence (Albuquerque Pai et al., 2024). Many organizations with values-driven executives engage in symbolic sustainability practices or make misleading environmental claims (Ren et al., 2023; Ivasciuc & Ispas, 2025). The disconnect suggests that sustainability leadership cannot be fully explained through the lens of individual-level ethics (Sajjad et al., 2024; Afshar Jahanshahi & Kalaycioglu, 2025). Leaders with strong moral commitments may operate within organizational systems characterized by information asymmetry, incentive distortion, and cultural pressures, which may unintentionally enable greenwashing (Miao & Nduneseokwu, 2025; Bîzoi & Bîzoi, 2025). Such limitations point to a need for a conceptual shift toward understanding sustainable leadership as a structural, rather than a purely personal, phenomenon (Sajjad et al., 2024; Wamsler et al., 2024). This observation raises an important question: If leaders are to guide their organization beyond greenwashing, what theoretical frameworks can explain the structural barriers they will face, and what governance solutions will be required?

Recent critiques have affirmed the gap and called for evolution. For instance, Maak et al. (2016) called for increased attention to be paid to the systemic and institutional dimensions of sustainability and asserted that leaders must be understood as ethical actors who are also designers of governance systems. The work suggested that sustainable outcomes will depend on the creation of policies and routines that are capable of translating values into verifiable practices. However, the shift has yet to be fully incorporated into the mainstream sustainability leadership literature, which remains largely focused on values-driven models. This theoretical shortfall can explain the systemic roots of greenwashing and why it persists, despite well-intentioned leadership. Therefore, it highlights the need to integrate governance theory into the core concept of sustainable leadership (Singh & Sangal, 2025; Jones et al., 2025).

Agency Theory and Accountability

Agency theory provides a compelling explanation for the persistence of greenwashing, even in organizations with genuine sustainability commitments. The agency problem stems from the separation of ownership and control. In the context of sustainability, this is evident when leaders prioritize short-term reputational gains or market expectations, which come at the expense of genuine long-term environmental and social impact (Warhurst, 2002). This rational divergence of interests can lead directly to greenwashing, including "green hushing" to avoid scrutiny or making vague, unsubstantiated claims to manage stakeholders' perceptions. According to Jensen and Meckling (1976), managerial behavior is fundamentally shaped by information asymmetry, monitoring limitations, and incentive misalignment between agents (managers) and principals (stakeholders and owners). This lens is relevant to sustainability contexts, in which environmental and social data are complex, nonfinancial, and difficult for external audiences and boards to verify independently. Such information transparency can create fertile ground for selective disclosure, which allows managers to highlight favorable metrics, frame marginal improvements as significant achievements, or omit contradictory data. Such behavior is not necessarily malicious and is typically a rational response to governance systems in which the cost of full transparency is high, but the benefits of symbolic reporting are immediate and visible.

The relationship between incentives and sustainability behavior shifts the problem from individual ethics to organizational structure (Gizyatova et al., 2024; Pfister et al., 2024). When executive compensation and performance prioritize short-term financial returns, material sustainability initiatives will typically be limited to symbolic gestures or high-visibility marketing efforts. Flammer (2021) examined sustainability-linked compensation and argued that genuine performance will improve when incentives are explicitly tied to verified ESG metrics. In the absence of mechanisms that align managerial rewards with long-term environmental outcomes, greenwashing will become a predictable byproduct of incentive misalignment. This incentive structure can also explain organizations' preference for narratives over substantive evidence. Reputational and market gains from sustainability branding can be captured quickly. By contrast, the cost of deep emissions cuts, supply chain reform, or responsible sourcing is substantial and can accumulate over time (Shao et al., 2025). Confronted with the timing and finance imbalance, managers will have a structural incentive to favor low-cost symbolic actions, such as "green" product lines or distant net-zero pledges, over costly, immediate commitments. Such practices may result in environments that could systematically enable and unintentionally encourage greenwashing.

Agency theory can explain why greenwashing may persist under ethical leadership. Well-intentioned leaders may genuinely pursue sustainability while challenging legacy structures, market pressures, and performance systems that reward superficial disclosure (Singh & Sangal, 2025). The absence of robust internal checks and balances and verification mechanisms may allow such practices to persist, thereby framing greenwashing as a governance issue (Free et al., 2025). Specifically, boards play a critical, yet underrealized, role in resolving the agency problem by enforcing transparent reporting, ethical scrutiny, and accountability. Therefore, agency theory leads to a clear conclusion: Transitioning from greenwashing to genuine sustainability requires going beyond ethical leadership; it demands the deliberate design of governance systems that can address the root causes of the agency problem. The deliberate design involves aligning incentives, reducing information asymmetry through rigorous auditing and transparency, and embedding accountability mechanisms that will make integrity the default organizational condition.

Governance and Organizational Integrity

Governance theory complements agency theory by explaining how organizational structures shape behavior. While agency theory clarifies why managers may misrepresent sustainability information, governance theory explains how organizations can design structures to prevent such behavior. Governance encompasses formal and informal mechanisms for guiding decision-making, allocating authority, overseeing performance, and ensuring compliance. In this context, sustainable governance refers to systems for embedding environmental and social commitments into organizational processes, rather than treating them as symbolic concerns. Institutional theory reinforces the perspective. "Institutional decoupling," which was conceptualized by Meyer and Rowan (1977), explains how organizations separate their formal public-facing structures from their actual day-to-day operations to maintain their legitimacy. In sustainability contexts, such decoupling can be observed in the proliferation of charters, green labels, certifications, and climate pledges, which are typically disconnected from internal operations. Therefore, greenwashing represents a specific form of institutional decoupling, that is, divergence between public environmental claims and operational realities.

Organizational integrity is a product of deliberately designed governance structures. Such structures can enforce consistency between claims and behavior through clear reporting, independent verification, cross-functional accountability, and authorized monitoring bodies. Without such mechanisms, organizations risk the ceremonial adoption of sustainability principles, in which policies exist formally but are weakly implemented. Extensive sustainability reporting infrastructures may facilitate incomplete or selective disclosure when the governance mechanisms lack rigor (Searcy & Buslovich, 2014). Moreover, governance breakdowns may occur for structural reasons. Specifically, boards may lack sustainability expertise and rely on management-provided information, sustainability functions may be isolated from core operations, fragmented data systems can undermine reporting reliability, and third-party certifications vary in their methodological rigor, which may enable reputational arbitrage, that is, an ethics–reputation trade-off. Such vulnerabilities can explain why greenwashing is best understood as a structural phenomenon, rather than an ethical lapse. The literature converges on a central insight: Sustainable leadership requires leaders who are capable of designing governance systems that can enforce integrity, reduce information asymmetry, strengthen accountability, and embed sustainability into decision-making processes. This insight provides the theoretical foundation of the governance architect model developed in this chapter.

Unpacking Greenwashing as a Governance Failure

Greenwashing, whether intentional or unintentional, can mislead stakeholders about an organization's environmental performance and benefits. Greenwashing ranges from minor exaggeration to the outright fabrication of claims or data. However, its consequences are severe, including reduced customer satisfaction, reputational damage, and legal penalties. The root of the problem is seldom a lack of good intentions and is typically weak internal controls and inadequate accountability mechanisms. Existing definitions of sustainable leadership emphasize vision, ethics, and long-term value creation. However, such definitions typically overlook the critical role of systemic governance in ensuring the credibility of sustainability claims. Traditional leadership models focus on short-term financial performance; thus, they are inadequate for addressing the complexities of sustainable development. Although the ESG framework represents progress, it typically prioritizes reporting over verifiable action, which can enable superficial compliance without substantive change. This reality necessitates a shift in perspectives: greenwashing is a systemic flaw that arises from inadequate governance and accountability structures. Therefore, it is a leadership challenge that requires more than rhetoric and demands organizationally embedded solutions (Free et al., 2025). Organizations rarely set out to deliberately mislead their stakeholders. However, they operate within governance environments that allow ambiguous claims, symbolic practices, and selective reporting.

Green Innovation Under Resource Constraints and the Phenomenon of Greenwashing on Green Innovation

Resources can comprise firm-specialized physical, human, and organizational assets (Wernerfelt, 1984) and form the cornerstone of organizational strategies. Whether tangible or intangible, resources can enable organizations to develop valuable capabilities, facilitate the creation of new or improved processes, products, and services, and provide organizations with a competitive advantage. When an organization's essential resources are inadequate to address challenges or seize opportunities, the organization can be described as experiencing "resource constraints." Given limited valuable resources, organizations must strategically acquire and deploy such resources. The world is facing resource shortages, compounded by developed nations' historical failure to adopt resource-constrained strategies, which has led to overproduction and environmental harm. Resource-constrained innovation is defined as the capability to "do better with fewer resources for more recipients" to create increased value while minimizing resource use (Liu et al., 2020). The characteristics of resource-constrained innovation include sustainability, minimalism, reuse of existing components, and user convenience (Prabhu, 2017). However, research provided limited insights into how firms leverage resource-constrained innovation and deploy resources to capture markets. Green innovation, which is a subset of resource-constrained innovation, is defined as sustainable innovation that can minimize environmental impact via technological, process, product, management, or service improvements. The green market has grown rapidly and given rise to greenwashing, in which firms with weak environmental performance communicate positive outcomes (De Freitas Netto et al., 2020). Greenwashing can reduce organizational green innovation inputs and thus weaken organizations' financial performance (Wang & Tian, 2025). This chapter examines strategies for curbing greenwashing.

From a governance perspective, internal systems can create conditions that can facilitate greenwashing. When sustainability reporting is fragmented across environmental, marketing, public relations, and supply chain units, no single function will be able to ensure data accuracy, which may generate inconsistencies that can be intentionally or unintentionally exploited. Sustainability data typically flows through informal channels instead of audited systems, which can increase the risk of unverified claims in reports or marketing (Jain & Mitra, 2025). Such systemic weaknesses can promote greenwashing, even without deceptive intent. Organizations typically lack robust verification mechanisms to improve clarity. Unlike financial reporting, sustainability data reporting lacks standardized audits. Thus, organizations rely on voluntary frameworks or industry metrics that vary in their scope and rigor. In addition, external assurance quality is inconsistent, and internal audit teams may lack ESG expertise. Hence, organizations can choose the least demanding metrics or selectively disclose favorable indicators to produce reports that appear comprehensive but are weak in verifiable substance. Organizational culture can also influence greenwashing. Public sustainability commitments can generate internal pressure to show rapid progress. Employees may present data highlighting successes while downplaying challenges to meet leadership expectations or protect their reputation. In such cultures, storytelling typically overrides accuracy, which can embed greenwashing into organizational routines.

The regulatory environment can also affect governance compliance. Historically, different jurisdictions have voluntary sustainability disclosure requirements, which can offer firms considerable reporting flexibility (Gunningham, 2009). However, despite the existence of regulations, enforcement is typically limited, which may result in minor consequences for making misleading claims. Such ambiguity can contribute to underdeveloped governance systems for environmental integrity (McIntyre, 2018). Corporate examples can illustrate this perspective. For instance, the H&M "Conscious Collection" controversy arose from the inadequate verification of recycled material claims. Meanwhile, oil and gas firms that announce climate commitments while expanding their fossil fuel assets highlight the gaps in governance structures. This can prevent alignment between pledges and practices (Curtin et al., 2019), and present greenwashing as governance fragility reframes sustainable leadership challenges. The problem is not leaders' intentions but their ability to design and enforce systems that can make sustainability claims credible. Ethical motivation alone is inadequate; leaders must actively shape governance (Avery & Bergsteiner, 2011). This approach underpins a governance-based redefinition of sustainable leadership.

4. TOWARD A GOVERNANCE-BASED DEFINITION OF SUSTAINABLE LEADERSHIP

The prevalence of greenwashing has exposed a critical flaw in normative definitions of sustainable leadership (Boeske, 2023). Governance deficit and deceptive claims necessitate redefining sustainable leadership beyond personal qualities. The concept must be comprehended as the capacity to design and enforce accountability systems that can maintain organizational sustainability (Metcalf & Benn, 2013). In other words, sustainable leadership must be a structural, governance-oriented role, rather than an individual characteristic or a rhetorical position (Liao, 2022). Sustainable leadership is redefined in this study as a structural role rooted in the design and enforcement of internal accountability mechanisms that can secure the integrity of environmental and social claims. This redefinition positions sustainable leaders as not merely a visionary or steward but a governance architect who is responsible for embedding sustainability through verifiable systems, transparent data flows, rigorous oversight bodies, and anti-greenwashing controls. Moreover, the governance architect role is not limited to building systems. In fact, he is an inspirational steward. Through designing strong structures, the leader signals that every employee's genuine environmental effort will be seen, measured, and valued, not neglected under corporate rhetoric. In this way, the architect's structural work becomes the foundation that inspires widespread stewardship across the organization: teams adopt pro-environmental behaviours because the system proves that integrity matters.

A governance-based definition will position leaders as the primary architects of organizational integrity. Moreover, instead of emphasizing inspiration, the redefinition focuses on designing and maintaining systems that can prevent misleading claims, including transparent data infrastructures, rigorous reporting protocols, independent verification, cross-functional accountability, and board-level oversight processes. Thus, leaders who are committed to authentic sustainability can serve as governance architects who can shape the conditions under which integrity will become the default (Avery & Bergsteiner, 2011). This idea aligns with broad corporate governance and sustainability reporting trends and the global shift toward mandatory audited disclosure through global frameworks. The reality reflects a move from voluntary narrative-driven reporting to standardized evidence-based accountability (Molnár et al., 2025). Ethical intention alone is inadequate. Technical systems, governance structures, and institutional safeguards are necessary to ensure that sustainability data are accurate, complete, and auditable.

Empirical evidence supports structural reform over ethical prompting to reduce greenwashing. Organizations that integrate sustainability into their risk management, link executives' compensation with verified ESG performance, and embed experts in boards will likely exhibit credible outcomes (Wang et al., 2023). Such governance-centered approaches are crucial for increasing regulatory compliance, investors' confidence, and meaningful contributions to frameworks, such as the UN SDGs (Bajra & Čadež, 2020). Illustrative examples can clarify this shift. For instance, a CEO who is committed to carbon neutrality but does not conduct supply chain emissions tracking will not be able to rely solely on their ethical intent. A leader who prioritizes rigorous reporting and board oversight will likely achieve sustainability, even with minimal rhetoric (Olu-Owolabi et al., 2024). Sustainable leaders can foster accountability through shared commitments to verifiable impact. This frames leadership as an institutional design that can embed sustainability into the organizational structure by laying the groundwork for a governance architect model, which is discussed in the next section.

Illustrative examples from different industries further demonstrate how sustainability failures often originate from governance weaknesses instead of ethical shortcomings. In the financial sector, ESG investment controversies such as DWS Group have highlighted weaknesses in sustainability verification and disclosure processes. In the fashion industry, sustainability-related product claims, such as H&M's Conscious Collection, have frequently been challenged because of insufficient supply-chain transparency. In the energy sector, firms such as Shell and BP have faced criticism for promoting ambitious climate commitments while continuing investments in carbon-intensive activities. Across these industries, the common pattern is not absence of ethical intentions but deficiencies in governance structures capable of verifying sustainability claims.

The continuous challenge of organizations to reduce greenwashing highlights a critical flaw in dominant sustainable leadership models. The literature emphasizes the ethical perspective of long-term orientation and stakeholder engagement. However, such approaches are inadequate to confirm credible sustainability claims (Rosário & Boechat, 2025). Greenwashing persists not because leaders lack ethical intent, but because organizational systems fail to translate values into verifiable action, which reflects deep governance and accountability deficiencies. The reality necessitates reconsidering sustainable leadership as a governance-centered function, rather than a moral role (Zioło et al., 2024).

In this reframing, sustainable leaders will be able to design, implement, and enforce institutional systems that can safeguard sustainability performance and disclosure (Olu-Owolabi et al., 2024). A governance architect will be able to emphasize structures that can constrain managerial discretion, reduce information asymmetry, and align incentives with verified sustainability outcomes, according to agency theory and corporate governance (Manning & Reinecke, 2016). Moreover, the role is systemic, rather than symbolic, and involves transparent reporting, verification similar to financial audits, cross-functional accountability across operations, finance, and risk management, and executive incentives tied to validated performance. Such mechanisms can collectively reduce the structural conditions under which greenwashing may become rational or routine. Ethical leadership remains essential; however, it can be implemented through structures that can ensure the alignment of sustainability claims with actual practices.

Figure 1. Central theme of a sustainable leader as a governance architect

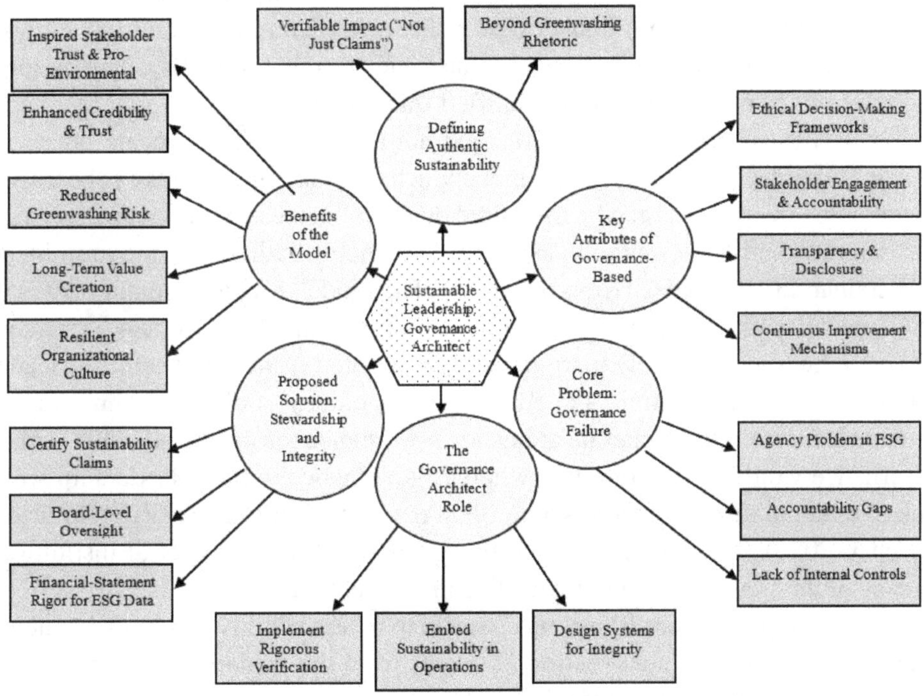

At the center of Figure 1 is governance failure, which is depicted as the primary structural condition under which greenwashing can emerge. This illustration reflects the argument that sustainability misrepresentation stems less from individual moral decay and more from systemic weaknesses in oversight, verification, and accountability mechanisms. Surrounding the core are the leadership responsibilities of the architect of governance, which emphasize the design of integrity systems that can embed sustainability into organizational decision-making, instead of isolating it as a marginal function. The figure illustrates causal logic, rather than a descriptive list of leadership traits. Governance noncompliance can generate accountability gaps and agency problems that, if unaddressed, can produce a sustainability rhetoric that is distinct from reality (Safari & Parker, 2024).

A governance architect can interfere by designing systems for integrity, including transparent data infrastructures and incentive alignment and board-level oversight mechanisms, to achieve verifiable performance. The explicit inclusion of a stewardship and integrity committee underscores the need for an oversight body to reinforce sustainability claims rigorously (Muneer et al., 2025). The outcomes depicted in the figure include enhanced credibility, reduced greenwashing risk, long-term value creation, and organizational resilience, which are consequences of the governance structure rather than aspirational leadership intentions. This framing reinforces the chapter's core proposition: Credible sustainability performance can be achieved through institutional design, rather than rhetorical commitment. Figure 1 depicts the conceptual model, which integrates leadership theory, agency logic, and governance mechanisms into a unified framework.

The governance architect has emerged not as a symbolic advocate but as the principal architect of the conditions under which sustainability claims can be trusted. Leadership effectiveness can be measured by the robustness of governance systems that can make misrepresentation structurally difficult, rather than the prominence of sustainability discourse (Ke et al., 2025). The model can also explain why many ESG and sustainability initiatives fail to prevent greenwashing. When governance is fragmented, operationally marginalized, or isolated from board scrutiny, leaders may face structural constraints in ensuring the accuracy of claims. Therefore, sustainable leadership must operate alongside governance reform. Leadership without governance would be aspirational, whereas governance without leadership would lack intentional design, which is a problem that the model can resolve (Rosário & Boechat, 2025). This rethinking sets the groundwork for the central institutional scheme of the stewardship and integrity committee. In designing integrity systems, leaders will require a board-level mechanism to independently verify sustainability claims rigorously. The next section advances from the leadership structure to the board structure and details how the committee can govern sustainable leadership and embed accountability at the highest organizational level.

5. STEWARDSHIP AND INTEGRITY COMMITTEE

Purpose and Governance-Based Rationale

Persistent greenwashing shows that sustainability breakdowns are ethical or managerial shortcomings and rooted in inadequate board-level oversight (Khan et al., 2026). Although many organizations have adopted sustainability policies, reporting, and initiatives, such measures typically lack the institutional authority and verification rigor required to ensure claims (Jamil & Wahyuni, 2024). In response to such limitations, this chapter proposes a stewardship and integrity committee as a dedicated board-level governance mechanism responsible for safeguarding the integrity of sustainability claims as rigorously as that of financial reports. The primary purpose of the committee is to transform sustainability from being a narrative-driven voluntary reporting activity into a formally governed organizational function. Positioned at the board level, the committee will serve as a guardian of sustainability integrity and ensure that environmental and social claims are auditable and aligned with recognized regulatory standards. Hence, the committee will address the structural conditions under which greenwashing can proliferate. This elevates sustainability as a core fiduciary concern, rather than merely a secondary corporate initiative.

Structure and Relationship to the Board of Directors

The stewardship and integrity committee is considered to be a permanent committee independent of the board of directors and mirrors the structural logic of audit and risk committees in financial governance. The committee members are expected to possess expertise in sustainability accounting, environmental science, risk management, ESG auditing, and regulatory compliance, and the committee must have the technical capacity to examine sustainability data and assurance processes effectively. The specialized composition distinguishes the committee from conventional ESG or CSR committees because the latter typically lack the authority, expertise, or procedural rigor necessary to prevent sustainability misrepresentation. The committee will operate as a specialized extension of the board's fiduciaries and report directly to the board and provide regular assessments of the sustainability integrity risks, assurance findings, and governance gaps. It will also serve as the board's technically informed arm for sustainability surveillance and embed verification and accountability into routine board deliberations. However, ultimate accountability will remain with the board of directors. The relationship can ensure the systematic integration of sustainability governance into organizational decision-making at the highest level.

Personal Qualities and Organizational Structures

Governance structures are central to the prevention of greenwashing. However, they cannot eliminate the importance of personal leadership qualities, such as ethical conviction, courage, and moral responsibility (Rihal et al., 2025). Sustainable leadership requires individuals who are willing to challenge misleading practices, can resist short-term pressures, and will prioritize long-term integrity over reputational symbolism. However, personal qualities alone will be inadequate to ensure sustainability integrity when organizational systems reward discretion, opacity, or selective disclosure. Nevertheless, personal qualities remain essential because they serve as incentive for structural reform. Leaders with strong moral courage and a commitment to transparency are more likely to challenge existing weak practices and champion governance practices that strengthen accountability. When a leader obviously leads by example and refuses to sign off on unverified claims and demands transparency even when it exposes short-term weaknesses, they send an unmistakable signal: integrity is non-negotiable. This personal conviction, when coupled with the structural change such as establishing a Stewardship and Integrity Committee against board resistance, becomes a primary inspirational force. Employees witness that their leader sacrifices comfort for credibility. In response, they adopt environmental stewardship as a shared moral commitment. Thus, personal ethical conviction does not replace governance systems; it enriches them. It transforms a compliance-driven entity into a steward-led organization where every member feels empowered to act.

Functions: Assurance, Audit, and Certification Oversight

The committee is designed to address the governance gaps that historically facilitated greenwashing. Its central function is the verification of sustainability claims through evidence-based internal reviews and independent assurance processes. The committee will evaluate the methodologies, assumptions, and data systems that underpin sustainability disclosures to ensure their audit readiness is comparable to financial standards. The committee will also oversee the organization's engagement with third-party certification bodies, such as the ISO 9000, and assess the credibility, independence, and methodological rigor of external assurance providers. Such oversight can mitigate "assurance shopping" risks, which can undermine the credibility of sustainability reporting. The committee will also integrate sustainability risks into enterprise risk management frameworks and identify material greenwashing risks, regulatory noncompliance, and reputational vulnerabilities associated with inaccurate or misleading claims (Deghaim et al., 2024). Other responsibilities of the committee include aligning the organization with frameworks such as the GRI and ISSB and reviewing ESG-linked compensation to ensure that rewards are based on verified performance, rather than self-reported data. Dedicated whistleblower channels for sustainability misconduct can further reinforce internal accountability mechanisms and detect integrity breaches in advance (Pratama et al., 2025).

Figure 2. Traditional approach versus governance-architect model

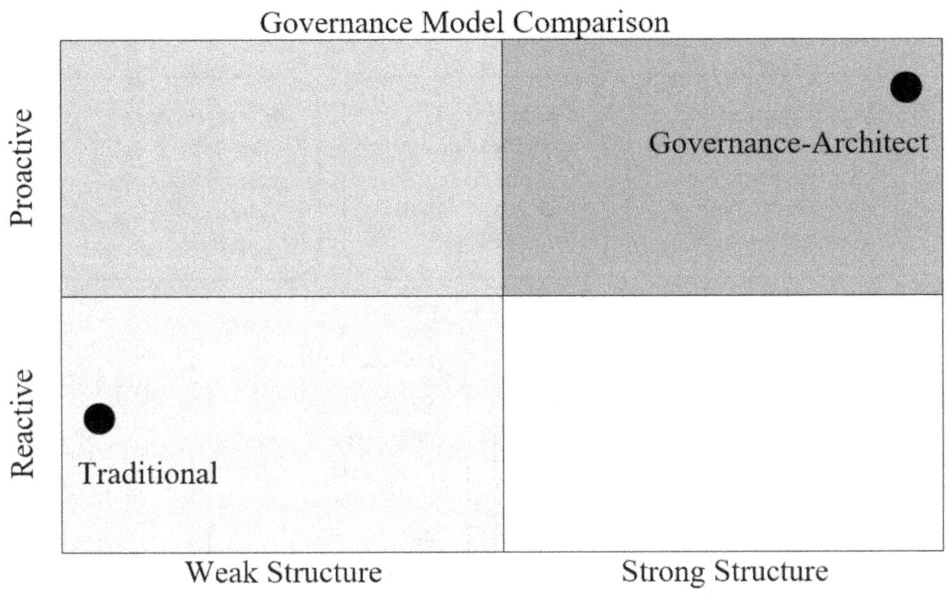

Figure 2 compares traditional sustainability governance approaches with the proposed model. Traditional approaches are characterized by weak structures and limited verification and score poorly on transparency, accountability, and assurance, which are optimal conditions for greenwashing (Shukla, 2025). By contrast, the architect model demonstrates superior performance across the dimensions. Specifically, it can embed sustainability oversight into the governance system, rather than relying on voluntary commitments or individual leadership ethics. The model reinforces the argument that sustainability integrity demands governance structures that can institutionalize accountability (Lee et al., 2025; Permatasari et al., 2025). Meanwhile, the stewardship and integrity committee serves as the board-level mechanism through which the governance approach can be deployed.

Figure 3. Responsibilities of the stewardship & integrity committee

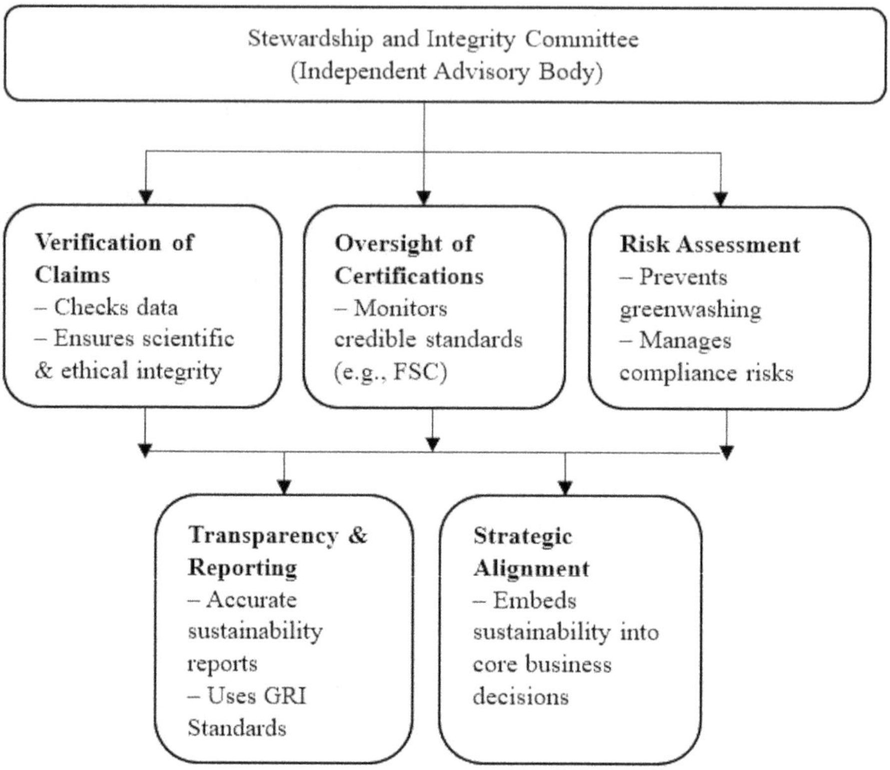

The committee's integrated system of responsibilities, including verification, assurance compliance, risk management, reporting integrity, and accountability, is illustrated in Figure 3. The system can reinforce sustainability disclosure reliability (Deghaim et al., 2024).

Figure 4. Core functions of the stewardship & integrity committee

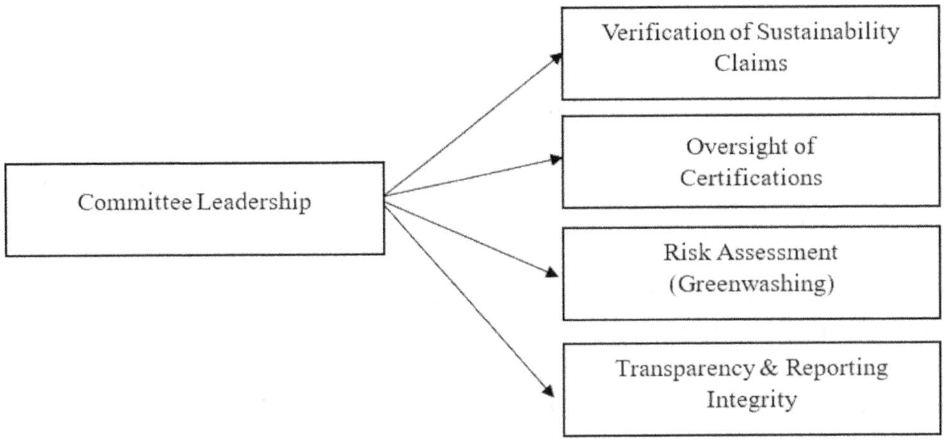

The committee's core functions are synthesized in Figure 4, which illustrates how verification, assurance oversight, risk management, reporting integrity, and accountability mechanisms interact as an integrated governance system. Rather than operating as individual controls, the functions can collectively increase disclosure reliability and reduce opportunities for selective reporting. By consolidating such control within a board-level committee, the model can avoid oversight fragmentation, which is common in sustainability governance. The model emphasizes that integrity will emerge from interacting systems and institutional authority. Moreover, the model highlights the governance architect's focus on structural design, rather than superficial action (Ye & Guo, 2025; Mei & Broyde, 2025).

Table 1. Contrasting traditional and governance-based leadership

Aspect	Traditional Sustainable Leadership	Governance-Based Sustainable Leadership
Focus	Rhetoric, communication, and individual values	Governance structures, systemic accountability, and integrity mechanisms
Risk of Greenwashing	High, due to a lack of verifiable oversight	Significantly reduced due to embedded controls
Role of Leader	Advocate, visionary, communicator	Architect of systems, guardian of integrity, accountability enforcer
Ultimate Outcome	Reputational goodwill, symbolic compliance	Verified Environmental Stewardship (collective, democratized accountability across all functions)

Table 1 clarifies the conceptual shift advanced by contrasting traditional sustainable leadership with the new governance-based definition. Traditional approaches prioritize communication, advocacy, and individual values. Hence, they may result in high exposure to greenwashing owing to weak verification mechanisms. By contrast, leaders, as guardians of integrity, can substantially reduce greenwashing risks through embedded controls and board-level accountability (Shukla, 2025). The comparison shows that leadership effectiveness in sustainability should be evaluated through governance robustness, which can ensure claim accuracy, rather than rhetoric. The suggested committee can handle the shift by providing the institutional infrastructure necessary to translate intent into verifiable organizational outcomes through formal verification mechanisms (Fasasi et al., 2021).

Why Traditional ESG and CSR Committees Are Insufficient

Many organizations have established ESG or CSR committees. However, such bodies typically lack the authority, independence, and technical expertise necessary to evaluate sustainability claims rigorously and are typically positioned as advisory or managerial bodies, rather than governance mechanisms with formal oversight power. Such shortcomings may limit their ability to prevent misrepresentation or enforce accountability. The stewardship and integrity committee can address the limitations by embedding sustainability verification into the board's formal governance structure to align it with financial oversight (Tin & Yong, 2025).

Board's Role in Facing Greenwashing

The stewardship and integrity committee can reframe the board of directors' role in sustainability governance. Boards have acknowledged that sustainability disclosures carry strategic, financial, reputational, and ethical implications. Through the committee's work, sustainability integrity will become a continuous governance responsibility, rather than a periodic reporting exercise, which can ultimately strengthen the board's capacity to detect, prevent, and respond to greenwashing risks (Hossain et al., 2025).

The practical relevance of the proposed committee can be illustrated through the cases discussed earlier. In the DWS Group controversy, allegations concerning overstated ESG practices revealed weaknesses in internal verification and board-level oversight. Under our Governance-Architect Model, a Stewardship and Integrity Committee would independently review sustainability methodologies, assess the evidence supporting ESG disclosures, and evaluate whether public sustainability claims precisely reflect organizational practices. Such mechanism could reduce the risk of contradictions between reported and actual ESG performance. Similarly,

H&M's Conscious Collection clash highlighted governance deficiencies related to sustainability verification within complex global supply chains. The Governance-Architect Model would require sustainability claims to undergo systematic validation through transparent reporting systems, independent assurance processes, and board-level review. These mechanisms could reduce the likelihood of misleading environmental representations.

Transition to the Governance-Architect Model

The stewardship and integrity committee can serve as an institutional bridge between leadership design and board oversight through its purpose, structure, functions, and relationship with the board of directors. The committee grounds the governance reforms required to operationalize the governance architect model and provides the foundation for a consolidated conceptual framework. The next section integrates several factors into the model to offer a unified explanation of how organizations can eliminate greenwashing through structural design, rather than commitments (Buttigieg & Pulis, 2024).

6. THE GOVERNANCE-ARCHITECT MODEL: A NEW CONCEPTUAL FRAMEWORK

The model redefines sustainable leaders as system designers, rather than merely advocates, who can build sustainability into their organization's structure. Unlike approaches that rely primarily on personal ethics or rhetorical commitments, the model emphasizes institutional design to ensure that claims are verifiable and auditable. By drawing on organizational design, institutional theory, and the responsible leadership view, the model demonstrates how leaders can translate sustainability values into structural mechanisms to assure accountability and integrity. At the model's core are interrelated dimensions that can prevent greenwashing. Systemic transparency relies on integrated data infrastructures that can secure accurate, consistent, and accessible environmental and social performance information. The combination of internal audits with external assurances can further validate high-risk claims. In addition, the model embeds sustainability responsibilities across operations, supply chains, strategies, and HR and risk management operations to reduce superficial efforts. Incentive structures can align rewards with verified performance. Meanwhile, a strong culture, ethical decision-making frameworks, and whistleblower protection channels can promote psychological safety and reporting without fear of harm. Furthermore, the model emphasizes alignment with globally recognized standards, such as the GRI, SASB, and ISSB, to ensure comparability, traceability, and credibility (Sneideriene & Legenzova, 2026; Varsha et al., 2024).

This cross-functional design goes beyond distribute responsibility as it democratizes stewardship. When sustainability metrics are embedded into the daily workflows of HR (e.g., green competency training), supply chain (e.g., supplier ESG scorecards), finance (e.g., carbon-adjusted budgets), and operations (e.g., waste-reduction targets), every employee becomes an environmental steward by default, not by exception. Hence, accountability becomes a shared, visible norm. In this structure, a warehouse worker who flags excessive packaging, a procurement officer who rejects a non-compliant supplier, or a financial analyst who questions a carbon offset claim are all acting as empowered stewards. Thus, the governance architect's most inspiring achievement is that it is a system that makes stewardship everyone's job.

Based on agency theory, the model acknowledges that managers may engage in impression management when oversight is weak or when incentives are misaligned. Governance architect systems are expected to constrain opportunistic behavior and reduce information asymmetry and enforce accountability, and leaders are expected to act as the architects of operational systems. Specifically, leaders can shape the conditions under which sustainability can be measured, verified, and disclosed. Although personal ethics and cultural norms are important, they pass through formal governance structures. The proposed approach can systematically embed integrity into organizational systems and ensure reliable ethical outcomes by avoiding reliance on the discretionary will of managers. Another central feature of the model is its board-level oversight through a stewardship and integrity committee, which will independently assess the operational and disclosure mechanisms. The operational systems and board oversight mechanism create a dual-governance structure that can align internal accountability with external regulatory expectations, such as CSRD and ISSB standards. The examination of sustainability claims as rigorous as financial standards can structurally control greenwashing and elevate sustainability integrity as a core governance responsibility (Noor et al., 2025).

The model can also address institutional separation when organizations adopt symbolic policies without substantive implementation. Responsibilities embedded across departments can reduce the gap between policy and practice, which can be reinforced by incentives tied to verified outcomes, whistleblower policies, and recognized reporting frameworks. Empirical studies showed that organizations with comprehensive governance systems produce superior disclosure reports that demonstrate high alignment between the stated policies and actual practices. The operationalization of the model will require a holistic approach. Therefore, data infrastructures must collect accurate and verifiable ESG information across the value chain. Moreover, reporting must adhere to global frameworks, and accountability must be distributed across organizational functions. Stakeholder engagement channels should enable scrutiny and the use of ethical frameworks, and whistleblower protection mechanisms must guide decision-making. Furthermore, internal and external audits should validate ESG claims, and executives' compensation should be tied to verified ESG performance to align incentives with long-term value (Marta & Rake, 2025; Sklavos et al., 2026). An additional outcome of this model is inspired stakeholder action. Governance systems verify sustainability claims through transparent reporting, independent assurance, and accountable oversight. Hence, stakeholders gain confidence in the credibility of organizational commitments that encourages employees, investors, customers, and communities to move beyond passive approval. Stakeholders will actively support sustainability initiatives and foster collective stewardship and strengthen long-term sustainability outcomes.

Table 2. Key functions of a governance architect within an organization

Governance Architect Function	Description	Impact on Greenwashing Prevention
Robust Data Infrastructure	Establishing systems for accurate, consistent, and verifiable collection of environmental and social data across the value chain.	Ensures claims are based on solid evidence, reducing the risk of unsubstantiated assertions.
Independent Verification Processes	Implementing internal and external audit mechanisms for sustainability reports and claims, akin to financial audits.	Provides an objective assessment of data and claims, increasing credibility and detecting potential misrepresentations.
Clear Reporting Standards	Adopting and adhering to globally recognized sustainability reporting frameworks (e.g., GRI, SASB) to ensure comprehensive and comparable disclosure.	Standardizes reporting, making it difficult to hide or selectively disclose information, thereby fostering transparency.
Cross-Functional Accountability	Assigning clear responsibilities and accountability for sustainability performance across different departments and leadership levels.	Distributes ownership of sustainability, preventing it from being isolated in a single department and ensuring collective responsibility.
Stakeholder Engagement Mechanisms	Establishing formal channels for engaging with stakeholders (employees, customers, investors, and communities) on sustainability issues and feedback.	Increases external scrutiny and feedback, helping to identify and correct any discrepancies or perceived greenwashing.
Ethical Decision-Making Frameworks	Integrating ethical considerations and sustainability impact assessments into all significant business decisions.	Guides leaders to make choices that genuinely advance sustainability rather than just appearing to do so.

The model can strengthen internal controls and enhance transparency by embedding such functions into a unified structure. Integration can foster a culture of evidence-based decision-making across an organization (Patil et al., 2025) and transform sustainability from a voluntary or symbolic initiative into a strategic capability that is operational, measurable, and accountable. Through the integration of leadership, governance, and organizational design, the model can provide a conceptual and practical framework for eliminating the structural conditions under which greenwashing can emerge.

The Governance-Architect Model can be further strengthened through emerging digital technologies that enhance mechanisms of organizational governance systems. Particularly, artificial intelligence can support the continuous monitoring of ESG performance, automated anomaly detection, and the identification of inconsistencies between sustainability claims and operational data. Therefore, AI assists leaders and oversight bodies in detecting potential greenwashing risks at early stages. Similarly, blockchain-verified ESG systems can strengthen the model's emphasis on verifiable performance by creating traceable records of environmental and social data across organizational value chains. Hence, they reduce opportunities for selective disclosure or information manipulation (Saberi et al., 2019; Kouhizadeh & Sarkis, 2018). Furthermore, real-time sustainability reporting platforms can enhance the transparency objectives of the Governance-Architect Model. They provide continuous access to sustainability metrics for boards, auditors, regulators, and stakeholders that moves beyond periodic reporting toward ongoing accountability. These technologies do not replace governance mechanisms, but they digitally reinforce the stewardship and integrity committee's verification role and strengthen the governance architect's capacity to ensure that sustainability claims remain aligned with actual organizational performance.

The foundation sets the stage for the next section, which explores the practical pathways through which organizations can implement the governance architect model and achieve sustainable and verifiable outcomes (P. et al., 2025; Tumpa & Naeni, 2025).

7. IMPLEMENTATION PATHWAYS FOR ORGANIZATIONS

The implementation of the governance architect model will require structured pathways that can align governance reforms with operational realities. The process begins with the systematic diagnosis of the governance gaps that can expose organizations to sustainability misrepresentation. Firms may underestimate how fragmented data systems, inconsistent reporting, and siloed responsibilities can increase greenwashing risk. The sustainability reporting literature shows that organizations typically overestimate the robustness of their internal controls because they assume that voluntary ESG teams can provide adequate monitoring. A diagnostic assessment is a crucial first step to examine data sources, reporting controls, verification practices, incentive structures, and organizational culture to identify misalignments between claims and evidence. Unless such assessments demonstrate clarity, reforms will remain symbolic and fail to impact real-world decision-making (Islam, 2025; Indrawati et al., 2025).

Once vulnerabilities are identified, organizations can design oversight systems that can embed sustainability integrity into the structural model. This phase manages the governance architect's role through transparent reporting, robust data governance, cross-departmental accountability, and formal verification workflows. Oversight systems must match regulations such as the CSRD, which requires equivalent assurances for sustainability disclosures. An effective system can integrate sustainability metrics into risk management mechanisms to ensure that environmental and social risks undergo the same evaluation as financial indicators. Organizations that embed sustainability into their core governance typically demonstrate high reporting reliability and experience reduced reputational and regulatory exposure from selective disclosure (Amin et al., 2025).

Implementation will also require alignment between the organizational culture, incentive structure, and governance to support the integrity mechanisms. Short-term-driven cultures can generate pressure that can undermine accountability. Initiatives will fail when employees lack psychological safety to report inconsistencies or when executives face conflicting incentives that reward appearances over verified outcomes. Alignment will require communicating the strategic value of sustainability integrity, establishing clear expectations for ethical reporting, and normalizing transparency and internal controls. Management incentives must reward independently verified sustainability performance and penalize practices that can compromise data accuracy or governance. When culture, incentives, and governance reinforce one another, sustainability integrity will become an organizational norm, rather than merely an aspirational goal. At the leadership level, the alignment can be managed through distinct competencies and behaviors characteristic of the architect role (Nashrullah et al., 2025).

Table 3. Competencies and behaviours for the governance-architect leadership model to practice

Educate and empower	Develop a deep understanding of sustainability complexities and empower teams to integrate sustainable practices into their daily operations. This includes investing in specialized leadership training (López et al., 2025).
Lead by Example	Demonstrate unwavering commitment to transparency and ethical conduct to foster a culture where integrity is paramount (Brown et al., 2005).
Champion Structural Change	Advocate for and implement governance structures like the stewardship and integrity committee (Aguilera et al., 2008).

continued on following page

Table 3. Continued

Embrace Technology	Utilize data and analytics to track progress, ensure accuracy, and provide transparent reporting on sustainability metrics (Rosário & Boechat, 2025).
Foster Collaboration	Engage with internal and external stakeholders, including regulators, NGOs, and industry peers, to align on best practices and build collective accountability (Harrison et al., 2010).
Narrative Integrity	Communicate an inspiring sustainability vision grounded in verified evidence. Sustainability narratives accurately reflect organizational performance, which motivates stakeholders to engage in authentic environmental stewardship.

As summarized in Table 3, effective governance architects combine governance literacy with ethical credibility, technological capability, and collaborative capacity. Leaders must gain a deep understanding of sustainability complexity and governance requirements, then embed such knowledge into managerial and operational roles, instead of limiting it to specialist units. Moreover, they should invest in targeted leadership development to enhance systems thinking and sustainability assurance competence. Leadership credibility will be further strengthened when executives show a consistent commitment to transparency and ethical conduct to emphasize that integrity is nonnegotiable, rather than discretionary. Most importantly, governance architects can drive structural reforms by governing mechanisms such as a stewardship and integrity committee and establishing sustainability oversight systems at the board level (Collevecchio et al., 2025; Yong et al., 2024). Technology also plays a key role in the model. Leaders can leverage data analytics and integrated reporting systems to ensure the traceability, accuracy, and audit readiness of sustainability metrics (Thawani et al., 2025). Furthermore, effective implementation will require collaboration beyond organizational boundaries, which will involve engaging regulators, assurance providers, civil society, and industry peers to align internal governance with evolving standards and accountability norms (Job & Khanna, 2024).

To enhance empirical application, the Governance-Architect Model could be operationalized using a handful of governance and sustainability dimensions. Governance architect effectiveness may be evaluated through the degree of sustainability integration into organizational decision-making, the existence of verified reporting systems, and the alignment of executive incentives with ESG performance. The effectiveness of a Stewardship and Integrity Committee may be assessed through indicators such as the frequency of sustainability assurance reviews, the proportion of sustainability claims independently verified, the identification and mitigation of greenwashing risks, and board engagement with sustainability oversight. Transparency systems may be evaluated through disclosure quality, reporting completeness, assurance coverage, and stakeholder access to sustainability information. Accountability mechanisms could be measured through whistleblower utilization, corrective action implementation, ESG-linked performance evaluation systems, and compliance with recognized reporting standards. Collectively, these dimensions provide an initial framework for evaluating the effectiveness of governance-centred sustainable leadership and offer a foundation for future empirical research.

Figure 5. Impact of ad-hoc sustainability efforts versus the governance-architect model

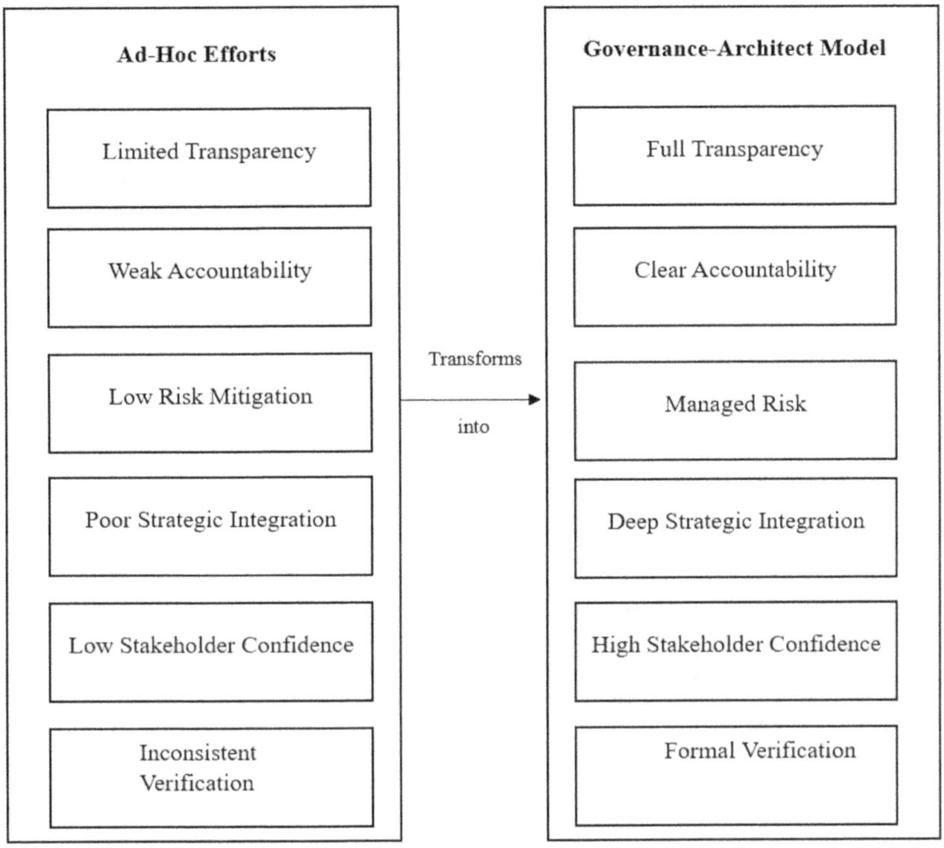

The contrast between ad hoc initiatives and governance-embedded implementation can strengthen the argument for the latter approach, as shown in Figure 5. Although ad hoc efforts can facilitate modest improvements, the governance model can produce stronger outcomes in transparency, accountability, risk mitigation, strategic integration, and stakeholder confidence (Bui & Fifka, 2025). The figure shows that material sustainability performance can be achieved not from isolated initiatives but from governance structures that can enforce verification and accountability (Pérez-Durán, 2024). Such implementation pathways can transform the model from a conceptual framework into an operational governance strategy. Organizations can create the institutional conditions necessary to eliminate greenwashing by diagnosing the governance gaps. Then, they can design rigorous oversight systems and align leadership incentives with their organizational culture (Ghitti et al., 2024). The linked three-stage pathway will emerge across reforms, which integrates diagnostic assessment, board-level reform, and leadership-driven system design.

Figure 6. Action plan of the governance architect implementation framework

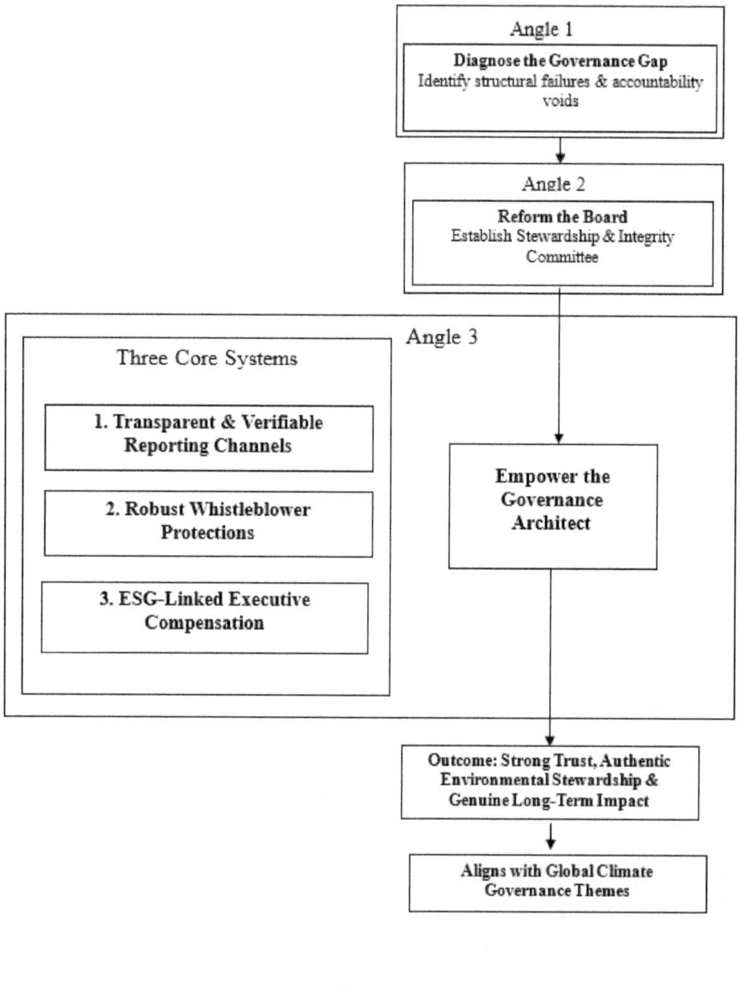

Figure 6 aggregates the chapter's main arguments and shows how organizations progress from diagnosing their governance failures to reforming their board oversight mechanism through a dedicated committee. Subsequently, leaders are empowered as governance architects who can build integrated integrity systems. The structured pathway can restore stakeholders' trust and enable verifiable impact and aligns with global imperatives. Thus, credible sustainability performance can be achieved not through rhetoric but through a deliberate governance structure.

The model implementation may face challenges or counterarguments. Naturally, organizations might resist governance reforms for several reasons. First, smaller firms may lack resources for independent verification and board-level expertise because of cost and capacity constraints. Firms can mitigate this obstacle by adopting governance reforms gradually, beginning with the most significant sustainability claims and expanding oversight as resources allow. Second, board and executive may perceive the committee as a threat to their autonomy. Demonstrating regulatory alignment (e.g., with CSRD or ISSB) and long-term risk reduction can build support. Third, data fragmentation is a threat as many firms lack integrated ESG data systems. Organizations can address this barrier by adopting simple digital reporting tools and gradually integrating sustainability data across the organization. Fourth, governance complexity may arise as sustainability oversight responsibilities would overlap with existing audit, risk, compliance, or ESG committees. Therefore, clear governance charters, defined reporting structures, and coordination mechanisms are crucial to prevent duplication and ambiguity. Fifth, regulatory frames differ across jurisdictions. While some regions mandate sustainability assurance and disclosure, others follow voluntary frameworks. Hence, organizations should adapt the Governance-Architect Model to local institutional roles while preserving its core principles. Finally, cultural inertia may arise because employees accustomed to symbolic reporting may resist rigorous verification. Leadership storytelling around "Narrative Integrity," as conceptualized in Table 3, can transform this resistance into ownership. Acknowledging these challenges does not weaken the model; it strengthens its practical credibility.

8. CONTRIBUTION AND INNOVATION

This chapter's contributions stem from three interrelated innovations that can advance sustainable leadership theory and practice. First, the accountable stewardship framework eliminates sustainability rhetoric and requires organizational claims to be real, measurable, and independently verifiable. Agency theory explains why well-intentioned leaders may make misleading claims when incentive structures reward appearances rather than integrity (Sasaki, 2025). The accountability framework provides a conceptual foundation for credible sustainability oversight mechanisms. Hence, this chapter introduces the stewardship and integrity committee, which is a novel board-level body that can certify sustainability claims. The committee can address systemic governance gaps by overseeing independent verifications, mitigating greenwashing risks, evaluating sustainability-related compensation, and ensuring the alignment of ESG data with global standards. Thus, the committee can transform sustainability from a voluntary aspiration into a verifiable organizational practice (Bu et al., 2024; Cohen et al., 2023).

Second, this chapter proposes a governance architect model of sustainable leadership and defines leaders as designers of systems, rather than merely proponents of values. Leaders can build structures that can prevent greenwashing, such as transparent reporting measures, cross-functional accountability systems, compensation packages linked with verified outcomes, whistleblower anonymity channels, independent audits, and unified data structures (Benveniste & Mizrahi, 2023). The model embeds sustainability into the organizational DNA by shifting the focus from beliefs to systemic design (Ma et al., 2026), which may have major implications for global climate governance. ESG regulators, investors, and civil society are demanding rigorous disclosure that can influence capital, markets, and policies (Goerzen et al., 2025). Thus, the model aligns with major regulatory reforms, including the EU's Corporate Sustainability Reporting Directive, the ISSB's IFRS S1 and S2, and the U.S. SEC's climate disclosure rules (Gadinis & Fregoni, 2026). Together, they signal the broad shift toward treating sustainability information with the seriousness of financial regulations.

Greenwashing remains a major obstacle to sustainable development, notably in sectors such as agriculture (Abouarab et al., 2025; Koch & Denner, 2025). This observation highlights the urgency of international collaboration. Organizations such as the FAO, World Bank, and ISO are advancing standardized ESG disclosure guidelines. Moreover, technologies, such as blockchain, have emerged to enhance data integrity and independent audits (Almadadha, 2024). Initiatives such as the WTO-managed "greenwashing blacklist" further highlight the growing intolerance for misleading claims. Such developments have exposed the limitations of voluntary ESG reporting, which typically suffers from inconsistent methodologies, weak verification, and selective disclosure (Bernini & La Rosa, 2024). The governance architect model can directly address the gaps by embedding transparency and accountability into governance structures, which can enable firms to meet emerging regulatory expectations while mitigating legal, financial, and reputational risks.

The model can also strengthen global climate action by supporting accurate emissions tracking, evaluating national commitments, informing mitigation strategies, and mobilizing climate finance. Investors rely on verified ESG data to assess risks and identify credible transition pathways. Institutionalized integrity mechanisms, such as board-level certification and independent assurances, can rebuild trust in corporate sustainability claims, which can increase investors' confidence, stakeholders' engagement, and the legitimacy of global climate initiatives (Gipper et al., 2025; Krasodomska et al., 2025). Such features can also improve ESG investment. Investors can distinguish genuinely sustainable firms from those that engage in performative disclosure. Such distinctions can reduce greenwashing risk and improve market efficiency.

The model also carries global weight pertaining the United Nations Sustainable Development Goals (SDGs) by strengthening the governance foundations necessary for credible sustainability action. In particular, the Stewardship and Integrity Committee promotes integrity which support SDG 13 (Climate Action) and SDG 16 (Strong Institutions). The model suggests that meaningful environmental stewardship at organizational and global levels requires more than aspirational commitments. Rather, it depends on robust technical systems, independent verification mechanisms, and institutional safeguards that ensure sustainability decisions are grounded in reliable evidence. The Governance-Architect Model provides a governance pathway through which organizations can contribute more credibly to global sustainability and climate objectives.

This chapter makes substantive scholarly and practical contributions to multiple domains. Specifically, to leadership studies, this study contributes by shifting sustainable leadership from individual ethics to institutional structures. To corporate governance and ESG research, this study contributes by introducing a novel board-level oversight mechanism to strengthen assurance and disclosure integrity. To environmental policy and global climate governance, this study contributes by providing a structural explanation for persistent greenwashing and institutional fragmentation. Insights from organizational behavior and institutional theory support and ground the model described in this chapter. Furthermore, the model translates into an actionable framework for executives, policymakers, and practitioners. The framework has the potential to influence academic discourse, regulatory development, and professional standards and shape executive education, board governance, and cross-sectoral ESG implementation.

9. CONCLUSION AND FUTURE RESEARCH DIRECTIONS

Greenwashing is a persistent challenge and a symptom of deep governance weakness (Sundarasen et al., 2024). This chapter argues that credible sustainability in a climate-challenged world will require the reconceptualization of leadership. New leaders must be governance architects who can design systems to institutionalize accountability and integrity (Bernini & La Rosa, 2024). Existing leadership models, which focus on values and stakeholder engagement, have proven themselves to be inadequate. Specifically, they have failed to prevent the separation of sustainability commitments from actual performance. By evaluating greenwashing with agency theory and institutional integrity, this analysis demonstrated that credibility depends on robust measurement, disclosure, and oversight structures. The governance architect model reframes leadership and focuses on creating transparent reporting systems, conducting independent verification, aligning incentives, and designing protected channels for accountability. The model can be institutionalized through a board-level committee tasked with certifying sustainability claims with the rigor of financial auditing. The mechanisms can build an organizational structure that can make greenwashing structurally difficult and shift sustainability from a symbolic commitment to an embedded practice.

The implications of the model extend across theory and practice. For leadership studies, the model expands sustainable leadership from personal morality to institutional design. For corporate governance and ESG, the model offers a concrete framework for enhancing disclosure reliability and aligning investors' expectations with verifiable outcomes. For environmental policy and climate governance, the model addresses the accountability gaps by supporting the transition from narrative-based reporting to evidence-based performance. Yet the ultimate research question is not merely empirical, it is moral: Will organizations and their leaders build the governance that makes greenwashing unthinkable? The answer begins not in a study, but in a choice.

Why This Chapter Matters

This chapter provides an academically rigorous, yet actionable, response to the contemporary crisis of trust in corporate sustainability. This chapter moves beyond a greenwashing diagnosis toward a governance-centered blueprint for credible leadership and bridges leadership theory, governance theory, and institutional integrity to open new scholarly pathways. Moreover, this chapter provides executives, board members, and policymakers with a concrete model to operationalize integrity through verifiable systems. In an era when climate credibility has become nonnegotiable, leadership must be judged by its developed accountability structures (Bernini & La Rosa, 2024). In addition, the model reframes sustainable leadership as a matter of governance design, verification, and systemic accountability and offers a pathway to render greenwashing structurally obsolete, with the goal of fostering resilient, trustworthy, and genuinely sustainable organizations.

The governance architect model also opens avenues for future research. First, scholars should empirically examine how specific governance structures, such as board-level oversight committees, can shape sustainability performance and reduce greenwashing. Specifically, longitudinal studies can identify effective configurations. Second, future research should explore how leaders implement governance reforms in practice, which will require focusing on competencies such as governance literacy, systems thinking, and risk management. Comparative studies can highlight the influence of regulatory contexts, cultural norms, and organizational structures.

Third, future work should assess interactions between evolving global standards and internal governance reforms. Such evaluations can measure how governance-centered approaches can enhance compliance, reduce reporting inconsistencies, and strengthen investors' confidence. Finally, scholars should investigate how governance-based leadership can contribute to broad climate governance, including how firm-level integrity in sustainability disclosure can support climate modeling, policy consistency, capital allocation, and the alignment of organizational governance with national and international climate objectives.

Future research may also examine some propositions derived from the model. First, organizations that establish dedicated Stewardship and Integrity Committees may face lower levels of greenwashing risk or higher sustainability disclosure credibility compared to organizations relying on conventional ESG committees. Second, the integration of independent sustainability assurance mechanisms may revive the relationship between sustainability commitments and actual sustainability performance. Third, the alignment of executive compensation with verified ESG outcomes may seize information asymmetry and thus improve accountability.

Several research questions also emerge from the model. How does governance-architect leadership influence organizational sustainability performance? Which governance mechanisms are most effective in reducing greenwashing risk? How do regulatory environments moderate the effectiveness of stewardship and integrity committees? To what extent do emerging technologies such as AI and blockchain empower governance-based sustainability oversight? These questions provide a foundation for future empirical studies that can evaluate, refine, and extend the Governance-Architect Model across different industries and institutional contexts.

REFERENCES

Abouarab, R., Mishra, T., & Wolfe, S. (2025). Spotting Portfolio Greenwashing in Environmental Funds. *Journal of Business Ethics, 197*(4), 811–839. DOI: 10.1007/s10551-024-05783-z

Afshar Jahanshahi, A., & Kalaycioglu, O. (2025). The Dual Faces of Leadership: CEOs' Dark and Light Traits in Shaping Business Ethical Climate and Sustainability. *Journal of Business Ethics*, •••, 1–21.

Aguilera, R. V., Filatotchev, I., Gospel, H., & Jackson, G. (2008). An organizational approach to comparative corporate governance: Costs, contingencies, and complementarities. *Organization Science, 19*(3), 475–492. DOI: 10.1287/orsc.1070.0322

Ahmed, A. (2024). Sustainable leadership: Integrating environmental responsibility into corporate governance. *Journal of Reproducible Research, 2*(2), 45–54. https://journalrrsite.com/index.php/Myjrr/article/view/69

Albuquerque Pai, A., Anand, A., Pazhoothundathil, N., & Ashok, L. (2024). Leadership perspectives on resilience capabilities for navigating disruption. *Journal of Asia Business Studies, 18*(1), 103–123. DOI: 10.1108/JABS-03-2023-0081

Almadadha, R. (2024). Blockchain Technology in Financial Accounting: Enhancing Transparency, Security, and ESG Reporting. *Blockchains, 2*(3), 312–333. DOI: 10.3390/blockchains2030015

Amin, A., Wiyarni, W., & Kadarusman, K. (2025). Corporate governance principles and their influence on sustainability reporting in Indonesian public companies: The moderating role of associations and regulators. *Edelweiss Applied Science and Technology, 9*(10), 1459–1471. DOI: 10.55214/2576-8484.v9i10.10680

Apostol, O., Dey, C., & Thomson, I. (2025). Introducing a Themed Issue on Sustainability Reporting Regulation in Europe. *Social and Environmental Accountability Journal, 45*(2), 89–94. DOI: 10.1080/0969160X.2025.2537994

Avery, G. C., & Bergsteiner, H. (2011). Sustainable leadership practices for enhancing business resilience and performance. *Strategy and Leadership, 39*(3), 5–15. DOI: 10.1108/10878571111128766

Bajra, U., & Čadež, S. (2020). Alternative regulatory policies, compliance, and corporate governance quality. *Baltic Journal of Management, 15*(1), 42–60. DOI: 10.1108/BJM-11-2018-0373

Benveniste, J., & Mizrahi, S. (2023). Accountability, governance, and institutional change in intergovernmental relations: A theoretical framework. *Politics & Policy, 51*(2), 242–255. DOI: 10.1111/polp.12520

Bernini, F., & La Rosa, F. (2024). Research in the greenwashing field: Concepts, theories, and potential impacts on economic and social value. *The Journal of Management and Governance, 28*(2), 405–444. DOI: 10.1007/s10997-023-09686-5

Bîzoi, A. C., & Bîzoi, C. G. (2025). Essence and Appearance: A Critical Exploration of Corporate Greenwashing Through Ethical Dilemmas in Global Supply Chains. *Green Wealth: Navigating towards a Sustainable Future*, 211-237.

Boeske, J. (2023). Leadership towards sustainability: A review of sustainable, sustainability, and environmental leadership. *Sustainability (Basel), 15*(16), 12626. DOI: 10.3390/su151612626

Brown, M. E., Treviño, L. K., & Harrison, D. A. (2005). Ethical leadership: A social learning perspective for construct development and testing. *Organizational Behavior and Human Decision Processes, 97*(2), 117–134. DOI: 10.1016/j.obhdp.2005.03.002

Bu, M., Liu, X., Zhang, B., Hazaea, S. A., Fan, R., & Wang, Z. (2024). Governance of Corporate Greenwashing through ESG Assurance. *Systems, 12*(9), 365. DOI: 10.3390/systems12090365

Bui, T. M. C., & Fifka, M. (2025). *From adversaries to allies: Cross-sector partnerships for sustainability between businesses and civil society organisations—a systematic literature review and future research avenues.* Management Review Quarterly., DOI: 10.1007/s11301-025-00573-x

Buttigieg, C. P., & Pulis, S. (2024). Strategic initiatives to address greenwashing. *ERA Forum, 25*, 327–337. https://doi.org/DOI: 10.1007/s12027-024-00815-7

Christensen, D. M., Serafeim, G., & Sikochi, A. (2022). Why is Corporate Virtue in the Eye of The Beholder? The Case of ESG Ratings. *The Accounting Review, 97*(1), 147–175. DOI: 10.2308/TAR-2019-0506

Cohen, S., Kadach, I., Ormazabal, G., & Reichelstein, S. (2023). Executive Compensation Tied to ESG Performance: International Evidence. *Journal of Accounting Research, 61*(3), 805–853. DOI: 10.1111/1475-679X.12481

Collevecchio, F., Temperini, V., Barba-Sanchez, V., & Meseguer-Martinez, A. (2025). Sustainable Governance: Board Sustainability Experience and the Interplay with Board Age for Firm Sustainability. *Journal of Business Ethics, 197*(2), 371–389. DOI: 10.1007/s10551-024-05739-3

Curtin, J., McInerney, C., Gallachóir, B. Ó., Hickey, C., Deane, P., & Deeney, P. (2019). Quantifying stranding risk for fossil fuel assets and implications for renewable energy investment: A review of the literature. *Renewable & Sustainable Energy Reviews, 116*, 109402. DOI: 10.1016/j.rser.2019.109402

Deghaim, S., Albanaqi, M., Albathali, M. A., Saad, N., Alotaibi, T., Ghazai, B., Alotaibi, G., Ali, W., Otayn, M., Awad, F., Alsharari, M., Shtewi, S., Alsharari, Q., Awad, A., Alsharari, S., Tabia, F., Alsharari, Q., Alenazi, J. H. A., Hathal, M., & Alotaibi, H. (2024). Strategic data stewardship in modern healthcare: An integrated governance framework for ensuring quality, privacy, and ethical disclosure across clinical, operational, and analytical domains. *Saudi Journal of Medicine and Public Health, 1*(2), 1619–1626. DOI: 10.64483/202412450

Du Toit, E. (2024). Thirty Years of Sustainability Reporting: Insights, Gaps and an Agenda for Future Research Through a Systematic Literature Review. *Sustainability (Basel), 16*(23), 10750. DOI: 10.3390/su162310750

European Commission. (2021). *Screening of websites for "greenwashing": Analysis of environmental claims in the EU consumer market.* European Commission. https://environment.ec.europa.eu/topics/circular-economy/green-claims_en

Fasasi, S. T., Adebowale, O. J., Abdulsalam, A., & Nwokediegwu, Z. Q. S. (2021). Framework for integrating methane monitoring data into carbon credit verification systems. *International Journal of Multidisciplinary Research and Growth Evaluation, 2*(2), 538–543. DOI: 10.54660/.IJMRGE.2021.2.2.538-543

Ferdig, M. A. (2007). Sustainability leadership: Co-creating a sustainable future. *Journal of Change Management, 7*(1), 25–35. DOI: 10.1080/14697010701233809

Flammer, C. (2021). Short-termism and executive compensation. [München: ifo Institut-Leibniz-Institut für Wirtschaftsforschung an der Universität München.]. *CESifo Forum, 22*(3), 17–19.

Forliano, C., Battisti, E., de Bernardi, P., & Kliestik, T. (2025). Mapping the greenwashing research landscape: A theoretical and field analysis. *Review of Managerial Science, 19*(11), 3407–3456. DOI: 10.1007/s11846-025-00856-3

Free, C., Jones, S., & Tremblay, M. S. (2025). Greenwashing and sustainability assurance: A review and call for future research. *Journal of Accounting Literature, 47*(4), 739–764. DOI: 10.1108/JAL-11-2023-0201

Gadinis, S., & Fregoni, S. (2026). Beyond the Brussels Effect: The surprising rise of the International Sustainability Standards Board. *Journal of International Economic Law*, •••, jgaf049. Advance online publication. DOI: 10.1093/jiel/jgaf049

Ghitti, M., Gianfrate, G., & Palma, L. (2024). The agency of greenwashing. *The Journal of Management and Governance, 28*(3), 905–941. DOI: 10.1007/s10997-023-09683-8

Gipper, B., Ross, S., & Shi, S. X. (2025). ESG assurance in the United States. *Review of Accounting Studies, 30*(2), 1753–1803. DOI: 10.1007/s11142-024-09856-2

Gizyatova, A., Dlusskaya, V., & Bisultanova, A. (2024). Behavioural economics and sustainability: Aligning human incentives with green growth strategies. *Reliability: Theory & Applications, 19*(SI 6 (81)), 1145-1150.

Goerzen, A., Van Assche, A., Zhan, J. X., & Zhang, L. (2025). From the editors: Global sustainability reporting standards and the future of international business. *Journal of International Business Policy, 8*(2), 125-136. https://doi.org/DOI: 10.1057/s42214-025-00213-x

Gunningham, N. (2009). Environment Law, Regulation and Governance: Shifting Architectures. *Journal of Environmental Law, 21*(2), 179–212. DOI: 10.1093/jel/eqp011

Hao, X., Tian, T., Dong, L., Wong, C. W. Y., & Lai, K. (2025). Unmasking greenwashing in ESG disclosure: Insights from evolutionary game analysis. *Annals of Operations Research, 356*(1), 667–695. DOI: 10.1007/s10479-025-06538-3

Harrison, J. S., Bosse, D. A., & Phillips, R. A. (2010). Managing for stakeholders, stakeholder utility functions, and competitive advantage. *Strategic Management Journal, 31*(1), 58–74. DOI: 10.1002/smj.801

Hossain, M. Z., Rashed, J. R., & Hredoy, A. I. (2025). Greenwashing in corporate sustainability: Implications for financial reporting and accounting practices. *European Journal of Innovative Studies and Sustainability, 1*(6), 45–57. DOI: 10.59324/ejiss.2025.1(6).05

Indrawati, N., Sensuse, D., Sumirat, D., & Purwaningsih, E. H. (2025). Blockchain governance assessment tools: A conceptual model from a business sustainability perspective. *Journal of Information Systems Engineering & Management, 10*(25), 828–850. DOI: 10.52783/jisem.v10i25s.4173

Islam, F. A. S. (2025). Artificial intelligence's hidden footprint in environmental engineering: A life-cycle risk assessment and the AI-ERAF governance framework. *American Journal of Data Science and Artificial Intelligence, 1*(2), 40–57. DOI: 10.54536/ajdsai.v1i2.6293

Ivasciuc, I. S., & Ispas, A. (2025). Communicating Sustainability in Hospitality: A Multi-Layer Analysis of Transparency, Green Claims, and Corporate Value Construction. *Sustainability (Basel), 18*(1), 172. DOI: 10.3390/su18010172

Jain, V., & Mitra, A. (2025). *Navigating Environmental Regulations in Digital Marketing: Ensuring Compliance While Promoting Green Claims*. Regulatory Frameworks and Digital Compliance in Green Marketing, 125-164. IGI Global. DOI: 10.4018/979-8-3693-9831-9.ch005

Jamil, N. N., & Wahyuni, E. T. (2024). Unveiling greenwashing: Risks in sustainability and ESG reporting. *International Journal of Accounting, Finance and Business, 9*(57), 30-42. https://academicinspired.com/ijafb/article/view/2608

Janik, A., & Ryszko, A. (2026). Greenwashing in Sustainability Reporting: A Systematic Literature Review of Strategic Typologies and Content-Analysis-Based Measurement Approaches. *Sustainability (Basel), 18*(1), 17. DOI: 10.3390/su18010017

Jensen, M. C., & Meckling, W. H. (1976). Theory of the firm: Managerial behavior, agency costs and ownership structure. *Journal of Financial Economics, 3*(4), 305–360. DOI: 10.1016/0304-405X(76)90026-X

Job, J. S., & Khanna, S. (2024). Building a resilient future: Collaborative sustainability regulation. *Clean Energy, 8*(3), 223–237. DOI: 10.1093/ce/zkae026

Jones, G. G., Lopes, T. D. S., Pananond, P., van Tulder, R., Sinkovics, N., & Sinkovics, R. R. (2025). Transitioning from responsible and reactive to deeply responsible and proactive international business. *Critical Perspectives on International Business, 21*(2), 196–225. DOI: 10.1108/cpoib-08-2024-0092

Ke, Y., Wen, Y., & Teng, L. (2025). Chain Leader Policy and Corporate Environmental Sustainability: A Multi-Level Analysis of Greenwashing Mitigation Mechanisms. *Sustainability (Basel), 17*(19), 8871. DOI: 10.3390/su17198871

Khan, N., Jabeen, S., Qureshi, M. I., & Falahat, M. (2026). Greenwashing Communication Deficiencies and Consumer Response: Implications for Sustainable Technology Entrepreneurship. *Sustainable Technology and Entrepreneurship, 100132*(2), 100132. Advance online publication. DOI: 10.1016/j.stae.2026.100132

Koch, T., & Denner, N. (2025). What is greenwashing – a scoping review of greenwashing definitions and development of the Need-for-Balance Model. *Journal of Sustainable Business, 10*(1), 17. DOI: 10.1186/s40991-025-00124-3

Kouhizadeh, M., & Sarkis, J. (2018). Blockchain practices, potentials, and perspectives in greening supply chains. *Sustainability (Basel), 10*(10), 3652. DOI: 10.3390/su10103652

Krasodomska, J., Zarzycka, E., Street, D. L., & Grabowski, W. (2025). The impact of companies' trust-building efforts on sustainability reporting assurance quality: Insights from Europe. *Meditari Accountancy Research*, *33*(7), 246–279. DOI: 10.1108/MEDAR-05-2024-2496

Kwon, K., Lee, J., Wang, C., & Diwanji, V. S. (2024). From green advertising to greenwashing: Content analysis of global corporations' green advertising on social media. *International Journal of Advertising*, *43*(1), 97–124. DOI: 10.1080/02650487.2023.2208489

Lee, S., Isa, M., & Shaiban, M. (2025). Corporate governance and greenwashing: The moderating role of political connection and sensitive industries. *SAGE Open*, *15*(3), 1–15. DOI: 10.1177/21582440251369200

Li, S., & Liu, C. (2025). Corporate strategic greenwashing under ESG disclosure uncertainty: Financing incentives and nonlinear effects. *Journal of Environmental Management*, *394*, 127473. DOI: 10.1016/j.jenvman.2025.127473 PMID: 41038102

Liao, Y. (2022). Sustainable leadership: A literature review and prospects for future research. *Frontiers in Psychology*, *13*, 1045570. DOI: 10.3389/fpsyg.2022.1045570 PMID: 36420377

Liu, Y., Wang, H., Chen, J., Zhang, X., Yue, X., Ke, J., Wang, B., & Peng, C. (2020). Emergency management of nursing human resources and supplies to respond to coronavirus disease 2019 epidemic. *International Journal of Nursing Sciences*, *7*(2), 135–138. DOI: 10.1016/j.ijnss.2020.03.011 PMID: 32292632

López, S. M. C., Guerra, I. R., & Pernas, J. B. (2025). Sustainable Leadership as a Governance Mechanism in the ESG Era: A Systematic Review of Organizational Transformation in the Hospitality Sector. *Merits*, *5*(4), 22. DOI: 10.3390/merits5040022

Ma, X., Ma, X., Fei, W., Jiang, Q., & Zhang, Y. (2026). Stop Where it Should Stop: Board Faultlines, Stakeholder Supervision and Greenwashing Behaviors of Firms. *Journal of Business Ethics*, *203*(2), 313–340. DOI: 10.1007/s10551-025-06046-1

Maak, T., Pless, N. M., & Voegtlin, C. (2016). Business statesman or shareholder advocate? CEO responsible leadership styles and the micro-foundations of political CSR. *Journal of Management Studies*, *53*(3), 463–493. DOI: 10.1111/joms.12195

Manning, S., & Reinecke, J. (2016). A modular governance architecture in-the-making: How transnational standard-setters govern sustainability transitions. *Research Policy*, *45*(3), 618–633. DOI: 10.1016/j.respol.2015.11.007

Marta, D. J., & Rake, P. (2025). The success of the whistleblowing system in preventing corruption in village financial management in Indonesia. *Jurnal Inspektorat, 1*(2), 72–88. DOI: 10.64527/inspektorat.v1i2.39

McIntyre, O. (2018). Transnational environmental regulation and the normativisation of global environmental governance standards: The promise of order from chaos? *Journal of Property. Planning and Environmental Law, 10*(2), 92–112. DOI: 10.1108/JPPEL-01-2018-0003

Mei, Y., & Broyde, M. J. (2025). Reclaiming constitutional authority of algorithmic power. *arXiv.* https://arxiv.org/abs/2508.11699 DOI: 10.2139/ssrn.5389563

Metcalf, L., & Benn, S. (2013). Leadership for sustainability: An evolution of leadership ability. *Journal of Business Ethics, 112*(3), 369–384. DOI: 10.1007/s10551-012-1278-6

Meyer, J. W., & Rowan, B. (1977). Institutionalized organizations: Formal structure as myth and ceremony. *American Journal of Sociology, 83*(2), 340–363. DOI: 10.1086/226550

Miao, Q., & Nduneseokwu, C. (2025). *Environmental leadership in Private organizations. Environmental leadership in a VUCA Era: An interdisciplinary handbook.* Springer Nature.

Molnár, P., Lukács, B., & Tóth, Á. (2025). Toward Social Disclosure Alignment: Evaluating the Interoperability of ISSB S2 with ESRS and GRI Standards. *Societies (Basel, Switzerland), 15*(10), 273. DOI: 10.3390/soc15100273

Muneer, S., Singh, A., Choudhary, M. H., Alshammari, A. S., & Butt, N. A. (2025). Does Environmental Disclosure and Corporate Governance Ensure the Financial Sustainability of Islamic Banks? *Administrative Sciences, 15*(2), 54. DOI: 10.3390/admsci15020054

Nashrullah, H., Maemunah, M., & Purnamasari, P. (2025). Building integrity from within: The role of organizational culture and commitment in preventing fraud in student organizations at Universitas Islam Bandung. *Journal of Economics and Business UBS, 14*(5), 1313–1323. DOI: 10.52644/n2ydyt42

Noor, F. A. M., Said, R., Arsad, S., & Shafie, N. A. (2025). Bridging governance and transparency: The role of board age diversity in shaping integrity disclosure. *International Journal of Research and Innovation in Social Science, 9*(8), 4682–4690. DOI: 10.47772/IJRISS.2025.908000375

Olu-Owolabi, F. E., Oladoyin, A. M., Odukoya, O., & Osimen, G. U. (2024). Harnessing traditional institutions for sustainable leadership and good governance in Nigeria. *Journal of Infrastructure. Policy and Development*, *8*(11), 6951. DOI: 10.24294/jipd.v8i11.6951

P., R., Gupta, A. K., Satyavathi, M., Chaitanya, M. S. K., Parameswari, M., & Kalpana, R. (2025). Sustainable IT governance models using blockchain for transparent decision-making. *Proceedings of the 4th International Conference on Innovative Mechanisms for Industry Applications (ICIMIA 2025)*, 1948–1954. https://doi.org/ DOI: 10.1109/ICIMIA67127.2025.11200918

Patil, P. A., Jadhav, R. S., Jagtap, P., Thorat, K. D., & Patil, H. K. (2025). Citizen-led AI audit platform for transparency and accountability in automated decision-making. *International Journal on Advanced Computer Engineering and Communication Technology*, *14*(1), 676–681. DOI: 10.65521/ijacect.v14i1.751

Pérez-Durán, I. (2024). Twenty-five years of accountability research in public administration: Authorship, themes, methods, and future trends. *International Review of Administrative Sciences*, *90*(3), 546–562. DOI: 10.1177/00208523231211751

Permatasari, Y., Rosyidah, M. A., Sabaruddin, L., Embuningtiyas, S. S., & Yuniawati, R. A. (2025). Sustainability governance and greenwashing: The role of sustainability committees in ASEAN-5 countries. *Asian Review of Accounting*, •••, 1–26. DOI: 10.1108/ARA-03-2025-0067

Pfister, J. A., Otley, D., Ahrens, T., Dambrin, C., Darwin, S., Granlund, M., Jack, S. L., Lassila, E. M., Millo, Y., Peda, P., Sherman, Z., & Wilson, D. S. (2024). Performance management in the prosocial market economy: A new paradigm for economic performance and sustainability. *Qualitative Research in Accounting & Management*, *21*(5), 397–443. DOI: 10.1108/QRAM-02-2024-0031

Pless, N. M., & Maak, T. (2011). Responsible leadership: Pathways to the future. *Journal of Business Ethics*, *98*(1), 3–13. DOI: 10.1007/s10551-011-1114-4

Prabhu, J. (2017). Frugal innovation: Doing more with less for more. *Philosophical Transactions. Series A, Mathematical, Physical, and Engineering Sciences*, *375*(2095), 20160372. DOI: 10.1098/rsta.2016.0372 PMID: 28461436

Pratama, A., Yusoff, H., Yadiati, W., & Jaenudin, E. (2025). Sustainability-related corporate governance and greenwashing practices: Preliminary evidence from Southeast Asian companies. *Meditari Accountancy Research*, *33*(3), 853–885. DOI: 10.1108/MEDAR-07-2024-2579

Ren, S., Cooke, F. L., Stahl, G. K., Fan, D., & Timming, A. R. (2023). Advancing the sustainability agenda through strategic human resource management: Insights and suggestions for future research. *Human Resource Management, 62*(3), 251–265. DOI: 10.1002/hrm.22169

Rihal, J., Alzubi, A., Aljuhmani, H. Y., & Berberoğlu, A. (2025). From ethics to ecology: How ethical leadership drives environmental performance through green organizational identity and culture. *PLoS One, 20*(11), e0336608. Advance online publication. DOI: 10.1371/journal.pone.0336608 PMID: 41231815

Rosário, A. T., & Boechat, A. C. (2025). How Sustainable Leadership Can Leverage Sustainable Development. *Sustainability (Basel), 17*(8), 3499. DOI: 10.3390/su17083499

Saberi, S., Kouhizadeh, M., Sarkis, J., & Shen, L. (2019). Blockchain technology and its relationships to sustainable supply chain management. *International Journal of Production Research, 57*(7), 2117–2135. DOI: 10.1080/00207543.2018.1533261

Safari, M., & Parker, L. D. (2024). Understanding multiple accountability logics within corporate governance policy discourse: Resistance, compromise, or selective coupling? *European Accounting Review, 33*(4), 1467–1496. DOI: 10.1080/09638180.2023.2194028

Sahu, M., Mishra, A., Alahdal, W. M., & Sami, M. (2025). ESG performance and audit committee expertise: Advancing sustainable development goals in leading nations. *International Review of Economics & Finance, 103*, 104445. DOI: 10.1016/j.iref.2025.104445

Sajjad, A., Eweje, G., & Raziq, M. M. (2024). Sustainability leadership: An integrative review and conceptual synthesis. *Business Strategy and the Environment, 33*(4), 2849–2867. DOI: 10.1002/bse.3631

Sánchez, V. P. D., Zorio-Grima, A., & Merello, P. (2026). Sustainability reporting assurance: Knowledge, lags, levels, and providers in leading global companies. *Journal of Innovation & Knowledge, 12*, 100899. DOI: 10.1016/j.jik.2025.100899

Sasaki, N. (2025). Addressing scandals and greenwashing in carbon offset markets: A framework for reform. *Global Transitions, 7*, 375–382. DOI: 10.1016/j.glt.2025.06.003

Searcy, C., & Buslovich, R. (2014). Corporate perspectives on the development and use of sustainability reports. *Journal of Business Ethics, 121*(2), 149–169. DOI: 10.1007/s10551-013-1701-7

Shao, M., Liu, Y., Zhao, G., Sun, H., & Zhao, P. (2025). Mitigating Involutionary Competition Through Corporate ESG Adoption: Evidence from the Consumer Electronics Manufacturing Industry. *Sustainability (Basel), 17*(20), 8998. DOI: 10.3390/su17208998

Shukla, T. (2025). Leveraging blockchain for ESG disclosure: Enhancing transparency and accountability in sustainability reporting. *International Journal of Engineering Technologies and Management Research, 12*(3). Advance online publication. DOI: 10.29121/ijetmr.v12.i3.2025.1546

Singh, V., & Sangal, S. (2025). *Evolving CSR: Ethical Leadership in Environmental Challenges and Sustainable Consumption*. Enhancing Social Sustainability in Manufacturing Supply Chains, 1-32. IGI Global Scientific Publishing.

Sklavos, G., Zournatzidou, G., & Sariannidis, N. (2026). Unveiling ESG controversy risks: A multi-criteria evaluation of whistleblowing performance in European financial institutions. *Risks, 14*(1), 10. DOI: 10.3390/risks14010010

Sneideriene, A., & Legenzova, R. (2026). A framework for mitigating greenwashing in sustainability reporting. *Sustainability (Basel), 18*(1), 524. DOI: 10.3390/su18010524

Spaniol, M. J., Danilova-Jensen, E., Nielsen, M., Rosdahl, C. G., & Schmidt, C. J. (2024). Defining greenwashing: A concept analysis. *Sustainability (Basel), 16*(20), 9055. DOI: 10.3390/su16209055

Sundarasen, S., Zyznarska-Dworczak, B., & Goel, S. (2024). Sustainability reporting and greenwashing: A bibliometrics assessment in G7 and non-G7 nations. *Cogent Business & Management, 11*(1), 2320812. DOI: 10.1080/23311975.2024.2320812

Taylor, R. M. (2024). Methodological reflections on normative case studies: What they are and why we need better quality criteria to inform their use. *Educational Theory, 74*(3), 301–311. DOI: 10.1111/edth.12555

Thawani, B., Bhatia, M., Panigrahi, T., & Mall, S. (2025). Research trends on use of technology in sustainability reporting: A keyword-based bibliometric analysis. *International Journal of Disclosure and Governance, 22*(4), 910–940. DOI: 10.1057/s41310-025-00315-1

Tin, S. T., & Yong, J. (2025). From promise to proof: GreenBrew Malaysia's sustainability challenge. *Semarak International Journal of Entrepreneurship, Economics, and Business Development, 4*(1), 25–48. DOI: 10.37934/sijeebd.4.1.2548

Tumpa, R. J., & Naeni, L. (2025). Improving decision-making and stakeholder engagement at project governance using digital technology for sustainable infrastructure projects. *Smart and Sustainable Built Environment, 14*(4), 1292–1329. DOI: 10.1108/SASBE-10-2024-0451

U.S. Securities and Exchange Commission (SEC). (2023). *SEC charges DWS Investment Management Americas Inc. with misstatements regarding ESG investments and anti-money laundering violations.* https://www.sec.gov/newsroom/press-releases/2023-194

Varsha, M., Dhanaraj, D. V. T., & Info, A. (2024). Analysis of the influence of accounting information on stock price with the extent of corporate social responsibility disclosures as a moderation. *African Journal of Commercial Studies, 5*(3), 157–163. DOI: 10.59413/ajocs/v5.i.3.2

Visser, W., & Courtice, P. (2011). Sustainability Leadership: Linking Theory and Practice, SSRN *Working Paper Series*, 2011. https://www.cisl.cam.ac.uk/resources/sustainability-leadership/sustainability-leadership-linking-theory-and-practice?utm_source=chatgpt.com DOI: 10.2139/ssrn.1947221

Wamsler, C., Osberg, G., Janss, J., & Stephan, L. (2024). Revolutionising sustainability leadership and education: Addressing the human dimension to support flourishing, culture and system transformation. *Climatic Change, 177*(1), 4. DOI: 10.1007/s10584-023-03636-8

Wang, W., Sun, Z., Zhu, W., Ma, L., Dong, Y., Sun, X., & Wu, F. (2023). How does multi-agent govern corporate greenwashing? A stakeholder engagement perspective from "common" to "collaborative" governance. *Corporate Social Responsibility and Environmental Management, 30*(1), 291–307. DOI: 10.1002/csr.2355

Wang, Z., & Tian, X. (2025). From Greenwashing to Sustainability: The Mediating Effect of Green Innovation in the Agribusiness Sector on Financial Performance. *Agriculture, 15*(12), 1316. DOI: 10.3390/agriculture15121316

Warhurst, A. (2002). Sustainability indicators and sustainability performance management. *Mining, Minerals and Sustainable Development [MMSD] project report, 43*, 129.

Wernerfelt, B. (1984). A resource-based view of the firm. *Strategic Management Journal, 5*(2), 171–180. DOI: 10.1002/smj.4250050207

Ye, M., & Guo, R. (2025). Regulatory overload or incentive deficit? Forty years of structural imbalances in China's marine environmental policy instruments. *Frontiers in Marine Science, 12*, 1674863. DOI: 10.3389/fmars.2025.1674863

Yong, C. K., Alfan, E., & Mustapha, M. Z. (2024). Sustainability Committee and Environmental, Social and Governance (ESG). *International Journal of Business and Society*, *25*(2), 484–508. DOI: 10.33736/ijbs.7603.2024

Zioło, M., Bąk, I., & Spoz, A. (2024). Literature review of greenwashing research: State of the art. *Corporate Social Responsibility and Environmental Management*, *31*(6), 5343–5356. DOI: 10.1002/csr.2842

KEY TERMS AND DEFINITIONS

Accountability Mechanisms: Mechanisms that can reduce greenwashing by establishing traceable responsibilities and consequences through organizational processes, policies, and structures that hold individuals and departments accountable for the accuracy, integrity, and performance of sustainability action and disclosure.

Agency Problem in Sustainability: A gap that can contribute to greenwashing when corporate leaders prioritize short-term reputational gains or profits over long-term environmental and social value, owing to misaligned incentives and inadequate oversight mechanisms.

Governance Architect: A sustainable leadership role focused on designing and enforcing governance systems, such as verification processes, transparent reporting methods, and integrity safeguards, to ensure that sustainability claims are accurate, verifiable, and resilient to manipulation.

Greenwashing: The practice of exaggerating, fabricating, or selectively disclosing environmental information to make an organization appear more sustainable than it actually is.

Stewardship and Integrity Committee: A proposed board-level committee that can verify sustainability claims with financial-reporting rigor and oversee data accuracy, third-party assurances, exposure to sustainability misconduct, and the alignment of executive incentives with verified ESG performance.

Sustainability Governance: Formal rules, structures, and accountability systems that can guide and monitor ESG commitments and thus reduce the structural conditions that can enable greenwashing.

Sustainability Reporting Integrity: The degree to which environmental and social disclosures can be described as being complete, accurate, consistent, and independently verifiable.

Sustainable Leadership: A leadership approach that can embed environmental and social integrity into organizational systems and emphasizes the creation of structures to make sustainability claims accurate and enforceable.

Transparency Systems: Organizational infrastructures, such as traceable data systems, open reporting channels, and audit-ready documentation, can ensure that sustainability information is accessible and difficult to manipulate.

Verification Processes: Internal and external audits that can confirm the accuracy and credibility of sustainability data and distinguish genuine performance from symbolic or exaggerated claims.

Chapter 3
Theocentric Sustainable Leadership:
A Faith-Driven Framework to Counter Greenwashing in Modern Organizations

Andika Andika
http://orcid.org/0009-0001-9569-4396
Brawijaya University, Indonesia

ABSTRACT

The phenomenon of greenwashing indicates that sustainability initiatives are often reduced to image-building efforts without genuine moral commitment. Contemporary environmental ethics—both anthropocentric and ecocentric—have proven insufficient in preventing information manipulation, as they fail to provide leaders with a profound ethical foundation. This chapter proposes theocentric sustainable leadership as a conceptual alternative rooted in spiritual values, particularly the principles of amanah (trust) and the human role as khalifah (steward) within Islamic tradition. A theocentric approach frames sustainability as a moral–transcendental responsibility, thereby strengthening integrity, transparency, and accountability in decision-making. By integrating spiritual values into organizational vision, governance, and culture, this model offers a more robust ethical framework to counter greenwashing and foster authentic sustainability for future generations.

DOI: 10.4018/979-8-3373-7877-0.ch003

INTRODUCTION

The dynamics of modern corporate organizations, which strongly emphasize productivity, have driven organizations to compete intensely in seeking efficient and effective strategies to gain competitive advantage. In the pursuit of resources and profits, organizations frequently marginalize moral values, leading to adverse consequences for the environment, governance, and society (Cole et al., 2021). In recent decades, growing concern over climate change and ecosystem degradation has emerged as a major global issue . This concern has intensified demands for corporations to adopt sustainability practices that are genuine, profound, and accountable. However, alongside the rising awareness of environmental responsibility, a troubling phenomenon has also emerged: greenwashing. This term refers to corporate practices that exaggerate or misrepresent environmental friendliness without being accompanied by concrete actions that generate meaningful change (Pratama et al., 2025; Zervoudi et al., 2025). Greenwashing not only deceives consumers and stakeholders but also undermines global efforts to address environmental degradation (Koch & Denner, 2025).

This phenomenon indicates that many contemporary corporate sustainability practices remain rooted in flawed paradigms. Numerous organizations adopt sustainability primarily as a marketing strategy or a tool for reputational negotiation rather than as a moral calling. This highlights the need to re-examine the fundamental perspectives that underpin sustainable leadership (Hoffman & Sandelands, 2005). The widely advocated paradigm shift from anthropocentrism to ecocentrism in global sustainability discourse, although progressive, has not yet fully addressed the root of the problem. Both paradigms remain confined within frameworks that either position humans as the central agents of control or treat nature as an entity governed by human-centered ethical considerations (Frantz et al., 2025). In contrast, authentic sustainability entails a deeper spiritual dimension—one that connects humanity to the Creator and provides a transcendent moral foundation for responsible leadership. Within this context, the theocentric sustainable leadership approach emerges as a relevant, robust, and contextually grounded framework, particularly for faith-based societies. A theocentric perspective places God at the center of moral orientation and as the ultimate purpose of human action, including the governance of organizations and the management of the environment. For Muslims, this approach resonates deeply with the foundational Islamic teaching of humans as *khalifah fil ardh*—entrusted stewards responsible for protecting, nurturing, and sustaining the earth. This role of stewardship is not merely a social or ecological function, but a spiritual mandate that entails accountability to God (Husamah et al., 2025; Rakhmat, 2022). Consequently, sustainability actions are no longer viewed as optional strategic choices, but as moral obligations embedded within religious identity and ethical responsibility.

A theocentric approach leads to the understanding that authentic sustainability arises from spiritual consciousness rather than merely from external pressures or regulatory compliance. If greenwashing emerges due to the absence of a strong ethical foundation, faith-based leadership offers an opposing orientation—one grounded in honesty, sincerity, and genuine commitment to producing meaningful ecological impact (Fry & Egel, 2021). From a theocentric perspective, leaders are accountable not only to shareholders, consumers, or regulatory bodies, but also to God, who entrusts the earth as a shared living space for all creation. It is this transcendent awareness that can serve as a moral safeguard against the manipulation of sustainability narratives. Such a fundamental shift requires a transformation in how leadership itself is understood. Leadership is no longer viewed merely as the function of managing people and resources, but as a spiritual practice that integrates vision, values, and actions in alignment with divine principles (Dhiman, 2017; Fry & Egel, 2021). Theocentric leaders recognize that every decision carries moral and ecological consequences. They perceive nature not simply as an economic asset, but as God's creation endowed with intrinsic value (Eom & Ng, 2023). Within this worldview, excessive exploitation of nature constitutes a betrayal of the stewardship mandate, while its protection and restoration represent acts of obedience and ethical responsibility.

The theocentric sustainable leadership framework also offers a compelling response to the contemporary dilemmas faced by modern organizations. On the one hand, companies are driven to grow and remain competitive; on the other, they are increasingly expected to safeguard their social and ecological impacts. A theocentric paradigm helps reconcile these competing demands by positioning sustainability not as a constraint on growth, but as an integral component of the organization's moral mission. Leaders guided by divine values seek to create harmony between economic achievement and environmental preservation without sacrificing either for short-term reputational gains. Moreover, this approach emphasizes that sustainability is not merely a technical project, but an ethical journey rooted in values, beliefs, and spirituality. Green technologies, ESG standards, and sustainability reports are merely instruments rather than ends in themselves. Without a strong moral foundation, such tools can easily be distorted into mechanisms of greenwashing (Dempere et al., 2024; Zervoudi et al., 2025). Theocentric sustainable leadership ensures that the entire sustainability process—from planning and implementation to reporting—is carried out with integrity, reverence for God, and an awareness that every action will ultimately be held accountable. Furthermore, although deeply grounded in specific religious traditions such as Islam, the theocentric approach embodies inclusive and universal principles. Most spiritual traditions emphasize respect for nature, human moral responsibility, and the importance of living in balance (Leal Filho et al., 2022). As such, theocentric sustainable leadership can serve as a bridge between modern sustainability practices and universal spiritual values, offering a deeper and more resilient ethical foundation for addressing global crises.

This chapter invites readers to consider that environmental crises and the phenomenon of greenwashing should not be understood solely as technical problems or regulatory shortcomings. Rather, they may be interpreted as manifestations of a deeper ethical disorientation underlying human approaches to environmental governance. From this perspective, effective responses require renewed attention to the moral and spiritual dimensions of leadership. Theocentric sustainable leadership is therefore proposed not merely as a conceptual alternative, but as an ethical orientation grounded in the understanding that the earth is not an object of ownership, but a trust that entails moral responsibility. The following sections examine the ethical challenges of greenwashing, the limitations of prevailing sustainability paradigms, the conceptual foundations of theocentric sustainable leadership, and potential pathways for its implementation within modern organizations. By integrating perspectives on spirituality, leadership, and sustainability, this chapter seeks to contribute to the development of leadership frameworks that are ethically grounded and responsive to long-term environmental responsibility.

The principal contribution of this chapter is the articulation of a systemic framework through which a theocentric spiritual worldview can be translated into institutional leadership mechanisms. Rather than treating faith as a private source of individual morality, the chapter conceptualizes spiritual values as organizationally embedded drivers of governance, accountability, stakeholder engagement, and sustainability-oriented decision-making. Through this perspective, Theocentric Sustainable Leadership is positioned as a mechanism for aligning organizational conduct with authentic sustainability objectives, thereby providing a normative and institutional response to the persistent challenge of greenwashing.

THE ETHICAL PROBLEM OF GREENWASHING

Greenwashing has emerged as one of the most significant ethical challenges in the contemporary sustainability landscape. The term, which gained prominence in the late 1980s, refers to corporate practices that present an environmentally friendly image—through advertising, reporting, or public campaigns—without being supported by substantive actions consistent with such claims (Bowen & Aragon-Correa, 2014). In an era in which organizations increasingly compete to demonstrate their commitment to sustainability, greenwashing functions as a veil that obscures underlying realities. It erodes public trust, misleads consumers, and, more fundamentally, hinders genuine environmental transformation. To understand the roots of this phenomenon, greenwashing should not be viewed merely as a communication error, but as an ethical issue embedded within flawed business paradigms. It manifests in various forms, including the use of "eco-friendly" labels lacking scientific validation, the selective disclosure of sustainability data, and ESG reports that emphasize positive achievements while downplaying or omitting negative environmental impacts (Sneideriene & Legenzova, 2025). In some cases, companies claim reductions in carbon footprints while merely relocating emission sources along their supply chains. In others, firms introduce so-called "green" products that differ only marginally from conventional alternatives, yet are packaged with nature-inspired aesthetics to create the impression of environmental responsibility. Despite their diverse expressions, these practices share a common characteristic: the instrumentalization of sustainability as a means of organizational legitimacy rather than as a genuine moral commitment.

To understand the drivers of greenwashing, it is necessary to examine the external pressures faced by corporations. Contemporary consumers are increasingly attentive to environmental issues, and a growing body of research indicates that younger generations—particularly Generation Z and millennials—are more inclined to prefer ethical and sustainable products. Regulatory frameworks have also become more stringent, with governments mandating greater transparency and environmental impact assessments. In addition, investors are progressively incorporating ESG criteria into their investment decisions. Within this environment, companies often feel compelled to project a "green" image, regardless of whether they have achieved substantive sustainability standards. Greenwashing thus emerges as an expedient response—less costly, faster, and easier than undertaking genuine operational transformation (Yang et al., 2020). However, the drivers of greenwashing are not solely external. Internal organizational factors also play a critical role, particularly incentive structures that prioritize short-term outcomes. In many firms, sustainability initiatives are designed primarily as public relations or marketing projects rather than as elements of strategic transformation. When sustainability responsibilities are assigned to communication departments instead of ethics, governance, or operational teams, the risk of narrative manipulation increases substantially. Executives whose performance is evaluated predominantly on short-term financial metrics may favor symbolic actions that enhance corporate image without requiring fundamental changes to business processes. Consequently, greenwashing arises from a misalignment between image and reality: organizations seek to appear responsible without committing to the substantive efforts required to become responsible.

This issue becomes increasingly complex when the roots of greenwashing are understood to lie in anthropocentric paradigms. Within this worldview, nature is positioned primarily as an instrument for human interests, and sustainability is framed as a strategy for maintaining business continuity rather than as a moral responsibility toward ecosystems (Hoffman & Sandelands, 2005). Sustainability is often perceived as a form of "added value" or a source of competitive differentiation, rather than as an ethical obligation. When sustainability ethics are detached from a deeper moral dimension, opportunities for manipulation expand significantly. For this reason, analyses of greenwashing cannot be confined to technical or regulatory considerations alone. Greenwashing represents a moral issue that reflects a deeper spiritual void within corporate leadership. The consequences of greenwashing are therefore extensive and profound.

The most immediate and visible consequence of greenwashing is the erosion of trust (Szabo & Webster, 2021). Consumers who feel deceived tend to become skeptical of sustainability claims more broadly. When a single company engages in greenwashing, the reputational damage often extends beyond the organization itself, casting doubt on entire industries that are genuinely striving to implement sustainable practices. Such skepticism is particularly harmful because it can foster public apathy or diminish the perceived urgency of environmental issues. As a result, collective efforts to address ecological crises are weakened. A second major consequence is the stagnation of meaningful change. Greenwashing creates an illusion of progress while masking the absence of substantive improvement. When organizations prioritize managing narratives over transforming operational practices, environmental problems remain fundamentally unresolved. In a global context marked by escalating climate risks, such inertia is especially dangerous. Moreover, greenwashing can generate unfair competition, whereby companies that invest substantial resources in environmentally responsible technologies are disadvantaged relative to those that merely simulate sustainability through symbolic actions. A third, equally significant consequence concerns the ethical impact on organizational culture. When greenwashing becomes normalized or even legitimized as a business strategy, organizational integrity gradually erodes. Employees who recognize the discrepancy between sustainability narratives and actual practices may experience moral dissonance, reduced motivation, or ultimately choose to disengage from or leave the organization. Over time, this dynamic undermines the ethical foundations of the organization and normalizes manipulative behavior. Organizational environments shaped by such conditions are unlikely to foster authentic and sustained innovation in sustainability.

The issue of greenwashing is also closely linked to intergenerational responsibility. Many sustainability claims prioritize short-term image management for immediate economic gains, despite the fact that ecological consequences are inherently long-term in nature. In this sense, greenwashing does not merely deceive the present generation; it constitutes a betrayal of future generations. It shifts ecological burdens onto the shoulders of those yet to come, while current actors benefit from an illusion of sustainability. Given this complexity, it becomes evident that greenwashing is fundamentally an ethical problem that requires a robust moral response. Technical solutions—such as enhanced reporting regulations, sustainability audits, or ESG standards—are undoubtedly important, yet they remain insufficient on their own. As long as internal motivations remain unchanged, organizations can continue to exploit loopholes and manipulate sustainability narratives. What is ultimately required is a deeper transformation of values: a reorientation that restores sustainability to its ethical foundation as a moral responsibility rather than a reputational instrument.

It is precisely at this point that the relevance of a theocentric approach becomes increasingly evident. By re-centering leadership on divine principles—where human actions are understood as a trust and ultimately accountable to God—the space for greenwashing is significantly reduced. Transcendental accountability provides an ethical safeguard that is arguably more robust than any formal standard or regulatory framework. Leaders guided by a theocentric orientation are less likely to instrumentalize sustainability for image construction, as deceptive practices are perceived not merely as violations of public trust, but as breaches of moral responsibility before God. In this sense, theocentric sustainable leadership offers both a conceptual and moral response to the problem of greenwashing.

LIMITATIONS OF ANTHROPOCENTRIC AND ECOCENTRIC PARADIGMS

Contemporary sustainability discourse is predominantly shaped by two ethical paradigms: anthropocentrism and ecocentrism (Kopnina et al., 2018; Washington & Maloney, 2020). Both have played a significant role in advancing environmental awareness across public policy, social movements, and organizational strategies. Anthropocentrism emphasizes human welfare as the primary moral reference, while ecocentrism challenges this hierarchy by recognizing the intrinsic value of nature. Despite their contributions, neither paradigm has proven sufficient in addressing persistent ethical failures such as greenwashing. The continued prevalence of misleading sustainability practices suggests that the problem cannot be explained solely by regulatory gaps or technical shortcomings, but rather by deeper philosophical limitations embedded within these ethical frameworks.

Anthropocentrism and the Instrumentalization of Nature

Anthropocentrism positions human interests at the center of moral consideration, evaluating environmental actions primarily in terms of their utility for human well-being (Kopnina et al., 2018). Within organizational and business contexts, this paradigm often translates into an instrumental view of nature—as a resource to be managed, optimized, and exploited in pursuit of economic objectives. Environmental responsibility is thus justified insofar as it contributes to profitability, competitiveness, or reputational advantage. This logic creates a critical moral vulnerability. When nature is valued only for its usefulness, environmental harm can be tolerated as long as it does not generate immediate or visible costs for human stakeholders. In such a context, greenwashing becomes not an ethical anomaly but a rational strategic response. By selectively communicating environmental achievements while concealing ecological harm, organizations can maintain legitimacy without undertaking substantive operational change. The absence of intrinsic moral limits allows sustainability narratives to be manipulated whenever doing so serves short-term organizational interests.

Moreover, anthropocentrism tends to reinforce short-termism. Market-driven pressures—quarterly performance targets, shareholder expectations, and rapid growth imperatives—often outweigh long-term ecological considerations. Even when leaders acknowledge the importance of sustainability, the dominant anthropocentric logic encourages solutions that deliver quick symbolic gains rather than deep structural transformation. Greenwashing thus emerges as a convenient compromise between environmental expectations and economic expediency. Crucially, moral reasoning within anthropocentrism lacks a transcendent reference point. Ethical standards are human-defined, context-dependent, and negotiable. As a result, moral boundaries can be reinterpreted or relaxed when they conflict with organizational goals. This moral relativity creates fertile ground for practices that appear sustainable in discourse but are ethically hollow in substance.

Ecocentrism and the Limits of Normative Enforcement

Ecocentrism arose as a corrective to anthropocentrism by asserting that nature possesses intrinsic value independent of human interests. From this perspective, humans are part of a broader ecological system and bear moral responsibility toward the environment. While this paradigm strengthens the ethical status of nature, it nevertheless encounters significant limitations when translated into organizational practice. One core challenge lies in the absence of absolute moral authority. Ecocentric values are typically grounded in scientific reasoning, ecological interdependence, or social consensus. While compelling, these foundations remain contestable and open to interpretation. Organizations can selectively appropriate ecological principles, redefining sustainability in ways that align with branding strategies rather than ecological reality. Competing metrics, indicators, and interpretations of environmental impact allow firms to construct favorable narratives without fundamentally altering harmful practices.

Additionally, ecocentrism relies heavily on scientific knowledge, which is often complex, probabilistic, and uncertain. This epistemic complexity provides opportunities for selective disclosure and data manipulation. Companies may highlight improvements in specific indicators while obscuring broader environmental degradation across supply chains or life cycles. Without a stronger moral anchor, scientific evidence alone lacks the normative force needed to restrain opportunistic behavior. Perhaps most importantly, ecocentrism assumes that increased awareness and information will naturally lead to ethical behavior. However, empirical evidence from organizational contexts suggests otherwise. Many leaders are fully aware of the environmental consequences of their actions, yet continue to engage in greenwashing due to powerful economic incentives. In the absence of spiritual or transcendent accountability, ecological values risk being reduced to aesthetic symbols or marketing tools rather than guiding moral commitments.

Why Existing Paradigms Fail to Contain Greenwashing

The persistence of greenwashing across industries indicates that both anthropocentric and ecocentric paradigms share structural weaknesses. In both cases, moral norms remain negotiable, enforcement mechanisms are external and limited, and ethical compliance depends largely on instrumental calculations. Sanctions for misleading sustainability claims are typically legal or reputational, and organizations often find ways to mitigate or avoid them. More fundamentally, neither paradigm adequately addresses the moral interiority of leadership. Ethical failures in sustainability are not solely the result of insufficient rules or metrics, but of leadership decisions shaped by values, beliefs, and moral commitments. When sustainability is disconnected from deeper moral or spiritual responsibility, it becomes vulnerable to instrumentalization.

Toward a Theocentric Reframing of Sustainable Leadership

Recent scholarship has increasingly emphasized sustainable leadership as a strategic mechanism for fostering long-term organizational performance while balancing economic, social, and environmental objectives. Among the leading contributors to this stream of research, Iqbal and colleagues have demonstrated how sustainable leadership influences a wide range of organizational outcomes, including sustainable performance, pro-environmental behavior, psychological empowerment, workforce localization, and green innovation. For example, sustainable leadership has been shown to enhance sustainable performance for SME's (Iqbal et al., 2020), while also encouraging pro-environmental behavior among employees through the conservation of resources mechanism (Piwowar-Sulej & Iqbal, 2025). Furthermore, recent studies highlight the role of leadership in translating sustainability-oriented policies into organizational green innovation and environmentally responsible practices (Iqbal et al., 2026). While existing sustainable leadership research provides substantial evidence regarding the positive effects of leadership on sustainability-related outcomes, the dominant focus remains on managerial capabilities, organizational processes, and behavioral mechanisms. However, less attention has been devoted to the deeper moral and spiritual foundations that shape leaders' sustainability commitments. Building upon these insights, this chapter extends the sustainable leadership literature by proposing Theocentric Sustainable Leadership as a faith-driven framework that embeds transcendent ethical accountability into organizational governance and sustainability decision-making. In doing so, the chapter complements existing sustainable leadership perspectives by addressing the moral vulnerabilities that often enable symbolic sustainability practices and greenwashing.

The limitations of anthropocentrism and ecocentrism point to the need for a more foundational ethical paradigm. A theocentric perspective reframes sustainability not as a strategic option or social expectation, but as a moral obligation rooted in accountability beyond human institutions. By understanding nature as a trust rather than an asset, and leadership as stewardship rather than control, theocentric sustainable leadership introduces a form of transcendental accountability that constrains opportunistic behavior. This paradigm does not reject scientific knowledge, regulatory mechanisms, or economic considerations. Rather, it situates them within a broader moral framework in which ethical responsibility is non-negotiable and sustainability is inseparable from integrity. In doing so, theocentric sustainable leadership offers a normative foundation capable of addressing greenwashing at its ethical roots—an argument that will be elaborated in the following chapter.

THEOCENTRIC STEWARDSHIP IN ISLAMIC THOUGHT

Within Islamic intellectual tradition, leadership is fundamentally inseparable from spirituality (Abeng, 1997). Leadership is not conceptualized merely as a managerial function, a position of authority, or a mechanism for coordinating resources, but rather as a moral responsibility rooted in the relationship between human beings and God. This perspective challenges dominant paradigms in modern organizational theory that tend to prioritize efficiency, control, and performance outcomes while marginalizing ethical and spiritual considerations. In Islamic thought, leadership is embedded within a theocentric worldview in which human actions are ultimately oriented toward divine accountability (Lahmar, 2024). This worldview provides the ethical foundation for what can be described as *theocentric stewardship*—a form of leadership that situates sustainability not as a strategic choice, but as a moral obligation.

Central to Islamic ethics is the concept of *khalifah*, which defines humanity's position in relation to the earth . The Qur'an describes human beings as God's vicegerents on earth (*khalīfatullāh fī al-ar*), a designation that confers responsibility rather than dominance (Glaser, 1998). Unlike anthropocentric paradigms that position humans as sovereign agents with the right to exploit natural resources for economic gain, the concept of *khalifah* emphasizes guardianship, care, and moral restraint. Humans are entrusted with maintaining balance, preventing harm, and ensuring that their actions do not disrupt the integrity of creation. Leadership, therefore, is inherently relational: it governs not only relationships among people within organizations, but also the relationship between humanity, nature, and God (Ismail & Berghout, 2024). This understanding of stewardship carries significant implications for contemporary organizations. When leaders view themselves as *khalifah*, organizational success cannot be defined solely in terms of profit maximization or market dominance. Environmental degradation, excessive resource extraction, and deceptive sustainability practices are not merely operational failures; they represent moral breaches of the stewardship mandate. From this perspective, greenwashing is not a neutral marketing tactic but a violation of the ethical responsibility entrusted to human beings. By misrepresenting environmental performance, organizations betray their role as caretakers of the earth and undermine the moral legitimacy of leadership itself.

Closely connected to the concept of *khalifah* is *amanah*, or trust, which constitutes one of the most profound ethical principles in Islamic thought. *Amanah* refers to a sacred trust bestowed by God upon humanity—a trust so weighty that, according to the Qur'anic narrative, the heavens, the earth, and the mountains declined to bear it (Herijanto, 2022; Islam & Samsudin, 2018). This imagery underscores the gravity of human responsibility, extending beyond personal morality to encompass social, economic, and ecological domains. Within organizational contexts, *amanah* reframes leadership as a spiritual contract rather than a purely contractual or instrumental role. Managing natural resources, producing sustainability reports, and communicating environmental commitments are thus acts of trust that demand honesty, transparency, and integrity. Practices such as selective disclosure, exaggerated environmental claims, or the concealment of ecological harm constitute breaches of *amanah*. These actions violate not only professional ethics but also spiritual accountability. From an Islamic perspective, ethical leadership cannot be compartmentalized; deception in sustainability reporting is ethically indistinguishable from deception in any other moral domain. This holistic moral vision directly challenges the normalization of greenwashing in modern corporate practice.

Islamic thought further strengthens the ethical foundation of stewardship through a rich ecological worldview that emphasizes harmony, justice, and balance. The Qur'an repeatedly affirms that creation is governed by *mīzān*, or balance, and that humans are prohibited from causing *fasād*, or corruption, on earth. Environmental harm, whether through pollution, excessive consumption, or unsustainable production, is understood as a disruption of this divinely ordained balance. Unlike instrumental approaches to sustainability that justify environmental protection only when it aligns with economic interests, Islamic ethics regard environmental care as intrinsically valuable. Nature, in Islamic theology, is not a passive object but a manifestation of divine signs (*āyāt*). Mountains, rivers, plants, and animals are portrayed as participants in a cosmic order that reflects divine wisdom. Consequently, environmental degradation or the manipulation of sustainability narratives is not merely a technical failure but a moral affront to the sanctity of creation. This theological framing introduces a depth of moral significance that extends beyond regulatory compliance or scientific calculation. Environmental responsibility becomes an expression of reverence rather than a function of risk management. Another critical ethical principle within Islamic stewardship is the prohibition of *isrāf*, or excess. Overconsumption, wastefulness, and extravagance are consistently condemned in Islamic teachings, as they reflect a lack of gratitude and moral discipline. This principle is particularly relevant in the context of contemporary sustainability challenges, where excessive consumption patterns and resource-intensive production systems drive ecological degradation. Greenwashing often operates by obscuring these structural problems, allowing organizations to maintain unsustainable practices while projecting an environmentally responsible image. The ethical logic of *isrāf* exposes this contradiction by demanding moderation, restraint, and genuine transformation rather than cosmetic change.

What ultimately distinguishes theocentric stewardship from anthropocentric and ecocentric paradigms is its emphasis on transcendental moral accountability. In Islamic belief, all human actions are subject to divine judgment, encompassing not only outcomes but intentions and processes. This form of accountability introduces a moral constraint that cannot be fully replicated by legal sanctions, market mechanisms, or reputational pressures. Leaders may evade regulatory scrutiny or manage public perception, but they cannot escape accountability before God. This awareness creates an internalized ethical discipline that discourages manipulation even in the absence of external enforcement. The concept of *ihsan*—acting with excellence and sincerity as though one stands before God—further deepens this moral orientation. In organizational leadership, *ihsan* translates into ethical consistency between rhetoric and reality. A leader guided by *ihsan* would reject greenwashing not merely because it is risky or unethical in business terms, but because it violates sincerity and truthfulness at a spiritual level. From this standpoint, greenwashing constitutes a form of *khiyānah*, or betrayal, as it involves deliberate deception regarding the care of a trust that ultimately belongs to God.

Taken together, the interrelated concepts of *khalifah*, *amanah*, Islamic ecological ethics, and spiritual accountability form a coherent framework of theocentric stewardship. This framework offers more than symbolic or cultural guidance; it provides a robust normative foundation capable of addressing the moral roots of greenwashing. By anchoring leadership in transcendental responsibility rather than strategic calculation alone, Islamic theocentric stewardship reframes sustainability as an ethical obligation embedded in faith and moral identity. Although deeply grounded in Islamic theology, the principles of theocentric stewardship resonate with broader spiritual and ethical traditions that emphasize humility, responsibility, and care for creation. As such, this framework does not seek to replace existing sustainability paradigms but to enrich them by addressing their moral limitations. In an era where sustainability is increasingly vulnerable to instrumentalization and symbolic compliance, theocentric stewardship offers a compelling alternative—one that restores integrity, sincerity, and moral depth to leadership in modern organizations.

THEOCENTRIC SUSTAINABLE LEADERSHIP

The emergence of Theocentric Sustainable Leadership represents a normative response to the deepening crisis of integrity in contemporary sustainability practices. As greenwashing proliferates across industries and as dominant paradigms—anthropocentric and naturalistic alike—struggle to provide compelling moral constraints, the need for a leadership model grounded in transcendent ethics becomes increasingly evident. Theocentric Sustainable Leadership does not merely offer an alternative managerial approach; it proposes a foundational reorientation of sustainability by anchoring leadership in spiritual accountability and moral responsibility toward God. At its core, this paradigm challenges the prevailing assumption that sustainability is primarily a matter of strategic alignment, reputational management, or regulatory compliance. Instead, it reframes sustainability as a moral obligation inherent in the human condition. From a theocentric perspective, human beings are not autonomous agents exercising unlimited discretion over natural resources, but moral stewards entrusted with the care of the earth (DeWeese, 2023). This ontological positioning fundamentally alters the ethical logic of leadership: sustainable action is no longer justified by instrumental outcomes alone, but by fidelity to a higher moral mandate.

Theocentric Sustainable Leadership may be defined as a leadership approach that places God at the center of moral orientation and treats spiritual accountability as the primary basis for decision-making in pursuit of ecological balance, social well-being, and long-term sustainability. Unlike secular models of sustainable leadership that typically focus on the economic, social, and environmental "triple bottom line," the theocentric framework introduces a fourth and decisive dimension: spirituality. This spiritual dimension does not function as an abstract value or symbolic add-on, but as the ethical engine that animates integrity, sincerity, and moral consistency in leadership practice. The inclusion of spirituality significantly strengthens the moral architecture of sustainability. Leaders guided by a theocentric orientation possess motivations that extend beyond compliance, incentives, or performance indicators (Wang et al., 2019). Their commitment to sustainability is grounded in the belief that actions are ultimately evaluated not only by stakeholders or institutions, but by God. As a result, ethical consistency is maintained even in contexts where regulation is weak, monitoring is limited, or reputational risks are minimal. Sustainability becomes a matter of conscience rather than convenience.

Central to this paradigm is the principle of *divine accountability*. In contrast to secular accountability structures, which are predominantly horizontal—directed toward shareholders, regulators, customers, or civil society—theocentric leadership introduces a vertical dimension of accountability. Leaders perceive their decisions as subject to moral scrutiny that transcends legal frameworks and market pressures. Every strategic choice, environmental policy, and public claim is understood as part of a broader spiritual responsibility. This transcendent accountability has profound implications for ethical conduct. Honesty, for instance, is no longer treated as a negotiable professional norm but as an absolute moral value. The misrepresentation of environmental performance, selective disclosure of sustainability data, or exaggeration of ecological achievements constitutes not merely unethical behavior, but a violation of trust before God. In this sense, greenwashing is fundamentally incompatible with theocentric leadership because it entails deliberate deception in matters that are morally sacred.

Equally important is the emphasis on coherence between intention, action, and impact. Theocentric Sustainable Leadership rejects symbolic compliance and cosmetic initiatives that serve reputational goals without producing substantive ecological improvement. Good intentions alone are insufficient; they must materialize in concrete practices that protect ecosystems, respect communities, and preserve resources for future generations. This insistence on moral congruence serves as a direct counterforce to the performative nature of sustainability often observed in contemporary corporate discourse. The conceptual strength of Theocentric Sustainable Leadership lies in its integration of four interrelated ethical orientations, which together form a coherent moral framework. First, leadership is guided by a sense of sacred purpose. Environmental stewardship is understood not as a discretionary organizational objective but as a divinely mandated responsibility. This sacred purpose cultivates a long-term orientation in which sustainability commitments transcend short-term market cycles and leadership tenures. Organizational visions informed by this orientation are less susceptible to opportunistic shifts driven by branding or public pressure.

Second, theocentric leadership emphasizes ethical guardianship. Leaders function as moral custodians who are responsible for safeguarding both environmental integrity and ethical transparency within the organization. This guardianship extends to ensuring truthful sustainability reporting, resisting narrative manipulation, and fostering accountability across supply chains. From this standpoint, greenwashing represents a serious moral breach rather than a technical or communicative error.

Third, the paradigm underscores the role of stewardship wisdom. Sustainability challenges are inherently complex and demand more than technical expertise. Stewardship wisdom involves the capacity to perceive the interconnectedness of ecological systems, human welfare, and spiritual responsibility. Leaders guided by such wisdom are attentive to long-term ecological consequences, cautious in innovation that may disrupt environmental balance, and deliberate in aligning business decisions with ethical values. This orientation fosters moral prudence, discouraging actions that may be legally permissible yet ethically destructive.

Fourth, Theocentric Sustainable Leadership is animated by transformative compassion. Compassion, understood as *rahmah* in Islamic ethics, extends beyond interpersonal relations to encompass concern for all forms of life and for future generations. Leadership informed by compassion prioritizes inclusive well-being, recognizes the vulnerability of affected communities, and seeks to minimize harm to ecosystems. This compassionate orientation transforms sustainability from a compliance-driven agenda into a moral commitment to shared flourishing.

For this framework to be meaningful, it must be operationalized within organizational systems. Theocentric Sustainable Leadership requires the integration of spiritual values into corporate vision, governance structures, and performance metrics. Sustainability indicators should not be limited to quantitative outputs but should also reflect ethical integrity, transparency, and the authenticity of environmental claims. Decision-making processes can be guided by moral reflection that asks not only whether an action is profitable or legal, but whether it fulfills the trust entrusted to humanity and avoids ecological harm. Moreover, this paradigm encourages collective moral responsibility rather than isolated ethical heroism. Leadership is viewed as a shared moral practice in which values are embedded across organizational levels. Educational initiatives, ethical dialogues, and reflective practices can help cultivate a culture where sustainability is perceived as both a professional duty and a spiritual responsibility.

One of the most significant contributions of Theocentric Sustainable Leadership lies in its capacity to counter greenwashing at its root. By grounding motivation in spiritual accountability rather than reputational gain, the framework eliminates the incentive to engage in deceptive sustainability practices. Ethical standards exceed regulatory minimums, transparency becomes an expression of trust, and environmental stewardship is embraced as a form of moral devotion rather than strategic positioning. Looking ahead, the relevance of this paradigm extends beyond faith-based organizations. In an era marked by ecological crisis and declining public trust, leadership models that integrate ethical depth with practical governance are increasingly necessary. Theocentric Sustainable Leadership offers a compelling vision of leadership for the future—one that restores moral seriousness to sustainability and reaffirms that the earth is not merely a resource to be managed, but a trust to be honored with integrity, humility, and responsibility.

To synthesize the chapter's central arguments, Table 1. presents the proposed Theocentric Sustainable Leadership framework. The table illustrates how core theocentric principles are translated into leadership orientations, institutional mechanisms, and organizational outcomes. Consistent with the chapter's primary contribution, it demonstrates the systemic translation of a spiritual worldview into organizational leadership practices capable of addressing the ethical roots of greenwashing.

Table 1. Theocentric sustainable leadership as a framework for countering greenwashing

Core Element	Theocentric Foundation	Leadership Orientation	Institutional Mechanism	Expected Organizational Outcome	Anti-Greenwashing Function
Sacred Purpose	Khalifah (stewardship) and divine responsibility	Sustainability as a moral obligation rather than a strategic option	Sustainability embedded in organizational vision and mission	Strong sustainability commitment and mission alignment	Reduces symbolic sustainability initiatives driven solely by reputation
Ethical Guardianship	Amanah (trust) and accountability before God	Leaders act as custodians of environmental integrity and ethical conduct	Ethical governance structures, oversight mechanisms, and truthful reporting systems	Greater integrity and ethical consistency	Prevents manipulation, selective disclosure, and misleading environmental claims
Stewardship Wisdom	Mizan (balance) and moral prudence	Long-term and holistic decision-making that considers ecological, social, and spiritual consequences	Values-based strategic decision-making and sustainability evaluation	Sustainable resource management and ecological responsibility	Discourages short-term opportunistic actions that enable greenwashing
Transformative Compassion	Rahmah (compassion) and care for creation	Commitment to the well-being of communities, ecosystems, and future generations	Stakeholder engagement, inclusive sustainability initiatives, and ethical human capital development	Social legitimacy and stakeholder trust	Reduces exploitative practices hidden behind sustainability narratives

Source: Author's own work

HOW THEOCENTRIC STEWARDSHIP COUNTERS GREENWASHING

Greenwashing endures not merely because of regulatory gaps or insufficient enforcement, but because it is embedded within a deeper ethical failure that characterizes many contemporary leadership paradigms. Organizations today are often highly capable of crafting sophisticated sustainability narratives, producing visually compelling ESG reports, and aligning themselves symbolically with environmental values, while substantive operational practices remain largely unchanged. This discrepancy highlights that greenwashing is not fundamentally a technical or informational problem; rather, it is a moral phenomenon rooted in leadership systems that lack deeply internalized ethical commitments.

Within dominant managerial logics, sustainability is frequently framed as a strategic instrument—valuable insofar as it enhances legitimacy, attracts investment, or strengthens brand reputation. Ethical considerations are thus subordinated to performance imperatives, market expectations, and short-term financial targets. In such an environment, misleading sustainability claims are often rationalized as pragmatic adaptations rather than moral failures. Greenwashing becomes normalized not because leaders are unaware of its consequences, but because prevailing leadership frameworks do not provide a sufficiently strong moral foundation to restrain opportunistic behavior when ethical integrity conflicts with organizational advantage. Theocentric stewardship offers a fundamentally different approach to this problem. Rather than emphasizing compliance, metrics, or external verification as the primary safeguards of sustainability, it begins with the moral identity of the leader. Leadership, in this paradigm, is understood as an entrusted responsibility (*amanah*) that carries accountability not only to stakeholders and institutions, but ultimately to a transcendent moral authority. By relocating accountability from external surveillance to internal moral conviction, theocentric stewardship addresses greenwashing at its ethical roots rather than merely mitigating its visible outcomes.

Moral Internalization: The Foundation of Anti-Deception

The persistence of greenwashing cannot be adequately explained without examining the role of moral internalization in leadership. In many contemporary organizational settings, morality is treated instrumentally—as a resource to be mobilized when it aligns with strategic interests. Ethical principles are often invoked rhetorically, yet remain weakly embedded in decision-making processes. This instrumentalization creates moral flexibility, enabling leaders to justify selective disclosure or exaggerated sustainability claims when such practices are perceived to serve organizational goals. Theocentric stewardship disrupts this logic by redefining morality as an intrinsic dimension of leadership identity. Values such as trust (*amanah*), moral consciousness (*taqwa*), and accountability (*hisab*) are not external constraints imposed by regulation, but internal commitments that shape how leaders understand their role and responsibilities. Within this framework, honesty is not a strategic choice contingent on reputational risk; it is a spiritual obligation inseparable from ethical leadership.

This internalization of moral values significantly reduces the propensity for deception. Leaders who perceive their actions as subject to divine accountability are less likely to engage in practices that misrepresent environmental performance, even in situations where manipulation is unlikely to be detected. Ethical self-regulation emerges not from fear of sanctions, but from a deeply embedded sense of moral responsibility. As a result, greenwashing is reframed from a reputational risk to a violation of trust that carries moral consequences beyond organizational boundaries. Importantly, moral internalization functions as a preventative mechanism. Decisions that may appear economically advantageous but ethically questionable are filtered out at the earliest stages of strategic deliberation. While sustainability standards, certifications, and audits remain relevant, they no longer serve as the primary sources of moral legitimacy. Instead, they operate as supporting instruments within a leadership framework already anchored in ethical conviction. In this sense, theocentric stewardship complements existing governance mechanisms while addressing their fundamental limitation: the absence of internal moral restraint.

Consider a hypothetical board meeting of a manufacturing company preparing its annual ESG report. During the reporting process, managers discover that a recently implemented emissions-reduction initiative has produced positive results in one production facility. However, the same reporting period also includes a significant increase in indirect emissions generated elsewhere in the supply chain. Legally, the company is permitted to highlight the successful initiative without prominently disclosing the broader emissions increase, thereby presenting a more favorable sustainability narrative to investors and the public.

From a conventional managerial perspective, emphasizing the positive achievement may be viewed as a legitimate communication strategy. However, within a theocentric leadership framework, the board initiates a session of *muhasabah* (moral self-reflection) before approving the report. Rather than asking only whether the disclosure complies with reporting requirements, board members ask deeper ethical questions: Does this report truthfully represent our environmental impact? Are we fulfilling our amanah as stewards of creation? Would this communication remain acceptable if subjected not only to stakeholder scrutiny but also to divine accountability?

Through this process of collective reflection, the board concludes that selective disclosure would create a misleading impression of the organization's sustainability performance. Consequently, the report is revised to include both achievements and shortcomings, accompanied by a transparent action plan for addressing supply-chain emissions. Although this decision may reduce short-term reputational benefits, it strengthens long-term credibility, organizational integrity, and trust among stakeholders. This example illustrates how *muhasabah* functions as an internal ethical mechanism that discourages greenwashing by aligning sustainability communication with moral responsibility rather than image management.

Transparency: From Regulatory Compliance to Moral Obligation

Greenwashing often thrives in organizational environments characterized by opacity, information asymmetry, and hierarchical control over narratives. In such contexts, transparency is frequently reduced to regulatory compliance—meeting disclosure requirements without fostering genuine openness. Sustainability reporting becomes an exercise in impression management, where selective data presentation and ambiguous language are used to construct favorable images rather than convey substantive truth. Theocentric stewardship reframes transparency as a moral obligation rather than a procedural requirement. Leaders who view themselves as stewards of divine trust perceive information—particularly environmental and social data—as part of a public trust that must be safeguarded. Concealment, distortion, or selective disclosure is therefore understood not merely as a governance failure, but as an ethical transgression. Transparency extends beyond numerical accuracy to encompass sincerity in communication and integrity in decision-making processes.

This moral orientation toward transparency has important organizational implications. When leaders consistently demonstrate ethical openness, they contribute to the formation of cultures in which honesty is normalized and expected. Employees are more likely to feel morally and psychologically safe in raising concerns, reporting inconsistencies, or questioning sustainability claims that do not align with operational realities. Such cultures structurally constrain opportunities for greenwashing by reducing the likelihood that deceptive practices will remain unchallenged. Furthermore, theocentric stewardship expands the scope of oversight beyond formal audit mechanisms. While compliance structures remain important, ethical vigilance becomes distributed across organizational levels. Individuals perceive themselves not merely as procedural actors fulfilling assigned roles, but as moral agents responsible for upholding organizational integrity. Transparency thus becomes embedded in everyday practices rather than confined to periodic reporting cycles, making greenwashing increasingly difficult to sustain.

Long-Term Responsibility: Overcoming Short-Termism

Short-term orientation is widely recognized as a structural driver of greenwashing. Pressures to demonstrate immediate results—whether in financial performance or sustainability metrics—often incentivize symbolic actions that generate positive narratives without producing meaningful environmental impact. In such conditions, organizations may prioritize cosmetic initiatives that are visible and marketable, while deferring substantive transformation that requires time, investment, and organizational change. Theocentric stewardship introduces an alternative temporal framework grounded in long-term moral responsibility. By emphasizing accountability that extends beyond immediate organizational outcomes to include intergenerational and spiritual dimensions, it weakens the appeal of short-term deception. Leaders are encouraged to evaluate decisions not only in terms of present gains, but also in relation to their enduring consequences for ecosystems, communities, and future generations.

This long-term orientation redefines the meaning of sustainability itself. Success is no longer measured by the sophistication of sustainability narratives, but by the depth and durability of ecological impact. Investments in systemic change—such as restructuring supply chains, reducing absolute emissions, or transforming production processes—are valued even when they yield slow or uncertain returns. Greenwashing, by contrast, becomes increasingly incompatible with leadership grounded in long-term moral accountability. Moreover, theocentric stewardship reframes the relationship between organizations, nature, and society. Rather than treating the environment as a resource to be managed for competitive advantage, it is understood as part of a moral trust that must be preserved. This relational perspective discourages practices that externalize environmental harm or shift ecological burdens onto future generations. Sustainability thus becomes an expression of ethical stewardship rather than image management.

Integrative Perspective

When moral internalization, transparency, and long-term responsibility operate simultaneously, they form a coherent and multilayered defense against greenwashing. Internalized morality reduces the motivation to deceive; transparency reduces opportunities for concealment; and long-term responsibility alleviates pressure to pursue symbolic shortcuts. Together, these mechanisms address not only the visible symptoms of greenwashing, but also its psychological, moral, and structural foundations. Unlike technocratic approaches that rely primarily on regulation and measurement, theocentric stewardship offers a values-based framework capable of sustaining ethical conduct even in ambiguous or weakly regulated environments. It does not reject existing sustainability instruments, but situates them within a deeper moral architecture. By restoring leadership to its ethical and spiritual foundations, theocentric stewardship provides a robust conceptual response to greenwashing as a moral crisis rather than a mere reporting failure.

IMPLICATIONS FOR ORGANIZATIONS AND LEADERSHIP PRACTICE

Integrating theocentric values into organizational life should not be understood as a symbolic spiritual gesture or a purely normative aspiration. Rather, it represents a strategic and ethical necessity for organizations seeking to achieve authentic sustainability while avoiding the growing crisis of credibility caused by greenwashing. As public skepticism toward corporate sustainability claims intensifies, organizations are increasingly compelled to reflect not only on *what* sustainability initiatives they pursue, but also on *why* they pursue them and *how* such initiatives are enacted in practice. Within this context, leadership grounded in divine trust (*amanah*) offers a distinctive and morally robust foundation for organizational sustainability. This section discusses the practical implications of theocentric sustainable leadership for contemporary organizations. It demonstrates how spiritual values can be translated into concrete organizational arrangements, including vision formulation, governance structures, decision-making processes, human capital development, sustainability reporting, and stakeholder engagement. The central argument advanced here is that spiritual values must be institutionalized within organizational systems if they are to meaningfully counter greenwashing rather than remain rhetorical commitments.

Reframing Organizational Vision and Purpose

The shift toward theocentric sustainable leadership begins with a fundamental reorientation of organizational vision and purpose. Many organizations articulate sustainability through technocratic language focused on efficiency, emissions reduction, or compliance with environmental standards. While such objectives are important, they often lack moral depth and therefore fail to inspire ethical transformation. A theocentric perspective challenges organizations to redefine their purpose by recognizing that they are not merely economic actors, but moral agents entrusted with public and ecological responsibility. When *amanah* is embedded in organizational vision, sustainability is no longer framed as an external obligation or reputational asset, but as a moral commitment to safeguard creation. This reframing transforms sustainability from a peripheral agenda into a core element of organizational identity. Consequently, ecological justice and social well-being are positioned not as secondary outcomes of economic success, but as intrinsic purposes that guide organizational conduct.

Embedding Theocentric Ethics in Governance Structures

Governance systems play a critical role in shaping organizational behavior, yet many contemporary governance models remain narrowly focused on shareholder interests and short-term performance. Theocentric sustainable leadership requires governance arrangements that explicitly recognize moral and spiritual accountability alongside economic responsibility. By incorporating ethical stewardship into governance mechanisms, organizations can create institutional safeguards against greenwashing. Ethical oversight bodies, moral evaluation processes, and spiritually informed codes of conduct function as counterweights to purely instrumental decision-making. Importantly, theocentric governance introduces a dual conception of accountability: leaders are answerable not only to regulators and stakeholders, but also to a transcendent moral authority. This expanded accountability discourages opportunistic behavior even in contexts where regulatory enforcement is weak or public scrutiny is limited.

Integrating Theocentric Values into Decision-Making Processes

Greenwashing often emerges at critical moments of strategic decision-making, particularly when organizations face tension between economic pressures and environmental responsibility. In the absence of strong moral anchors, leaders may rationalize deceptive practices as necessary compromises. Theocentric sustainable leadership addresses this vulnerability by embedding *amanah* directly into decision-making logic. Decisions are evaluated not solely on financial outcomes or reputational benefits, but on their moral intention, ecological impact, and spiritual accountability. This approach encourages leaders to reflect on whether their actions genuinely serve the common good or merely create the appearance of responsibility. Over time, such reflective decision-making fosters organizational habits that prioritize ethical consistency over symbolic compliance.

Human Capital Development: From Competence to Character

A defining implication of theocentric sustainable leadership lies in its approach to human capital development. Conventional leadership development programs tend to emphasize technical expertise, strategic capability, and performance optimization, often overlooking moral character formation. The theocentric perspective challenges this imbalance by asserting that sustainable leadership ultimately depends on who leaders are, not merely on what they can do. By integrating values such as trust, honesty, and ecological responsibility into leadership development, organizations cultivate moral resilience among their members. Leaders who internalize these values are more likely to resist pressures toward manipulation and short-termism. Moreover, ethical role modeling by senior leaders plays a decisive role in shaping organizational culture, signaling that integrity and responsibility are not negotiable traits but foundational expectations.

Reorienting Sustainability Reporting Practices

Sustainability reporting has become one of the primary arenas in which greenwashing occurs, as organizations selectively present favorable data while obscuring environmental harm. Theocentric sustainable leadership calls for a fundamental reorientation of reporting practices, grounded in the moral imperative of truthfulness. Within a theocentric framework, reporting is understood as an act of moral testimony rather than a branding exercise. Accuracy, transparency, and consistency between claims and actions are treated as expressions of *amanah*. By expanding reporting systems to include ethical dimensions—such as honesty in disclosure and integrity in decision-making—organizations move beyond performative transparency toward substantive accountability. Public participation and community verification further reinforce this orientation by positioning society as a legitimate moral stakeholder.

Transforming Stakeholder Relationships

Another significant implication of theocentric stewardship concerns the nature of stakeholder relationships. Whereas many organizations engage stakeholders transactionally, the theocentric paradigm reframes these relationships as moral trusts that require care, fairness, and long-term commitment. This relational orientation discourages practices that externalize environmental harm onto suppliers or marginalized communities. It also demands honest communication with consumers, rejecting exaggerated sustainability claims that mislead public perception. By engaging stakeholders in genuine moral dialogue, organizations strengthen social legitimacy and align their sustainability efforts with broader ethical expectations.

Cultivating a Culture of Divine Accountability

Ultimately, the effectiveness of theocentric sustainable leadership depends on cultural transformation. Organizational culture determines whether spiritual values are lived realities or mere symbolic references. A culture grounded in divine accountability nurtures ethical self-regulation, where individuals act responsibly not because they are monitored, but because they recognize their moral obligations. Practices that encourage reflection, reward integrity, and normalize honesty—especially in acknowledging sustainability failures—play a critical role in sustaining this culture.

Such an environment reduces fear-driven concealment and enables continuous ethical improvement.

CONCLUSIONS

The widespread phenomenon of greenwashing in the modern era has revealed a critical insight: sustainability cannot rest solely on technocratic, procedural, or regulatory foundations. No matter how meticulously a sustainability report is designed or how comprehensive the certifications obtained, as long as sustainability is not supported by a strong moral foundation, the space for manipulation remains open. This is the central irony of contemporary sustainability practices: as environmental issues gain prominence, they simultaneously create new opportunities for organizations to exploit green rhetoric for reputational purposes. It is precisely through this irony that the need for a new paradigm—or more accurately, for a paradigm that returns to its ethical roots—becomes increasingly evident.

This book chapter emerges from a deep concern over the widening gap between sustainability rhetoric and organizational reality. Many corporations perceive sustainability primarily as an administrative obligation rather than a moral trust. As a result, sustainability is reduced to a marketing slogan instead of being embraced as a deeply rooted spiritual consciousness. Under such conditions, greenwashing becomes a normalized phenomenon and, in some cases, is even regarded as a legitimate strategic practice. As previously discussed, however, greenwashing is not merely an isolated ethical lapse; it is a reflection of a flawed paradigm governing the relationship between humanity, nature, society, and God.

The preceding chapters have demonstrated that anthropocentric and naturalistic paradigms, while contributing important insights to modern environmental ethics, possess significant limitations. Neither is capable of addressing the true root of greenwashing: the absence of an internal moral mechanism that binds individuals to principles of truthfulness, honesty, and long-term responsibility. Anthropocentrism fails because it prioritizes human interests and economic efficiency, reducing nature to a mere resource. Naturalistic paradigms fail because they implicitly ascribe moral agency to nature itself, despite the fact that morality resides solely within human consciousness. Neither paradigm offers a sufficiently compelling moral rationale to resist manipulation when corporate interests are at stake.

As an alternative, this chapter proposes a theocentric paradigm as the foundation of *theocentric sustainable leadership*, a leadership model grounded in spiritual values—particularly the concepts of *amanah* (trust), *khalifah* (stewardship), and moral accountability within the Islamic tradition. This approach positions humans not as absolute dominators, but as custodians of creation entrusted with maintaining balance. Humans are granted the freedom to manage resources, yet this freedom is inseparable from profound responsibility—accountability not only to other human beings, but ultimately to God, who is ever-watchful. This belief forms the core of strong moral internalization: the conviction that no action, however concealed or insignificant, escapes divine evaluation.

The concept of *khalifah* teaches that every leader is a bearer of responsibility. Leadership is not merely a privilege, but a moral burden. The principle of *amanah* reinforces the understanding that nothing under human control—whether nature, organizations, or fellow human beings—is owned absolutely; all are entrusted responsibilities that must be managed with honesty and justice. Meanwhile, *muhasabah* cultivates reflective moral awareness, emphasizing that every action carries ethical consequences. Through these foundations, theocentric leadership offers a deeper ethical framework than secular ethics—one that asks not only what is right according to human judgment, but what is right before God.

Building upon this theoretical foundation, the framework of theocentric sustainable leadership is developed as a synthesis of spiritual values and modern organizational practices. This model illustrates how divine-centered values can be translated into managerial principles such as long-term orientation, transparency, honesty, and ecological justice. More importantly, it demonstrates that theocentric values are not limited to religious organizational contexts; they possess universal relevance as moral principles that prioritize integrity, responsibility, and the preservation of life.

This chapter further explains how theocentric leadership directly addresses the problem of greenwashing. The central mechanism lies in moral internalization. When sustainability is driven by spiritual motivation rather than administrative compliance, manipulation loses its appeal. Greenwashing becomes morally untenable because leaders recognize that false claims constitute a betrayal of trust. In addition, the theocentric paradigm reinforces transparency through a culture of honesty. It encourages organizations to develop reporting systems that are truthful and non-selective, while rejecting green narratives that do not correspond to operational realities. Equally important, this paradigm expands the horizon of responsibility across generations, compelling organizations to think beyond short-term profit considerations.

Ultimately, this chapter offers practical recommendations for integrating theocentric ethics into organizational structures. These include reformulating organizational vision, establishing ethical councils, embedding *amanah*-based values into decision-making systems, and developing character-based human capital. Together, these recommendations affirm that theocentric leadership is not a utopian ideal. Rather, it is a practicable model that can be institutionalized through deliberate structures, systems, and organizational cultures—enabling organizations to operate not only efficiently, but also with genuine moral integrity.

REFERENCES

Abeng, T. (1997). Business Ethics in Islamic Context: Perspectives of a Muslim. *Business Ethics Quarterly*, *7*(3), 47–54. DOI: 10.2307/3857312

Bowen, F., & Aragon-Correa, J. A. (2014). Greenwashing in Corporate Environmentalism Research and Practice: The Importance of What We Say and Do. In *Organization and Environment* (Vol. 27, Number 2, pp. 107–112). SAGE Publications Inc. DOI: 10.1177/1086026614537078

Cole, R., Johan, S., & Schweizer, D. (2021). Corporate failures: Declines, collapses, and scandals. *Journal of Corporate Finance*, *67*, 101872. Advance online publication. DOI: 10.1016/j.jcorpfin.2020.101872

Dempere, J., Alamash, E., & Mattos, P. (2024). Unveiling the truth: greenwashing in sustainable finance. In *Frontiers in Sustainability* (Vol. 5). Frontiers Media SA. DOI: 10.3389/frsus.2024.1362051

DeWeese, G. J. (2023). A Theocentric Environmental Ethic. *Religions*, *14*(7), 913. Advance online publication. DOI: 10.3390/rel14070913

Dhiman, S. (2017). Leadership and Sustainability. In Marques, J., & Dhiman, S. (Eds.), *Leadership Today: Practices for Personal and Professional Performance* (pp. 161–177). Springer International Publishing., DOI: 10.1007/978-3-319-31036-7_9

Eom, K., & Ng, S. T. (2023). The Potential of Religion for Promoting Sustainability: The Role of Stewardship. *Topics in Cognitive Science*, *15*(3), 480–499. DOI: 10.1111/tops.12641 PMID: 36780334

Frantz, P., Rego, F., & Barbas, S. (2025). Ecocentrism vs. Anthropocentrism: To the Core of the Dilemma to Overcome It. *The Linacre Quarterly*, *92*(4), 449–459. DOI: 10.1177/00243639251339844 PMID: 40510561

Fry, L. W., & Egel, E. (2021). Global leadership for sustainability. *Sustainability (Basel)*, *13*(11), 6360. Advance online publication. DOI: 10.3390/su13116360

Glaser, I. (1998). Roles and relationships: Reflections on the khalifah and the image of God. *Transformation (Durban)*, *15*(1), 18–23. DOI: 10.1177/026537889801500108

Herijanto, H. (2022). Al amanah in al qur'an vs trust: A comparative study. *International Journal of Ethics and Systems*, *38*(4), 549–575. DOI: 10.1108/IJOES-03-2021-0064

Hoffman, A. J., & Sandelands, L. E. (2005). Getting right with nature: Anthropocentrism, ecocentrism, and theocentrism. *Organization & Environment, 18*(2), 141–162. DOI: 10.1177/1086026605276197

Husamah, R. A., Permana, T. I., & Lestari, N. (2025). Islam and sustainability issues, how far has the relationship progressed? A bibliometric analysis. In *Social Sciences and Humanities Open* (Vol. 12). Elsevier Ltd. DOI: 10.1016/j.ssaho.2025.101703

Iqbal, Q., Ahmad, N. H., Nasim, A., & Khan, S. A. R. (2020). A moderated-mediation analysis of psychological empowerment: Sustainable leadership and sustainable performance. *Journal of Cleaner Production, 262*, 121429. Advance online publication. DOI: 10.1016/j.jclepro.2020.121429

Iqbal, Q., Piwowar-Sulej, K., & Andersén, J. (2026). From Eco-Policy to Green Practice: Government Eco-Initiatives, Managerial Attention and Frugal Green Innovation in Saudi Arabia. *Business Strategy and the Environment*, bse.70821. Advance online publication. DOI: 10.1002/bse.70821

Islam, M. S., & Samsudin, S. (2018). Interpretations of Al-amanah Among Muslim Scholars and Its Role in Establishing Peace in Society. *Social Change, 48*(3), 437–450. DOI: 10.1177/0049085718781689

Ismail, H. W., & Berghout, A. (2024). *AL-SHAJARAH ISTAC Journal of Islamic Thought and Civilization.*

Koch, T., & Denner, N. (2025). Different shades of green deception. Greenwashing's adverse effects on corporate image and credibility. *Public Relations Review, 51*(1), 102521. Advance online publication. DOI: 10.1016/j.pubrev.2024.102521

Kopnina, H., Washington, H., Taylor, B., & Piccolo, J. J. (2018). Anthropocentrism: More than Just a Misunderstood Problem. In *Journal of Agricultural and Environmental Ethics* (Vol. 31, Number 1, pp. 109–127). Springer Netherlands. DOI: 10.1007/s10806-018-9711-1

Lahmar, F. (2024). Redefining Leadership: The Role of Spirituality and Motherhood in Muslim Women's Educational Leadership. *Religions, 15*(12), 1565. Advance online publication. DOI: 10.3390/rel15121565

Leal Filho, W., Salvia, A. L., Ulluwishewa, R., Abubakar, I. R., Mifsud, M., LeVasseur, T. J., Correia, V., Consorte-McCrea, A., Paço, A., Fritzen, B., Ray, S., Gordon, N., Luetz, J. M., Borsari, B., Venkatesan, M., Mukul, S. A., Carp, R. M., Begum, H., Nunoo, E. K., & Farrugia, E. (2022). Linking sustainability and spirituality: A preliminary assessment in pursuit of a sustainable and ethically correct world. *Journal of Cleaner Production, 380*, 135091. Advance online publication. DOI: 10.1016/j.jclepro.2022.135091

Piwowar-Sulej, K., & Iqbal, Q. (2025). *Sustainability-focused leadership and pro-environmental behavior in SMEs: through the lens of conservation of resources theory.* Central European Management Journal., DOI: 10.1108/CEMJ-11-2024-0375

Pratama, A., Yusoff, H., Yadiati, W., & Jaenudin, E. (2025). Sustainability-related corporate governance and greenwashing practices: Preliminary evidence from southeast Asian companies. *Meditari Accountancy Research, 33*(3), 853–885. DOI: 10.1108/MEDAR-07-2024-2579

Rakhmat, A. (2022). Islamic Ecotheology: Understanding the Concept of Khalifah and the Ethical Responsibility of The Environment. *Academic Journal of Islamic Principles and Philosophy, 3*(1), 1–24. DOI: 10.22515/ajipp.v3i1.5104

Sneideriene, A., & Legenzova, R. (2025). Greenwashing prevention in environmental, social, and governance (ESG) disclosures: A bibliometric analysis. *Research in International Business and Finance, 74*, 102720. Advance online publication. DOI: 10.1016/j.ribaf.2024.102720

Szabo, S., & Webster, J. (2021). Perceived Greenwashing: The Effects of Green Marketing on Environmental and Product Perceptions. *Journal of Business Ethics, 171*(4), 719–739. DOI: 10.1007/s10551-020-04461-0

Wang, M., Guo, T., Ni, Y., Shang, S., & Tang, Z. (2019). The effect of spiritual leadership on employee effectiveness: An intrinsic motivation perspective. *Frontiers in Psychology, 9*(JAN), 2627. Advance online publication. DOI: 10.3389/fpsyg.2018.02627 PMID: 30662419

Washington, H., & Maloney, M. (2020). The need for ecological ethics in a new ecological economics. In *Ecological Economics* (Vol. 169). Elsevier B.V., DOI: 10.1016/j.ecolecon.2019.106478

Yang, Z., Nguyen, T. T. H., Nguyen, H. N., Nguyen, T. T. N., & Cao, T. T. (2020). Greenwashing behaviours: Causes, taxonomy and consequences based on a systematic literature review. In *Journal of Business Economics and Management* (Vol. 21, Number 5, pp. 1486–1507). VGTU. DOI: 10.3846/jbem.2020.13225

Zervoudi, E. K., Moschos, N., & Christopoulos, A. G. (2025). From the Corporate Social Responsibility (CSR) and the Environmental, Social and Governance (ESG) Criteria to the Greenwashing Phenomenon: A Comprehensive Literature Review About the Causes, Consequences and Solutions of the Phenomenon with Specific Case Studies. In *Sustainability (Switzerland)* (Vol. 17, Number 5). Multidisciplinary Digital Publishing Institute (MDPI). DOI: 10.3390/su17052222

KEY TERMS AND DEFINITIONS

Amanah (Trust): A sacred trust bestowed by God upon humanity, referring to moral responsibility in managing resources, authority, and relationships. In organizational contexts, amanah implies that leadership roles, environmental resources, and stakeholder relationships are entrusted responsibilities requiring honesty, transparency, and accountability.

Anthropocentrism: An ethical paradigm that positions human interests as the central criterion for moral consideration. Within sustainability discourse, anthropocentrism often treats nature instrumentally—as valuable primarily for its utility to human welfare and economic growth.

Divine Accountability Culture: An organizational culture in which individuals act ethically not merely due to monitoring, but because they perceive moral responsibility before God.

Ecocentrism: An environmental ethical perspective that recognizes the intrinsic value of nature independent of human interests. While ecocentrism challenges human-centered dominance, it may lack a transcendent moral anchor capable of restraining opportunistic behavior.

Ethical Guardianship: The responsibility of leaders to protect organizational integrity, ensure truthful sustainability communication, and prevent manipulative practices.

Greenwashing: Corporate practices that exaggerate, misrepresent, or selectively disclose environmental performance to create an appearance of sustainability without substantive ecological transformation. It represents a moral and communicative failure rooted in instrumental leadership paradigms.

Ihsan: An Islamic ethical principle referring to excellence and sincerity in action, performed with awareness of divine presence. In leadership, ihsan demands consistency between rhetoric and reality.

Israf (Excess): A prohibition against wastefulness and overconsumption. In sustainability discourse, israf challenges consumerism and resource-intensive production systems.

Khalifah (Stewardship): An Islamic concept describing humanity's role as vicegerent or steward of the earth. It emphasizes responsibility, guardianship, and moral restraint rather than domination or exploitation.

Mizan (Balance): A Qur'anic concept describing the divinely ordained balance in creation. Environmental degradation and excessive exploitation are viewed as disruptions of this balance.

Moral Congruence: Consistency between intention, communication, and action in sustainability practices.

Moral Internalization: The process by which ethical values become embedded within a leader's identity and decision-making logic, reducing reliance on external enforcement mechanisms.

Muhasabah: Self-reflection and moral evaluation of one's actions. In organizational leadership, it refers to continuous ethical assessment of decisions and sustainability practices.

Sacred Purpose: The understanding that environmental stewardship is not optional or strategic alone, but a moral calling rooted in spiritual identity.

Short-Termism: A managerial orientation that prioritizes immediate financial gains or reputational benefits over long-term ecological responsibility. It is a structural driver of greenwashing.

Spiritual Dimension of Sustainability: The recognition that authentic sustainability requires moral depth, conscience, and accountability beyond technical compliance or economic calculation.

Sustainability Integrity: The alignment between sustainability rhetoric, operational practices, and measurable ecological impact. Integrity eliminates the gap that enables greenwashing.

Theocentric Sustainable Leadership: A leadership framework that integrates spiritual accountability, moral integrity, and ecological responsibility. It positions sustainability as a divine trust and rejects symbolic compliance in favor of authentic, ethically grounded practice.

Theocentrism: A worldview that places God at the center of moral orientation and ultimate accountability. In environmental ethics, theocentrism understands nature as divine creation and human beings as morally responsible stewards accountable to God.

Transcendental Accountability: A form of accountability that extends beyond regulatory or stakeholder oversight to include moral responsibility before God. It functions as an internal ethical constraint against deception and opportunism.

Chapter 4
Sustainable Leadership for Environmental Stewardship:
A Framework for Responsible Organizational Practice

A. Sharmila
http://orcid.org/0009-0001-8110-2548
Saveetha Institute of Medical and Technical Sciences, Chennai, India

Raja Shree K. C.
http://orcid.org/0009-0009-1029-910X
Saveetha School of Management, Saveetha Institute of Medical and Technical Sciences, Chennai, India

Gigi G. S.
Saveetha Institute of Medical and Technical Sciences, Chennai, India

Praveen Kumar P.
http://orcid.org/0000-0002-0707-039X
Srm Institute of Science and Technology, Vadapalani, India

ABSTRACT

As businesses deal with escalating environmental issues and sustainability demands, sustainable leadership has grown in significance. There is still little empirical data connecting sustainable leadership to environmental stewardship, despite the fact that leadership and sustainability have been discussed extensively. The purpose of this study is to investigate how sustainable leadership contributes to the advancement of ecologically conscious business practices. Utilising survey information gathered

DOI: 10.4018/979-8-3373-7877-0.ch004

from managerial staff, a quantitative research methodology was used. The results show that through increased resource efficiency and green initiatives, sustainable leadership greatly improves environmental stewardship. The study adds to the body of knowledge on sustainability by putting forth an empirically supported framework that connects environmental outcomes and leadership behaviours. In practical terms, the findings offer direction to policymakers and leaders who want to improve ESG-driven organisational processes.

1. INTRODUCTION

In recent decades, organizations across the globe have been confronted with unprecedented environmental, social, and economic challenges that demand a fundamental rethinking of leadership and organizational responsibility. Climate change, environmental degradation, resource scarcity, social inequality, and increasing regulatory and stakeholder pressures have significantly altered the landscape in which organizations operate (United Nations, 2015; World Economic Forum, 2022). These challenges have intensified expectations for organizations to move beyond short-term profit maximization toward more responsible, ethical, and sustainable forms of value creation.

Consequently, leadership paradigms are evolving from traditional, economically driven models to approaches that emphasize long-term sustainability, environmental stewardship, and stakeholder well-being (Suriyankietkaew &Avery, 2014; Pless et al., 2012).Organizations are no longer evaluated solely on their financial performance; rather, they are increasingly assessed based on their environmental and social impact. Investors, regulators, customers, employees, and communities are demanding greater transparency, accountability, and responsibility from organizations regarding their environmental practices (Freeman & Boeker, 1984); (Eccles et al., 2014).

Leaders influence organizational vision, strategy, culture, and decision-making processes, thereby determining whether sustainability initiatives are symbolic or genuinely embedded in organizational practice (Metz et al., 2020). Unlike traditional leadership models that prioritize short-term financial outcomes, sustainable leadership focuses on long-term organizational success by balancing economic performance with environmental protection and social responsibility (Avery & Bergsteiner, 2011). Sustainable leaders adopt a systems-thinking perspective, recognizing the interconnectedness of organizational activities, environmental systems, and societal well-being (Liao, 2022).

1.1 History and Evolution of Sustainable Leadership

The concept of sustainable leadership emerged from the convergence of sustainability and leadership research in response to growing concerns regarding environmental degradation, social inequality, and long-term organizational survival. Traditional leadership theories, including trait, behavioural, and contingency approaches, primarily emphasized leader characteristics, managerial effectiveness, and organizational performance outcomes (Badshah, 2012). While these theories contributed significantly to leadership studies, they offered limited guidance for addressing complex sustainability challenges facing modern organizations. As global concerns related to climate change, resource scarcity, and social responsibility intensified, scholars recognized the need for a leadership approach capable of balancing economic, environmental, and social objectives (Egri & Herman, 2000).

Consequently, sustainable leadership evolved as an integrative paradigm that draws upon transformational leadership, ethical leadership, responsible leadership, and stakeholder theory. Transformational leadership contributes the vision and inspiration necessary for organizational change (Bass & Avolio, 1994), while ethical leadership emphasizes integrity, fairness, and accountability in decision-making (Brown et al., 2005). Stakeholder theory further expands leadership responsibility beyond shareholders to include employees, customers, communities, suppliers, and the natural environment (Freeman, 1984). Sustainable leadership synthesizes these perspectives by promoting long-term value creation, organizational resilience, stakeholder engagement, and environmental stewardship (Avery & Bergsteiner, 2011).

Recent developments in sustainability research have further strengthened the relevance of sustainable leadership through its alignment with Environmental, Social, and Governance (ESG) principles, the United Nations Sustainable Development Goals (SDGs), and stakeholder capitalism. Contemporary studies suggest that sustainable leadership enables organizations to navigate uncertainty, foster innovation, and enhance resilience in increasingly dynamic business environments (Hallinger & Suriyankietkaew, 2021; Suriyankietkaew & Avery, 2022). The COVID-19 pandemic further highlighted the importance of leadership approaches that balance organizational performance with societal and environmental responsibilities, thereby accelerating scholarly interest in sustainable leadership as a strategic imperative for long-term success.

Figure 1. History and Evolution of Sustainable Leadership

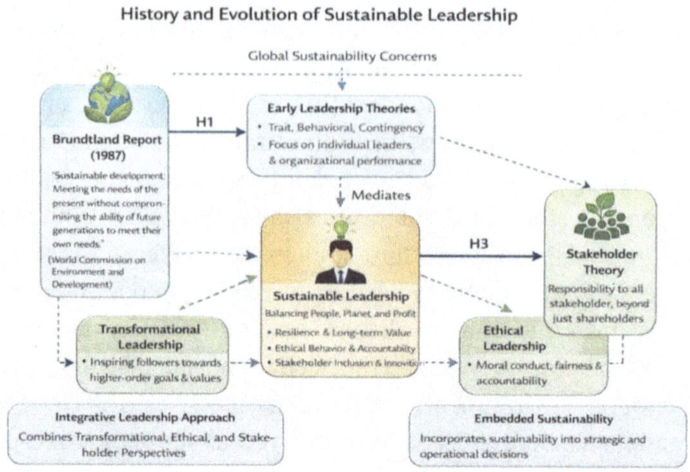

(Egri & Herman, 2000; Pless et al., 2012; Brown et al., 2005; Freeman & Boeker, 1984; Avery & Bergsteiner, 2011)

1.2 Environmental Challenges and Organizational Responsibility

Organizations today operate within an increasingly complex environmental landscape characterized by climate change, biodiversity loss, resource depletion, pollution, and ecological degradation. Scientific evidence indicates that industrial activities, unsustainable production systems, and global supply chains are among the primary contributors to environmental deterioration and greenhouse gas emissions (Hart, 1995; Steffen et al., 2015). These environmental challenges have intensified pressure on organizations to reconsider their operational practices and adopt more sustainable approaches to business management. Governments and regulatory agencies worldwide have introduced stricter environmental regulations related to carbon emissions, waste management, environmental reporting, and resource conservation, compelling organizations to integrate sustainability considerations into strategic and operational decision-making (Aguilera et al., 2007). At the same time, stakeholder expectations have evolved significantly.

Consumers, investors, employees, local communities, and non-governmental organizations increasingly demand that organizations demonstrate environmental responsibility through sustainable sourcing, renewable energy adoption, carbon footprint reduction, circular economy initiatives, and transparent sustainability reporting (Porter & van der Linde, 1995; Bansal & Roth, 2000). Consequently, environmental responsibility is no longer viewed merely as a compliance requirement but as a strategic necessity that can enhance organizational reputation, stakeholder trust, and competitive advantage. The emergence of Environmental, Social, and Governance (ESG) frameworks has further reinforced the connection between sustainability performance and organizational success by providing measurable indicators of environmental and social responsibility (Eccles et al., 2014).

Recent studies suggest that organizations with strong ESG performance tend to exhibit superior risk management, greater financial resilience, and enhanced long-term value creation (Friede et al., 2015; Eccles et al., 2023). Moreover, environmental stewardship has evolved from a reactive compliance-oriented approach to a proactive strategy focused on preserving natural resources, promoting green innovation, and creating sustainable business models (Aragón-Correa & Sharma, 2003). As environmental risks continue to intensify, organizations increasingly recognize that responsible environmental practices are essential not only for ecological sustainability but also for long-term organizational performance and stakeholder value creation (World Economic Forum, 2024; UNEP, 2023).

Figure 2. Environmental challenges and organizational responsibility

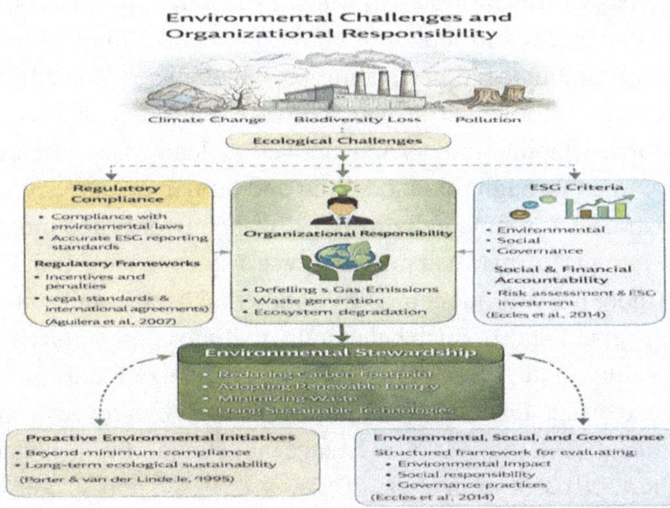

(Rockström et al., 2009; Hart, 1995; Aguilera et al., 2007; Eccles et al., 2014; Avery & Bergsteiner, 2011; Shrivastava, 1995)

1.3 Role of Leadership in Environmental Stewardship

Leadership plays a pivotal role in shaping how organizations perceive, interpret, and respond to environmental challenges. Leaders influence strategic direction, allocate financial and human resources, establish performance expectations, and cultivate organizational culture, all of which are critical determinants of effective environmental stewardship (Waldman & Siegel, 2008). In the absence of strong leadership commitment, sustainability initiatives often remain fragmented, symbolic, or short-lived, lacking integration with core organizational processes and long-term strategy (Linnenluecke & Griffiths, 2010);(Banerjee, 2002).Sustainable leadership emphasizes the integration of environmental responsibility into organizational vision, mission, and strategic planning. Sustainable leaders articulate a clear and compelling sustainability-oriented vision that aligns environmental objectives with business goals, thereby embedding stewardship principles into organizational strategy rather than treating them as peripheral concerns (Avery & Bergsteiner, 2011; Epstein, 2018).

By encouraging experimentation, supporting risk-taking, and investing in sustainable technologies, leaders create an enabling environment for eco-innovation, such as the development of environmentally friendly products, cleaner production processes, and energy-efficient systems (Aragón-Correa & Sharma, 2003). Empirical studies suggest that leadership support for environmental initiatives enhances organizational learning and innovation capabilities, which in turn contribute to improve environmental and economic performance (Porter & Linde, 1995; Hart & Dowell, 2011).In addition to shaping strategy and innovation, leaders significantly influence employee behaviours through role modelling and human resource practices. When leaders demonstrate environmentally responsible behaviours and integrate sustainability criteria into performance appraisal, reward systems, and training programs, employees are more likely to adopt pro-environmental behaviours at work (Renwick et al 2013; Daily et al., 2009). Such behaviours include energy conservation, waste reduction, recycling, and active participation in organizational green initiatives. Research indicates that leadership support enhances employees' environmental commitment and intrinsic motivation to engage in sustainability-oriented behaviours (Ones & Dilchert, 2012; Norton et al., 2015).

Stakeholder theory suggests that proactive engagement with these groups enables organizations to identify environmental risks, respond to stakeholder expectations, and co-create sustainable solutions (Freeman & Boeker, 1984; Donaldson & Preston, 1995). Through ethical governance, transparency, and accountability, leaders enhance organizational legitimacy and stakeholder trust, which are essential for sustaining long-term environmental initiatives (Suchman, 1995; Aguilera et al., 2007). Transformational and responsible leadership styles have been shown to positively influence environmental performance by fostering values-based leadership, employee empowerment, and collective responsibility for sustainability outcomes (Waldman et al., 2006; Pless et al., 2012).

Figure 3. Role of leadership in environmental stewardship

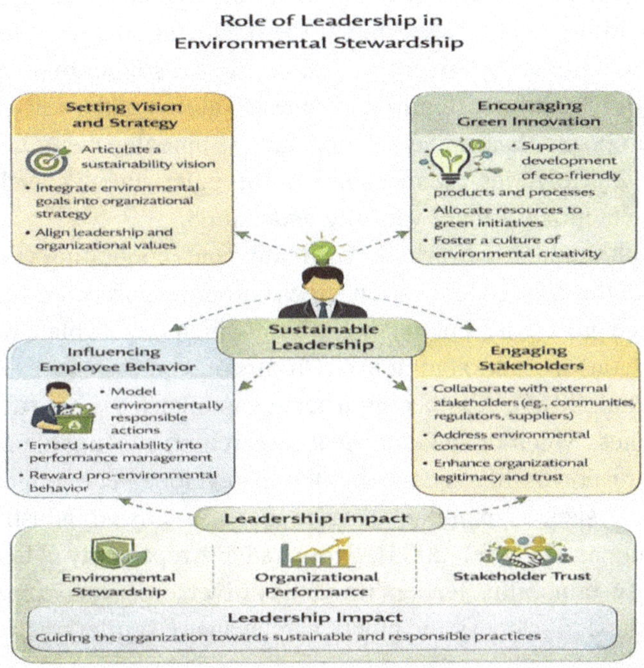

(Linnenluecke & Griffiths, 2010; Banerjee, 2002; Pless et al., 2012; Freeman & Boeker,1984; Suchman, 1995)

1.4 Research Gap

Despite the growing scholarly attention devoted to sustainability, environmental management, and leadership, several important gaps persist in the existing literature. A substantial body of research has examined sustainability initiatives, environmental management systems, and corporate environmental responsibility from strategic or operational perspectives, often without explicitly incorporating leadership processes as a central explanatory factor (Hart, 1995; Aragón-Correa & Sharma, 2003). Conversely, leadership research has traditionally focused on outcomes such as organizational performance, employee motivation, job satisfaction, and change management, with limited attention to environmental stewardship or sustainability-related outcomes (Yukl et al., 2013; Bass & Riggio, 2006). Even when sustainability is addressed, it is frequently treated as a peripheral concern rather than as a core leadership responsibility influencing long-term organizational resilience and ecological impact (Banerjee, 2002; Waldman & Siegel, 2008).

Although conceptual models of sustainable leadership and responsible leadership are well articulated in the literature (Avery & Bergsteiner, 2011) empirical investigations that systematically examine how sustainable leadership practices translate into measurable environmental stewardship at the organizational level remain limited. Existing empirical studies often rely on qualitative case studies or descriptive analyses, which, while insightful, offer limited generalizability and insufficient empirical validation of proposed theoretical relationships (Linnenluecke & Griffiths, 2010; Pless et al., 2012).Moreover, the mechanisms through which leadership influences environmental outcomes are not yet fully understood.

Prior research frequently overlooks mediating and moderating variables such as organizational culture, green innovation, green human resource management practices, and employee pro-environmental behaviour, which may explain how leadership commitment is converted into tangible environmental performance (Renwick et al., 2013; Norton et al.,2015). Another significant gap relates to contextual limitations in existing studies. Much of the empirical research on sustainable leadership and environmental responsibility is concentrated in developed economies and specific industrial sectors, such as manufacturing or energy-intensive industries (Bansal & Roth, 2000; Delmas & Toffel, 2008). Scholars have repeatedly called for research that examines sustainability leadership across diverse organizational contexts to enhance theoretical robustness and practical relevance (Aguilera et al., 2007; Jamali & Karam,2018).

1.5 Objectives of the Study

In response to the identified research gaps, the primary objective of this study is to empirically examine the impact of sustainable leadership on environmental stewardship within organizations. Specifically, the study aims to:

1. To Assess the influence of sustainable leadership practices on organizational environmental performance;
2. To identify key leadership behaviors that promote environmental responsibility and stewardship; and
3. To develop and test a conceptual framework linking sustainable leadership with responsible organizational practices and environmental outcomes.

By achieving these objectives, the study seeks to provide empirical evidence that strengthens the theoretical and practical understanding of sustainable leadership and its role in environmental stewardship.

1.6 Significance of the Study

- This study makes important contributions to both theory and practice. From a theoretical perspective, it extends leadership and sustainability literature by empirically validating the relationship between sustainable leadership and environmental stewardship.
- To integrating perspectives from stakeholder theory, resource-based view, and leadership theory, the study provides a holistic framework that explains how leadership capabilities translate into environmental value creation.
- The findings offer actionable insights for organizational leaders, policymakers, and sustainability practitioners. By identifying leadership behaviors and organizational mechanisms that enhance environmental stewardship, the study supports organisations in designing effective sustainability strategies.

2. LITERATURE REVIEW

The literature on sustainable leadership and environmental stewardship has expanded significantly over the past two decades, reflecting growing academic and managerial interest in responsible organisational practices. This section reviews key concepts, theoretical foundations, empirical findings, and existing research gaps to establish a robust foundation for the present study.

2.1 Concept of Sustainable Leadership

2.1.1 Definition and Evolution

Sustainable leadership refers to leadership practices that balance economic performance with social equity and environmental responsibility to ensure long-term organizational viability (Avery & Bergsteiner, 2011). Unlike traditional leadership models focused on short-term financial outcomes, sustainable leadership adopts a long-term and holistic perspective by integrating sustainability principles into strategic decision-making (Hargreaves & Fink, 2006).

2.1.2 Sustainable vs. Traditional Leadership

Short-term orientation, centralised decision-making, and a major focus on shareholder value are common characteristics of traditional leadership techniques (Dickson, 2023). On the other hand, long-term value generation for a variety of stakeholders, such as society and the environment, is emphasised by sustainable leadership (Avery & Bergsteiner, 2011).

Table 1. Sustainable vs. traditional leadership

S.No	Dimension	Traditional Leadership	Sustainable Leadership
1	Time Horizon	Short-term focus on immediate financial results	Long-term focus on creating value for multiple stakeholders
2	Decision-Making	Centralized and hierarchical	Participative and inclusive, involving diverse stakeholders
3	Primary Focus	Shareholder value and profit maximization	People, planet, and profit; emphasizes environmental and social responsibility
4	Innovation & Creativity	Primarily driven by efficiency and operational performance	High priority on innovation, creativity, and sustainable solutions
5	Resource Management	Utilitarian approach with emphasis on cost reduction and resource exploitation	Prudent, ethical, and sustainable utilization of resources
6	Organizational Culture	Sustainability is peripheral, symbolic, or compliance-driven	Sustainability is embedded in the vision, mission, values, and daily practices
7	Leadership Role	Directive, control-oriented, and authority-based	Change agent, ethical role model, and sustainability champion

Source: (Dickson, 2023) & (Avery & Bergsteiner, 2011)

2.2 Environmental Stewardship in Organisations

2.2.1 Concept and Dimensions of Environmental Stewardship

Environmental stewardship refers to the proactive responsibility of organisations to protect and conserve the natural environment through ethical decision-making and sustainable business practices (Daily & Huang, 2001). It involves integrating environmental considerations into organizational strategy and operations, ensuring accountability for environmental impacts across the value chain rather than treating them as externalities. Scholars conceptualize environmental stewardship as a multidimensional construct. The natural resource–based view identifies pollution prevention, product stewardship, and sustainable development as strategic dimensions that enable organizations to reduce environmental harm while achieving competitive advantage (Hart, 1995). Similarly, proactive environmental strategies such as resource efficiency, cleaner technologies, and green innovation support long-term sustainability and responsiveness to stakeholder and regulatory pressures (Aragón-Correa & Sharma, 2003).Importantly, environmental stewardship extends beyond regulatory compliance by encouraging voluntary, forward-looking environmental practices that enhance ecological sustainability and reinforce ethical business conduct (Bansal & Roth, 2000; Aragón-Correa, et al., 2008).

2.2.2 Corporate Environmental Responsibility

Corporate Environmental Responsibility (CER) is a core element of environmental stewardship and refers to an organization's obligation to minimize environmental harm while supporting sustainable development (Shrivastav,1995;Banerjee,2002). CER involves integrating environmental considerations into corporate strategies and operations through initiatives such as energy conservation, waste reduction, emissions control, and the adoption of environmentally friendly technologies, which can improve efficiency and regulatory compliance (Delmas & Toffel, 2008).Beyond operational benefits, CER enhances organizational performance by strengthening stakeholder trust, corporate reputation, and legitimacy, thereby contributing to sustained competitive advantage (Fombrun,2005; Aguilera, et al., 2007; Hart, & Dowell, 2011). By embedding environmental responsibility into organizational culture and leadership practices, CER also supports innovation, employee engagement, and long-term value creation in response to growing sustainability challenges (Bansal, 2005).

2.2.3 Institutional Theory

Institutional theory explains how organizational practices are shaped by external pressures arising from regulatory requirements, normative expectations, and cognitive frameworks within the institutional environment (Powell & DiMaggio, 1983). Organizations often adopt similar structures and practices in response to these pressures to gain legitimacy, stability, and acceptance within their institutional context. Environmental stewardship is strongly influenced by institutional pressures such as environmental regulations, industry standards, professional norms, and societal expectations for sustainable development (Delmas & Toffel, 2008). Regulatory pressures compel organizations to comply with environmental laws, while normative pressures from stakeholders and industry associations encourage the adoption of best practices in sustainability. Cognitive pressures further shape organizational beliefs regarding the importance of environmental responsibility as a legitimate and necessary organizational goal (Jennings & Zandbergen, 1995).Leaders play a crucial role in interpreting and responding to these institutional pressures. By translating external expectations into internal policies, structures, and routines, leaders facilitate the institutionalization of environmental stewardship within organizations (Bansal & Clelland, 2004).

2.2.4 Resource-Based View (RBV)

The resource-based view (RBV) argues that organizations can achieve sustained competitive advantage by developing and leveraging resources that are valuable, rare, inimitable, and non-substitutable (Barney et al., 2001). These resources may include tangible assets, such as technology, as well as intangible assets, such as organizational culture, leadership capabilities, and knowledge. Within the sustainability context, environmental capabilities such as green innovation, environmental management expertise, and sustainability-oriented leadership can function as strategic resources that enhance long-term organizational performance (Hart, 1995; Hart & Dowell, 2011). Sustainable leadership, in particular, represents a critical intangible resource that enables organizations to integrate environmental stewardship into strategy and operations, thereby creating difficult-to-imitate competitive advantages. The natural resource based view extends RBV by emphasizing how environmental stewardship contributes to competitive advantage through pollution prevention, product stewardship, and sustainable development strategies (Hart, 1995). Organizations that invest in sustainability-focused leadership and environmental capabilities are better positioned to respond to environmental challenges, reduce risks, and capitalize on emerging market opportunities related to green products and services (Porter & Linde, 1995).

2.3 Key Outcomes of Sustainable Leadership

- Sustainable leadership is associated with a range of positive environmental and organizational outcomes. One of the most significant outcomes is the reduction of environmental impact through improved resource efficiency, waste reduction, and lower emissions (Zhu et al., 2013; Porter & Linde, 1995).
- Embedding sustainability into operational processes, organizations can achieve both environmental and economic benefits, supporting the notion of "eco-efficiency." Another key outcome is enhanced stakeholder trust and organizational legitimacy.
- Organizations led by sustainability-oriented leaders are perceived as more responsible and ethical, which strengthens relationships with stakeholders such as customers, investors, regulators, and local communities (Aguilera, et al., 2007; Fombrun, C. J. 2005).

- Improved environmental reputation further contributes to competitive advantage by differentiating organizations in increasingly sustainability-conscious markets (Bansal, & Clelland, 2004).

2.4 Mediating and Moderating Variables

Understanding how sustainable leadership influences environmental performance requires examining the mediating and moderating mechanisms that shape this relationship. Prior research suggests that organizational culture, green innovation, and employee environmental behaviours play critical roles in translating leadership values into environmental outcomes.

2.4.1 Organizational Culture

Organizational culture acts as a key mediating mechanism through which sustainable leadership influences environmental stewardship. Culture reflects shared values, norms, and assumptions that guide employee behaviour and decision-making within organizations .When leaders emphasize sustainability, they contribute to the development of a sustainability-oriented culture that reinforces pro-environmental norms, ethical responsibility, and collective commitment to environmental goals (Avery & Bergsteiner, 2011).Empirical studies indicate that organizations with strong sustainability-oriented cultures exhibit higher levels of environmental performance, as employees internalize environmental values and consistently engage in environmentally responsible practices (Norton et al., 2015; Linnenluecke & Griffiths, 2010).

2.4.2 Green Innovation

Green innovation serves as a critical mediating variable linking sustainable leadership to environmental performance. Green innovation refers to the development and implementation of eco-friendly products, processes, and technologies that reduce environmental harm and improve resource efficiency (Su et al., (2020). Leaders who support sustainability create conditions that encourage creativity, experimentation, and investment in environmentally innovative solutions. Research suggests that leadership commitment to sustainability positively influences green innovation by promoting knowledge sharing, cross-functional collaboration, and long-term investment in sustainable technologies (Jabbour et al., 2010).

2.4.3 Employee Environmental Behaviour

Employee environmental behaviour functions as both a mediating and moderating variable in the relationship between sustainable leadership and environmental performance. Employee environmental behaviour includes voluntary and discretionary actions such as energy conservation, recycling, waste reduction, and participation in organizational environmental initiatives (Ones & Dilchert, 2012). Leadership support for sustainability enhances employees' environmental awareness, motivation, and sense of responsibility, thereby encouraging such behaviours (Robertson & Barling, 2013).As a mediating variable, employee environmental behaviour explains how leadership influences environmental performance by translating leadership commitment into daily operational practices. As a moderating variable, the strength of this relationship may vary depending on employees' environmental values, engagement levels, and perceived organizational support for sustainability (Norton et al., 2015)

3. CONCEPTUAL FRAMEWORK AND HYPOTHESES

3.1 Conceptual Model Development

This study's conceptual model is created by combining ideas from environmental management literature, stakeholder theory, and sustainable leadership theory. According to the model, organisational environmental stewardship is influenced by sustainable leadership, which in turn has an impact on organisational success. Leadership behaviours that prioritise moral decision-making, long-term planning, stakeholder participation, and environmental responsibility are referred to as sustainable leadership.

According to stakeholder theory (Freeman & Boeker, 1984), companies run by sustainable leaders are more likely to strike a balance between societal and environmental expectations and economic goals, which strengthens environmental stewardship programs. Additionally, as a mediating mechanism, the model integrates green organisational culture. Shared norms, values, and beliefs that direct employee behaviour are reflected in organisational culture. The effectiveness of environmental stewardship activities is increased when a culture that places a high priority on environmental responsibility is shaped by sustainable leaders. This mediating link is consistent with transformational leadership theory, which highlights the impact of leaders on the attitudes and values of followers (Bass & Avolio, 1994).

Furthermore, it is believed that environmental stewardship enhances organisational performance, which includes both environmental and non-environmental objectives including stakeholder trust, operational efficiency, and reputation. According to the resource-based perspective (Barney et al., 2001), environmental skills developed by sustainable leadership can be strategic assets that improve an organization's performance over the long run.

3.2 Hypotheses Formulation

Based on the conceptual framework and prior empirical evidence, the following hypotheses are proposed:

> **H1**: Sustainable leadership has a positive and significant influence on environmental stewardship.

Sustainable leaders are expected to promote environmentally responsible strategies and allocate resources toward environmental initiatives, thereby strengthening organizational stewardship practices.

> **H2:** Environmental stewardship has a positive effect on organizational performance.

Organizations that actively engage in environmental stewardship are likely to experience improved efficiency, enhanced reputation, and long-term competitive advantage.

> **H3:** Sustainable leadership positively influences organizational performance.

Leadership commitment to sustainability is expected to directly contribute to organizational outcomes by fostering ethical governance, innovation, and stakeholder trust.

> **H4:** Green organizational culture mediates the relationship between sustainable leadership and environmental stewardship.

Sustainable leadership is expected to cultivate a culture that values environmental responsibility, which in turn enhances stewardship practices.

> **H5:** Green innovation mediates the relationship between environmental stewardship and organizational performance.

Organizations with strong stewardship practices are more likely to invest in green innovation, leading to improved performance outcomes.

H6: Employee environmental behavior moderates the relationship between sustainable leadership and environmental stewardship.

The positive effect of sustainable leadership on environmental stewardship is expected to be stronger when employees actively engage in pro-environmental behaviours.

Figure 4. Hypotheses formulation

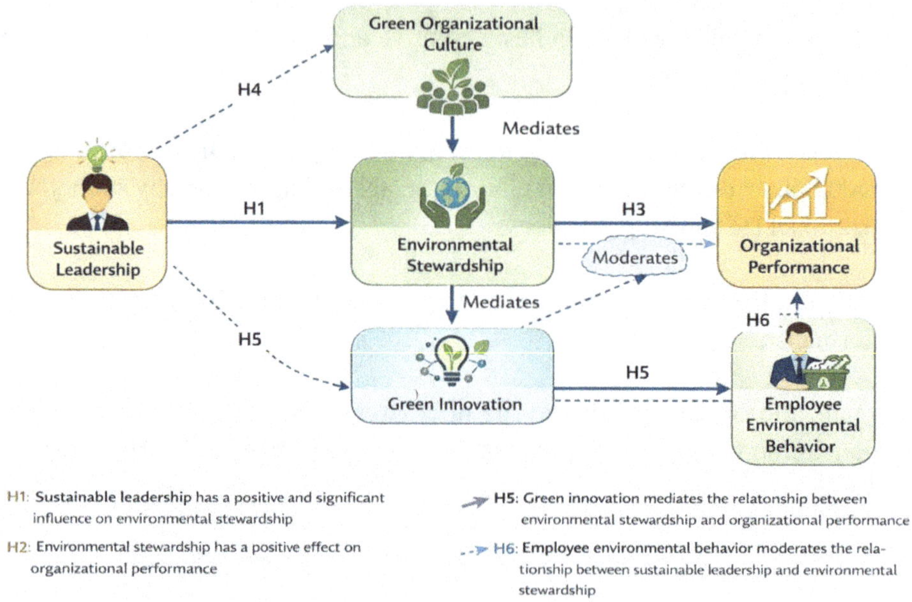

(Freeman & Boeker, 1984; Bass & Avolio, 1994; Barney et al., 2001)

Summary of the Conceptual Model

Independent Variable: Sustainable Leadership
Mediators: Green Organizational Culture, Green Innovation
Moderator: Employee Environmental Behaviour
Dependent Variables: Environmental Stewardship, Organizational Performance

This framework provides a comprehensive structure for empirically examining how leadership behaviors translate into environmentally responsible organizational practices and performance outcomes.

4. RESEARCH METHODOLOGY

The research design, demographic and sample processes, data collection methodologies, variable measurement, data analysis methods, and ethical considerations used in the study are all described in this section. The methodological approach is intended to investigate the linkages suggested in the conceptual framework through empirical means.

4.1 Research Design

The study uses a quantitative research approach to investigate the connections between organisational success, environmental stewardship, and sustainable leadership. Because it enables objective measurement, hypothesis testing, and the generalisation of results across organisational contexts, a quantitative approach is suitable. The study uses a cross-sectional methodology, gathering information at one particular moment in order to document attitudes and behaviours pertaining to environmental stewardship and sustainable leadership. This design's capacity to analyse intricate interactions between several characteristics has led to its widespread application in leadership and sustainability research.

4.2 Population and Sample

4.2.1 Target Population

Employees at the managerial and supervisory levels of medium-sized and large organisations that have implemented environmental management or sustainability efforts make up the study's target group.

4.2.2 Sample Size and Sampling Technique

To guarantee sufficient statistical power for multivariate analysis, especially structural equation modelling, a sample size of 300 respondents was used. The study used a stratified random sampling technique in which respondents were chosen at random within each stratum after organisations were first classified according to the kind of industry. This method lessens sampling bias and improves representativeness.

4.3 Data Collection Methods

A structured questionnaire survey, which is appropriate for obtaining standardised responses from a sizable sample, was the main method used to collect data. Closed-ended questions with a five-point Likert scale ranging from "strongly disagree" to "strongly agree" made up the questionnaire. To support the contextual understanding of organisational environmental practices, secondary data, including sustainability reports and environmental policy documents, were examined in addition to survey data. Informal conversations with a subset of respondents helped elucidate survey replies and contextual subtleties, even though the study is mostly quantitative.

4.4 Ethical Considerations

Ethical standards were strictly adhered to throughout the research process. Participation in the study was voluntary, and informed consent was obtained from all respondents. Confidentiality and anonymity of responses were ensured, and data were used solely for academic purposes. The study complied with institutional ethical guidelines and avoided any form of coercion or misrepresentation.

5. DATA ANALYSIS AND RESULTS

The findings of the empirical investigation into the connections between environmental stewardship, green organisational culture, sustainable leadership, and organisational performance are presented in this section. Respondent profiles, descriptive data, validity and reliability evaluations, hypothesis testing, and a synopsis of significant empirical findings are all included in the analysis.

5.1 Respondent Profile

The respondents demographic profile guarantees sufficient representation for the study and gives a summary of the sample characteristics. 300 managerial and supervisory staff members from businesses actively involved in environmental management and sustainability projects made up the final sample.

Figure 5. Demographic profile of respondents

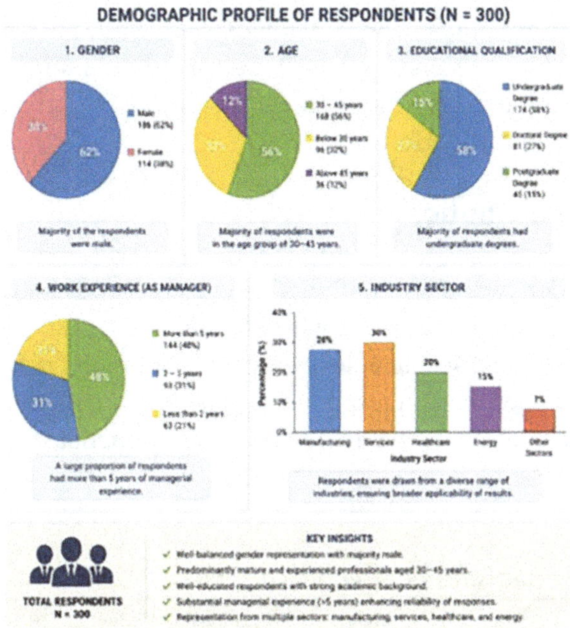

The demographic profile of the respondents indicates that the study sample was adequately representative of managerial and supervisory employees working in organizations engaged in environmental management and sustainability initiatives. Among the 300 respondents, 62% were male and 38% were female, reflecting the gender composition of managerial personnel in the participating organizations. The majority of respondents (56%) belonged to the 30–45 years age group, while 32% were below 30 years and 12% were above 45 years. This suggests that most participants were mature professionals with sufficient experience to provide informed opinions on sustainability and leadership practices.

The educational qualifications of the respondents reveal a highly educated sample. A majority of the participants held undergraduate degrees (58%), followed by doctoral degrees (27%) and postgraduate degrees (15%). Furthermore, the respondent's possessed considerable managerial experience, with 48% reporting more than five years of experience, 31% having between two and five years of experience, and 21% having less than two years of experience. The high level of education and professional experience among respondents enhances the reliability and credibility of the information collected. The respondents were drawn from diverse industry sectors, thereby increasing the generalizability of the study findings. The largest proportion of participants represented the services sector (30%), followed by manufacturing

(28%), healthcare (20%), energy (15%), and other sectors (7%). This broad industry representation provides a comprehensive perspective on sustainability-related practices across different organizational contexts. Overall, the demographic characteristics indicate that the sample was well-suited for examining the relationships among sustainable leadership, environmental stewardship, green organizational culture, and organizational performance.

5.2 Descriptive Statistics

The central tendency and variability of the research variables were examined using descriptive statistics. The mean values for environmental stewardship, green organisational culture, and sustainable leadership were found to be above the midpoint of the scale, suggesting that respondents generally thought their organisations were moderately to highly involve in environmental practices and sustainability-oriented leadership.

Table 2. Descriptive statistics of research variables

Variable	Mean	Standard Deviation	Interpretation
Sustainable Leadership	4.12	0.68	High
Environmental Stewardship	4.05	0.72	High
Green Organizational Culture	3.98	0.75	Moderate to High
Organizational Performance	4.18	0.65	High

Interpretation: Table 2 presents the descriptive statistics of the study variables. The mean scores for Sustainable Leadership (M = 4.12), Environmental Stewardship (M = 4.05), Green Organizational Culture (M = 3.98), and Organizational Performance (M = 4.18) were all above the midpoint of the scale, indicating that respondents generally perceived a moderate to high level of sustainability-oriented leadership, environmental responsibility, and organizational effectiveness within their organizations. The standard deviation values ranged from 0.65 to 0.75, suggesting adequate variability in responses and indicating that the data were not excessively concentrated around the mean. These findings imply that while sustainability-related practices are evident across the sampled organizations, variations exist in the extent of their implementation. Overall, the descriptive statistics provide preliminary support for further inferential analyses to test the proposed hypotheses and relationships among the study constructs.

5.3 Reliability and Validity Analysis

5.3.1 Reliability and convergent validity Analysis

Reliability and convergent validity analyses were conducted to evaluate the adequacy of the measurement model. Reliability refers to the consistency and stability of the measurement items used to assess the study constructs, while convergent validity examines the extent to which the indicators of a construct are related and measure the same underlying concept. Reliability was assessed using Cronbach's Alpha and Composite Reliability (CR), with values greater than 0.70 indicating satisfactory internal consistency. Convergent validity was evaluated using Average Variance Extracted (AVE), where values exceeding 0.50 demonstrate that the construct explains more than half of the variance of its indicators.

Table 3. Reliability and convergent validity analysis

Construct	Cronbach's Alpha	Composite Reliability (CR)	Average Variance Extracted (AVE)
Sustainable Leadership	0.874	0.901	0.604
Environmental Stewardship	0.842	0.886	0.567
Green Organizational Culture	0.889	0.914	0.639
Organizational Performance	0.861	0.895	0.588

Interpretation: Table 2 shows that the Cronbach's alpha values ranged from 0.842 to 0.889, exceeding the recommended threshold of 0.70, indicating satisfactory internal consistency. Composite Reliability (CR) values ranged from 0.886 to 0.914, confirming construct reliability. Furthermore, all AVE values exceeded 0.50, demonstrating adequate convergent validity.

5.3.2 Discriminant Validity (Fornell-Larcker Criterion)

Discriminant validity was assessed using the Fornell-Larcker criterion to determine whether the study constructs were empirically distinct from one another. According to this criterion, the square root of the Average Variance Extracted (AVE) for each construct should be greater than its correlations with other constructs.

Table 4. Discriminant validity (Fornell-Larcker criterion)

Construct	SL	ES	GOC	OP
Sustainable Leadership (SL)	**0.777**			
Environmental Stewardship (ES)	0.532	**0.753**		
Green Organizational Culture (GOC)	0.615	0.581	**0.799**	
Organizational Performance (OP)	0.648	0.604	0.692	**0.767**

Interpretation: The square root of AVE for each construct was greater than its correlations with other constructs, satisfying the Fornell-Larcker criterion and confirming discriminant validity. Therefore, the constructs were empirically distinct and measured different theoretical concepts.

5.4 Hypotheses Testing

Hypotheses were tested using structural equation modeling (SEM), which allows simultaneous examination of multiple relationships and mediating effects. The results indicated that sustainable leadership has a significant positive effect on environmental stewardship, providing support

Hypothesis 1 - Leaders who demonstrate sustainability-oriented behaviors were found to significantly enhance organizational commitment to environmental practices.Environmental stewardship was found to have a positive and significant influence on organizational performance, supporting
Hypothesis 2 - Organizations actively engaged in environmental initiatives reported improved operational efficiency, reputation, and stakeholder trust. The direct relationship between sustainable leadership and organizational performance was also significant, supporting
Hypothesis 3-This suggests that leadership commitment to sustainability contributes to organizational outcomes beyond environmental practices alone. Mediation analysis revealed that green organizational culture partially mediates the relationship between sustainable leadership and environmental stewardship, supporting.
Hypothesis 4 -Sustainable leaders foster a culture that values environmental responsibility, which in turn strengthens stewardship practices.Furthermore, green innovation was found to mediate the relationship between environmental stewardship and organizational performance, supporting
Hypothesis 5. Finally, moderation analysis showed that employee environmental behavior strengthens the relationship between sustainable leadership and environmental stewardship.

Figure 6. Hypotheses testing

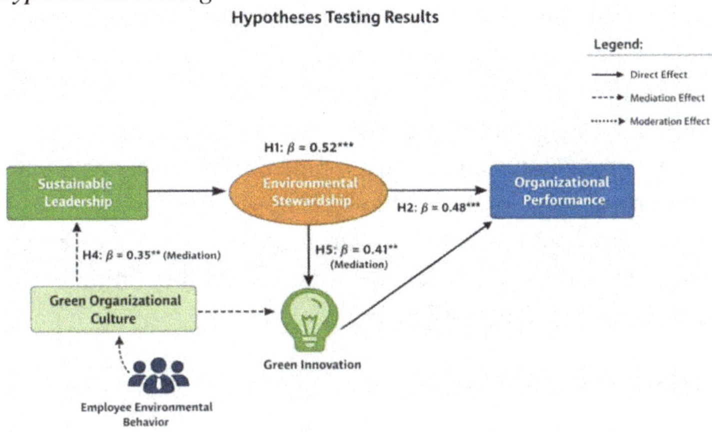

5.5 Key Empirical Findings

The study's empirical results provide a number of significant insights. First, environmental stewardship is found to be significantly influenced by sustainable leadership, demonstrating the crucial role that leadership plays in pushing sustainability objectives. Second, environmental stewardship makes a substantial contribution to organisational performance, proving that commercial success and environmental responsibility are complimentary rather than antagonistic goals. Third, the results highlight how crucial internal organisational mechanisms like employee behaviour, culture, and innovation are in converting leadership goals into observable environmental results. When taken as a whole, the findings offer solid empirical backing for the suggested conceptual framework and emphasise the strategic importance of sustainable leadership for ethical business practices.

6. DISCUSSION

6.1 Interpretation of Results

The results of this study offer solid empirical proof that sustainable leadership is essential for improving environmental stewardship in businesses. Stronger commitment to environmental practices is fostered by leaders who exhibit long-term orientation, ethical decision-making, and care for social and environmental effects (Avery & Bergsteiner, 2011). According to (Porter & Linde, 1995 and Hart, 1995), environmental stewardship has a substantial direct impact on organisational performance, indicating that eco-friendly policies improve stakeholder trust, operational efficiency, and reputation. Pro-environmental behaviours are reinforced by shared values and norms that are shaped by sustainable leadership, as evidenced by the partly mediating role of green organisational culture. Furthermore, when businesses use eco-innovative procedures and goods, the mediating effect of green innovation validates that environmental endeavours result in better performance (Ahuja et al., 2023). Employee environmental behaviour's moderating effect emphasises that when staff members actively participate in sustainability initiatives, leadership effectiveness is increased (Ones & Dilchert, 2012).

6.2 Theoretical Implications

Sustainable leadership can be better understood through the integration of the Resource-Based View (RBV), Transformational Leadership Theory, and Stakeholder Theory, which collectively explain how leadership contributes to sustainability-driven organizational outcomes. Stakeholder Theory emphasizes that organizations have responsibilities not only to shareholders but also to a broader range of stakeholders, including employees, customers, suppliers, communities, governments, and the natural environment (Freeman & Boeker, 1984). Sustainable leaders recognize these diverse stakeholder interests and incorporate environmental and social considerations into strategic decision-making processes. By fostering stakeholder engagement and balancing competing interests, sustainable leadership enhances organizational legitimacy, trust, and long-term stakeholder relationships, which are essential for sustainable business success.

From the perspective of the Resource-Based View (RBV), sustainable leadership facilitates the development of valuable, rare, inimitable, and non-substitutable (VRIN) organizational resources that serve as sources of sustained competitive advantage (Barney et al., 1991). Environmental stewardship, sustainability-oriented organizational culture, green innovation capabilities, and environmentally responsible practices are strategic resources that competitors find difficult to replicate. (Hart, 1995) Natural Resource-Based View further extends RBV by arguing that environmental capabilities such as pollution prevention, product stewardship, and sustainable development contribute significantly to long-term organizational performance. Sustainable leaders play a critical role in nurturing these capabilities by allocating resources toward environmental initiatives, promoting green technologies, and embedding sustainability into organizational strategies. Consequently, organizations that effectively leverage these sustainability-related resources are better positioned to achieve superior environmental and economic performance.

Transformational Leadership Theory provides another important theoretical foundation for understanding sustainable leadership. Transformational leaders inspire and motivate followers to transcend short-term self-interests and pursue collective organizational goals through vision, intellectual stimulation, individualized consideration, and inspirational motivation (Bass & Avolio, 1994). Sustainable leadership extends these transformational principles by explicitly incorporating environmental responsibility and long-term sustainability objectives into the leadership agenda. Through their behaviours and actions, sustainable leaders encourage employees to adopt pro-environmental values, engage in innovative problem-solving, and support sustainability initiatives. They create a culture that values learning, experimentation, and continuous improvement, thereby fostering green innovation and environmental stewardship across the organization.

The integration of these three theoretical perspectives offers a comprehensive explanation of how sustainable leadership influences organizational sustainability outcomes. Stakeholder Theory explains why organizations must address environmental and social concerns; RBV explains how sustainability-related capabilities become strategic resources that generate competitive advantage; and Transformational Leadership Theory explains the behavioural and cultural mechanisms through which leaders inspire employees to embrace sustainability-oriented values and practices. Together, these theories suggest that sustainable leadership is not merely an ethical or environmental responsibility but also a strategic capability that enhances organizational resilience, innovation, stakeholder satisfaction, and long-term competitive performance. Therefore, sustainable leadership serves as a critical link between environmental stewardship and sustainable organizational success in increasingly complex and dynamic business environments.

6.3 Practical Implications for Leaders and Organizations

The findings of this study have important implications for both policymakers and organizational managers seeking to enhance environmental sustainability and long-term organizational performance. Organizations should invest in developing sustainable leadership competencies through leadership development programs, executive training, and capacity-building initiatives that emphasize ethical decision-making, stakeholder engagement, and environmental responsibility (Avery & Bergsteiner, 2011). For example, Unilever has embedded sustainability into its leadership development programs, encouraging managers to integrate environmental and social objectives into strategic decision-making while pursuing business growth. Similarly,

Patagonia promotes values-based leadership by empowering leaders to prioritize environmental stewardship and corporate responsibility.

Organizations can further strengthen sustainability by cultivating a robust green organizational culture through environmentally supportive human resource practices, sustainability-focused communication, and employee recognition programs (Renwick et al., 2013). A notable example is Google, which encourages employees to participate in sustainability initiatives through internal environmental campaigns, green workplace programs, and employee-led sustainability projects. Such initiatives help embed environmental values into everyday organizational activities and foster a shared commitment to sustainability.

Leaders should also actively promote green innovation by investing in environmentally friendly technologies, sustainable production processes, and renewable energy solutions. For instance, Tesla has demonstrated how leadership commitment to innovation can accelerate the development of sustainable technologies while simultaneously creating competitive advantage. Likewise, Interface transformed its business model by investing in circular manufacturing processes and sustainable product design, significantly reducing its environmental footprint while improving operational efficiency.

Furthermore, environmental stewardship can be substantially enhanced by encouraging employee environmental behaviour through participation, empowerment, and recognition programs (Ones & Dilchert, 2012). For example, Toyota Motor Corporation engages employees in continuous environmental improvement initiatives through its Kaizen philosophy, enabling staff at all levels to contribute ideas for reducing waste, conserving resources, and improving environmental performance. Similarly, IKEA involves employees in sustainability campaigns focused on energy conservation, waste reduction, and responsible resource utilization, thereby fostering environmentally responsible behaviour throughout the organization.

These real-world examples demonstrate that sustainable leadership extends beyond policy formulation and plays a crucial role in embedding sustainability into organizational culture, driving green innovation, and encouraging pro-environmental employee behaviour. By integrating these practices, organizations can strengthen their environmental stewardship efforts, improve ESG performance, enhance stakeholder trust, and achieve long-term sustainable growth and competitive advantage.

7. PROPOSED FRAMEWORK FOR SUSTAINABLE LEADERSHIP

7.1 Dimensions of the Framework

Vision and Values

Vision and values form the foundation of sustainable leadership by providing long-term direction and purpose beyond short-term financial goals. Sustainable leaders articulate a clear sustainability-oriented vision that integrates environmental protection, social responsibility, and economic viability (Avery & Bergsteiner, 2011). Such values guide strategic decision-making and influence employee behavior by embedding sustainability into the organizational identity. A strong sustainability vision also aligns organizational goals with global agendas such as the Sustainable Development Goals (SDGs) and ESG principles (UN Global Compact, 2019).

Ethical Governance

Ethical governance emphasizes transparency, accountability, and responsible decision-making in leadership practices. Sustainable leadership promotes ethical standards that ensure compliance with environmental regulations and voluntary environmental initiatives (Pless et al., 2012). Ethical governance strengthens trust among stakeholders and reduces environmental and reputational risks (Freeman & Boeker, 1984). By integrating ethics into governance structures, organizations ensure that sustainability commitments are consistently translated into policies, monitoring systems, and performance evaluations (Aguilera et al., 2007).

Environmental Innovation

Environmental innovation refers to the development and implementation of eco-friendly products, services, and processes that reduce environmental impact while enhancing organizational performance. Sustainable leaders play a critical role in fostering a climate that supports experimentation, learning, and green innovation. From a resource-based perspective, environmental innovation represents a strategic capability that contributes to long-term competitive advantage (Hart, 1995).

Stakeholder Engagement

Stakeholder engagement involves actively involving internal and external stakeholders such as employees, customers, suppliers, communities, and regulators in sustainability initiatives. Sustainable leadership recognizes that environmental stewardship is a collective responsibility and requires collaboration across organizational boundaries (Freeman & Boeker, 1984). Effective engagement enhances legitimacy, improves decision quality, and strengthens stakeholder trust (Donaldson & Preston, 1995).

7.2 Implementation Strategies

To operationalize the proposed framework, organizations should integrate sustainability into leadership development and strategic planning processes. Leadership training programs should emphasize ethical decision-making, systems thinking, and environmental awareness (Avery& Bergsteiner, 2011). Organizations can embed sustainability into governance structures through clear policies, performance metrics, and ESG reporting mechanisms (Aguilera et al., 2007). Encouraging cross-functional collaboration and investing in green technologies can further support environmental innovation. Additionally, transparent communication and participatory mechanisms such as sustainability committees and stakeholder forums can enhance stakeholder engagement and accountability (Linnenluecke & Griffiths, 2010).

7.3 Expected Organizational Outcomes

The proposed framework is expected to generate multiple positive organizational outcomes. Environmentally, organizations are likely to achieve improved environmental stewardship through reduced emissions, efficient resource utilization, and compliance with environmental standards (Hart, 1995). Strategically, sustainable leadership can enhance organizational performance by fostering innovation, strengthening reputation, and building long-term competitive advantage (Porter & Linde, 1995). Socially, stronger stakeholder relationships and ethical governance contribute to trust, legitimacy, and employee commitment (Freeman & Boeker, 1984).

8. MANAGERIAL AND POLICY IMPLICATIONS

8.1 Implications for Top Management

The findings of this study highlight the critical responsibility of top management in embedding sustainability into organizational strategy and culture. Senior leaders should act as role models by demonstrating commitment to environmental stewardship through ethical decision-making, transparent governance, and long-term strategic thinking (Avery & Bergsteiner, 2011). Integrating sustainability objectives into corporate vision, mission statements, and performance management systems can ensure alignment between leadership intent and organizational action .Top management should also invest in leadership development programs that cultivate sustainability-oriented competencies, such as systems thinking and stakeholder engagement (Pless et al., 2012). Furthermore, encouraging green innovation by allocating resources to environmentally friendly technologies and processes can enhance both environmental and financial performance. By fostering a supportive green organizational culture, leaders can motivate employees to actively participate in sustainability initiatives, thereby strengthening environmental stewardship across all organizational levels (Renwick et al., 2013).

8.2 Implications for Policymakers

For policymakers, the study underscores the importance of creating an institutional environment that actively promotes sustainable leadership and environmentally responsible organizational practices. Governments and regulatory agencies can encourage environmental stewardship by implementing comprehensive environmental regulations, sustainability reporting standards, and ESG disclosure requirements that hold organizations accountable for their environmental performance (Aguilera et al., 2007). Beyond regulatory compliance, policymakers can develop incentive-based mechanisms such as tax benefits, low-interest green financing, subsidies, and research grants to encourage organizations to invest in environmentally sustainable technologies and green innovation initiatives (Porter & van der Linde, 1995). For example, government-supported renewable energy programs and green technology funding schemes have enabled organizations in several countries to reduce carbon emissions while improving operational efficiency.

Policymakers can also facilitate the development of sustainable leadership competencies by incorporating sustainability and environmental management modules into executive education, professional certification programs, and industry-specific leadership training initiatives. Such programs can equip current and future leaders with the knowledge and skills required to address environmental challenges and implement sustainability-oriented strategies. In addition, public-sector agencies can establish sustainability excellence awards, benchmarking frameworks, and recognition programs to encourage organizations to adopt best practices in environmental stewardship.

Furthermore, policymakers can strengthen collaboration among government institutions, industry associations, academic institutions, and civil society organizations to promote knowledge sharing and the diffusion of sustainable practices. In sectors such as healthcare, manufacturing, and energy, sector-specific sustainability guidelines can be developed to assist organizations in integrating environmental stewardship into their operational and strategic activities. By creating supportive regulatory frameworks, financial incentives, and leadership development opportunities, policymakers can accelerate the adoption of sustainable leadership practices, enhance environmental performance, and contribute to the achievement of national sustainability and climate-related goals.

8.3 Alignment with SDGs and ESG Goals

The proposed framework and empirical findings are closely aligned with global sustainability agendas, particularly the United Nations Sustainable Development Goals (SDGs) and Environmental, Social, and Governance (ESG) frameworks. Sustainable leadership contributes directly to SDG 12 (Responsible Consumption and Production) by encouraging organizations to adopt resource-efficient practices, reduce waste generation, and promote environmentally sustainable operations. It also supports SDG 13 (Climate Action) by fostering environmental stewardship, green innovation, and proactive measures to mitigate environmental impacts (United Nations, 2015). Through the promotion of ethical behaviour, transparency, and stakeholder engagement, sustainable leadership further advances SDG 16 (Peace, Justice, and Strong Institutions) by strengthening organizational accountability and responsible governance practices.

The findings also have important implications for organizations seeking to improve their ESG performance. From the Environmental (E) perspective, sustainable leadership facilitates environmental stewardship and green innovation, enabling organizations to reduce resource consumption, lower emissions, and enhance environmental performance. From the Social (S) perspective, sustainable leadership promotes employee engagement, collaboration, stakeholder trust, and socially responsible organizational practices. These outcomes contribute to stronger relationships with employees, customers, communities, and other stakeholders. From the Governance (G) perspective, sustainable leadership encourages ethical decision-making, transparency, accountability, and responsible management practices, which are essential components of effective corporate governance (Eccles et al., 2014).

The practical significance of these findings is particularly relevant for policymakers and organizational leaders who are increasingly required to demonstrate measurable progress toward sustainability objectives. Organizations can utilize the proposed framework as a strategic guide for integrating sustainability into leadership development programs, environmental management systems, and ESG reporting initiatives. Similarly, policymakers can use these findings to design sustainability regulations, leadership development schemes, and ESG disclosure frameworks that encourage organizations to align their operations with national and international sustainability targets. By strengthening sustainable leadership capabilities, organizations can simultaneously enhance ESG performance, contribute to SDG achievement, and create long-term economic, social, and environmental value.

9. LIMITATIONS AND FUTURE RESEARCH DIRECTIONS

9.1 Methodological Limitations

Despite its contributions, this study has certain methodological limitations that should be acknowledged. First, the use of a cross-sectional research design limits the ability to infer causal relationships among sustainable leadership, environmental stewardship, and organizational performance. Longitudinal designs would provide stronger evidence of causality and allow researchers to examine changes in sustainability practices over time (Podsakoff et al., 2003). Second, the study relies primarily on self-reported survey data, which may be subject to common method bias and social desirability effects. Although statistical remedies were applied, future studies could incorporate objective performance indicators or multi-source data to enhance validity (Podsakoff et al., 2014). Third, while SEM offers robust analytical capabilities, model specification and measurement choices may influence results, suggesting the need for replication using alternative analytical techniques.

9.2 Contextual Limitations

The generalizability of the findings may be constrained by contextual factors such as industry type, organizational size, and geographical setting. Sustainability challenges and leadership practices may differ significantly across sectors, particularly between manufacturing, services, and public organizations (Linnenluecke & Griffiths, 2010). Cultural and institutional differences across countries may also influence leadership behaviors and environmental priorities, limiting the applicability of the results to other national or regional contexts (Hofstede, 2011).

9.3 Directions for Future Studies

Future research should address these limitations by adopting longitudinal and mixed-method approaches to deepen understanding of sustainable leadership dynamics. Comparative studies across industries and countries would help identify contextual factors that shape the effectiveness of sustainable leadership (Egri & Herman, 2000). Researchers could also explore additional mediating and moderating variables, such as green human resource management, digital technologies, or regulatory pressure, to further refine the proposed framework (Renwick, et al., 2013).

10. SUMMARY OF FINDINGS

This study examined the role of sustainable leadership in promoting environmental stewardship and organizational performance. The results show that sustainable leadership has a significant positive influence on environmental stewardship, confirming that leaders who emphasize long-term vision, ethical governance, and stakeholder responsibility drive environmentally responsible practices (Avery & Bergsteiner, 2011; Daily et al., 2012). Environmental stewardship was also found to positively affect organizational performance by enhancing operational efficiency, reputation, and stakeholder trust, supporting the natural resource–based view that proactive environmental strategies contribute to sustained competitive advantage (Hart, 1995; Hart & Dowell,2011).The findings further indicate that green organizational culture and green innovation partially mediate the relationship between sustainable leadership and environmental stewardship, highlighting the role of shared values and eco-innovation in achieving environmental outcomes (Metz Ilies & Metz, 2020). Additionally, employee environmental behavior strengthens this relationship, underscoring the importance of employee engagement in translating leadership vision into effective environmental action (Ones & Dilchert, 2012; Norton et al., 2015).

11. CONCLUSION

This study contributes to the leadership and sustainability literature by developing and empirically validating an integrated framework that links sustainable leadership, environmental stewardship, green organizational culture, green innovation, and organizational performance. Drawing on stakeholder theory (Freeman & Boeker, 1984), the resource-based view (Barney, et al.,2001; Hart, 1995), and transformational leadership theory (Rolfe, 2011) the research demonstrates how leadership capabilities translate into environmental and strategic value creation (Avery & Bergsteiner, 2011).The inclusion of mediating and moderating mechanisms advances existing research beyond direct-effect models by explaining how sustainable leadership operates through green organizational culture, green innovation, and employee environmental behaviour (Ones & Dilchert, 2012).Finally, the study contributes to emerging ESG and sustainability research by empirically linking leadership practices with environmental stewardship and organizational performance outcomes, highlighting sustainable leadership as a key driver of long-term organizational resilience and competitive advantage (Eccles et al., 2014; Flammer, 2021).

In conclusion, sustainable leadership emerges as a critical driver of responsible organizational practice in an era of escalating environmental and social challenges. Organizations that invest in sustainability-oriented leadership are better positioned to respond to regulatory pressures, stakeholder expectations, and global sustainability agendas such as the Sustainable Development Goals (SDGs) and ESG frameworks (Freeman & Boeker,1984); Eccles et al., 2014, Ioannou & Serafeim, 2015; United Nations, 2015). Sustainable leadership not only supports environmental stewardship but also fosters innovation, organizational resilience, and long-term value creation (Avery & Bergsteiner, 2011; Maak et al., 2016). As environmental concerns continue to intensify, the adoption of sustainable leadership practices will be essential for organizations seeking enduring success while contributing positively to society and the natural environment (Hart, 1995; Flammer, 2021).

REFERENCES

Aguilera, R. V., Rupp, D. E., Williams, C. A., & Ganapathi, J. (2007). Putting the S back in corporate social responsibility: A multilevel theory of social change in organizations. *Academy of Management Review*, *32*(3), 836–863. DOI: 10.5465/amr.2007.25275678

Ahuja, J., Yadav, M., & Sergio, R. P. (2023). Green leadership and pro-environmental behaviour: A moderated mediation model with rewards, self-efficacy and training. *International Journal of Ethics and Systems*, *39*(2), 481–501. DOI: 10.1108/IJOES-02-2022-0041

Aragón-Correa, J. A., Hurtado-Torres, N., Sharma, S., & García-Morales, V. J. (2008). Environmental strategy and performance in small firms: A resource-based perspective. *Journal of Environmental Management*, *86*(1), 88–103. DOI: 10.1016/j.jenvman.2006.11.022 PMID: 17239519

Aragón-Correa, J. A., & Sharma, S. (2003). A contingent resource-based view of proactive corporate environmental strategy. *Academy of Management Review*, *28*(1), 71–88. DOI: 10.5465/amr.2003.8925233

Avery, G. C., & Bergsteiner, H. (2011). Sustainable leadership practices for enhancing business resilience and performance. *Strategy and Leadership*, *39*(3), 5–15. DOI: 10.1108/10878571111128766

Badshah, S. (2012). Historical study of leadership theories. *Journal of Strategic Human Resource Management*, *1*(1), 49–59. DOI: 10.18533/journal.v8i2.1582

Banerjee, S. B. (2002). Corporate environmentalism: The construct and its measurement. *Journal of Business Research*, *55*(3), 177–191. DOI: 10.1016/S0148-2963(00)00135-1

Bansal, P. (2005). Evolving sustainably: A longitudinal study of corporate sustainable development. *Strategic Management Journal*, *26*(3), 197–218. DOI: 10.1002/smj.441

Bansal, P., & Clelland, I. (2004). Talking trash: Legitimacy, impression management, and unsystematic risk in the context of the natural environment. *Academy of Management Journal*, *47*(1), 93–103. DOI: 10.2307/20159562

Bansal, P., & Roth, K. (2000). Why companies go green: A model of ecological responsiveness. *Academy of Management Journal*, *43*(4), 717–736. DOI: 10.2307/1556363

Barney, J., Wright, M., & Ketchen, D. J.Jr. (2001). The resource-based view of the firm: Ten years after 1991. *Journal of Management, 27*(6), 625–641. DOI: 10.1177/014920630102700601

Bass, B. M., & Avolio, B. J. (1994). Transformational leadership and organizational culture. *International Journal of Public Administration, 17*(3-4), 541–554. DOI: 10.1080/01900699408524907

Bass, B. M., & Riggio, R. E. (2006). *Transformational leadership*. Psychology press., DOI: 10.4324/9781410617095

Brown, M. E., Treviño, L. K., & Harrison, D. A. (2005). Ethical leadership: A social learning perspective for construct development and testing. *Organizational Behavior and Human Decision Processes, 97*(2), 117–134. DOI: 10.1016/j.obhdp.2005.03.002

Daily, B. F., Bishop, J. W., & Govindarajulu, N. (2009). A conceptual model for organizational citizenship behavior directed toward the environment. *Business & Society, 48*(2), 243–256. DOI: 10.1177/0007650308315439

Daily, B. F., Bishop, J. W., & Massoud, J. A. (2012). The role of training and empowerment in environmental performance: A study of the Mexican maquiladora industry. *International Journal of Operations & Production Management, 32*(5), 631–647. DOI: 10.1108/01443571211226524

Daily, B. F., & Huang, S. C. (2001). Achieving sustainability through attention to human resource factors in environmental management. *International Journal of Operations & Production Management, 21*(12), 1539–1552. DOI: 10.1108/01443570110410892

Delmas, M. A., & Toffel, M. W. (2008). Organizational responses to environmental demands: Opening the black box. *Strategic Management Journal, 29*(10), 1027–1055. DOI: 10.1002/smj.701

Dickson, R. K. (2023, August). Analysis of the traditional leadership theories: A review of contemporary leadership approaches and management effectiveness. In Information and Knowledge Management (Vol. 13, No. 5, pp. 9-23). DOI: DOI: 10.7176/IKM/13-5-02

Donaldson, T., & Preston, L. E. (1995). The stakeholder theory of the corporation: Concepts, evidence, and implications. *Academy of Management Review, 20*(1), 65–91. DOI: 10.2307/258887

Eccles, R. G., Ioannou, I., & Serafeim, G. (2014). The impact of corporate sustainability on organizational processes and performance. *Management Science, 60*(11), 2835–2857. DOI: 10.1287/mnsc.2014.1984

Egri, C. P., & Herman, S. (2000). Leadership in the North American environmental sector: Values, leadership styles, and contexts of environmental leaders and their organizations. *Academy of Management Journal, 43*(4), 571–604. DOI: 10.2307/1556356

Epstein, M. J. (2018). *Making sustainability work: Best practices in managing and measuring corporate social, environmental and economic impacts*. Routledge., DOI: 10.4324/9781351280129

Flammer, C. (2021). Corporate green bonds. *Journal of Financial Economics, 142*(2), 499–516. DOI: 10.1016/j.jfineco.2021.01.010

Fombrun, C. J. (2005). Corporate reputations as economic assets. The Blackwell handbook of strategic management, 285-308.DOI: 10.1111/b.9780631218616.2006.00011.x

Freeman, J., & Boeker, W. (1984). The ecological analysis of business strategy. *California Management Review, 26*(3), 73–86. DOI: 10.2307/41165081

Friede, G., Busch, T., & Bassen, A. (2015). ESG and financial performance: Aggregated evidence from more than 2000 empirical studies. *Journal of Sustainable Finance & Investment, 5*(4), 210–233. DOI: 10.1080/20430795.2015.1118917

Hargreaves, A., & Fink, D. (2006). Redistributed leadership for sustainable professional learning communities. *Journal of School Leadership, 16*(5), 550–565. DOI: 10.1177/105268460601600507

Hart, S. L. (1995). A natural-resource-based view of the firm. *Academy of Management Review, 20*(4), 986–1014. DOI: 10.2307/258963

Hart, S. L., & Dowell, G. (2011). Invited editorial: A natural-resource-based view of the firm: Fifteen years after. *Journal of Management, 37*(5), 1464–1479. DOI: 10.1177/0149206310390219

Hofstede, G. (2011). Dimensionalizing cultures: The Hofstede model in context. *Online Readings in Psychology and Culture, 2*(1), 8. DOI: 10.9707/2307-0919.1014

Ioannou, I., & Serafeim, G. (2015). The impact of corporate social responsibility on investment recommendations: Analysts' perceptions and shifting institutional logics. *Strategic Management Journal, 36*(7), 1053–1081. DOI: 10.1002/smj.2268

Jabbour, C. J. C., Santos, F. C. A., & Nagano, M. S. (2010). Contributions of HRM throughout the stages of environmental management: Methodological triangulation applied to companies in Brazil. *International Journal of Human Resource Management, 21*(7), 1049–1089. DOI: 10.1080/09585191003783512

Jamali, D., & Karam, C. (2018). Corporate social responsibility in developing countries as an emerging field of study. *International Journal of Management Reviews*, *20*(1), 32–61. DOI: 10.1111/ijmr.12112

Jennings, P. D., & Zandbergen, P. A. (1995). Ecologically sustainable organizations: An institutional approach. *Academy of Management Review*, *20*(4), 1015–1052. DOI: 10.2307/258964

Liao, Y. (2022). Sustainable leadership: A literature review and prospects for future research. *Frontiers in Psychology*, *13*, 1045570. DOI: 10.3389/fpsyg.2022.1045570 PMID: 36420377

Linnenluecke, M. K., & Griffiths, A. (2010). Corporate sustainability and organizational culture. *Journal of World Business*, *45*(4), 357–366. DOI: 10.1016/j.jwb.2009.08.006

Maak, T., Pless, N. M., & Voegtlin, C. (2016). Business statesman or shareholder advocate? CEO responsible leadership styles and the micro-foundations of political CSR. *Journal of Management Studies*, *53*(3), 463–493. DOI: 10.1111/joms.12195

Metz, D., Ilies, L., & Metz, M. (2020). The impact of organisational culture on leadership effectiveness and performance. In Proceedings of the 14th International Management Conference" Managing Sustainable Organizations (pp. 776-788).DOI: DOI: 10.24818/IMC/2020/04.09

Norton, T. A., Parker, S. L., Zacher, H., & Ashkanasy, N. M. (2015). Employee green behavior: A theoretical framework, multilevel review, and future research agenda. *Organization & Environment*, *28*(1), 103–125. DOI: 10.1177/1086026615575773

Ones, D. S., & Dilchert, S. (2012). Environmental sustainability at work: A call to action. *Industrial and Organizational Psychology: Perspectives on Science and Practice*, *5*(4), 444–466. DOI: 10.1111/j.1754-9434.2012.01478.x

Pless, N. M., Maak, T., & Waldman, D. A. (2012). Different approaches toward doing the right thing: Mapping the responsibility orientations of leaders. *The Academy of Management Perspectives*, *26*(4), 51–65. DOI: 10.5465/amp.2012.0028

Podsakoff, N. P., Podsakoff, P. M., MacKenzie, S. B., Maynes, T. D., & Spoelma, T. M. (2014). Consequences of unit-level organizational citizenship behaviors: A review and recommendations for future research. *Journal of Organizational Behavior*, *35*(S1), S87–S119. DOI: 10.1002/job.1911

Podsakoff, P. M., MacKenzie, S. B., Podsakoff, N. P., & Lee, J. Y. (2003). The mismeasure of man (agement) and its implications for leadership research. *The Leadership Quarterly*, *14*(6), 615–656. DOI: 10.1016/j.leaqua.2003.08.002

Porter, M. E., & Linde, C. V. D. (1995). Toward a new conception of the environment-competitiveness relationship. *The Journal of Economic Perspectives*, *9*(4), 97–118. DOI: 10.1257/jep.9.4.97

Powell, W. W., & DiMaggio, P. J. (1983). The iron cage revisited: Institutional isomorphism and collective rationality in organizational fields. *American Sociological Review*, *48*(2), 147–160. DOI: 10.2307/2095101

Renwick, D. W., Redman, T., & Maguire, S. (2013). Green human resource management: A review and research agenda. *International Journal of Management Reviews*, *15*(1), 1–14. DOI: 10.1111/j.1468-2370.2011.00328.x

Robertson, J. L., & Barling, J. (2013). Greening organizations through leaders' influence on employees' pro-environmental behaviors. *Journal of Organizational Behavior*, *34*(2), 176–194. DOI: 10.1002/job.1820

Rockström, J., Falkenmark, M., Karlberg, L., Hoff, H., Rost, S., & Gerten, D. (2009). Future water availability for global food production: The potential of green water for increasing resilience to global change. *Water Resources Research*, *45*(7), 2007WR006767. Advance online publication. DOI: 10.1029/2007WR006767

Rolfe, P. (2011). Transformational leadership theory: What every leader needs to know. *Nurse Leader*, *9*(2), 54–57. DOI: 10.1016/j.mnl.2011.01.014

Shrivastava, P. (1995). The role of corporations in achieving ecological sustainability. *Academy of Management Review*, *20*(4), 936–960. DOI: 10.2307/258961

Steffen, W., Richardson, K., Rockström, J., Cornell, S. E., Fetzer, I., Bennett, E. M., ... & Sörlin, S. (2015). Planetary boundaries: Guiding human development on a changing planet. science, 347(6223), 1259855.DOI: DOI: 10.1126/science.1259855

Su, X., Xu, A., Lin, W., Chen, Y., Liu, S., & Xu, W. (2020). Environmental leadership, green innovation practices, environmental knowledge learning, and firm performance. *SAGE Open*, *10*(2), 2158244020922909. DOI: 10.1177/2158244020922909

Suchman, M. C. (1995). Managing legitimacy: Strategic and institutional approaches. *Academy of Management Review*, *20*(3), 571–610. DOI: 10.2307/258788

Suriyankietkaew, S., & Avery, G. C. (2014). Leadership practices influencing stakeholder satisfaction in Thai SMEs. *Asia-Pacific Journal of Business Administration*, *6*(3), 247–261. DOI: 10.1108/APJBA-01-2014-0010

Waldman, D. A., & Siegel, D. (2008). Defining the socially responsible leader. *The Leadership Quarterly*, *19*(1), 117–131. DOI: 10.1016/j.leaqua.2007.12.008

Waldman, D. A., Siegel, D. S., & Javidan, M. (2006). Components of CEO transformational leadership and corporate social responsibility. *Journal of Management Studies, 43*(8), 1703–1725. DOI: 10.1111/j.1467-6486.2006.00642.x

Xie, X., Huo, J., & Zou, H. (2019). Green process innovation, green product innovation, and corporate financial performance: A content analysis method. *Journal of Business Research, 101*, 697–706. DOI: 10.1016/j.jbusres.2019.01.010

Yukl, G., Mahsud, R., Hassan, S., & Prussia, G. E. (2013). An improved measure of ethical leadership. *Journal of Leadership & Organizational Studies, 20*(1), 38–48. DOI: 10.1177/1548051811429352

Zhu, Q., Sarkis, J., & Lai, K. H. (2013). Institutional-based antecedents and performance outcomes of internal and external green supply chain management practices. *Journal of Purchasing and Supply Management, 19*(2), 106–117. DOI: 10.1016/j.pursup.2012.12.001

Chapter 5
Sustainable Leadership for Institutional Environmental Stewardship:
A Behavioral and Systems Approach for Transforming Organizations

Sarina Asif
School of Management and Commerce, K.R. Mangalam University, Gurugram, India

Nitish Kumar Minz
http://orcid.org/0009-0000-4770-0336
VIT Business School, Vellore Institute of Technology, Vellore, India

Anumeha Mathur
School of Management and Commerce, K.R. Mangalam University, Gurugram, India

ABSTRACT

Environmental sustainability remains a strategic priority for organizations, yet a persistent gap exists between formal sustainability commitments and actual environmental outcomes. This chapter introduces Sustainable Stewardship Leadership (SSL) as a behavioral and systems-based framework explaining how leaders can embed environmental stewardship into institutional culture and organizational systems. Drawing on sustainability leadership, green transformational leadership,

DOI: 10.4018/979-8-3373-7877-0.ch005

stewardship theory, and systems thinking, the chapter demonstrates how leadership influences environmental outcomes through value-based influence, behavioral modeling, structural alignment, and community empowerment. By integrating behavioral and institutional perspectives, the chapter offers actionable insights for leaders and policymakers seeking to move beyond compliance-driven sustainability toward enduring environmental transformation and long-term ecological resilience.

1. INTRODUCTION

The demands on organizations to focus on strategies of sustainability are amplified by environmental degradation, climate volatility, and systemic ecological risk, which is shifting towards an organization focused on being environmentally responsible rather than merely complying with the regulations (Rosário and Boechat, 2025; Farag, 2026; Udin et al., 2025). Even with popular practices in institutional sustainability policies, environmental management systems, and Environmental, Social, and Governance frameworks, there is still a gap between the institutional sustainability commitments and actual perceived environmental outcomes. Such an implementation gap indicates that the challenge of sustainability is not really technical or procedural, but fundamentally behavioral and leadership-oriented (Buhr and Weissbrod, 2026; Reyes et al., 2026; Twesigye et al., 2026).

Despite the widespread adoption of sustainability frameworks, many institutions continue to experience difficulties translating environmental commitments into measurable and enduring environmental outcomes. This sustainability implementation gap suggests that environmental transformation depends not only on policies, technologies, or governance mechanisms but also on the behavioural and leadership processes that shape organizational action (Senge, 2006).

1.1 The Need for Sustainable Stewardship Leadership

As we now know, sustainability scholarship is no longer the sole domain of compliance, reporting, and technological fixes. These approaches play a role in environmental accountability and environmental performance measurement but do not entirely account for the successful institutionalization of sustainability in some organizations when compared with other organizations relying on isolated initiatives, short-term projects, or individual leadership champions (Althnayan et al., 2022; Hanif et al., 2023; Haile, 2025). As a result, sustainability activities can become disjointed and compartmentalized, weakly embedded in organizational culture, and are easily lost when other initiatives are prioritized or when external pressures surface.

The restrictions indicate that sustainability can be thought about not just as a management challenge, but also as a stewardship one. Environmental stewardship goes beyond strict adherence to regulations; it's about shared responsibility for the long-term preservation of ecological systems, and embedding sustainability practices within the culture of the organisation. In this sense, sustainability is achieved when members of the organization see environmental responsibility not as an outside mandate but as a part of the organizational values. To achieve such stewardship, there is a need for leadership approaches that will impact behavioural norms, increase ethical responsibility, and promote shared ownership of sustainability goals across the institution (Ismail, 2025; Miao & Nduneseokwu, 2025; Parvez et al., 2026).

In addition, environmental sustainability is a systemic issue, with a need for synergy between individual behavior, organizational culture, governance, and stakeholder relationships. Leadership is therefore a crucial component not just for setting environmental priorities, but for institutionalising sustainability into the key processes that are likely to lead to responsible behaviours over time (Farag, 2026; Liu et al., 2025; Sadiq et al., 2026). The expectations that leadership models are dominated by the outcomes of performance, regulatory compliance, or technological solutions may fall short of the behavioural and cultural mechanisms that can drive the institutionalization of environmental responsibility (Ivanova et al., 2026; Umar et al., 2025). Thus, sustainability is a long-term issue, and the real issue is whether organizations can develop stewardship as a lasting capacity.

Prior research has examined sustainable leadership, green transformational leadership, stewardship theory, and environmental management as important yet largely independent perspectives for understanding sustainability outcomes. While these approaches provide valuable insights, limited attention has been devoted to explaining how leadership behaviour, organizational culture, governance systems, and stakeholder engagement interact to institutionalize environmental stewardship across organizational contexts. This fragmentation has constrained the development of integrative explanations for how sustainability commitments are translated into enduring environmental practices.

To address this gap, this chapter proposes Sustainable Stewardship Leadership (SSL) as a behavioural and systems-oriented framework for institutional environmental transformation. SSL conceptualizes environmental stewardship as an organizational capability that emerges through the alignment of ethical leadership, behavioural reinforcement, organizational systems, and stakeholder participation. Drawing upon insights from sustainability leadership, organizational behaviour, systems thinking, stewardship theory, and sustainability governance, the framework explains how institutions can bridge the gap between sustainability intentions and environmental action while strengthening their capacity for long-term environmental stewardship (Farag, 2026; Miao & Nduneseokwu, 2025; Sposato & Dittmar, 2025).

2. THEORETICAL FOUNDATIONS OF SUSTAINABLE STEWARDSHIP LEADERSHIP

Sustainable institutional change should be guided by leadership paradigms that are no longer based on operational efficiency and regulatory compliance but on value orientation and behaviorally embedded stewardship orientations (Alemu, 2025; Rosário and Boechat, 2025; Farag, 2026). The modern challenges of sustainability are multidimensional and systemic in nature and require multidimensional leadership strategies that combine ethical responsibility, affective behavior, and organizational systems alignment. In this regard, the concept of SSL is proposed as an integrative phenomenon and has its foundations in a variety of theoretical frameworks, such as sustainable leadership, green transformational leadership, stewardship theory, and systems thinking (Miao and Nduneseokwu, 2025; Vargas-Hernández, 2022; Udin et al., 2025).

2.1 Sustainable Leadership as a Strategic and Ethical Imperative

The theory of sustainable leadership emphasizes long-term value creation through the alignment of organizational purpose with environmental, social, and governance objectives. Unlike short-term performance-oriented leadership approaches, sustainable leadership prioritizes intergenerational responsibility, ethical decision-making, and organizational resilience in response to ecological uncertainty. Research suggests that leaders who frame sustainability as both a moral responsibility and a strategic priority are more successful in integrating environmental considerations into organizational decision-making processes (Rosário & Boechat, 2025; Farag, 2026; Buhr & Weissbrod, 2026).

A central contribution of sustainable leadership is its redefinition of organizational success beyond financial and operational outcomes to include environmental performance and long-term societal value. In doing so, sustainability becomes a core managerial responsibility rather than a peripheral organizational initiative. This perspective provides an important normative foundation for stewardship-oriented leadership by emphasizing that environmental responsibility should be reflected in leadership behaviour, organizational culture, and institutional decision-making systems (Alemu, 2025; Liu et al., 2025; Twesigye et al., 2026).

A significant limitation of conventional sustainability leadership models, however, is their tendency to operationalize sustainability primarily through strategic management instruments rather than through institutional behavioural transformation. While policy alignment and ESG integration remain important, sustainability initiatives often struggle to achieve lasting impact when environmental responsibility is not internalized within organizational identity and leadership culture. Sustainable leadership, therefore, becomes most effective when ethical stewardship is translated into behavioural norms, institutional routines, and long-term organizational learning processes (Senge, 2006; Avery & Bergsteiner, 2011).

Sustainable leadership highlights the ethical and strategic significance of environmental responsibility, but it is mainly about leadership's influence and the vision of the organisation. It offers little explanation of how sustainability values are institutionalised within governance structures, relationships, and day-to-day organisational practices, however. This restriction indicates a need for broader frameworks that can link leadership influence and long-term stewardship processes.

2.2 Green Transformational Leadership and Pro-Environmental Behavior

Green transformational leadership (GTL) is a continuation of transformational leadership theory that specifically includes emphasis on the values of the environment, behaviors, and performance results (Althnayan et al., 2022; Hanif et al., 2023; Udin et al., 2025). It has been proven in the literature that GTL has a positive impact on environmental organizational citizenship behavior, green innovation, and sustainability performance in all industries (Ismail, 2025; Kebe et al., 2025; Haile, 2025). GTL leaders define an impressive vision of the environment, challenge employees to innovate with a sustainable approach, and model environmental responsibility (Althnayan et al., 2022; Nduneseokwu and Harder, 2023; Seema et al., 2025).

Notably, GTL can be used to promote stewardship practices through emotional commitment and moral identification to environmental objectives instead of using formal controls or incentives only (Parvez et al., 2026; Hanif et al., 2023). This behavioral activation is essential in operationalizing sustainability strategies into everyday operation, especially in institutes that are complex, where direct observation is restricted (Ismail, 2025; Kebe et al., 2025; Twesigye et al., 2026).

While green transformational leadership has contributed significantly to environmental awareness and environmentally responsible employee behavior, the lasting impact of this strategy may be jeopardized due to its reliance on charismatic individuals for implementing sustainable practices, as opposed to integrating such practices into institutional operations. Many companies demonstrate temporary interest in sustainability issues during the reign of inspiring leaders; however, they often struggle to maintain this level of behavior once their priorities have shifted or organizational demands have increased. Therefore, the practice of sustainability transformation requires not only motivational leadership practices but also governance, accountability, and institutional learning procedures.

One more drawback of sustainability transformations based on transformational approaches refers to the focus of such practices on visionary leadership and neglecting structural factors and institutional complexities. Sustainability tends to become a component of the leader's rhetoric rather than something that is integrated into institutional structures and practices.

The problem above can be addressed through Sustainable Stewardship Leadership (SSL), which highlights long-term behavioral integration, systemic governance, and institutional accountability. Unlike the previous paradigm, SSL recognizes environmental stewardship as an organizational ability embedded into governance, organizational culture, and stakeholder relations instead of a task performed by inspirational leaders.

Green transformational leadership emphasizes the role of inspiration, motivation, and behaviour in inducing environmental responsibility. But when the sustainability program is primarily based on transformational influence, it is possible that it is still susceptible to leadership change and shifting priorities within the organization. To ensure long-term environmental stewardship, mechanisms are therefore needed that operate beyond the leader/follower dynamic, and become institutionalised and routinised.

2.3 Stewardship Theory and Environmental Responsibility

Stewardship theory provides an important theoretical foundation for understanding environmental responsibility by viewing organizational members as actors who are motivated to pursue collective and long-term interests rather than solely short-term personal gains. Within environmental contexts, stewardship is reflected in voluntary and discretionary behaviours that contribute to the protection and improvement of ecological systems, even in the absence of formal requirements or direct incentives (Twesigye et al., 2026; Buhr & Weissbrod, 2026; Nduneseokwu & Harder, 2023).

Leadership plays a critical role in activating stewardship orientations by fostering ethical climates, strengthening shared responsibility, and legitimizing pro-environmental initiative-taking. When stewardship values become integrated into leadership practices, employees are more likely to internalize sustainability as part of their professional identity, resulting in stronger environmental commitment and greater behavioural consistency across organizational contexts (Miao & Nduneseokwu, 2025; Vargas-Hernández, 2022; Parvez et al., 2026).

The concept of stewardship theory brings a valuable normative point of view, focusing on responsibility, trust, and the collective good. However, theory provides little information on how stewardship values can be intentionally developed, consolidated, and maintained throughout complex systems of organizations. This gap is what makes considering stewardship principles alongside behavioural and institutional perspectives so important.

2.4 Systems Thinking and Institutional Sustainability

Sustainability issues are manifested in and through interwoven organizational, social, and ecological systems, and systems thinking is an important theoretical foundation to sustainable stewardship leadership (Farag, 2026; Ivanova et al., 2026; Liu et al., 2025). Systems-based leadership focuses on alignment of strategy, structure, incentives, and culture to make sure that sustainability is strengthened by institutional operations (Sposato and Dittmar, 2025; Umar et al., 2025; Vargas-Hernández, 2022).

Systems thinking contributes significantly to sustainability leadership by recognizing that environmental challenges are interconnected rather than isolated institutional problems. Organizational sustainability outcomes are influenced by the interaction between governance structures, employee behavior, technological systems, resource management, stakeholder participation, and institutional culture. Consequently, fragmented sustainability initiatives frequently fail because they address environmental concerns independently without considering broader institutional interdependencies.

Environmental degradation, climate instability, and resource inefficiency often emerge through cumulative organizational processes rather than isolated operational failures. Institutions that focus only on technical sustainability interventions while neglecting behavioral and structural dimensions may therefore struggle to achieve long-term environmental transformation. Systems-oriented leadership becomes essential because it allows institutions to identify relationships between organizational decisions, governance systems, environmental performance, and stakeholder behavior.

Within the SSL framework, systems thinking strengthens institutional environmental stewardship by encouraging cross-functional collaboration, organizational learning, and long-term strategic alignment. This perspective enables sustainability to become integrated into institutional decision-making processes rather than remaining confined to isolated environmental programs or compliance-oriented sustainability policies.

A lack of systemic alignment implies a fragile position of sustainability efforts, which can be seen as being torn apart and symbolic (Buhr and Weissbrod, 2026; Reyes et al., 2026; Udin et al., 2025). Sustainable stewardship leadership thus combines behavioral, structural, and governance systems to entrench environmental responsibility and not rely on champions (Farag, 2026; Twesigye et al., 2026; Sposato and Dittmar, 2025).

Systems thinking offers a useful approach to thinking about sustainability issues that are interrelated. Systems-based explanations, however, tend to focus on structural connections and relationships and fail to sufficiently account for leadership behaviors and cultural processes that trigger institutional change. Systemic alignment and stewardship-oriented leadership are needed together to achieve effective sustainability transformation.

Table 1. Theoretical foundations informing sustainable stewardship leadership

Theory	Core Focus	Leadership Role	Environmental Outcome
Sustainable Leadership	Long-term value, ethics, ESG integration	Strategic and moral anchor	Institutional sustainability resilience
Green Transformational Leadership	Vision, inspiration, modeling	Behavioral activation	Green innovation and performance
Stewardship Theory	Intrinsic motivation, collective interest	Ethical climate builder	Voluntary pro-environmental behavior
Systems Thinking	Structural and process alignment	System integrator	Durable environmental transformation

2.5 Toward an Integrated Perspective of Institutional Environmental Stewardship

The theoretical perspectives reviewed in this section collectively demonstrate that sustainability leadership cannot be adequately understood through a single conceptual lens. Sustainable leadership emphasizes long-term value creation and environmental responsibility, green transformational leadership highlights behavioural influence, stewardship theory focuses on ethical responsibility and collective accountability, and systems thinking explains the interconnected nature of organizational and environmental processes (Althnayan et al., 2022; Hanif et al., 2023; Kebe et al., 2025; Senge, 2006).

Despite their contributions, these perspectives provide only partial explanations of how environmental stewardship becomes institutionalized across organizations. Leadership-based approaches explain influence and motivation, while stewardship and systems perspectives emphasize responsibility and complexity; however, they offer limited insight into how behavioural, cultural, governance, and stakeholder dimensions interact to sustain environmental responsibility over time.

This limitation highlights the need for a more integrated perspective on sustainability leadership. Building on the complementary insights of these theories, this chapter proposes Sustainable Stewardship Leadership (SSL) as a behavioural and systems-oriented framework that explains how environmental stewardship becomes embedded within organizational culture, governance systems, and stakeholder relationships. From this perspective, long-term sustainability is viewed not merely as an environmental outcome but as an institutional capability shaped through ethical leadership, behavioural reinforcement, and organizational alignment (Alemu, 2025; Buhr & Weissbrod, 2026; Rosário & Boechat, 2025).

3. BEHAVIORAL FOUNDATIONS OF ENVIRONMENTAL STEWARDSHIP

Although the preceding passage laid out the theoretical basis of Sustainable Stewardship Leadership, an effective analysis of institutional change in the environmental context must also consider the behavioral approach to leadership, whereby it influences organizational behavior and environmental stewardship.

The phenomenon of environmental stewardship in institutions is essentially a behavioral phenomenon, the timing of which depends on the values of individuals, collective norms, and sensemaking processes that are driven by the leaders (Alemu, 2025; Buhr and Weissbrod, 2026; Twesigye et al., 2026). Although organizational sustainability strategies usually focus on the systems, policies, and technologies, the long-term environmental outcomes are based on whether people internalize the ecological responsibility as a professional and moral identity (Parvez et al., 2026; Ismail, 2025; Nduneseokwu and Harder, 2023). This section studies the mechanisms of behavior that sustainable stewardship leadership triggers pro-environmental behavior within institutional settings.

3.1 Values, Beliefs, and Ethical Motivation

The cognitive and moral basis of environmental stewardship is values and beliefs, which determine the interpretation of sustainability objectives and reaction to leadership indicators by the organizational members (Alemu, 2025; Buhr and Weissbrod, 2026). Individuals tend to internalize the stewardship values and behave in ways that are not specified by their job title when leaders make a consistent effort to communicate the importance of environmental responsibility as a mutual ethical duty, but not as a mandatory compliance measure (Miao and Nduneseokwu, 2025; Reyes et al., 2026). The studies show that organizational values that aim at achieving sustainability enhance moral responsibility and decrease the opposition to organizational environmental change efforts (Alemu, 2025; Ivanova et al., 2026).

Further ethical motivation is enhanced as the alignment of sustainability messages with institutional purpose and societal impact by leadership allows employees to make environmental stewardship a point of purposeful work instead of imposed responsibility (Rosário and Boechat, 2025; Farag, 2026). Value-based framing can lead to intrinsic motivation and increase commitment to long-term sustainability in knowledge-intensive organizations, especially universities and state agencies (Umar et al., 2025; Ivanova et al., 2026).

In fact, "behavioral sustainability" is an important factor in the institutionalization of environmental initiatives in organizational systems. The translation of environmental expectations into organizational practice and individual reinforcement in daily organizational practices is the key to sustainability transformation, although this process is formalized in sustainability policies and governance frameworks. Thus, sustainability-oriented behavioural mechanisms are a critical basis for understanding how environmental stewardship is integrated in institutional culture and operational decision-making.

3.2 Pro-Environmental Organizational Citizenship Behavior

The term pro-environmental organizational citizenship behavior (OCB-E) describes the voluntary activities of employees undertaken to protect and improve the natural environment beyond their formal job responsibilities. Empirical evidence consistently associates green transformational leadership with higher levels of OCB-E through the development of environmental commitment, trust, and psychological ownership among organizational members. In this context, leaders serve as visible role models whose behaviours establish environmental norms that encourage voluntary sustainability initiatives (Althnayan et al., 2022; Hanif et al., 2023; Ismail, 2025; Nduneseokwu & Harder, 2023).

Organizational Citizenship Behavior for the Environment (OCB-E) is one of the most important behavioural dimensions of long-term sustainability transformation, as it reflects environmental actions that extend beyond formal institutional requirements. Employees demonstrating OCB-E often engage in resource conservation, waste reduction, sustainability advocacy, and encouraging environmentally responsible practices among colleagues without direct organizational enforcement.

The significance of OCB-E lies in its ability to strengthen sustainability culture through informal behavioural influence. Environmental responsibility becomes more deeply embedded within the workplace when sustainability is viewed as an integral part of employees' roles rather than as an externally imposed obligation. As a result, environmentally responsible behaviours become incorporated into organizational culture and contribute to long-term environmental performance.

Leadership plays a crucial role in promoting OCB-E by shaping ethical norms, organizational trust, and collective environmental responsibility. Participatory governance, behavioural modelling, and sustainability-oriented leadership practices can increase voluntary support for environmental initiatives and reinforce the integration of sustainability values into everyday organizational activities. Consequently, behavioural sustainability emerges as an important institutional mechanism through which organizations translate environmental commitments into routine practice (Parvez et al., 2026; Haile, 2025; Twesigye et al., 2026).

OCB-E also serves as a critical behavioural link between leadership intent and organizational environmental performance, particularly in complex systems where formal monitoring is limited. When environmental stewardship becomes embedded in daily activities such as energy conservation, waste minimization, and green innovation, organizations generate sustainability benefits that extend beyond isolated environmental interventions (Hanif et al., 2023; Kebe et al., 2025; Sposato & Dittmar, 2025).

3.3 Psychological Ownership and Green Identity

Environmental stewardship is further reinforced through psychological ownership and green organizational identity, both of which contribute to the development of individual responsibility for ecological performance. Individuals are more likely to engage in sustained pro-environmental behaviour when sustainability is perceived as a shared responsibility rather than solely as a management initiative. Leadership contributes to the development of this ownership by involving employees in sustainability-related decision-making processes and recognizing environmental contributions (Parvez et al., 2026; Buhr & Weissbrod, 2026; Miao & Nduneseokwu, 2025).

Green identity formation is particularly relevant within institutional settings such as higher education and public-sector organizations, where professional values and behavioural norms are strongly influenced by mission-oriented objectives. When leaders embed sustainability within institutional narratives, symbols, and organizational practices, they strengthen a shared environmental identity that can sustain stewardship behaviours despite resource constraints or leadership transitions (Umar et al., 2025; Ivanova et al., 2026; Rosário & Boechat, 2025).

From an SSL perspective, psychological ownership and green identity strengthen the behavioural foundations of environmental stewardship by transforming sustainability from an organizational objective into a shared institutional value. This internalization increases the likelihood that stewardship-oriented behaviours will persist over time and remain resilient under changing organizational conditions (Nduneseokwu & Harder, 2023; Alemu, 2025).

Figure 1. Behavioral pathways to environmental stewardship through sustainable stewardship leadership

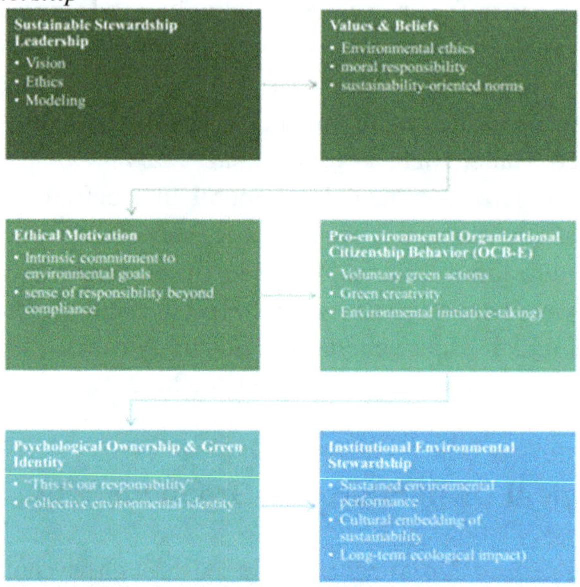

Developed by the author based on Alemu (2025), Buhr and Weissbrod (2026), Twesigye et al. (2026), and Miao and Nduneseokwu (2025)

Figure 1 illustrates the behavioral mechanisms through which Sustainable Stewardship Leadership influences institutional environmental outcomes. Leadership-driven values and ethical framing shape employees' beliefs, which activate intrinsic environmental motivation. This motivation fosters pro-environmental organizational citizenship behaviors and strengthens psychological ownership and green identity, ultimately leading to sustained institutional environmental stewardship.

4. SUSTAINABLE STEWARDSHIP LEADERSHIP (SSL) FRAMEWORK

An integrative leadership model termed Sustainable Stewardship Leadership (SSL) is proposed to explain how environmental stewardship can be embedded within organizational behaviour and institutional systems through leadership action. Unlike traditional sustainability leadership approaches that primarily emphasize policy adoption or environmental management practices, SSL focuses on the behavioural and systemic pathways through which leaders translate sustainability intentions into enduring environmental outcomes. The framework views leadership as a mechanism for aligning values, behaviours, organizational structures, and stakeholder engagement around long-term ecological responsibility (Alemu, 2025; Rosário & Boechat, 2025; Farag, 2026).

At its core, SSL is founded on the premise that environmental sustainability is not achieved through isolated initiatives but through the continuous reinforcement of stewardship-oriented values and practices across organizational systems. Leaders play a central role in shaping environmental responsibility, facilitating its implementation, and ensuring its institutionalization within everyday decision-making processes. To explain how this transformation occurs, the SSL framework identifies four interrelated mechanisms: value-based influence, behavioural modelling, structural alignment, and community empowerment (Althnayan et al., 2022; Kebe et al., 2025).

Value-based influence represents the ethical foundation of SSL and focuses on the leader's role in cultivating ecological values and shared moral responsibility throughout the organization. Rather than presenting sustainability as a compliance requirement or reputational strategy, leaders frame environmental responsibility as an institutional value connected to organizational purpose and societal contribution. This value-oriented approach influences employee attitudes and ethical decision-making, increasing the likelihood that sustainability considerations become integrated into both operational and strategic activities (Buhr & Weissbrod, 2026; Ivanova et al., 2026; Parvez et al., 2026).

The second mechanism, behavioural modelling, highlights the symbolic influence of visible leadership actions. Through responsible resource use, ethical decision-making, and support for environmental innovation, leaders establish behavioural norms that legitimize stewardship practices across organizational levels. Such modelling can strengthen employee engagement in voluntary environmental activities by reducing uncertainty and reinforcing psychological ownership of sustainability outcomes (Nduneseokwu & Harder, 2023; Ismail, 2025; Seema et al., 2025).

Structural alignment represents the systems-oriented dimension of SSL and emphasizes the integration of sustainability into organizational processes, governance structures, and performance systems. Leaders operationalize stewardship by embedding environmental objectives within resource allocation, evaluation systems, and decision-making procedures. This alignment helps ensure consistency between organizational values and practices while strengthening institutional resilience during periods of environmental uncertainty or leadership transition (Farag, 2026; Hanif et al., 2023; Vargas-Hernández, 2022).

The fourth mechanism, community empowerment, extends stewardship beyond formal organizational boundaries by encouraging employees, partners, students, and other stakeholders to become active participants in environmental responsibility. Leaders promote empowerment through participation, collaboration, and shared ownership of sustainability initiatives. By decentralizing stewardship responsibilities, organizations can strengthen collective efficacy, enhance environmental impact, and improve institutional legitimacy (Twesigye et al., 2026; Miao & Nduneseokwu, 2025; Reyes et al., 2026).

Taken together, these four mechanisms position SSL as both a behavioural and systems-based framework for institutional environmental transformation. Rather than viewing sustainability as a strategic objective alone, SSL conceptualizes stewardship as an organizational capability that is reinforced through ethical influence, behavioural example, structural support, and stakeholder participation. This integrative perspective provides a pathway through which environmental responsibility can become embedded within organizational culture, governance systems, and everyday practice (Udin et al., 2025; Buhr & Weissbrod, 2026).

4.1 Institutional Challenges that Restrict Environmental Stewardship

Many institutions are struggling to move environmental ideals into sustainable stewardship, despite the increase in organizational support for sustainability. Although environmental policies, sustainability reporting frameworks, and ESG initiatives have enhanced accountability and visibility, they do not always create lasting behavioral and organizational change (Alemu, 2025; Sposato & Dittmar, 2025; Vargas-Hernández, 2022). This often means sustainability initiatives are confined to particular departments, strategic initiatives, or leadership agendas, and are not integrated across the institution.

One of the common issues is the key role played by individual leaders in sustainability results. Although influential leaders can create a lot of momentum, the sustainability functions can get weakened when priorities shift or when a change in leadership occurs (Buhr & Weissbrod, 2026; Reyes et al., 2026). Similarly, many organizations suffer from disconnects among sustainability strategies, governance systems, employee behaviors, and stakeholder engagement processes, which hinder the effectiveness of environmental initiatives (Farag, 2026; Liu et al., 2025; Sadiq et al., 2026).

These issues indicate that strong environmental policy and leadership are not enough to manage long-term environmental good stewardship. There is a need for leadership strategies that can integrate behavioural, cultural, structural, and governance aspects of sustainability in institutions. In this context, Sustainable Stewardship Leadership (SSL) is proposed as an integrative framework for institutional environmental transformation.

Figure 2. Conceptual framework of sustainable stewardship leadership (ssl) linking ethical leadership, behavioral change, environmental stewardship, and long-term sustainability

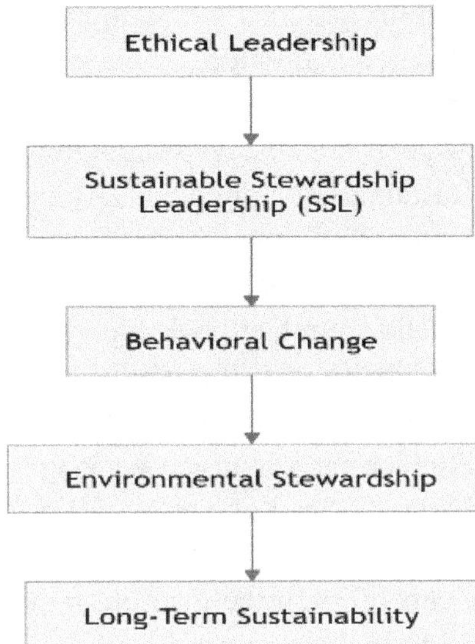

Developed by the author based on sustainability leadership and systems thinking literature

Within the Sustainable Stewardship Leadership framework, the relationship between ethical leadership, sustainability-oriented organizational behaviour, environmental stewardship, and long-term institutional sustainability happens sequentially as depicted in Figure 2. The model highlights that the transformation towards sustainability starts with ethical leadership, which can make environmental responsibility a value of the institution. These leadership practices promote the adoption of practices that lead to behavioral change, such as fostering accountability, engaging stakeholders, and promoting a sense of environmental responsibility throughout organizational systems.

The framework also illustrates how sustainability-driven behaviour change leads to environmental stewardship, through enhanced institutional commitment for responsible governance and sustainability-based decision making. Stewardship becomes an ingrained way of doing things in the organization and, therefore, an institution is better positioned to deliver long-term sustainability outcomes, namely ecological resilience, organizational legitimacy, and systems-oriented environmental governance. Hence, the model accentuates that sustainability can be achieved only in a situation of integrated leadership, behavioural and institutional transformation processes, and not by exclusive environmental activities.

4.2 Conceptual Logic of the SSL Framework

The Sustainable Stewardship Leadership (SSL) framework is designed to address a recurring issue in sustainability practice—how to make the environment part of permanent organizational behavior. Previous studies have focused on leadership influence, environmental management systems, governance, and stakeholder engagement; however, these aspects are analyzed separately, even though they are all interrelated and influence sustainability outcomes (Althnayan et al., 2022; Hanif et al., 2023; Vargas-Hernández, 2022).

The SSL offers a framework for institutional environmental stewardship that embodies a combination of behavioural, ethical, and systems-oriented perspectives and thereby supports the leadership in sustainability scholarship. The framework suggests that a combination of leadership behaviour, organisational culture, governance systems, and stakeholder participation, working synergistically, reinforces long-term sustainability (Farag, 2026; Kebe et al., 2025; Senge, 2006). SSL does not think about environmental stewardship just as an environmental result; instead, it thinks about becoming a capability of the institution that is embedded in organizational practices and decision-making processes.

In this view, sustainability transition requires not only environmental policy-making and conformance systems, but also institutional capacity to develop a sense of stewardship that will last beyond change and leadership turnover.

4.3 Theoretical Contribution of Sustainable Stewardship Leadership

The key strength of SSL is the combination of leadership, stewardship, behavioural, and systems perspectives in one framework for environmental change. Most of the current sustainability literature has investigated these aspects separately, leading to disjointed accounts of the emergence and maintenance of environmental responsibility in organisations (Althnayan et al., 2022; Hanif et al., 2023; Senge, 2006). SSL overcomes this by providing a description of how leadership influence, behavioural reinforcement, organizational alignment, and stakeholder engagement all work together to ensure institutional stewardship.

A key difference from an approach that merely concentrates on environmental performance results is that SSL takes account of the processes of institutionalization that involve integrating sustainability into the culture and governance of organizations. This approach focuses on building a capability for stewardship that can outlast changes in leadership, organizational structure, and changing environmental issues (Buhr & Weissbrod, 2026; Rosário & Boechat, 2025).

Therefore, it is important to see SSL not only as a leadership model for sustainability, but also as a tool for institutionalising environmental responsibility by achieving congruence between behavioural and systemic mechanisms. The integrative perspective is the central theoretical contribution of the chapter.

5. THE USE AND APPLICATION OF SUSTAINABLE STEWARDSHIP LEADERSHIP IN AN INSTITUTIONAL SETTING

Sustainable Stewardship Leadership comes to life in looking at how sustainability principles are adopted across different organizational contexts. Each institutional environment faces varied sustainability challenges determined by governance arrangements, operational priorities, stakeholder expectations, and available resources. Therefore, an adaptive organizational approach that can address the different challenges of the environment and institutions is essential in the implementation of stewardship-oriented leadership. Considering the use of SSL in organizational settings thus enhances knowledge of the leadership system, long-term environmental responsibility, and institutional sustainability.

Having provided a basis for the theory, this chapter shows how the theoretical construct works within different organizational settings defined by varied governance frameworks and sustainability concerns.

Due to the differences in mission orientation, governance structure, stakeholder expectations, and resource constraints, different institutional contexts exercise Sustainable Stewardship Leadership (SSL) differently (Farag, 2026; Reyes et al., 2026; Rosário and Boechat, 2025). Although the fundamental principles of the system include value-based influence, behavior modeling, structural alignment, and community empowerment, all these aspects have to be used in a situation-specific manner to make them relevant and efficient (Miao and Nduneseokwu, 2025; Vargas-Hernández, 2022; Udin et al., 2025). This section analyses the way that the activity of the SSL occurs in three key institutional contexts, such as those in institutions of higher education, corporate organizations, and the government sector.

5.1 Higher Education Institutions

Higher education institutions (HEIs) are at a privileged place in transformation within sustainability since they are at the same time producers of knowledge, role models within society, and anchors (Umar et al., 2025; Tapsir et al., 2025; Ivanova et al., 2026). In HEIs, value-based influence is the focus of the art of creating environmental stewardship and aligning it with the academic missions, ethical education, and societal responsibility (Alemu, 2025; Rosário and Boechat, 2025; Buhr and

Weissbrod, 2026). University leaders are central to integrating sustainability into learning programs, research priorities, and institutional practices, and they, therefore, determine the behavior of individuals and institutional identity (Umar et al., 2025; Ivanova et al., 2026; Reyes et al., 2026).

Academic and administrative leaders should act as role models and exercise stewardship through behavioral modeling, which provides stability to sustainable practices like responsible resource usage, green research practices, and environmentally friendly procurement (Tapsir et al., 2025; Vargas-Hernández, 2022; Nduneseokwu and Harder, 2023). Continuity across the different leadership periods in the organization is created through structural alignment necessary in the integration of sustainability into the governance systems, accreditation standards, and performance evaluation systems (Farag, 2026; Umar et al., 2025; Sposato and Dittmar, 2025). Community empowerment provided in HEIs also facilitates the involvement of students, faculty, and external partners as environmental stewards to increase the sustainability outputs by engaging collectively (Ivanova et al., 2026; Rosairo and Boechat, 2025; Twesigye et al., 2026).

5.2 Corporate Organizations

On the corporate level, SSL overcomes the ongoing dilemma of making environmental responsibility, competition, and financial goals meet (Asad, 2024; Hanif et al., 2023; Haile, 2025). The concept of value creation in firms presupposes that the leaders should present sustainability as a part of long-term value creation rather than a cost or a reputational risk (Rosário and Boechat, 2025; Farag, 2026; Buhr and Weissbrod, 2026). This type of framing helps increase the commitment of employees and enables internalization of sustainability objectives between functional areas (Alemu, 2025; Ismail, 2025; Parvez et al., 2026).

Senior executive behavior is especially significant in corporate environments due to the fact that behaviors by leaders support green innovation, ethics, and risk-taking associated with the environment (Althnayan et al., 2022; Kebe et al., 2025; Subaşi et al., 2026). The structural alignment is carried out by use of environmental management systems, green performance measures, incentive systems, and reward systems based on sustainability that incorporate stewardship into daily organizational practices (Hanif et al., 2023; Liu et al., 2025; Sposato and Dittmar, 2025). Moreover, community empowerment expands the scope of SSL beyond the organizational scope as it involves suppliers, customers, and local communities in common sustainability activities, which builds legitimacy and environmental impact (Parvez et al., 2026; Kebe et al., 2025; Reyes et al., 2026).

Many companies are making green and sustainable commitments, but they have been criticized for symbolic environmentalism and greenwashing. Communication is often used to boost institutional "reputation", strengthen stakeholder perception of sustainability, and create ESG legitimacy, but without the necessary operational shift. When this is the case, sustainability is performative and not actually institutionalised.

Greenwashing undermines institutional trust because organizations take steps to demonstrate to the public how "environmentally responsible" they are while maintaining unsustainable governance structures, practices,s and resource use. Such a gap between what is said on the topic of sustainability and how the organization acts may raise doubts on the part of stakeholders and weaken the legitimacy of the company's environmental efforts.

In this context, Sustainable Stewardship Leadership focuses on behavioural commitment, ethical alignment,t and governance integration of sustainability and institutional behaviour. In the SSL context, environmental stewardship cannot only be materialized through sustainability reporting and symbolic environmental branding, but also through measurable transformation of the organization, stakeholder engagement,t and sustainability-oriented governance systems. Thus, the framework adds to the institutional legitimacy by calling for the organisations to integrate the sustainability values and thus to behave and structure themselves in a long-term manner.

This indicates that establishing environmental policies is not enough to make sustainability efforts successful; organisations need to be able to integrate the principle of stewardship into leadership practices, governance systems, and employee behaviour. A key aspect of environmental responsibility from an SSL perspective is that it should become a part of the organizational culture and not a standalone compliance requirement.

5.3 Public Sector and Government Institutions

Bureaucratic systems, political responsibility, and the lack of resources expose institutions in the public sector to unique sustainability issues (Twesigye et al., 2026; Reyes et al., 2026; Vargas-Hernández, 2022). The ethics of stewardship and creation of value to society are the key characteristics of the development of the concept of the role of an environmentally responsible entity in the government (SSL), making environmental responsibility a communal duty of society instead of a choice of an organization (Alemu, 2025; Rosário and Boechat, 2025; Farag, 2026). The leaders can contribute to the environment by aligning the priorities of sustainability with the missions of the public services and developmental goals (Twesigye et al., 2026; Miao and Nduneseokwu, 2025; Buhr and Weissbrod, 2026).

The legitimacy and trust provided by the model behavior among public leaders is reinforced by the situations when the success of environmental policy requires the compliance and participation of citizens (Nduneseokwu & Harder, 2023; Reyes et al.,

2026; Udin et al., 2025). Policy integration, regulatory frameworks, and systems of inter-agency coordination institutionalize sustainability across governance systems to achieve structural alignment (Farag, 2026; Vargas-Hernández, 2022; Sposato and Dittmar, 2025). The ideas of community empowerment are very important in the context of the public institutions mobilizing citizens, NGOs, and private actors as co-stewards of the environmental resources, extending stewardship beyond the formal organizational level (Twesigye et al., 2026; Rosário and Boechat, 2025; Ivanova et al., 2026).

Public institutions are in a privileged position, not only from the point of environmental protection but also from the perspective of practicing SSL, to deepen and reinforce the efforts towards environmental protection through the inclusion of sustainability goals, accountability mechanisms, and citizen engagement processes. This alignment can strengthen the institutional legitimacy and assist long-term environmental governance.

6. ENABLERS OF ENVIRONMENTAL STEWARDSHIP IN INSTITUTIONS

Leadership behaviour alone is not enough to make Sustainable Stewardship Leadership effective; the institutional environment in which they operate is also important in facilitating the long-term environmental change. Institutional environmental stewardship is strengthened through sustainability-oriented governance structures, organizational learning systems, involvement of stakeholders, and technological innovation. In the absence of enablers, sustainability efforts may be disjointed, inconsistent, or reliant on short-term leadership priorities. The institutional enablers of SSL must therefore be known to understand how environmental responsibility can be maintained in the organization in different operational and governance contexts.

In addition to leadership behavior, environmental stewardship within the organization relies on institutional support that fosters sustainable thinking in governance arrangements and institutional learning processes.

SSL is primarily hinged on the intentional and behavioral attributes of leadership as well as a collection of enabling conditions that enhance stewardship through organizational systems and networks of stakeholders (Farag, 2026; Rosário and Boechat, 2025; Reyes et al., 2026). When the leadership practices are reinforced with the institutional enablers that help to convert the values into action, behavioral habits into routines, and commitment into impact, the most likely to maintain environmental stewardship (Alemu, 2025; Buhr and Weissbrod, 2026; Twesigye et al., 2026). This section discusses three pivotal facilitators of stewardship, which include digital transformation, education and learning mechanisms, and governance and accountability structures.

6.1 Digital Transformation as a Stewardship Enabler

Digital transformation is becoming a more important element of boosting environmental stewardship by improving institutional system transparency, measurement, and coordination (Liu et al., 2025; Farag, 2026; Sposato and Dittmar, 2025). Digital tools help leaders to incorporate environmental information into the strategic decision-making process, track the sustainability performance, and share the environmental priorities that are communicated uniformly across organizational levels (Hanif et al., 2023; Asad, 2024; Udin et al., 2025). Digital systems are seen as stewardship enablers as opposed to compliance tools when aligned with the principles of SSL (Rosário and Boechat, 2025; Vargas-Hernández, 2022; Buhr and Weissbrod, 2026).

Digital transformation and AI hold great promise for enhancing environmental governance, but they also create key ethical and institutional issues. While AI systems can help optimize resources, monitor the environment, and predict sustainability issues, an overreliance on technological control could raise questions of transparency, accountability, and organizational monitoring.

Institutions that have decided to adopt systems for digital sustainability should thus keep technological innovation in mind in combination with ethical leadership and involvement of stakeholders. The sustainability governance process that only focuses on automated systems might lead to a low level of participative environmental responsibility and dampen employee commitment if institutional sustainability is seen as a technology-driven process, rather than an organizational commitment shared by employees.

Another difficulty is the disconnect in the technological readiness of institutions and regions. Institutional sustainability disparities could grow when organizations running in resource-poor settings don't have the infrastructure to support advanced sustainability technologies. Sustainable Stewardship Leadership responds to these concerns by proposing that digital sustainability systems can and should be used alongside other approaches to ethics, learning, and collaborative environmental governance.

Nevertheless, digital transformation stewardship is subject to framing and governance by leadership. Leaders who focus on learning, improvement, and shared responsibility stimulate employees to proactively utilize digital sustainability tools that will foster engagement and innovation (Liu et al., 2025; Kebe et al., 2025; Subaşi et al., 2026). Conversely, highly surveillance-based digital controls are likely to compromise voluntary environmental behavior and intrinsic motivations (Alemu, 2025; Parvez et al., 2026; Twesigye et al., 2026). SSL thus needs digital systems to be constructed and implemented in a manner that strengthens ethical accountability and shared custodianship (Farag, 2026; Sposato and Dittmar, 2025; Reyes et al., 2026).

6.2 Education, Learning, and Capability Development

Education and lifelong learning are part of the enabling factors of environmental stewardship as they develop mindsets, skills, and ethical consciousness (Ivanova et al., 2026; Umar et al., 2025; Rosário and Boechat, 2025). Leaders can contribute positively to the results of stewardship by integrating sustainability education in professional development, leadership training, and organizational learning systems (Miao and Nduneseokwu, 2025; Nduneseokwu and Harder, 2023; Tapsir et al., 2025). These educational interventions enhance the ability of the employees to view the environmental challenges systemically and be responsive (Buhr and Weissbrod, 2026; Farag, 2026; Reyes et al., 2026).

Experiential and arts-based pedagogies also contribute to the improvement of stewardship as they engage emotional, ethical, and reflective aspects of learning, which increases the commitment to the sustainability agenda (Ivanova et al., 2026; Alemu, 2025; Rosário and Boechat, 2025). Learning environments supported by leaders stimulate experimentation, discussion, and interdisciplinary collaboration so that individuals are able to convert sustainability knowledge into action (Umar et al., 2025; Tapsir et al., 2025; Vargas-Hernández, 2022). SSL develops stewardship as an organizational capability that is constantly changing and not fixed, and remains in line with education and learning (Buhr and Weissbrod, 2026; Twesigye et al., 2026; Reyes et al., 2026).

This underscores the need for stewardship-based leadership in the development of environmental values and environmental responsibilities among future generations. In addition to environmental awareness, educational institutions can play a role in building sustainability cultures that carry stewardship behaviours to other systems in society.

6.3 Governance, Accountability, and Institutional Support

The enablers of environmental stewardship include governance and accountability mechanisms, which are crucial to ensuring the reinforcement of leadership values and behavioral norms by means of institutional frameworks (Farag, 2026; Sposato & Dittmar, 2025; Vargas-Hernández, 2022). A successful implementation of sustainability in formal governance, such as policy frameworks, performance evaluation, and resource allocation systems (Hanif et al., 2023; Liu et al., 2025; Umar et al., 2025), is involved. This kind of alignment minimizes the charisma of individual leadership through enforcing stewardship into rules and routines (Buhr and Weissbrod, 2026; Reyes et al., 2026; Twesigye et al., 2026).

Mechanisms of accountability are also important to legitimize the stewardship activities and keep stakeholders in trust (Rosário and Boechat, 2025; Farag, 2026; Parvez et al., 2026). Open reporting, participatory assessment, and ethical control can help organizations to monitor environmental development, as well as to strengthen the collective responsibility (Alemu, 2025; Sadiq et al., 2026; Kebe et al., 2025). Governance systems that are more conducive to learning and improvement, as opposed to punitive control, enhance intrinsic motivation and commitment to sustainability in the long term (Buhr and Weissbrod, 2026; Twesigye et al., 2026; Reyes et al., 2026).

7. MEASUREMENT AND EVALUATION OF INSTITUTIONAL ENVIRONMENTAL STEWARDSHIP

The question of whether environmental stewardship is sustainable at institutions and, even, sustainability relies not only on intent and behavioral activation of leaders but also on the fact that it is possible to measure, evaluate, and consolidate the results of stewardship over the years (Farag, 2026; Rosário and Boechat, 2025; Reyes et al., 2026). Measurement is a very important process by which Sustainable Stewardship Leadership (SSL) transforms values and behaviors into institutional accountability and ongoing improvement (Alemu, 2025; Buhr and Weissbrod, 2026; Twesigye et al., 2026). Devoid of sound assessment systems, stewardship efforts may be left only as symbols, disjointed, or susceptible to leadership transitions and financial limitations (Vargas-Hernández, 2022; Sposato & Dittmar, 2025; Udin et al., 2025).

Environmental stewardship needs to be measured beyond output-based measurements of sustainability, to reflect behavioral, cultural, and systemic aspects of sustainability (Althnayan et al., 2022; Hanif et al., 2023; Ismail, 2025). Voluntary behaviors, ethical decision-making, and collective responsibility are those leadership-based results for which the multidimensional assessment methods are necessary (Parvez et al., 2026; Nduneseokwu and Harder, 2023; Seema et al., 2025). Thus, the evaluation models that are consistent with SSL focus on the tangible performance of the environment and intangible behavioral change (Alemu, 2025; Buhr and Weissbrod, 2026; Twesigye et al., 2026).

At the behavioral level, the measurement of stewardship includes organizational citizenship behavior of a pro-environmental nature, engagement of employees in sustainability programs, and internalization of environmental value (Althnayan et al., 2022; Kebe et al., 2025; Parvez et al., 2026). Empirical research shows that the green transformational leadership reinforces voluntary green behaviors, which are initial predictors of the institutional stewardship capacity (Ismail, 2025; Nduneseokwu and Harder, 2023; Subaşi et al., 2026). The quantification of such behaviors helps leaders to determine whether sustainability is already present in daily practice, but not limited to formal policies (Hanif et al., 2023; Twesigye et al., 2026; Udin et al., 2025).

The indicators based on cultural and identity add to the understanding of stewardship further by reflecting on common beliefs, norms, and organizational identity in terms of sustainability (Buhr and Weissbrod, 2026; Parvez et al., 2026; Alemu, 2025). Green organizational identity, ethical climate, and stewardship orientation scales can offer us information about the level of environmental commitment and indicate its sustainability (Rosário and Boechat, 2025; Reyes et al., 2026; Vargas-Hernández, 2022). These measures are especially helpful in the institutional context, in which the results of environmental impacts are materialized not through direct change in the operations but through a cultural reinforcing process (Umar et al., 2025; Ivanova et al., 2026; Twesigye et al., 2026).

System-level measurement is the complement of behavioral indicators as it evaluates how stewardship is institutionalized by measuring governance, processes, and resource allocation (Farag, 2026; Sposat and Dittmar, 2025; Liu et al., 2025). This level of assessment encompasses the aspect of structural alignment of the sustainability dimension of the SS in strategic planning, performance management, and incentive structures (Hanif et al., 2023; Umar et al., 2025; Vargas-Hernández, 2022). Systemic indicators also make sure that stewardship is cemented by organizational practices so that individual leadership influence can be minimized (Buhr and Weissbrod, 2026; Reyes et al., 2026; Twesigye et al., 2026).

Notably, stewardship assessment must be constructed as developmental and not punitive, which focuses on learning, adaptation, and perpetual improvement (Alemu, 2025; Rosário and Boechat, 2025; Farag, 2026). Sustainability-oriented feedback processes backed by the leadership promote reflection and innovation, which help institutions to optimize sustainability strategies to the environmental complexity and demands of the stakeholders (Ivanova et al., 2026; Kebe et al., 2025; Reyes et al., 2026). When measurement systems are given in line with the principles of SSL, they strengthen intrinsic motivation, ethical responsibility, and shared ownership of the environmental results (Buhr and Weissbrod, 2026; Twesigye et al., 2026; Parvez et al., 2026).

Overall, the measurement and evaluation of the institutional environmental stewardship needs a comprehensive strategy that incorporates behavioral, cultural, and systemic measures through the leadership of sustainable stewardship (Alemu, 2025; Farag, 2026; Rosário and Boechat, 2025). These multidimensional assessment systems allow companies to go beyond the symbolic sustainability pronouncement to evidence-based environmental change that is resilient, responsible, and firmly embedded in the organizational life (Reyes et al., 2026; Vargas-Hernández, 2022; Udin et al., 2025).

7.1 Challenges in Implementing Sustainable Stewardship Leadership

There is increasing institutional interest in sustainability leadership, but there are still major obstacles for organizations when implementing environmental stewardship initiatives. Resistance from institutions is often a result of hierarchical systems of governance, pressures to perform in the short term, constraints on resources, and cultures of organisation that resist behavioural change. Sustainability initiatives may also fail when environmental responsibility is seen as a symbolic or compliance-based obligation rather than a system of behaviors embedded institutionally. Hence, long-term stewardship requires not only leadership commitment but also governance integration, stakeholder participation, and organizational learning mechanisms.

7.2 Future Directions for Sustainable Stewardship Leadership

Future investigations of sustainability leadership must increasingly focus on how emergent technologies, governance frameworks, and institutional change contribute to environmental stewardship in different organizational settings. AI, sustainability prediction analytics, and digital environmental monitoring systems have great potential for contributing to enhanced institutional accountability, effective resource management, and better environmental decision-making. Such tools can assist organizations in monitoring their emissions, decreasing waste generation, and improving their sustainability reporting mechanisms.

However, digital sustainability solutions also bring about a number of serious ethical implications concerning transparency, surveillance, accountability, and unequal access to technological tools. It is necessary to investigate the ways through which institutions may be able to incorporate digital governance practices without compromising stakeholder involvement, ethical leadership, and institutional trust.

Finally, an interesting future research path may lie in the examination of the applicability of sustainability leadership in the context of developing economies. Contemporary discussions of sustainability leadership are predominantly confined to developed organizational environments, which means that little information exists about how stewardship-oriented leadership functions in different institutional contexts. Therefore, future work must continue exploring cross-sector collaboration, participatory governance, and institutional relationships that would contribute to improved environmental stewardship.

7.3 Cross-Sector Relevance of Sustainable Stewardship Leadership

While the way Sustainable Stewardship Leadership is practiced can differ in each context, the principles are generally applicable. Corporate organizations can focus on governance alignment and sustainability performance, public institutions on accountability and environmental legitimacy, and educational institutions on stewardship learning and cultural transformation. The framework is consistent in emphasizing the need to link together leadership behavior, organizational systems, and stakeholder involvement for the institutionalization of environmental responsibility, despite differences in context. It shows that SSL is not just a model for specific sectors but a framework for environmental stewardship that can be applied in various institutional contexts (Alemu, 2025; Farag, 2026; Udin et al., 2025).

8. DISCUSSION AND IMPLICATIONS

The main thesis of this chapter is that environmental sustainability cannot only be seen as an organizational result but also as an institutional capacity. Although sustainability scholarship has mainly concentrated on the environmental performance, the governance mechanisms, and leadership effectiveness, less attention has been directed to understanding how environmental responsibility becomes part of the systems and is sustained across different institutional contexts. The Sustainable Stewardship Leadership framework tackles this challenge by making stewardship the behavioural and systemic vehicle to turn sustainability pledges into lasting environmental action.

8.1 Theoretical Implications

Theoretically, this chapter is relevant to the field of sustainability and leadership, as stewardship is not viewed as an individual attribute or a single act, but as a resultant quality of leadership-based organizational systems (Miao and Nduneseokwu, 2025; Vargas-Hernández, 2022; Udin et al., 2025). The current research on green transformational leadership and environmental management usually focuses on direct performance impacts, but the views of the values, identity, and behavioral norms as mediating factors in the generation of lasting environmental performance are highlighted by SSL (Althnayan et al., 2022; Hanif et al., 2023; Kebe et al., 2025). The integrative thinking unites the micro-level theories on behavior with macro-level institutional sustainability studies (Buhr and Weissbrod, 2026; Reyes et al., 2026; Twesigye et al., 2026).

The stewardship theory is also applied through the scope of the SSL framework that incorporates it into the current organizational environment that is complex, digitalized, and has multi-stakeholder ownership (Farag, 2026; Liu et al., 2025; Sposato and Dittmar, 2025). The framework explains the influence of leadership on the environmental behavior of time and institutional layers more comprehensively by valuing the interdependence of value-based influence, behavioral modeling, structural alignment, and community empowerment (Alemu, 2025; Rosário and Boechat, 2025; Ivanova et al., 2026).

8.2 Managerial Implications

To the organizational leaders and practitioners, the chapter highlights that it is not possible to outsource environmental stewardship to sustainability units nor to inscribe it within formal management systems only (Alemu, 2025; Hanif et al., 2023; Sposato and Dittmar, 2025). Instead, the SSL needs leaders on all levels to set forth to model environmental values, integrate incentives with the aim of sustainability, and work towards a shared sense of ownership of environmental performance (Ismail, 2025; Parvez et al., 2026; Seema et al., 2025). It is stated that managers should consider stewardship a daily decision-making and interaction practice, along with role expectations, instead of a periodic initiative (Rosário and Boechat, 2025; Buhr and Weissbrod, 2026; Twesigye et al., 2026).

The discussion also recommends that sustainability ethics, behavioral change strategies, and systems thinking should be incorporated into the leadership development programs to develop stewardship capability within the institutions (Miao and Nduneseokwu, 2025; Nduneseokwu and Harder, 2023; Ivanova et al., 2026). Strategic clarity and ethical consistency are widely accepted as the qualities of leaders who are more likely to maintain the engagement of employees and voluntary pro-environmental behavior in the situation of ambiguity and conflicting organizational priorities (Kebe et al., 2025; Subaşi et al., 2026; Reyes et al., 2026).

8.3 Policy and Institutional Implications

On the policy level, the results indicate the need to shift the system beyond the compliance-based sustainability requirements to models that facilitate leadership responsibility and behavioural change (Farag, 2026; Vargas-Hernández, 2022; Rosário and Boechat, 2025). Policymakers and regulators can help promote the use of SSL by rewarding leadership-based sustainability, encouraging transparency, and acknowledging institutions that show entrenched stewardship, instead of just following standards (Sposato and Dittmar, 2025; Udin et al., 2025; Reyes et al., 2026).

Educational institutions, including government and social organizations, will have the opportunity to use the particular benefits of the mentioned tool to align sustainability objectives with the development agendas of societies and community participation programs (Umar et al., 2025; Ivanova et al., 2026; Twesigye et al., 2026). Participatory governance policies and cross-sector collaboration, coupled with sustainability education, help enhance the systemic state where the stewardship-based leadership is likely to thrive (Alemu, 2025; Rosário and Boechat, 2025; Buhr and Weissbrod, 2026).

Altogether, the discussion supports the idea that Sustainable Stewardship Leadership may become the feasible course of implementing environmental responsibility not only as a strategic goal within the institution but as a permanent organizational phenomenon (Althnayan et al., 2022; Farag, 2026; Udin et al., 2025). Through coordinated leadership action bringing together organizational systems, leadership behavior, and engagement with stakeholders, SSL creates a sound basis of ecological resilience, institutional legitimacy, and future societal prosperity (Reyes et al., 2026; Twesigye et al., 2026; Rosário and Boechat, 2025).

The discussion overall indicates that sustainability issues are not solvable by single-handedly intervening in policy compliance, environmental reporting, and technological innovation. To change the environment in the long-term, it is necessary to align leadership behaviour, the organisational culture, governance,e and the stakeholder engagement process. In this context, the importance of SSL is its capacity to encompass these aspects in a unified institutional structure that fosters sustainable environmental management.

9. CONCLUSION

Today's environmental challenges are more complex, and leadership strategies are more essential to achieving sustainability in the organization. As more institutions implement sustainability policies, governance,e and environmental efforts, many still face challenges in moving towards long-term environmental stewardship. In this chapter, it has been suggested that sustainability is not just a technical and procedural problem, but is a behavioural, cultural, and institutional one.

In response to this challenge, the chapter developed a Sustainable Stewardship Leadership (SSL) framework as a way of working toward an institutional environmental transformation in an integrative manner. It highlights that ethical leadership, behavioural stewardship, organizational systems, and stakeholder participation are vital factors in enhancing environmental responsibility, irrespective of the context of the organization. Instead of seeing sustainability as a series of stand-alone projects, SSL views sustainability as a capacity of the institution, which can be developed into the organizational culture, governance, and decision-making process.

The chapter also shows that sustainability results will be more sustainable when the environment is part of the organisation's culture and not an obligation imposed from outside. Leadership behaviour, institutional learning, and systems-oriented governance can help organisations build their capacity to live and act environmentally responsibly through times of change, uncertainty, and leadership transition.

In summary, Sustainable Stewardship Leadership offers a way to comprehend how organisations can mature beyond a symbolic sustainability commitment to a more resilient, adaptive,e and enduring approach to environmental stewardship. The framework provides a theory and a concrete roadmap to institutions that want to integrate sustainability into key aspects of institutional life.

10. FUTURE RESEARCH DIRECTIONS

Although Sustainable Stewardship Leadership offers a conceptual framework to think about institutional environmental transformation, a number of opportunities exist for future research. Empirical research is required in the first place to validate the framework and explore the connections between the core dimensions of the framework in various organizational settings. This research would contribute to the implementation of the framework's applicability and explanatory power.

Second, longitudinal studies might investigate the effects of stewardship leadership on sustainability outcomes over time. The nature and process of developing, reinforcing, and sustaining environmental stewardship throughout an organization's lifecycle would give important clues to the effectiveness of the framework over time.

Third, comparative studies of other sectors/industries and within other national contexts could investigate the influence of cultural, institutional, and regulatory factors in the implementation of Sustainable Stewardship Leadership. Such investigations would help to develop a more sophisticated understanding of the ways in which stewardship practices change in different environments.

There is also potential for future studies to focus on how new technologies such as artificial intelligence, digital governance systems, and sustainability analytics can boost stewardship leadership and environmental decision-making. As more organizations turn to digital tools to power their sustainability efforts, the relationship between technological innovation and stewardship practices will be of growing significance.

Lastly, scientists can investigate ways in which the SSL framework can be used to tackle current challenges in the fields of climate adaptation, environmental resilience, transition towards a circular economy, and stakeholder-driven sustainability governance. The extension of the framework in these new domains would make a significant contribution to both theory and progress in sustainability leadership research.

REFERENCES

Alemu, B. A. (2025). Cultivating a Culture of Sustainability: The Role of Organizational Values and Leadership in Driving Sustainable Practices. *Business Ethics and Leadership*, *9*(1), 79–94. DOI: 10.61093/bel.9(1).79-94.2025

Althnayan, S., Alarifi, A., Bajaba, S., & Alsabban, A. (2022). Linking Environmental Transformational Leadership, Environmental Organizational Citizenship Behavior, and Organizational Sustainability Performance: A Moderated Mediation Model. *Sustainability (Basel)*, *14*(14), 8779. DOI: 10.3390/su14148779

Asad, M. (2024). Impact of Environmental Management on Sustainable Performance of Pakistani Entrepreneurial Firms: The Mediating Role of Green Product Innovation and the Moderating Effect of Transformational Leadership. *Sustainability (Basel)*, *16*(24), 10935. DOI: 10.3390/su162410935

Avery, G. C., & Bergsteiner, H. (2011). Sustainable leadership practices for enhancing business resilience and performance. *Strategy and Leadership*, *39*(3), 5–15. DOI: 10.1108/10878571111128766

Buhr, M., & Weissbrod, I. (2026). The Role of Individual Beliefs for Corporate Sustainability Transformations. *Business Ethics, the Environment & Responsibility*, *35*(1), 340–366. DOI: 10.1111/beer.12783

Farag, M. I. H. (2026). Sustainability as a Management Strategy: Integrating Environmental, Social, and Governance Practices into Business Administration. *Management Science Advances*, *3*(1), 20–44. DOI: 10.31181/msa31202630

Haile, E. A. (2025). The nexus between green transformational leadership and financial performance of hotels: The mediating effect of environmental management practices. *Journal of Hospitality and Tourism Insights*, *8*(4), 1416–1435. DOI: 10.1108/JHTI-05-2024-0524

Hanif, S., Ahmed, A., & Younas, N. (2023). Examining the impact of environmental management accounting practices and green transformational leadership on corporate environmental performance: The mediating role of green process innovation. *Journal of Cleaner Production*, *414*, 137584. DOI: 10.1016/j.jclepro.2023.137584

Ismail, I. J. (2025). Transforming small business ventures for a greener tomorrow: The interplay between green transformational leadership, organizational green culture, and environmental sustainability. *Cogent Social Sciences*, *11*(1), 2441400. Advance online publication. DOI: 10.1080/23311886.2024.2441400

Ivanova, E., Rimanoczy, I., & Singhal, D. (Eds.). (2026). *The Power of Arts-Based Pedagogies: Transforming Mindsets to Shape Our Future*. Taylor & Francis. DOI: 10.4324/9781003666325

Kebe, I. A., Liu, Y., & Kahl, C. (2025). Greener solutions: Green transformational leadership and sustainable environmental outcomes within Sierra Leone's telecom sector. *Journal of Cleaner Production*, *517*, 145811. DOI: 10.1016/j.jclepro.2025.145811

Liu, M., Zhao, J., & Liu, H. (2025). Digital transformation, employee and executive compensation, and sustained green innovation. *International Review of Financial Analysis*, *97*, 103873. DOI: 10.1016/j.irfa.2024.103873

Miao, Q., & Nduneseokwu, C. (2025). Applications of leadership theories and concepts in environmental leadership. In *Environmental leadership in a VUCA era: An interdisciplinary handbook* (pp. 237–275). Springer Nature Singapore.

Nduneseokwu, C. K., & Harder, M. K. (2023). Developing environmental transformational leadership with training: Leaders and subordinates' environmental behaviour outcomes. *Journal of Cleaner Production*, *403*, 136790. DOI: 10.1016/j.jclepro.2023.136790

Parvez, M. O., Kim, W. G., Patwary, A. K., & Zainol, N. A. (2026). Active participation matters: Impacts of green organizational identity on employees' voluntary green behaviors. *International Journal of Hospitality Management*, *133*, 104447. DOI: 10.1016/j.ijhm.2025.104447

Reyes, J. P., Malabanan, M., & Monterroyo, C. B. (2026). A Contemporary Evidence Synthesis of Leadership Styles and Their Impact on Organizational Performance: A Systematic Review. *Journal of Interdisciplinary Perspectives*, *4*(1), 38–45. DOI: 10.69569/jip.2025.707

Rosário, A. T., & Boechat, A. C. (2025). How Sustainable Leadership Can Leverage Sustainable Development. *Sustainability (Basel)*, *17*(8), 3499. DOI: 10.3390/su17083499

Sadiq, M., Leong, M. K., Jiayao, Z., & Siddiqui, K. (2026). The effect of perceived firm diversity, equity,y and inclusion (DEI) on employee wellbeing and green performance: A moderated mediated model. *Corporate Social Responsibility and Environmental Management*, *33*(1), 500–518. DOI: 10.1002/csr.70186

Seema, S., Saini, G., Mittal, A., & Dhankar, N. (2025). Green transformational leadership and green management practices-based environmental stewardship: A SEM model with mediation and parallel moderation effects. *Global Knowledge, Memory and Communication*. Advance online publication. DOI: 10.1108/GKMC-07-2024-0445

Senge, P. M. (2006). *The fifth discipline: The art and practice of the learning organization*. Broadway Business.

Sposato, M., & Carlos Dittmar, E. (2025). Leadership Strategies for Implementing Environmental Management Systems. In Ordóñez de Pablos, P., Lytras, M., Alkhaldi, A., & Malik, S. (Eds.), *Insights Into Digital Business, Human Resource Management, and Competitiveness* (pp. 277–300). IGI Global Scientific Publishing., DOI: 10.4018/979-8-3693-9440-3.ch012

Subaşı, E., Karatepe, O. M., Rezapouraghdam, H., & Kim, T. T. (2026). Green transformational leadership and green recovery performance: Green adaptability and green creativity as mediators and emotional intelligence as a moderator. *International Journal of Hospitality Management*, *133*, 104508. DOI: 10.1016/j.ijhm.2025.104508

Tapsir, S. H., Yusof, R., & Hashmi, A. (2026). Transformational leadership, Green HRM and top management commitment: Pathways to sustainability in Malaysian higher education. *Journal of Applied Research in Higher Education*, *18*(2), 478–494. DOI: 10.1108/JARHE-09-2024-0493

Twesigye, M., Obedgiu, V., Nkurunziza, G., & Oceng, P. (2026). The mediating role of stewardship behaviour on leadership competencies and sustainability of government projects in developing economies. *The International Journal of Organizational Analysis*, *34*(1), 143–164. DOI: 10.1108/IJOA-08-2023-3935

Udin, U., Chanthes, S., & Dananjoyo, R. (2025). Green transformational leadership and environmental performance: Insights from bibliometric analysis for future research agenda. *International Journal of Sustainable Development and Planning*, *20*(6), 2331. DOI: 10.18280/ijsdp.200606

Umar, S. B., Ahmad, J., Mohd Bukhori, M. A. B., Ali, K. A. M., & Hussain, W. M. H. W. (2025). Transforming Higher-Education-Intitutes: Impact of Change Management on Sustainable Performance Through Transformational Leadership and Knowledge Management. *Sustainability (Basel)*, *17*(6), 2445. DOI: 10.3390/su17062445

Vargas-Hernández, J. G. (2022). Implications Between Environmental Management and Green Transformational Leadership in Greening Organizations. In Ortiz-Rodriguez, F., Tiwari, S., Iyer, S., & Medina-Quintero, J. (Eds.), *Futuristic Trends for Sustainable Development and Sustainable Ecosystems* (pp. 165–190). IGI Global Scientific Publishing., DOI: 10.4018/978-1-6684-4225-8.ch011

Wegner, R. da S. Estivalete, V. de F. B., Matheis, T. K., & Vieira, K. M. (2025). The influence of transformational leadership on environmental sustainability and organizational citizenship behaviors (Version 1.0) [Data set]. Harvard Dataverse. https://doi.org/DOI: 10.7910/DVN/BTSE7R

Chapter 6
Sustaining Trust:
Intergenerational Leadership and Environmental Stewardship

K. C. Raja Shree
http://orcid.org/0009-0009-1029-910X
Saveetha School of Management, Saveetha Institute of Medical and Technical Sciences, India

Gigi G. S.
http://orcid.org/0009-0000-0552-256X
Saveetha School of Management, Saveetha Institute of Medical and Technical Sciences, India

P. Praveen Kumar
http://orcid.org/0000-0002-0707-039X
SRM Institute of Science and Technology, Vadapalani, India

ABSTRACT

This chapter explores the intersection of generational leadership along with green and ethical consumption, focusing on how brands navigate value-based preferences and financial performance across Gen Alpha, Gen Z, Millennials, and Gen X. It examines the distinctive ethical priorities and consumption behaviors of each generation, highlighting Gen Z's demand for authenticity and social responsibility, Millennials' balance between ideals and affordability, Gen X's evolving attitudes, and Gen Alpha's emerging influence. The chapter analyzes how leaders can develop purpose-driven strategies and ethical organizational cultures to build trust and loyalty among these diverse cohorts. Emphasis is placed on the necessity for brands to integrate ethical commitments with accessible pricing and seamless dig-

DOI: 10.4018/979-8-3373-7877-0.ch006

ital experiences to sustain financial performance. The discussion provides insights into managing multi-generational expectations, ensuring that ethical values are authentically embedded in brand practices and communications.

INTRODUCTION

In recent times, businesses are reshaped by a remarkable shift in terms of ethical and sustainable practices (Karuppiah et al., 2024). Consumers make moral choices based on the business practices followed by every brand (Kuokkanen & Sun, 2020). This global shift is rooted in the generational dynamics by shaping the behaviour, values, and leadership of individuals across generations. Unlike past eras, social media has a profound influence on the younger generations, who are typically inundated with a vast amount of information about everything. Consequently, the demand for eco-friendly and socially fair business practices by businesses is highly amplified by social media (Abbas et al., 2019). To understand this, it is essential to explore the generational layers that shape the global economy. Generation X acts as a catalyst between traditional and modern values.

In contrast, millennials advocate for sustainability, followed by Generation Z, who stand for ethical issues, and all these behaviors and values are furthermore deeply rooted in Generation Alpha. Each of these generational groups contributes towards unique ethical values, purchasing and consumption patterns, as well as leadership orientations leading to ethical branding towards environmental stewardship (Bucic et al., 2012). Considering these factors, this chapter explores the influence of generational leadership on aligning branding in the market, thereby contributing to financial metrics and growth. Therefore, this chapter focuses on the significance of an organisation being strong and financially viable, leading to better economic performance.

Theoretical Framework

This chapter encompasses ethical leadership, value systems based on generations, and brand financial outcomes, which are drawn from multiple theoretical foundations. The transformational leadership theory given by Bass, 2015 focuses on leaders' ability to motivate purpose and instill ethical values for driving collective vision. Ethical leadership catalyzes redefining the organizational culture, which is influenced by the theory of planned behaviour and explains consumers' attitudes and consumption patterns. Furthermore, Stakeholder theory propounded by Freeman, 2010 explains about multigenerational stakeholder needs and how the generational aspects possess various expectations in terms of social responsibility and trust towards organisations. Strauss-Howe Generational cohort theory propounded by William Strauss and Neil Howe. demonstrates these differences by advocating for the distinct value systems. The consumer perceptions can be drawn from customer-based brand equity by Keller, 1993, and the triple bottom approach by Elkington, 1997 embeds the sustainable outcomes by aligning the three p's: people, planet, and profit for societal good as well as financial viability.

Conceptual Framework

Figure 1. Conceptual model

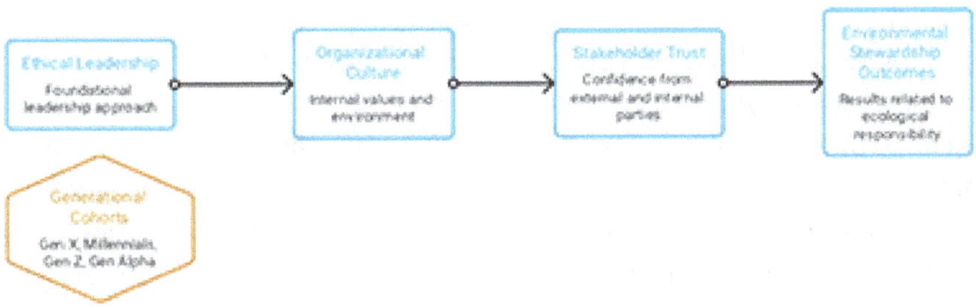

The proposed model illustrates the relationship between ethical leadership and ethical branding outcomes, mediated by organizational culture and stakeholder trust. This framework explains ethical values within leadership behavior, stakeholder perceptions, and ethical brand performance with financial sustainability. The model explains about ethical leadership and leaders' fairness, empathy and transparency in making decisions. Several transformational leadership theories say that leaders' motivational approach is aligned with moral values in par with organizational objectives, which result in teams' cohesiveness and brand image. These changes are mainly transferred through organizational culture, which is a mediator. Progressive cultural infrastructure flows through all channels vertically and horizontally, which demonstrates consistent moral behavior, open communication and authenticity. Another mediator, digital transformation strengthens real time transparency through AI, well documented data analytics and block chain. Consumers can easily verify their claims regarding sustainability and inclusivity, that are major factors of ethical branding. These mediators ensures trust and credibility. Psychological Contract Theory and Equity Theory say that fairness and reciprocity are important in gaining trust. The another mediator, stakeholder trust, is a major outcome due to internal ethical culture. It creates confidence among all affected people positively and visible results can be noted through ethical actions. Employee loyalty, advocacy and their engagement bring favorable outcomes such as enhanced brand equity, customer lifetime value, return on ethical investment (RoEI), and favorable ESG performance. Last mediator, generational cohorts interpret ethical leadership and brand communication through their respective their value orientations and technological familiarity. To be precise, Gen X looks on authenticity and reliability; while Millennials value purpose and inclusivity. Gen Z gives importance to activism, diversity, and digital transparency; while Gen Alpha seeks for ethical personalization through technology. All these variables clearly explain how brands can bring ethical intent and values in different generations.

METHODOLOGY

This chapter includes a conceptual research using narrative review employing literatures based on ethical leadership, generational cohorts, environmental stewardship and ethical branding. Also, relevant literatures have been studies from various databases to understand the emerging scope, theoretical perspectives as well as managerial implications. Therefore, this chapter develop an integrative concept associating the generational leadership, environmental stewardship, ethical branding and financial sustainability outcomes.

Generational Profiles and Environmental Values

Generational differences evolve according to various aspects, and they highly reflect the ethical values, leadership style they choose and purchasing behavior of individuals towards environmental stewardship (Vasanicova & Melnyk, 2025). Existing literature (Jorgensen, 2003) highlights that Generation X experienced a lot of economic fluctuations and rapid globalization; therefore, they show balance in pragmatism, which focuses on the application of ideas into actions after proper testing. Other literature (Lissitsa & Kol, 2016, 2021; Dhanapal et al., 2015; Parment, 2013) suggests that Generation X makes only calculative purchases, expecting long-term value for the products and services they purchase; moreover, they prefer in-store experiences over online shopping for major purchases. As leaders, they emphasize accountability and responsibility, giving more focus on fairness, expecting transitional corporate discipline, and they show more practicality than an idealistic view and the Generation X period ranges from 1965 - 1980 (DePew & Gonzales, 2020). Studies (DePew & Gonzales, 2020; Binu et al., 2024) advocate that the millennial era was from 1981 to 1996, and they showcase a value-driven ethical compass focusing on sustainability, transparency, and also inclusivity. Past literature (Jasrotia et al., 2023; Chatzopoulou & Kiewiet, 2021) highlights that they are digitally fluent and research-oriented, actively engaging in conscious purchasing and prioritizing ethical brands and eco-friendly product lines towards environmental stewardship. As leaders, they are at mid-level or senior levels. They are the ones reshaping the entire corporate governance, and they believe in flat hierarchies and collaboration within the organizations, and they embrace environmental as well as social impact in all the activities (Gopal & Pitts, 2025). The Gen Z era spanned from 1997 to 2012, characterized by a heightened awareness of the global community due to the pervasive connection and information-rich environment (Confetto et al., 2023). Various authors (Coman et al., 2022) recommend that they are activist-oriented based on the social media influences and have exposure to diversified aspects from a younger age. Studies (Theocharis & Tsekouropoulos, 2025; Confetto et al., 2023) suggest that consumers have zero tolerance for green washing and prefer brands that take visible stances. Greenwashing signifies a structural ethical catastrophe because it distorts environmental commitments, destabilizes stakeholder trust, and deteriorates organizational accountability. Their purchasing behavior is significantly influenced by social media information, but they expect ethical considerations to be a priority in all aspects. As leaders, they are still in their early careers, but they have high levels of networking and they try to redefine leadership through growing online communities (Hidayat et al., 2023; Gabrielova & Buchko, 2021). Generation Alpha is the current ongoing generation, from 2013. Studies (Knihova, 2024) suggest that Gen Alpha is growing in the AI world, where they are exposed to immersive media from childhood. Consequently, all their activities are based on the technology aspects they experience. The purchase pattern is highly influenced by AI-driven personalization and eco-innovation. As leaders, they would be human AI-integrated decision makers in the future, which would be normalized, and they would focus more on both environmental beneficence and humanitarian aspects (Oktay & Acaralp, 2025). Research (Stewart et al., 2017) demonstrates that businesses should understand generational behaviors to accelerate and serve in the evolving marketplace.

Figure 2. Generational shifts

Environmental Stewardship and Ethical Branding

Aligning the ethical values of an organisation with the financial goals transforms the purpose into a measurable performance. Ethical branding refers to building of brand based on dimensions such as honesty, sustainability, fairness, and social responsibility. It ensures the brand's values align with moral principles and stakeholders' well-being. Ethical branding reduces risk and fosters trust in brands, thereby enhancing resilience and revenue (Kang et al., 2023). Studies (Ali & Shahn, 2024) suggest that customer lifetime value (CLV) demonstrates the discounted profit an organisation expects from individual customers across relationships, and ethical practice increases the retention. Literature (Antonaras et al., 2011; Zouari-Hadiji & Chouaibi, 2021) advocates that return on ethical investment (RoEI) explains the net financial gains drawn from ethically driven initiatives and their cost. Authors (Ramirez et al., 2022; Pavlakova Docekalova & Kocmanova, 2018) highlight that ESG metrics explore the environment, social, and governance performance into decision-making indicators, which supports cost of capital advantages. Other studies (Garcia-Sanchez et al., 2019) depict that sustainability-linked disclosures improve the quality aspects and credibility. Pricing power outrage arises when perceptions of ethical value exceed the parity, such as transparent supply chains, reduced carbon footprints, inclusive practices, justifying the premium tiers and stabilising the prices towards environmental stewardship (Eyo-Udo et al., 2024). The generational expectations completely shape this elasticity, where Gen Z and millennials expect sustainability and are willing to pay for faster adoption, as stated by Gomes et al. (2023) and Brand et al. (2022). Other studies (Mhatre & Conger, 2011; Chhabra,

2010) show that Gen X gives importance to authenticity as well as reliability and expect durability. Ultimately, ethical branding is considered a moral imperative and market opportunity, aligning the corporate purpose and compounding the financial value with better societal outcomes through loyalty and pricing power (Hunt, 2019).

Strategic Environmental Leadership Across Generations

Strategic environmental leadership across generations needs the integration of intergenerational leadership models that emphasise workforce, inclusivity, agility, clear goals, sustainable environment, and ethical equity (Huyler et al., 2025). To connect the generations from boomers and GenX to Millennials, Gen Z and the future Gen Alpha, companies must foster inclusive decision-making by implementing youth advisory boards, intergenerational councils (Zurba et al., 2023), as well as reverse mentoring, which may help to bring in the latest insights from experienced leaders (Browne, 2021). Through this mechanism, organisations can foster empathy, alleviate generational differences, and enhance shared ownership of the brand. Due to the volatility, the branding landscape needs adaptive strategies (Murillo & Atristain-Suárez, 2023; Moşteanu, 2024). Gen Z, for example, expects digital platforms for engagement (Theocharis et al., 2025), while the older generation values consistency and reliability, making it crucial for brands to create sustainable innovations (Cubukcu Cerasi & Balcioglu, 2024). Building cross-generational communication platforms, such as online review forums and brand labs, enables meaningful conversations and co-creation, ensuring both cohesion and authenticity (Chaudhuri et al., 2022). Several researchers highlight the increasing importance of authentic, value-driven storytelling, especially for Gen Z and Alpha, who are highly inclined to emotional brand storytelling as well as value alignment. Brands that demonstrate transparency, showcase real people, and uncover both success and struggle paths are more likely to gain deep loyalty (Chopdar & Paul, 2024). Organisations should plan segment-specific strategies to manage brand trust and loyalty across all generations (Rachbini et al., 2025). Younger cohorts resonate with value-based moral behaviour and collective engagement (Weber & Urick, 2023), while older generations are attracted to dependable service, quality products, and maintenance of privacy. Ethical communication is the main factor in the digital-first age, especially as Gen Z seeks clarity about influencers and AI applications (Aru & Srivastava, 2025). When organisations fail to provide or humanise these elements, trust is eroded. Instead, brands can leverage micro influencers, authenticated user-generated content, and peer reinforcement. Finally, brands should align short-term winnings from social media campaigns and trend-based tactics with the sustainable development of ethical brand equity focusing on environmental stewardship. By integrating brand dealership in governance, responsiveness, and ethical storytelling with long-term purpose, organisations can achieve endurance through generational brand equity.

Figure 3. Strategic environment leadership framework

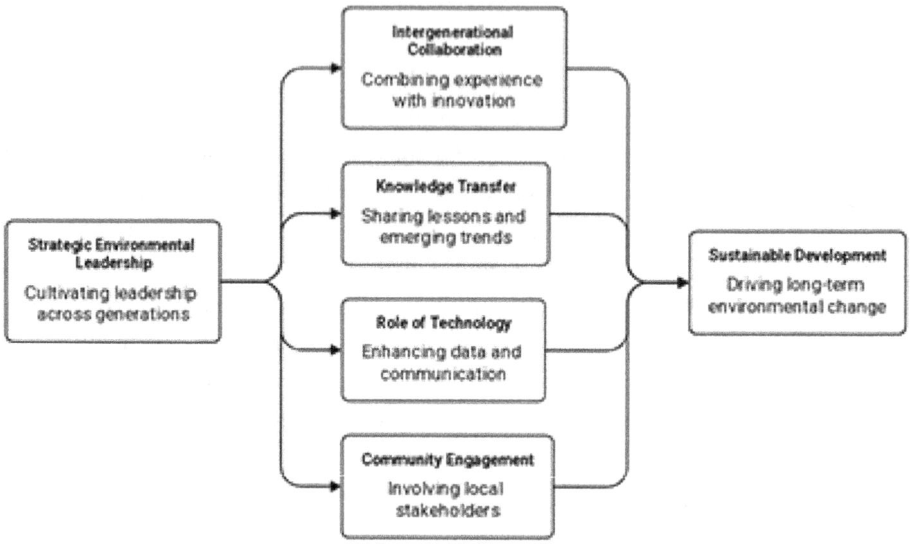

Digital Transformation and Environmental Transparency

Due to digital transformations, the dimensions of ethical consumption and consumer buying behaviour have been reframed across all generations (Djafarova & Foots, 2022; Joseph & Rao, 2022). From Baby Boomers to Gen Z, online platforms have become significant tools for branding and monitoring consumer brand ethics. Digital platforms enable consumers to evaluate, review, and share ethical information at a faster pace and on a larger scale (Tolegenov et al., 2024; Singh et al., 2024). Many researchers have found that ethical awareness can trigger brand reputational damage by exacerbating social and environmental issues and promoting peer-to-peer influence (Elshaer et al., 2025). However, social media platforms that publicize ethical awareness can also aggravate reputational damage via online exposure of deceitful practices (Chauhan et al., 2022). This twofold capacity of digital platforms puts brands under continuous digital scrutiny, with even minor errors potentially leading to significant criticism (Ong & Toh, 2023). Artificial Intelligence and advanced analytics using big data are now vital in brand communication in terms of ethical positioning (Mullins et al., 2021; Arora & Sharma, 2023). These digital tools allow advanced personalised messaging that connects with individual customer value systems, demographics, and psychographic behaviours (Theodorakopoulos & Theodoropoulou, 2024; Adeniran et al., 2024). Researchers have warned about privacy issues, data manipulation, and algorithmic bias resulting from increased personalisation and engagement (Bitra, 2025).

Young consumers are to be targeted with ethical applications in AI as they are tech-savvy and more responsive to the customised messages. But in reality, the risk of utilising psychological profiles for profit motive is higher than authentic value alignment continues as a primary concern (Williams et al., 2022). Simultaneously, digital transformation has increased the concept of cancel culture, particularly among Gen Z, as their purchase decisions are strongly tied to perceived brand values (Roldan et al., 2024; Costa & Azevedo, 2024; Djafarova & Foots, 2022). Several research studies show that while Gen Z customers are subject to negative environmental information regarding a brand, their purchase intentions decline severely (Zimand-Sheiner & Lissitsa, 2024; Sun & Xing, 2022; Tran et al., 2022). Gen Z also connects actively with eWOM, balancing both support and negative criticisms through online platforms (Perez-Aranda et al., 2024; Pillay, 2021). Considering these, brands should prioritize ethical consistency as an essential element (Akbar et al., 2024). To facilitate an ethical environment, many organisations are implementing real-time ethical declaration accelerated by blockchain and digital tools (Omisola et al., 2023; Udokwu et al., 2025). Research conducted in the manufacturing and retail sectors reveals that these technologies can improve eco-innovation, trust, and transparency (Di Chiacchio et al., 2024). Manufacturers show their supply chains as visible and transparent, and implement measurable sustainable goals, which demonstrate brands' accountability towards ethical values implementation (Mageto, 2021). This transparency creates credibility, especially among young consumers who demand real-time information and access. Considering these factors, companies must meet the ethical considerations of young generations (Djafarova & Foots, 2022; Hua et al., 2024).

While Gen Z looks for real-time interaction, story-driven content that focuses on inclusivity and activism (Rodrigues et al., 2024), older generations such as Gen X and Baby Boomers look for clarity, accessibility, and reliability (Sotelo-Duarte & Gonzalez-Cavazos, 2023; Charles & Duffett, 2025). Many studies show that Millennials and Gen Z give equal importance to ethical products, but Gen Z is more likely to express their value perception through digital platforms (Walters, 2021; Fraser et al., 2023). In reality, their financial limitations may limit their purchases ethically, showing a gap between values and real purchasing power that could reduce as the generation matures (Pokharel & Maharjan, 2024; Shaheen Hosany & Serdiuk, 2025). Ultimately, it can be argued that ethical consumption in the digital era largely hinges on brands' effective utilization of technology to cater to diverse generational expectations, maintain transparent communication, and foster authentic engagement (Lee, 2025; Halibas et al., 2025). Companies should build sustainable trust across all processes rather than only focusing on leveraging digital tools for marketing purposes towards environmental stewardship (Dadabaeva & Jamoliddinov, 2024; Gunduzyeli, 2024; Gopikrishna et al., 2023).

CROSS-CULTURAL AND REGIONAL PERSPECTIVES

The discussion on generational leadership and ethical consumption is conceptualized by Western-centric scholarship. Still, the global economy has cultural, socioeconomic, and political differences, which influence each generation's ethical values, their brand preferences and sustainability (Tkalac Vercic & Vercic, 2025). Even though Gen X, Millennials, Gen Z, and Gen Alpha generational categories have a general framework, their characteristics and behaviours towards the brands change across Western societies, Asian economies, and the Global South. These differences are caused by respective cultures, traditions, economic disparities, legal frameworks, as well as strategic developments in the markets. Learning these differences gives greater insights for brands to effectively direct cross-generational leadership and consumption globally.

Western Context

In Europe and North America, ethical consumption is connected with environmentalism, lifestyle preferences and political views (Hasan et al., 2023). Millennials and Gen Z of these geographical regions are greatly motivated by Fridays for Future movements, environmental strikes, which hold environmental sustainability as a moral obligation. Cultural foundations of these societies highlight consumer power, in the form of identity signaling. To consider the situation, European Millennials prefer locally sourced organic products that are fairly traded, which show environmental concerns and lifestyle alignment. Therefore, ethical leadership is highly needed for business organisations such as Unilever and IKEA, implementing circular economy models. In addition to this, the European Union's Green Deal and carbon disclosure requirements also emphasize upholding generational values. The United States is also not an exception; Gen Z consumers show sustainability-driven, corporate activism and inclusivity. These consumers expect the brands to be eco-friendly and also reflect on protecting racial justice, gender equality, and LGBTQ+ rights (Halibas et al., 2025). Nike, for instance, has gained the brand love of Gen Z through their marketing campaigns with Colin Kaepernick, which connect with social justice.

In contrast, Gen X, looks at pragmatic decision-making, sustainability, durability, reliability and authenticity rather than outspoken advocacy. In conclusion, Western countries' ethical consumption is closely knitted with personal identity, regulatory and sustainable frameworks, to ensure transparency-based sustainable business models and sustainable leadership.

Asian Context

Ethical consumption in Asian countries is driven by rapid modernization, collectivist cultural values, and economic diversity (Cui et al., 2022). Here, constant tension between traditional values and the globalization pressures is seen. Millennials and Gen Z's ethical consumption of Japan, South Korea, and China is drastically emerging and is also powered by technology and innovation. For instance, Chinese Gen Z consumers desire eco-innovative products like electric vehicles, while looking for digital transparency and blockchain-based supply chains. They also look for affordability and quality, as family responsibilities and financial security are the main social norms.

Ethical consumption in India has an outstanding social justice orientation. Millennials in this country mostly prefer affordability and cost-effectiveness, and also emphasise fair labour and social inclusivity. A classic example is the Tata Group, which earned consumer trust by incorporating corporate social responsibility, focusing on service values ("seva"). Simultaneously, Indian Gen Z is driven by social media activism to address unethical practices, especially gender equity, climate change and quick fashion, but their purchasing limitations do not end up in actual demand (Confetto et al., 2023). The case is different in Southeast Asia, especially Indonesia and Malaysia, where consumers' ethical leadership is formed through socio-cultural traditions, especially halal certification. Ethical consumption is the combination of both environmentalism and moral responsibility, which is driven by Islamic values. Gen Alpha in these countries are digitized yet rooted to cultural values, joining AI with traditional ethical norms. Asian Gen X is more conservative, looking for affordability and family well-being. They give more importance to long-term stability and community values. In conclusion, Asian generations have the culmination of traditional values, collectivism, affordability and innovation.

Global South Context

In Africa, Latin America, and parts of the Middle East, ethical consumption is framed by socioeconomic factors, namely economic inequality, low infrastructure and political instability. Generations give priorities to affordability and social justice, and give less importance to lifestyle ethical consumption (Seyfi et al.,

2025). Millennials and Gen Z in these countries possess high interest in social entrepreneurship and community-based consumption models. In these countries, Ethical consumption focuses more on product accessibility, an equal wage system and inclusivity rather than premiumness. For instance, Jumia's business practice is the integration of e-commerce fair trade along with local empowerment. In this context, basic social and economic factors are dominant over environmental factors and sustainable leadership.

Latin America focuses on activism and affordability. Gen Z and Millennials express their emotions through climate protests and anti-corruption, despite facing economic constraints. Consumers buy local products over global brands, giving importance to authenticity and regional identity. Thus, ethical leadership is more viewed as cultural respect and anti-exploitation operations rather than green branding. Coffee federations in Colombia and fair-trade campaigns in Brazil are the classic instances.

In the Middle East, religious values and governance factors dominate ethical consumption (Almalchi et al., 2024). Generational leadership preferences are dominated by cultural factors and charity (zakat), especially Millennials and Gen Z choose brands which are closely associated with religious endorsements. Recently, sustainable urban projects are on the rise in the Middle East, as Gen Z and Alpha prefer brands which are technology-driven connected to ethical frameworks. In the Global South, while choosing brands, Gen X is more cautious, stability, economic affordability and flexibility. The uncertainty towards global sustainability often reflects poverty and inequality, making the brands more realistic in managing ethics for survival.

COMPARATIVE INSIGHTS ACROSS REGIONS

There are striking contrasts and similarities in ethical consumption and leadership across generations. In western countries, consumers are connected closely with lifestyle, regulatory factors and activism, demanding transparency and collective action, while Asian societies are more realistic and collectivist culture, dominated by technology, culture, and family values, innovation as major factors. In the Global South, ethical consumption is deeply connected to social justice, product affordability, and ethics, focusing majorly on community empowerment over luxury sustainable products. In Western Millennials and Gen Z focus on corporate activism, while Asian consumers find a strong balance of innovation with collectivist ethics. Global South generational leaders majorly concentrate on flexibility and equitable growth, while Gen Alpha across all geographical regions shows strong digital attitude, and leadership in unpredictable ways.

Implications for Global Brand Strategy

By understanding these cross-cultural differences, brands should localize their ethical and sustainable strategies. Brands cannot go for a universal branding approach as it risks isolating certain consumer groups, especially not considering culture and economy. Multinational brands should frame global strategies by collaborating global ethical factors with local sensitivities. For example, European brands can incorporate carbon neutrality and sustainable innovation for Western Millennials, it must also prioritise fair pricing, regional employment, and community enhancing initiatives when doing business with African or Indian consumers. Similarly, Asian Gen Z demands tech-enabled ethical transparency, while Western Gen Z requires activism and fair accountability. By considering these regional distinctions, brands can form authentic cross-generational trust, improve financial viability, and bring ethical branding that resembles all cultural boundaries.

Challenges and Trade-Offs

Companies' primary challenge is to strike a balance between ethical premium pricing and affordability for Millennials and Gen Z, particularly for brands that aim to align social and economic purposes (Chatzopoulou & de Kiewiet, 2021). Customers' disposable income often prevents them from purchasing the ethical brands at higher prices, even though they have a strong interest in using sustainable fair trade products (Besharat et al., 2024; Chatzopoulou & de Kiewiet, 2021). Consumers are often unaware of the reasons behind higher product costs; therefore, companies should provide clear and transparent communication by explaining sustainable sourcing and fair labour practices (Chen, 2024; Turunen & Halme, 2021). To curb this issue, brands should be offered in tiered product lines and subscription-based selling models, which help in bridging the affordability gap and brand integrity (Prashad et al., 2025). However, fixing optimum pricing policies has become more complex in the digital marketplace, where surveillance over corporate responsibility is intensified. Consequently, one wrong turn may damage a brand's reputation (Piercy et al., 2010; Sanchez-Cartas & León, 2021; Seele et al., 2021). Another related challenge is curbing greenwashing accusations, where confused or exaggerated sustainability affirmations affect trust (Ben Mahjoub, 2025; Mousavi et al., 2024). Selective environmental promotions by covering up harmful practices result in low credibility that can damage brand equity (Barber et al., 2009; Yang et al., 2020). Gen Z is more environmentally informed and competent in identifying misleading remarks, creating transparency and independent verification needed for brand credibility (Ananda et al., 2025; Nazlim & Ozgen, 2024). This heightened mistrust must force brands to frame measurable sustainable metrics and communicate clearly to gain the confidence of the customers (Lee, 2025). These external pressures are due to resistance from traditional leadership that may be reluctant to move away from established cost-effective operational models. Implementation of new ethical frameworks often creates resistance to organizational change as there is friction between profitability and sustainability (Santos de Souza & Chimenti, 2024; Stouten et al., 2018). Companies also need adaptive leadership approaches for this approach (Ysoff, 2025). In some situations, senior leaders may undervalue the long-term advantages of practicing ethical policies, in favour of short-term economic gains, slowing down transformation efforts.

Added to these, cross-generational communication and brand positioning are also ongoing barriers. Each generation has different media habits, its value systems, and brand loyalty factors (Emini & Zeqiri, 2021). Designing a single logical brand perspective that represents across all these generations needs a multifaceted approach, adaptive messaging styles, and a proper understanding of generational imperatives (Aksoy, 2023). For example, older generations may respond more positively to traditional forms of advertising and culturally driven brand stories, while younger generations always seek interactive, open, and socially relevant messages. Breaking down in bridging these gaps risks undermining brand identity and weakening market positioning (Jaas, 2022). Thus, nurturing intergenerational collaboration and transparent evidence-based ethical branding practices becomes fundamental (Sergi, 2024). Considering all these brands should address affordability issues, eliminate greenwashing, reduce leadership resistance, and customize communication channels through different age groups, brands can overcome these challenges while upholding authenticity and building sustainable long-term relationships with inclusive consumer groups (Lee, 2025; Ines & Moreira, 2025).

Financial Models and Measurement Tools

Ethical branding has become an essential factor for long-term success in today's business world. Consumers expect businesses to show financial performance but also to showcase responsibility towards the environment and governance. Organisations need to frame robust, financial models, as well as measurement techniques that are connected with ethical practices and positive outcomes. These measurement tools provide the evaluation of sustainability, customer loyalty, and social impacts, and they also help to communicate value creation to its customers and communities. In this context, generational leadership has a prominent role in different age groups such as Baby Boomers, Generation X, Millennials, Generation Z, and Generation Alpha. The financial model, such as Return on Ethical Investment, Customer Lifetime Value, ESG scores, and frameworks like SASB, Integrated Reporting, and the Global Reporting Initiative, helps in understanding the space between ethical responsibility and financial accountability (Akpan & Oluwagbade, 2023).

Return on Ethical Investment (RoEI) is the traditional form of return on investment, by understanding that ethical practices bring both financial and tangible value. Conventional ROI measures the profitability in terms of costs, while RoEI also includes the investments such as sustainable procurement, fair labour, and carbon reduction benefits that are not immediately measurable but contribute to the sustainable equity and resilience (Ooi, 2025). To understand this, a brand that invests in sustainable energy sources may face high capital expenses, but they can reduce the future risks such as regulatory factors, fluctuating fuel prices and corporate damage. Likewise, a clothing company that has shifted its business strategy to organic cotton may initially forgo its margins, but receives high customer trust, brand, goodwill, pricing, power, and strengthened market positioning. Millennials and Gen Z value such brands that show commitments, ethical responsibilities; in turn, their loyalty is converted to measurable returns. Baby boomers and Gen X often view RoEI as a tool for risk mitigation, business stability and flexibility during vulnerable business cycles and sustainable leadership.

Another important tool in ethical branding is customer lifetime value, which quantifies the total financial contribution of a consumer throughout their connection with the brand. Conventionally, CLV is measured by considering average purchase value, frequency, and customer lifespan. In the current Global scenario, ethical branding is incorporated, and CLV is calculated by adding more loyalty factors to the existing ones, such as trust, shared values, and emotional connection. Businesses that integrate sustainability into their regular operations could see a drastic change in CLV because customers are engaged in a sustainable relationship aligned with the brand's values (Ali & Shabn, 2024). For instance, Patagonia customers have been highly committed to the brand for decades due to its environmental activism. They also act as brand ambassadors who extend their lifetime value through word-of-mouth marketing. Millennials and Gen Z, show higher CLV when ethical leadership is strong, as they remain committed to the businesses that value social and environmental factors. Boomers and Gen X initially favour product, quality and price, but once they receive authenticity, they become highly loyal.

ESG scores have become a prominent tool for evaluating corporate performance (Kao, 2023). ESG frameworks evaluate a company's operations, such as environmental governance, fair labour practices, diversity, and ethical transparency. These scores directly influence investors' decisions as it is related to both ethical and financial concerns. High ESG-rated companies have lower capital costs, efficient risk management, and a greater reputational image. For instance, Unilever ESG scores are consistently high due to its commitment towards long-term sustainability, which attracts all generational cohorts. In contrast, Tesla pinpoints the importance of ESG scoring as it performs environmental innovation, but encounters challenges in governance and fair labour practices. Millennials and Gen Z, ESG factors are not negotiable, while they prefer buying shares. Older investors are more vigilant and look at ESG scores as a risk mitigation tool and a long-term financial growth.

The Sustainability Accounting Standards Board has developed standards that give importance to sustainability factors and financial performance within each sector. For instance, the pharmaceutical sector metrics focus on access to medicine and patient safety. While in the textile sector, importance is given to supply chain, fair labour practices and raw materials sourcing. Investors who need complete knowledge of material risk and opportunities, rather than widened ethical statements, would prefer SASB guidelines. By implementing these standards, brands can upgrade their credibility with all stakeholders that addressing both ethical and financial concerns for generational leadership. This framework emphasizes measurable industry-specific metrics, especially for boomers and Gen X investors looking for accountability, while the younger generation demands transparency and authenticity.

Along with the above approaches, Integrated Reporting (IR) and Global Reporting Initiative (GRI) give a holistic view of value creation (Nicolo et al., 2023). IR includes financial, social, human, intellectual, and capital resources, which a company uses for creating value. This approach resembles Millennials and Gen Z, who place value on financial outcomes and society's well-being. For instance, Novo Nordisk uses integrated reporting to connect with its diabetes care initiatives for long-term financial growth and reputation. On the other hand, GRI focuses on multi-stakeholder accountability, through guidelines from fair labour practices to biodiversity. GRI reporting of Starbucks on their coffee sourcing and employee welfare shows the company's customer trust, investor satisfaction, revealing the younger generation's need for transparency and the older generation's need for compliance and benchmarking (Baruah et al., 2024).

These measurement tools are closely connected with evolving investor trends. 90% of millennials take investment decisions based on financial returns and social or environmental impact, while 60% of boomers do it. There is a high growth in the purchase of green and sustainable bonds, which shows a structural shift in businesses towards supporting projects such as sustainable energy, affordable housing, and access towards clean water. Green bonds are bought for an estimated amount of more than $2 trillion, especially by young investors who seek sustainable climate change. Millennials and Gen Z prefer transparent and technology-driven platforms providing ESG funds, while boomers and Gen X look for sustainable assets that manage long-term risks. Along with these advantages, these tools come with challenges as well. Greenwashing is the growing problem where overemphasizing ethical commitments reduces trust, especially among Gen Z customers, who are quicker in identifying inconsistencies. Even though regulatory frameworks such as SASB, IR and GRI are formed, issues still exist due to incomparability and standardization, making it more tough for stakeholders to make decisions across sectors and geographical regions. Another problem is the temporal dimension; ethical investments are of high cost in the short term, as investors have to show more patience to generate long-term value, which is a crucial issue. To address these gaps, effective ethical leadership strategies are to be framed, including generational cohorts. Furthermore, technology and generational convergence will influence financial models in ethical branding. Block chain technology provides measures for supply and transparency, and ESG claims, while AI delivers improved predictive analytics in risk assessment. Ethical financial measures will become standard expectations as Generation Alpha matures in a digital-first and climate-conscious environment. Ethical branding will become an essential factor for organizations for market survival, for quick adaptation to the constant changes and sustainable leadership.

Operational Insights

The alignment of generational leadership and ethical consumption shows a gradual shift for companies looking to align value-based brand choices with long-term financial performance. Enhancing cross-generational leadership boards by corporate leaders is essential, as it merges the activism and sustainability objectives of Gen Z and Gen Alpha with the long-term foresight and financial utilitarianism of Millennials and Gen X. This open-minded leadership approach strengthens decision-making patterns and ensures organizations remain agile, while being ethically accountable in fast-moving markets. Marketing experts thereby create sophisticated ethical policies that resonate across all generational cohorts. Digital-first gamified ethical campaigns engage Gen Alpha, while Gen Z looks on activism-based branding and corporate social responsibility. Meanwhile, Millennials look on authenticity as well as lifestyle factors, whereas Gen X focuses on trust, best quality, and value. Tailoring communication strategies based on these differences improves their trustworthiness and brand loyalty. Financial and regulatory policies also should be evolved to accommodate diverse generations. Companies should integrate environmental, Social, and Governance (ESG) criteria and Return on Ethical Investment (RoEI) with their traditional platforms, helping organizations to create equilibrium in social and economic motives for meeting the demands of ethically conscious generations. Strategic leaders can emphasize these policies to minimize greenwashing by enforcing mandatory transparent reporting to increase ethical sustainability. Finally, academicians and researchers play a crucial role by writing more literature on generational branding, sustainable socioeconomic policies, and ethical consumption, thereby connecting theory and practice to demonstrate the right path for business leaders in driving value-based consumer behavior. These integrated recommendations never undermine the importance of inclusivity, genuineness, and transparency in nurturing cross-generational relationships that enhance both authenticity and sustainable growth.

Practical Implications

This article provides many practical insights for policymakers, strategists, managers who aim at bringing sustainable ethical leadership across generational cohorts. Organizations can implement ethical leadership practices through interventions to strengthen ethical performance and sustainable brand equity. For this, organizations should cultivate ethical leadership practices in all managerial levels by conducting leadership training programs such as reflective practices, real time business case studies and ethical simulations. Organizations should embed ethical culture in to their systems through mission statement, HR policies and technologically enabled communication platforms. Cross-functional ethics committees or "integrity councils" can also be implemented to ensure authenticity in business operations. Organizations can reduce the gap between organizational ethics and brand success by building transparent and fair stakeholder trust, this can be achieved through ethical audits, grievance redressal mechanisms and open stakeholder communication. Leverage digital transformation to ensure ethical visibility by using block chain technologies, AI and social media platforms. Organizations can create digital dashboards that connect with performance metrics namely carbon footprint reduction, gender equality, community development, to increase brand reputation. Adding to this, adapt strategies for generational diversity, integrate ethical metrics in to brand performance evaluation and proper decision making during ethical ambiguity are to be done by organizations to ensure fairness across generational cohorts. By incorporating all these measures, organizations can attain sustainable brand success in the global marketplace.

CONCLUSION

The future of business is evolving not only by defining innovation and profits, but also by its capability to extend its goals towards ethical values across generational expectations. Companies should consider generational alignment as a strategic imperative and understand the expectations, ethical sensitivities, as well as consumption patterns. To sustain themselves in the competitive market, businesses should design themselves to cater to all groups through inclusive ethical frameworks. Ethical branding and consumption have emerged as a moral commitment and also a market opportunity. Once, this was an idealistic view, but the latest generations expect businesses to be eco-friendly and socially fair. Therefore, organisations need to create shared values, ethical agility, as well as adaptive leadership for reinforcing sustainable and equitable business practices. This would yield them profitability along with brand trust. Furthermore, together, a sustainable global economy can be created. In conclusion, businesses should utilize this strategic pathway to enhance both their operations and societal well-being.

REFERENCES

Abbas, J., Mahmood, S., Ali, H., Ali Raza, M., Ali, G., Aman, J., Bano, S., & Nurunnabi, M. (2019). The effects of corporate social responsibility practices and environmental factors through a moderating role of social media marketing on sustainable performance of business firms. *Sustainability (Basel), 11*(12), 3434. DOI: 10.3390/su11123434

Adeniran, I. A., Efunniyi, C. P., Osundare, O. S., & Abhulimen, A. O.Ibrahim Adedeji AdeniranChristianah Pelumi EfunniyiOlajide Soji OsundareAngela Omozele Abhulimen. (2024). Transforming marketing strategies with data analytics: A study on customer behavior and personalization. *International Journal of Management & Entrepreneurship Research*, *6*(8), 41–51. DOI: 10.56781/ijsret.2024.4.1.0022

Akbar, M. A., Khan, A. A., Mahmood, S., Rafi, S., & Demi, S. (2024). Trustworthy artificial intelligence: A decision-making taxonomy of potential challenges. *Software, Practice & Experience*, *54*(9), 1621–1650. DOI: 10.1002/spe.3216

Akpan, J. U., & Oluwagbade, O. (2023). Social and environmental responsibility in accounting: Beyond financial metrics. *International Journal of social sciences and management research, 9*(9), 163-188. https://doi.org/DOI: 10.56201/ijssmr.v9.no9.2023.pg148.173

Aksoy, C. (2023). Digital business ecosystems: An environment of collaboration, innovation, and value creation in the digital age. *Journal of business and trade, 4*(2), 156-180. DOI: 10.58767/joinbat.1358560

Ali, N., & Shabn, O. S. (2024). Customer lifetime value (CLV) insights for strategic marketing success and its impact on organizational financial performance. *Cogent Business & Management, 11*(1), 2361321. DOI: 10.1080/23311975.2024.2361321

Almalchi, H. N. H., Abedin, B., Safari, M., Sani, M. A., & Mat, A. (2024). Identifying Factors Affecting Ethical Consumption among Customers: Evidence from Iraq. *Pakistan Journal of Life and Social Sciences*, *22*(2). Advance online publication. DOI: 10.57239/PJLSS-2024-22.2.00190

Ananda, A. R., Putra, M. N. T., & Azhar, M. I. Y. (2025). Virtual Influencers, Real Impact: A Narrative Review on Credibility, Generational Trust, and Purchase Intention in Digital Marketing. *Sinergi International Journal of Communication Sciences*, *3*(2), 99–112. DOI: 10.61194/ijcs.v3i2.690

Antonaras, A., Iacovidou, M., & Memtsa, C. (2011). Measuring social return on investment using the EBEN GR Business Ethics Excellence Model. *Current Issues of Business and Law*, *6*(1), 69–89. DOI: 10.5200/1822-9530.2011.04

Arora, M., & Sharma, R. L. (2023). Artificial intelligence and big data: Ontological and communicative perspectives in multi-sectoral scenarios of modern businesses. *foresight, 25*(1), 126-143. DOI: 10.1108/FS-10-2021-0216

Aru, S., & Srivastava, A. (2025). The role of ai in shaping influencer marketing strategies for genz and millennials on instagram. *Learnovate-International*, 1-7. https://doi.org/DOI: 10.70604/learnint.v2i1.34

Barber, N., Taylor, C., & Strick, S. (2009). Wine consumers' environmental knowledge and attitudes: Influence on willingness to purchase. *International Journal of Wine Research*, 59-72. https://doi.org/DOI: 10.2147/IJWR.S4649

Baruah, A., Shaikh, M., Kumar, R. M., Shaikh, I. A. K., & Thomas, S. N. (2024, April). Artificial intelligence influence on leadership styles in human resource management for employee engagement. In *2024 Ninth International Conference on Science Technology Engineering and Mathematics (ICONSTEM)* (pp. 1-5). IEEE. DOI: 10.1109/ICONSTEM60960.2024.10568819

Ben Mahjoub, L. (2025). Greenwashing practices and ESG reporting: An international review. *The International Journal of Sociology and Social Policy*, *45*(1/2), 173–188. DOI: 10.1108/IJSSP-08-2024-0365

Besharat, A., Nardini, G., & Mesler, R. M. (2024). Bringing ethical consumption to the forefront in emerging markets: The role of product categorization. *Journal of Business Ethics*, *191*(4), 777–792. DOI: 10.1007/s10551-024-05667-2

Binu Raj, A., Subramani, A. K., & Jan, N. A. (2024). Faculty engagement, quality of work-life, organizational commitment and spiritual leadership–examining the mediation and moderation effects. *Industrial and Commercial Training*, *56*(1), 41–52. DOI: 10.1108/ICT-03-2023-0018

Bitra, S. K.Sai Kumar Bitra. (2025). Ethical AI and privacy in digital personalization: Balancing personalization and user trust. *World Journal of Advanced Engineering Technology and Sciences*, *15*(2), 774–779. DOI: 10.30574/wjaets.2025.15.2.0596

Brand, B. M., Rausch, T. M., & Brandel, J. (2022). The importance of sustainability aspects when purchasing online: Comparing generation X and generation Z. *Sustainability (Basel)*, *14*(9), 5689. DOI: 10.3390/su14095689

Browne, I. (2021). Exploring Reverse Mentoring;" Win-Win" Relationships in The Multi-Generational Workplace. *International Journal of Evidence Based Coaching and Mentoring*, *15*. Advance online publication. DOI: 10.3390/su14095689

Bucic, T., Harris, J., & Arli, D. (2012). Ethical consumers among the millennials: A cross-national study. *Journal of Business Ethics*, *110*(1), 113–131. DOI: 10.1007/s10551-011-1151-z

Charles, J., & Duffett, R. G. (2025). *The role of brand awareness and trust on purchase intent in Google shopping ads and demographic factor influences among Millennials and Generation Z*. BP International., DOI: 10.9734/bpi/mono/978-93-49238-37-4/CH9

Chatzopoulou, E., & de Kiewiet, A. (2021). Millennials' evaluation of corporate social responsibility: The wants and needs of the largest and most ethical generation. *Journal of Consumer Behaviour*, *20*(3), 521–534. DOI: 10.1002/cb.1882

Chaudhuri, S., Yoon, S. W., Hagen, M., & Legrand, A. (2022). Inter-generational developmental network and the impact of technology on knowledge creation and sharing. In *HRD Perspectives on Developmental Relationships: Connecting and Relating at Work* (pp. 123–150). Springer International Publishing., DOI: 10.1007/978-3-030-85033-3_6

Chauhan, R. S., Connelly, S., Howe, D. C., Soderberg, A. T., & Crisostomo, M. (2022). The danger of "fake news": How using social media for information dissemination can inhibit the ethical decision making process. *Ethics & Behavior*, *32*(4), 287–306. DOI: 10.1080/10508422.2021.1890598

Chen, C. W. (2024). Utilizing a hybrid approach to identify the importance of factors that influence consumer decision-making behavior in purchasing sustainable products. *Sustainability (Basel)*, *16*(11), 4432. DOI: 10.3390/su16114432

Chhabra, D. (2010). Back to the past: A sub-segment of Generation Y's perceptions of authenticity. *Journal of Sustainable Tourism*, *18*(6), 793–809. DOI: 10.1080/09669582.2010.483280

Chopdar, P. K., & Paul, J. (2024). The impact of brand transparency of food delivery apps in interactive brand communication. *Journal of Research in Interactive Marketing*, *18*(2), 238–256. DOI: 10.1108/JRIM-12-2022-0368

Coman, I. A., Yuan, S., & Tsai, J. Y. (2022). Toward an audience-centric framework of corporate social advocacy strategy: An exploratory study of young consumers from Generation Z. *Sustainability (Basel)*, *14*(7), 4099. DOI: 10.3390/su14074099

Confetto, M. G., Covucci, C., Addeo, F., & Normando, M. (2023). Sustainability advocacy antecedents: How social media content influences sustainable behaviours among Generation Z. *Journal of Consumer Marketing*, *40*(6), 758–774. DOI: 10.1108/JCM-11-2021-5038

Costa, C., & Azevedo, A. (2024). Antecedents and consequences of the "cancel culture" firestorm journey for brands: Is there a possibility for forgiveness? *Journal of Marketing Management, 40*(3-4), 289–312. DOI: 10.1080/0267257X.2023.2266465

Cubukcu Cerasi, C., & Balcioglu, Y. S. (2024). Bridging Generations and Values: Understanding Generation Z's Organizational Preferences and the Mediating Role of Sustainability and Innovation Attitudes in Turkey. *Administrative Sciences, 14*(9), 229. DOI: 10.3390/admsci14090229

Cui, G. Q., Zeng, J. Y., & Jin, C. H. (2022). The impact of vertical/horizontal individualism and collectivism on ethical consumption. *Sustainability (Basel), 14*(21), 14254. DOI: 10.3390/su142114254

Dadabaeva, R. A., & Jamoliddinov, F. S. (2024). Digital marketing as an element of sustainable development: Trends, challenges and opportunities. *Digital Models and Solutions, 3*(1), 65–79. DOI: 10.29141/2949-477X-2024-3-1-6

DePew, R., & Gonzales, G. (2020). Differences in health outcomes between millennials and generation X in the USA: Evidence from the National Health Interview Survey. *Population Research and Policy Review, 39*(3), 605–616. DOI: 10.1007/s11113-019-09563-w

Dhanapal, S., Vashu, D., & Subramaniam, T. (2015). Perceptions on the challenges of online purchasing: A study from "baby boomers", generation "X" and generation "Y" point of views. *Contaduría y Administración, 60*, 107–132. DOI: 10.1016/j.cya.2015.08.003

Di Chiacchio, L., Martínez-Caro, E., Cegarra-Navarro, J. G., & Garcia-Perez, A. (2024). Eco-innovation in the textile industry: The strategic importance of data privacy and environmental knowledge management. *Kybernetes, 53*(10), 3444–3461. DOI: 10.1108/K-02-2023-0222

Djafarova, E., & Foots, S. (2022). Exploring ethical consumption of generation Z: Theory of planned behaviour. *Young Consumers, 23*(3), 413–431. DOI: 10.1108/YC-10-2021-1405

Elshaer, I. A., Azazz, A. M., Fayyad, S., Kooli, C., Fouad, A. M., Hamdy, A., & Fathy, E. A. (2025). Consumer boycotts and fast-food chains: Economic consequences and reputational damage. *Societies (Basel, Switzerland), 15*(5), 114. DOI: 10.3390/soc15050114

Emini, A., & Zeqiri, J. (2021). Social media marketing and purchase intention: Evidence from Kosovo. *Ekonomska Misao i Praksa, 30*(2), 475–492. DOI: 10.17818/EMIP/2021/2.8

Eyo-Udo, N. L., Mokogwu, C., Olufemi-Phillips, A. Q., & Adewale, T. T. (2024). Developing ethical frameworks for sustainable food pricing through supply chain transparency. *International Journal of Innovation and Scientific Research, 11*(12), 919–947. DOI: 10.51244/IJRSI.2024.11120081

Fraser, J. R., Chung, M., & Cheon, H. J. (2023). Ethical consumption in the digital age: Analyzing benefit types, temporal distance, and normative factors for Gen Z. [GBFR]. *Global Business & Finance Review, 28*(3), 50–67. DOI: 10.17549/gbfr.2023.28.3.50

Gabrielova, K., & Buchko, A. A. (2021). Here comes Generation Z: Millennials as managers. *Business Horizons, 64*(4), 489–499. DOI: 10.1016/j.bushor.2021.02.013

Garcia-Perez, A. (2024). Eco-innovation in the textile industry: The strategic importance of data privacy and environmental knowledge management. *Kybernetes, 53*(10), 3444–3461. DOI: 10.1108/K-02-2023-0222

Garcia-Sanchez, I. M., Hussain, N., Martínez-Ferrero, J., & Ruiz-Barbadillo, E. (2019). Impact of disclosure and assurance quality of corporate sustainability reports on access to finance. *Corporate Social Responsibility and Environmental Management, 26*(4), 832–848. DOI: 10.1002/csr.1724

Gomes, S., Lopes, J. M., & Nogueira, S. (2023). Willingness to pay more for green products: A critical challenge for Gen Z. *Journal of Cleaner Production, 390*, 136092. DOI: 10.1016/j.jclepro.2023.136092

Gopal, S., & Pitts, J. (2025). ESG Integration: Unveiling Risk and Driving Innovation in Sustainable Finance. In *The FinTech Revolution: Bridging Geospatial Data Science, AI, and Sustainability* (pp. 35-81). Cham: Springer Nature Switzerland. DOI: 10.1007/978-3-031-74418-1_2

Gopikrishna, M., Burri, R. D., Kumar, N. M., Mahesh, D., Sankar, G. S., & Rao, N. K. K. (2023, October). Green Marketing Analysis for significant Investment Utilizing NLP Process. In *2023 International Conference on New Frontiers in Communication, Automation, Management and Security (ICCAMS)* (Vol. 1, pp. 1-8). IEEE. DOI: 10.1109/ICCAMS60113.2023.10525932

Gunduzyeli, B. (2024). Artificial Intelligence in Digital Marketing Within the Framework of Sustainable Management. *Sustainability (Basel), 16*(23), 10511. DOI: 10.3390/su162310511

Halibas, A., Akram, U., Hoang, A. P., & Thi Hoang, M. D. (2025). Unveiling the future of responsible, sustainable, and ethical consumption: A bibliometric study on Gen Z and young consumers. *Young Consumers*, *26*(7), 142–171. DOI: 10.1108/YC-11-2024-2327

Hasan, S., Wooliscroft, B., & Ganglmair-Wooliscroft, A. (2023). Drivers of ethical consumption: Insights from a developing country. *Journal of Macromarketing*, *43*(2), 175–189. DOI: 10.1177/02761467231168045

Hidayat, D., Tjandra, E. U., Herawati, N., & Simajuntak, M. R. A. (2023). Gen Z digital leadership through social media. *Widyakala Journal: Journal of Pembangunan Jaya University*, *10*(2), 62–71. DOI: 10.36262/widyakala.v10i2.779

Hua, S., Jin, S., & Jiang, S. (2024). The limitations and ethical considerations of chatgpt. *Data Intelligence*, *6*(1), 201–239. DOI: 10.1162/dint_a_00243

Hunt, S. D. (2019). The ethics of branding, customer-brand relationships, brand-equity strategy, and branding as a societal institution. *Journal of Business Research*, *95*, 408–416. DOI: 10.1016/j.jbusres.2018.07.044

Huyler, D., Gomez, L., Rocco, T. S., & Plakhotnik, M. S. (2025). Leading different generational cohorts in the workplace: Focus on situational and inclusive leadership. *New Horizons in Adult Education and Human Resource Development*, *37*(1), 6–19. DOI: 10.1177/19394225241297230

Ines, A., & Moreira, A. C. (2025). Exploring the Intersection of Environmental Sustainability and Anti-Consumption: A Review and Research Agenda. *Journal of Consumer Behaviour*, *24*(4), 2121–2142. Advance online publication. DOI: 10.1002/cb.2513

Jaas, A. (2022). E-marketing and its strategies: Digital opportunities and challenges. *Open Journal of Business and Management*, *10*(2), 822–845. DOI: 10.4236/ojbm.2022.102046

Jasrotia, S. S., Darda, P., & Pandey, S. (2023). Changing values of millennials and centennials towards responsible consumption and sustainable society. *Society and Business Review*, *18*(2), 244–263. DOI: 10.1108/SBR-01-2022-0013

Jorgensen, B. (2003). Baby Boomers, Generation X and Generation Y? Policy implications for defence forces in the modern era. *Foresight*, *5*(4), 41–49. DOI: 10.1108/14636680310494753

Joseph, S., & Rao, C. B. N. (2022). Transformation leadership and work performance: Mediating role of faculty engagement in higher educational institutions during COVID-19 pandemic. *International Journal of Management Education, 16*(6), 664–680. DOI: 10.1504/IJMIE.2022.126588

Kang, J., Faria, A. A., Lee, J., & Choi, W. J. (2023). Will consumers give us another chance to bounce back? Effects of precrisis commitments to social and product responsibility on brand resilience. *Journal of Product and Brand Management, 32*(6), 927–941. DOI: 10.1108/JPBM-03-2022-3899

Kao, F. C. (2023). How do ESG activities affect corporate performance? *MDE. Managerial and Decision Economics, 44*(7), 4099–4116. DOI: 10.1002/mde.3944

Karuppiah, K., Virmani, N., & Sindhwani, R. (2024). Toward a sustainable future: Integrating circular economy in the digitally advanced supply chain. *Journal of Business and Industrial Marketing, 39*(12), 2605–2619. DOI: 10.1108/JBIM-12-2023-0742

Knihova, L. (2024). AI-Powered Immersive Learning: Transforming Tertiary Education. DOI: 10.24132/jbt.2024.14.2.14_24

Kuokkanen, H., & Sun, W. (2020). Companies, meet ethical consumers: Strategic CSR management to impact consumer choice. *Journal of Business Ethics, 166*(2), 403–423. DOI: 10.1007/s10551-019-04145-4

Lee, G. K. S. (2025). Building Brands for a Sustainable Future: Integrating Strategy, Consumer Response and Ethical Communication. *Asian Journal of Economics. Business and Accounting, 25*(5), 10–9734. DOI: 10.9734/ajeba/2025/v25i51800

Lissitsa, S., & Kol, O. (2016). Generation X vs. Generation Y–A decade of online shopping. *Journal of Retailing and Consumer Services, 31*, 304–312. DOI: 10.1016/j.jretconser.2016.04.015

Lissitsa, S., & Kol, O. (2021). Four generational cohorts and hedonic m-shopping: Association between personality traits and purchase intention. *Electronic Commerce Research, 21*(2), 545–570. DOI: 10.1007/s10660-019-09381-4

Mageto, J. (2021). Big data analytics in sustainable supply chain management: A focus on manufacturing supply chains. *Sustainability (Basel), 13*(13), 7101. DOI: 10.3390/su13137101

Mhatre, K. H., & Conger, J. A. (2011). Bridging the gap between gen x and gen y: Lessons from authentic leadership. *Journal of Leadership Studies, 5*(3), 72–76. DOI: 10.1002/jls.20235

Moşteanu, N. R. (2024). Adapting to the unpredictable: Building resilience for business continuity in an ever-changing landscape. *European Journal of Theoretical and Applied Sciences*, *2*(1), 444–457. DOI: 10.59324/ejtas.2024.2(1).37

Mousavi, K., Kowsari, E., Ramakrishna, S., Chinnappan, A., & Gheibi, M. (2024). A comprehensive review of greenwashing in the textile industry (life cycle assessment, life cycle cost, and eco-labeling). *Environment, Development and Sustainability*, *27*(9), 1–41. DOI: 10.1007/s10668-024-04508-6

Mullins, M., Holland, C. P., & Cunneen, M. (2021). Creating ethics guidelines for artificial intelligence and big data analytics customers: The case of the consumer European insurance market. *Patterns (New York, N.Y.)*, *2*(10), 100362. Advance online publication. DOI: 10.1016/j.patter.2021.100362 PMID: 34693379

Murillo, E., & Atristain-Suárez, C. (2023). Leveraging Strategic Planning to Navigate Volatile Environments. In *Creating Economic Stability Amid Global Uncertainty: Post-Pandemic Recovery in Mexico's Emerging Economy* (pp. 189-217). Cham: Springer Nature Switzerland. DOI: 10.1007/978-3-031-41386-5_9

Nazlim, O. F., & Ozgen, N. (2024). Perceived information and disinformation of generation Z in the context of the changing consumer society: A research on news consumption behavior on social media. *Turkish Online Journal of Design Art and Communication*, *15*(1), 309–332. DOI: 10.7456/tojdac.1569297

Nicolo, G., Zanellato, G., Tiron-Tudor, A., & Tartaglia Polcini, P. (2023). Revealing the corporate contribution to sustainable development goals through integrated reporting: A worldwide perspective. *Social Responsibility Journal*, *19*(5), 829–857. DOI: 10.1108/SRJ-09-2021-0373

Oktay, D., & Acaralp, M. C. (2025). Leadership approach for generation alpha and a greener future: The health professionals of tomorrow. In *Generation Alpha and Next Generation Leadership for Greener Futures* (pp. 63–86). IGI Global Scientific Publishing. DOI: 10.4018/979-8-3693-9281-2.ch004

Omisola, J. O., Bihani, D., Daraojimba, A. I., Osho, G. O., Ubamadu, B. C., & Etukudoh, E. A. (2023). Blockchain in supply chain transparency: A conceptual framework for real-time data tracking and reporting using blockchain and AI. *International Journal of Multidisciplinary Research and Growth Evaluation*, *4*(1), 1238–1253. Advance online publication. DOI: 10.54660/.IJMRGE.2023.4.1.1238-1253

Ong, B., & Toh, D. J. (2023). Digital Dominance and Social Media Platforms: Are Competition Authorities Up to the Task? *IIC; International Review of Industrial Property and Copyright Law*, *54*(4), 527–572. DOI: 10.1007/s40319-023-01302-1 PMID: 37073397

Ooi, R. (2025). The Regenerative Wealth Economy: From ESG to EEoM in Transforming Capitalism via Multi-Capital Reinvestment & AI Governance. https://doi: DOI: 10.20944/preprints202503.1365.v2

Parment, A. (2013). Generation Y vs. Baby Boomers: Shopping behavior, buyer involvement and implications for retailing. *Journal of Retailing and Consumer Services, 20*(2), 189–199. DOI: 10.1016/j.jretconser.2012.12.001

Pavlakova Docekalova, M., & Kocmanova, A. (2018). Comparison of sustainable environmental, social, and corporate governance value added models for investors decision making. *Sustainability (Basel), 10*(3), 649. DOI: 10.3390/su10030649

Perez-Aranda, J., Tolkach, D., & Panchal, J. H. (2024). Reputation and eWOM in accommodation decision-making: Insights from Generation Z users. *Tourism Review*. Advance online publication. DOI: 10.1108/TR-03-2024-0185

Piercy, N. F., Cravens, D. W., & Lane, N. (2010). Thinking strategically about pricing decisions. *The Journal of Business Strategy, 31*(5), 38–48. DOI: 10.1108/02756661011076309

Pillay, S. (2021). The influence of electronic word-of-mouth adoption on brand love amongst Generation Z consumers. *Acta Commercii, 21*(1), 1–11. DOI: 10.4102/ac.v21i1.928

Pokharel, J., & Maharjan, I. (2024). Financial behavior of generation Z and Millennials. *Journal of Emerging Management Studies, 1*(2), 148–170. DOI: 10.3126/jems.v1i2.71522

Prashad, A., Rogers, H., & Srivastava, M. (2025, March). Pay-per-use business models as a driver for additive manufacturing adoption: supply chain implications. In *Supply Chain Forum: An International Journal* (pp. 1-22). Taylor & Francis. DOI: 10.1080/16258312.2025.2476383

Rachbini, W., Anggraeni, D., Burhan, A. B., Sovianti, R., & Rahmawati, E. (2025). Marketing Communications Across Generations: Insight from Indonesia's Online Marketplaces. *Jurnal The Messenger, 17*(1), 37–53. DOI: 10.26623/themessenger.v17i1.11832

Ramirez, A. G., Monsalve, J., González-Ruiz, J. D., Almonacid, P., & Peña, A. (2022). Relationship between the cost of capital and environmental, social, and governance scores: Evidence from Latin America. *Sustainability (Basel), 14*(9), 5012. DOI: 10.3390/su14095012

Rodrigues, P., Sousa, A., & Pinto Borges, A. (2024). Generation Z's attitudes toward virtual tourism: The impact of implicit and explicit attitudes. *EuroMed Journal of Business*, *19*(1), 21–45. DOI: 10.1108/EMJB-05-2023-0135

Roldan, C. J. L., Ong, A. K. S., & Tomas, D. Q. (2024). Cancel culture in a developing country: A belief in a just world behavioral analysis among generation Z. *Acta Psychologica*, *248*, 104378. DOI: 10.1016/j.actpsy.2024.104378 PMID: 38941914

Sanchez-Cartas, J. M., & León, G. (2021). Multisided platforms and markets: A survey of the theoretical literature. *Journal of Economic Surveys*, *35*(2), 452–487. DOI: 10.1111/joes.12409

Santos de Souza, F. D. O., & Chimenti, P. (2024). Emotions in organizational change: An integrative review. *Journal of Change Management*, *24*(2), 137–176. DOI: 10.1080/14697017.2024.2345078

Seele, P., Dierksmeier, C., Hofstetter, R., & Schultz, M. D. (2021). Mapping the ethicality of algorithmic pricing: A review of dynamic and personalized pricing. *Journal of Business Ethics*, *170*(4), 697–719. DOI: 10.1007/s10551-019-04371-w

Sergi, A. (2024). 'Familiness' diversification and recognition: Intergenerational changes in mobile 'ndrangheta families in Australia. *Trends in Organized Crime*, •••, 1–22. DOI: 10.1007/s12117-024-09537-1

Seyfi, S., Michael Hall, C., & Strzelecka, M. (2025). Gen Z–pioneers or paradoxes in sustainable tourism? *Journal of Sustainable Tourism*, *33*(6), 987–1015. DOI: 10.1080/09669582.2025.2491702

Shaheen Hosany, A. R., & Serdiuk, K. (2025). Understanding Gen Z Consumers: A Typology of (Un) sustainable Purchases. *Psychology and Marketing*, *42*(11), 2820–2832. Advance online publication. DOI: 10.1002/mar.70013

Singh, S., Subramani, A. K., David, R., & Jan, N. A. (2024). Workplace ostracism influencing turnover intentions: Moderating roles of perceptions of organizational virtuousness and authentic leadership. *Acta Psychologica*, *243*, 104136. DOI: 10.1016/j.actpsy.2024.104136 PMID: 38244371

Sotelo-Duarte, M., & Gonzalez-Cavazos, B. (2023). Young consumers influence by older generations: Developing the construct and scale to measure intergenerational brand influence. *Young Consumers*, *24*(3), 253–271. DOI: 10.1108/YC-06-2022-1539

Stewart, J. S., Oliver, E. G., Cravens, K. S., & Oishi, S. (2017). Managing millennials: Embracing generational differences. *Business Horizons*, *60*(1), 45–54. DOI: 10.1016/j.bushor.2016.08.011

Stouten, J., Rousseau, D. M., & De Cremer, D. (2018). Successful organizational change: Integrating the management practice and scholarly literatures. *The Academy of Management Annals, 12*(2), 752–788. DOI: 10.5465/annals.2016.0095

Sun, Y., & Xing, J. (2022). The impact of social media information sharing on the green purchase intention among generation Z. *Sustainability (Basel), 14*(11), 6879. DOI: 10.3390/su14116879

Theocharis, D., & Tsekouropoulos, G. (2025). Sustainable consumption and branding for Gen Z: How brand dimensions influence consumer behavior and adoption of newly launched technological products. *Sustainability (Basel), 17*(9), 4124. DOI: 10.3390/su17094124

Theocharis, D., Tsekouropoulos, G., & Athanasios, G. (2025). The Digital Generation: Branding and Consumer Behavior in Tech Adoption. *Rural and Regional Development, 3*(3), 10009. DOI: 10.70322/rrd.2025.10009

Theodorakopoulos, L., & Theodoropoulou, A. (2024). Leveraging big data analytics for understanding consumer behavior in digital marketing: A systematic review. *Human Behavior and Emerging Technologies, 2024*(1), 3641502. DOI: 10.1155/2024/3641502

Tkalac Vercic, A., & Vercic, D. (2025). Attitudes toward sustainable development and employer brands: Comparing generations X, Y and Z in two countries. *Corporate Communications, 30*(2), 355–371. DOI: 10.1108/CCIJ-11-2023-0161

Tolegenov, B., Rafif, K., Rafi, M., Herawati, R., Abdullah, A., & Sari, L. K. (2024). Digital Transparency and Consumer Awareness: Digital Platforms for Educating Consumers about Product Origins and Sustainability. *AL-MUZARA'AH, 12*(2), 277–287. DOI: 10.29244/jam.12.2.277-287

Tran, K., Nguyen, T., Tran, Y., Nguyen, A., Luu, K., & Nguyen, Y. (2022). Eco-friendly fashion among generation Z: Mixed-methods study on price value image, customer fulfillment, and pro-environmental behavior. *PLoS One, 17*(8), e0272789. DOI: 10.1371/journal.pone.0272789 PMID: 35972928

Turunen, L. L. M., & Halme, M. (2021). Communicating actionable sustainability information to consumers: The Shades of Green instrument for fashion. *Journal of Cleaner Production, 297*, 126605. DOI: 10.1016/j.jclepro.2021.126605

Udokwu, C., Voicu-Doroban u, R., Ogunyemi, A. A., Norta, A., Sturua, N., & Craß, S. (2025). Leveraging Blockchain for Ethical AI: Mitigating Digital Threats and Strengthening Societal Resilience. *Future Internet, 17*(7), 309. DOI: 10.3390/fi17070309

Vasanicova, P., & Melnyk, K. (2025). Generational Differences in Motivational Drivers and Travel Preferences: An Exploration of International Travel Behavior of Slovak Women. *Tourism and Hospitality*, *6*(2), 117. DOI: 10.3390/tourhosp6020117

Walters, P. (2021). Are generation Z ethical consumers? In *Generation Z marketing and management in tourism and hospitality: The future of the industry* (pp. 303–325). Springer International Publishing., DOI: 10.1007/978-3-030-70695-1_12

Weber, J., & Urick, M. J. (2023). Searching for a stronger generational understanding than just age: A multi-country analysis of millennials' personal values and moral reasoning. *Journal of Intergenerational Relationships*, *21*(1), 110–135. DOI: 10.1080/15350770.2021.1951920

Williams, P., Escalas, J. E., & Morningstar, A. (2022). Conceptualizing brand purpose and considering its implications for consumer eudaimonic well-being. *Journal of Consumer Psychology*, *32*(4), 699–723. DOI: 10.1002/jcpy.1324

Yang, Z., Huong, N. T. T., Nam, N. H., Nga, N. T. T., & Thanh, C. T. (2020). Greenwashing behaviours: Causes, taxonomy and consequences based on a systematic literature review. [JBEM]. *Journal of Business Economics and Management*, *21*(5), 1486–1507. DOI: 10.3846/jbem.2020.13225

Zimand-Sheiner, D., & Lissitsa, S. (2024). Generation Z-factors predicting decline in purchase intentions after receiving negative environmental information: Fast fashion brand SHEIN as a case study. *Journal of Retailing and Consumer Services*, *81*, 103999. DOI: 10.1016/j.jretconser.2024.103999

Zouari-Hadiji, R., & Chouaibi, Y. (2021). Corporate ethical behavior and the cost of equity capital: Evidence from the world's most ethical companies. *Journal of Financial Reporting and Accounting*, *19*(5), 939–964. DOI: 10.1108/JFRA-08-2020-0223

Zurba, M., Dhyani, S., Mwaura, G., Sivadas, D., Elegbede, I., & Williamson, D. F. (2023). Pathway to mainstream youth engagement and intergenerational partnership in nature conservation. *Frontiers in Ecology and the Environment*, *21*(4), 175–181. DOI: 10.1002/fee.2612

KEY TERMS AND DEFINITIONS

Baby Boomers: They were born during post world war II and popular for hard work, loyalty and their discipline. They prefer long term value and face to face communication.

Digital Transformation: It refers to the employment of digital technologies in business processes for improving the efficiency and customer experience.

Ethical Branding: Ethical branding refers to building of brand based on dimensions such as honesty, sustainability, fairness, and social responsibility.

Ethical Consumption: It demonstrates the practice of purchasing products and services from businesses that do production responsibly with consideration for environment, fair labour, animal welfare and sustainability.

Generation Alpha: The youngest and latest generation, they are children of Millennials. They are exposed to high loads of information and artificial intelligence. They are predicted to be ultra smart generation with highest education and upgraded global knowledge.

Generation X: They are known as middle child and are independent, skeptical and exposure towards early modernisation. They value work life balance.

Generation Z: They are true digital natives, born with digitally equipped world and are inclusive with global awareness.

Millennials: They were the first generation to grow with internet as well as mobile phones. They are tech-savvy, and collaborative. They value experiences more than possessions.

Strategic Brand Leadership: Strategic Brand Leadership demonstrates leading a brand in such a way that it build its long-term value, strength, loyalty, trust along with competitive advantage in customer minds.

Chapter 7
From Vision to Adaptation:
Environmental Leadership in Indonesia and Beyond

Masyhuri Masyhuri
http://orcid.org/0009-0009-3976-1521
Institute Technology and Business Widya Gama Lumajang, Indonesia

ABSTRACT

Environmental challenges such as climate change, biodiversity loss, and resource depletion demand leadership that goes beyond technical solutions. This chapter explores how green transformational leadership and adaptive leadership shape environmental performance across sectors. Green transformational leadership mobilizes employees through a compelling sustainability vision, role modelling, and green HRM alignment, while adaptive leadership enables stakeholders to address uncertainty, conflicting interests, and long-term ecological risks. Drawing from international studies and Indonesian cases—including state-owned enterprises, multinational firms, tourism villages, and ecopesantren—the chapter illustrates how leaders integrate environmental goals into governance, innovation, and community engagement. Although Indonesia has advanced national commitments such as Net Zero Emissions 2060 and FOLU Net Sink 2030, significant challenges persist, including policy–implementation gaps, economic pressures, and equity concerns.

1. INTRODUCTION

For much of the twentieth century, organizational leadership was primarily oriented toward economic growth, operational efficiency, and shareholder value

DOI: 10.4018/979-8-3373-7877-0.ch007

maximization, reflecting dominant industrial-era management paradigms. Environmental concerns were largely treated as externalities rather than integral components of organizational strategy. However, this perspective began to shift significantly following the publication of the Brundtland Report, which introduced sustainable development as development that meets present needs without compromising the ability of future generations to meet their own needs (Brundtland Commission, 1987). This transformation fundamentally altered expectations toward organizations, requiring them to balance economic performance with environmental stewardship and broader social responsibility. The emergence of the triple bottom line framework further reinforced this shift by emphasizing that organizational success should no longer be evaluated solely through financial outcomes, but also through social and environmental performance (Elkington, 1997).

The acceleration of climate change, biodiversity loss, pollution, and resource depletion has moved environmental issues from the periphery to the core of organizational and policy agendas. Technical solutions such as renewable energy, low-carbon technologies, and circular business models are necessary but not sufficient. Without leadership that can mobilize diverse stakeholders, reframe incentives, and embed sustainability into everyday decision-making, environmental initiatives often remain symbolic or fragmented.

Leadership in environmental contexts differs from traditional leadership by its time horizon, scope of stakeholders, and ethical stakes. Leaders are called to balance short-term economic performance with long-term ecological resilience, navigate trade-offs between local and global interests, and take responsibility not only for current but also future generations. This chapter discusses key theoretical perspectives particularly green transformational leadership and adaptive leadership and illustrates how they operate in practice through international and Indonesian cases.

The growing recognition of environmental degradation and social inequality also challenged conventional assumptions regarding the role of business and leadership in society. Stakeholder theory provided an important conceptual foundation by arguing that organizations bear responsibility not only to shareholders but also to employees, customers, governments, communities, and broader ecological systems that sustain long-term organizational viability (Freeman, 1984). This perspective encouraged a broader understanding of organizational accountability, where leaders increasingly became responsible for balancing competing stakeholder interests while addressing long-term sustainability challenges. As sustainability issues became more central to organizational decision-making, leadership itself gradually evolved beyond traditional managerial functions toward more socially and environmentally embedded forms of strategic leadership.

Beyond these core perspectives, the urgency of environmental degradation has led scholars to emphasize that sustainability transformations require leaders capable of integrating ethical judgment, stakeholder engagement, and systemic thinking into strategic and operational processes. Environmental problems today manifest as "adaptive challenges"—issues that do not have straightforward technical solutions and instead demand shifts in values, behaviors, and institutional arrangements. Thus, leadership must extend beyond the conventional domains of efficiency and performance to encompass moral accountability, long-term ecological stewardship, and collaborative governance.

In many contexts, particularly emerging economies such as Indonesia, environmental challenges are deeply intertwined with socio-economic development pressures. Natural resource exploitation, deforestation, industrial pollution, and climate-induced disasters present complex governance dilemmas that cannot be addressed through fragmented or sector-specific interventions. Leaders across government, business, and civil society are increasingly expected to exercise foresight, promote inclusive decision-making, and cultivate cross-sectoral partnerships that support sustainable development pathways.

Recent scholarship also highlights the importance of responsible leadership as a complementary lens that enriches existing models such as green transformational leadership and adaptive leadership. Responsible leadership integrates ethical and values-based considerations with stakeholder-oriented decision-making, thereby strengthening the legitimacy and effectiveness of environmental initiatives. This is particularly relevant in settings where environmental issues intersect with community livelihoods, cultural values, and intergenerational equity concerns. While the primary focus of this chapter remains on green transformational and adaptive leadership, responsible leadership provides a useful umbrella for understanding how environmental commitments can be translated into socially grounded and ethically informed action.

Furthermore, the rise of sustainability assessment frameworks, global reporting standards, and the Sustainable Development Goals (SDGs) has expanded expectations for leaders to align organizational strategies with planetary boundaries and environmental justice principles. In this regard, sustainability-oriented leadership is not simply an extension of traditional management practices; it represents a paradigmatic shift toward holistic, multi-level, and future-oriented governance. As Hasan et al. (2024) observe, sustainability leadership requires leaders to operate within complex adaptive systems where global supply chains, regulatory regimes, social movements, and ecological feedbacks interact in often unpredictable ways. Such complexity makes environmental leadership inherently cross-boundary and multilevel by nature, spanning local communities, corporations, government agencies, and transnational networks.

These evolving expectations underscore why the core characteristics of environmental leadership—systems thinking, long-term orientation, ethical and values-based decision-making, collaborative orientation, and the capacity to navigate uncertainty and conflict—are indispensable. They embody the shift from conventional managerialism to transformative, sustainability-driven leadership capable of guiding organizations through ecological turbulence and institutional change.

Taken together, these dynamics demonstrate that environmental leadership is not merely a functional competency but a strategic imperative for organizations operating in an era defined by climate risk, environmental constraints, and social demand for accountability. The theoretical lenses discussed in this chapter—green transformational leadership and adaptive leadership—offer robust foundations for understanding how leaders can mobilize people, navigate complexity, and institutionalize sustainability. In addition, examples from international and Indonesian contexts demonstrate how these forms of leadership take shape in practice, shaping institutional reform, community engagement, and environmental innovation.

2. METHODOLOGICAL APPROACH

This chapter adopts a conceptual review approach integrating theoretical synthesis with illustrative case studies to examine the role of leadership in addressing contemporary environmental sustainability challenges. The conceptual review method is particularly appropriate for exploring emerging interdisciplinary issues where theoretical integration and contextual interpretation are necessary to build comprehensive understanding.

The chapter synthesizes interdisciplinary scholarship from leadership studies, sustainability management, environmental governance, organizational behavior, and climate transition literature. Particular emphasis is placed on two complementary theoretical frameworks, namely Green Transformational Leadership (GTL) and Adaptive Leadership, as these perspectives provide important insights into how leaders mobilize environmental commitment, influence organizational behavior, and respond to complex sustainability challenges under conditions of uncertainty.

To complement theoretical analysis, this chapter incorporates illustrative case studies drawn from both international and Indonesian contexts. The international cases provide broader comparative perspectives regarding how environmental leadership practices emerge across organizational and sectoral settings, while Indonesian cases are intentionally selected to reflect the complexity of sustainability transitions in emerging economies characterized by rapid development pressures, institutional fragmentation, and diverse socio-ecological challenges.

The analytical process focuses on identifying recurring leadership mechanisms, environmental governance challenges, organizational transformation processes, stakeholder engagement patterns, and adaptive responses observed across both theoretical and empirical contexts. Through integrating theoretical synthesis with illustrative case analysis, this chapter develops a comprehensive framework for understanding how leadership contributes to environmental sustainability transitions across multiple organizational and institutional levels.

3. SUSTAINABLE LEADERSHIP

3.1 From Traditional to Sustainability Leadership

Traditional leadership models in management tended to emphasize organizational performance, efficiency, and competitive advantage. These models were largely shaped by industrial-era assumptions in which economic growth, productivity, and market competition were prioritized as primary indicators of organizational success. Within such paradigms, environmental or social considerations were often treated as externalities—issues that organizations acknowledged only insofar as they affected profitability, compliance, or risk management.

The evolution toward sustainability-oriented leadership reflects broader transformations in organizational theory, corporate governance, and societal expectations regarding the role of organizations in addressing increasingly complex global challenges. For much of the twentieth century, leadership theory developed primarily within the context of industrial-era management paradigms, where organizational success was strongly associated with operational efficiency, hierarchical control, productivity maximization, and the effective coordination of human resources to achieve economic objectives. During this period, leadership effectiveness was largely evaluated based on the ability to improve internal organizational performance, strengthen managerial authority, and sustain competitive advantage in increasingly industrialized markets (Burns, 1978).

This traditional management paradigm was closely aligned with shareholder-centered capitalism, which positioned financial performance and shareholder wealth maximization as the central objectives of organizational decision-making. Organizations were primarily expected to generate economic returns for investors, while broader environmental and social consequences of business activities often remained secondary considerations. Leadership within this paradigm therefore emphasized performance management, cost efficiency, strategic control, and organizational growth, with relatively limited attention given to long-term ecological sustainability or broader societal impact (Bass, 1985).

However, growing awareness of environmental degradation, resource scarcity, industrial pollution, and widening social inequality gradually exposed the limitations of purely shareholder-oriented organizational models. A major conceptual shift emerged through stakeholder theory, which fundamentally challenged the assumption that organizations exist solely to serve shareholder interests. Freeman (1984) argued that organizations operate within broader networks of relationships and therefore bear responsibility not only to shareholders, but also to employees, consumers, governments, local communities, and other stakeholders whose interests influence long-term organizational sustainability. This perspective significantly expanded the scope of leadership responsibility by recognizing that organizational success depends on balancing diverse stakeholder expectations rather than focusing exclusively on short-term financial outcomes.

The sustainability discourse gained further momentum following the publication of the Brundtland Report, which introduced sustainable development as development that meets present needs without compromising the ability of future generations to meet their own needs (Brundtland Commission, 1987). This framework fundamentally altered the relationship between organizations and society by establishing sustainability as a long-term strategic imperative rather than a peripheral social concern. Building upon this development, Elkington (1997) introduced the Triple Bottom Line framework, emphasizing that organizational performance should no longer be measured solely through economic profitability, but also through environmental stewardship and social responsibility. The emergence of these frameworks significantly redefined leadership expectations by requiring organizational leaders to integrate financial performance, environmental accountability, and broader societal well-being into strategic decision-making processes.

As sustainability challenges became increasingly complex, traditional leadership models were gradually recognized as insufficient for addressing multidimensional problems characterized by uncertainty, systemic interdependence, and long-term ecological consequences. This realization contributed to the emergence of sustainability leadership as a distinct field of scholarship. Early sustainability leadership literature emphasized that effective leadership in contemporary organizations requires systems thinking, long-term orientation, ethical responsibility, stakeholder collaboration, and the capacity to align organizational strategy with broader sustainability objectives. Avery (2005) and Ferdig (2007) argued that leadership can no longer be understood solely as the capacity to direct organizational performance, but increasingly involves the ability to facilitate organizational transformation while simultaneously balancing economic, environmental, and social responsibilities.

More recent scholarship further conceptualizes sustainability leadership as a leadership paradigm specifically designed to address complex socio-ecological systems where organizational boundaries increasingly intersect with environmental governance, climate transition, and global sustainability challenges. In this perspective, leadership effectiveness depends not only on the ability to inspire organizational commitment, but also on the capacity to navigate uncertainty, coordinate multiple stakeholders, facilitate institutional learning, and create adaptive responses to rapidly changing environmental conditions. Scholars increasingly argue that sustainability leadership requires leaders capable of integrating organizational strategy with broader ecological resilience while fostering long-term adaptive capacity across institutions and stakeholder networks (Visser & Courtice, 2011; Shriberg & MacDonald, 2013; Robertson & Barling, 2013). The emergence of environmental leadership models, including Green Transformational Leadership and Adaptive Leadership, can therefore be understood as a continuation of this broader intellectual evolution toward leadership frameworks capable of supporting long-term sustainability transitions across increasingly complex organizational and institutional environments.

Early leadership scholarship largely focused on hierarchical authority, managerial control, and organizational performance, where effective leadership was primarily associated with the ability to direct subordinates and maximize productivity outcomes. Foundational leadership theory later evolved through transformational leadership, which emphasized the leader's capacity to inspire followers, shape organizational vision, and drive long-term organizational change (Burns, 1978; Bass, 1985). As global sustainability challenges intensified, scholars increasingly recognized that traditional leadership models were insufficient to address complex environmental and social problems characterized by uncertainty, interdependence, and long-term systemic consequences. This recognition contributed to the emergence of sustainability leadership, a leadership perspective emphasizing systems thinking, ethical responsibility, long-term orientation, and the integration of economic, social, and environmental objectives in organizational decision-making (Avery, 2005; Ferdig, 2007).

Sustainability-oriented leadership extends this mandate by explicitly including environmental integrity and social justice as core objectives alongside financial outcomes. In contemporary scholarship, this evolution reflects a broader recognition that long-term organizational viability depends on the health of the ecological and social systems in which organizations operate. As environmental pressures intensify, leaders are increasingly required to integrate sustainability concerns into vision-setting, strategic planning, and operational management.

Authors increasingly use terms such as sustainability leadership, environmental leadership, green leadership, and climate leadership to describe leaders who intentionally align organizational strategies with planetary boundaries and Sustainable Development Goals (SDGs) (Hasan et al., 2024). These leadership forms acknowledge that sustainability is not merely a technical challenge but also a moral and strategic imperative. Leaders must therefore engage in sense-making to interpret environmental risks, communicate sustainability priorities, and encourage cultural shifts within their organizations.

Such leaders operate within a complex adaptive system: global supply chains, regulatory regimes, social movements, and ecological feedbacks interact in often unpredictable ways. This interconnectedness means that environmental leadership cannot be confined to a single sector or administrative silo but must extend across diverse domains. For instance, leaders in government agencies must coordinate with private-sector innovators; community leaders must collaborate with policymakers; and corporate executives must engage with civil society organizations to address shared environmental challenges.

More recent scholarship has further expanded sustainability leadership as a distinct field of inquiry by emphasizing the need for leaders capable of navigating complex socio-ecological systems while facilitating collective learning, stakeholder collaboration, and institutional transformation. Sustainability leadership increasingly requires leaders to move beyond conventional organizational boundaries and adopt integrative approaches that connect environmental governance, ethical responsibility, innovation, and adaptive organizational change. Scholars argue that effective sustainability leadership is fundamentally characterized by the ability to align long-term ecological resilience with organizational strategy, while simultaneously responding to dynamic environmental challenges that cut across sectors and institutional contexts (Visser & Courtice, 2011; Shriberg & MacDonald, 2013; Robertson & Barling, 2013).

This complexity makes environmental leadership cross-boundary and multi-level by nature, spanning local communities, corporations, government agencies, and transnational networks. Leaders must therefore cultivate the ability to manage competing interests, navigate power asymmetries, and anticipate long-term consequences. Unlike traditional leadership, which often relies on hierarchical control or technical expertise, sustainability leadership emphasizes adaptability, inclusiveness, and the capacity to mobilize distributed knowledge across varied stakeholders.

3.2 Core Characteristics

Across diverse models, several characteristics recur in descriptions of effective leadership in environmental contexts:

1. Systems thinking – the ability to see interconnections between economic activities, ecological conditions, and social structures. Leaders must understand how decisions in one domain—such as supply chain management or land-use planning—have cascading effects on environmental outcomes and community well-being. Systems thinking allows leaders to anticipate unintended consequences, identify leverage points, and design integrated solutions.

2. Long-term orientation – prioritizing intergenerational equity and resilience rather than only short-term gains. Sustainability leadership demands awareness that environmental degradation unfolds over decades or centuries. Leaders must therefore adopt temporal horizons that extend beyond immediate financial cycles, considering how current actions will influence future generations and ecosystem stability.
3. Ethical and values-based decision-making – integrity, transparency, and justice in how environmental burdens and benefits are distributed. Environmental decisions often involve moral dilemmas, such as balancing economic growth with ecological protection or determining which communities bear the cost of environmental harm. Values-based leadership ensures that such decisions honor fairness and accountability.
4. Collaborative orientation – building coalitions across sectors and levels of governance. Because environmental problems are inherently multi-dimensional, no single organization can address them alone. Effective leaders cultivate partnerships, facilitate dialogue, and integrate perspectives from scientists, policymakers, businesses, and local communities. Collaboration enhances innovation, legitimacy, and shared ownership of environmental outcomes.
5. Capacity to navigate uncertainty and conflict – engaging stakeholders with divergent interests, knowledge, and power. Environmental issues frequently involve scientific uncertainty, conflicting priorities, and contested values. Leaders must be skilled in negotiation, conflict resolution, and adaptive learning to guide stakeholders through complexity without polarizing or fragmenting collective action.

These characteristics are at the heart of two particularly relevant theoretical lenses: green transformational leadership and adaptive leadership. Together, they illuminate how leaders can articulate compelling environmental visions, empower stakeholders, and guide organizations through the uncertainties inherent in ecological transformation.

4. THEORETICAL FOUNDATIONS

Early leadership theory historically emphasized organizational efficiency, productivity, hierarchical control, and shareholder value maximization. Classical management paradigms emerging from the industrial era largely positioned economic growth and organizational performance as primary indicators of success, while environmental and social concerns were frequently treated as secondary externalities. However, the global sustainability discourse fundamentally shifted following the publication of the Brundtland Report, which introduced the concept of sustainable development as development that meets present needs without compromising the ability of future generations to meet their own needs (Brundtland Commission, 1987). This paradigm shift significantly expanded organizational responsibility beyond financial performance toward broader environmental stewardship and social accountability (Avery, 2005).

The emergence of stakeholder theory further reinforced this transformation by arguing that organizations bear responsibility not only to shareholders but also to employees, communities, governments, consumers, and wider ecological systems (Freeman, 1984). As sustainability challenges intensified, leadership scholarship gradually evolved toward sustainability-oriented leadership models emphasizing systems thinking, long-term orientation, ethical responsibility, stakeholder engagement, and cross-sector collaboration. Scholars increasingly argue that effective leadership in sustainability transitions requires leaders who can integrate economic, environmental, and social objectives while navigating institutional complexity and long-term uncertainty (Ferdig, 2007; Visser & Courtice, 2011; Shriberg & MacDonald, 2013).

4.1 Transformational and Green Transformational Leadership

Transformational leadership, as articulated by Bass and Riggio, focuses on leaders who inspire followers to transcend self-interest for collective goals, through idealized influence, inspirational motivation, intellectual stimulation, and individualized consideration. This style is often associated with organizational change, innovation, and cultural transformation because it emphasizes motivation through purpose and identity rather than control or transactional exchange. Transformational leaders establish a compelling organizational vision, challenge established assumptions, and empower followers to engage in continuous learning and experimentation.

Green transformational leadership (GTL) is a contextualized extension of transformational leadership where the "vision" and "idealized influence" are explicitly oriented toward environmental goals such as emissions reduction, resource efficiency, and circular economy practices. GTL applies the core transformational mechanisms to sustainability challenges, emphasizing the strategic role of leaders in shaping pro-environmental culture, enabling green innovation, and influencing employee behaviors toward environmental responsibility.

Recent empirical studies show that GTL can:

(1) strengthen employees' environmental commitment and green performance,
(2) encourage green creativity and innovation,
(3) and enhance environmental performance at organizational level (Çop et al., 2021).

These findings align with broader research demonstrating that leadership that emphasizes environmental values can significantly shape organizational outcomes. For example, green transformational leaders can influence employees' green self-efficacy, which in turn improves green creativity and voluntary environmental participation (Chen et al., 2020). Other studies have shown that GTL enhances both environmental citizenship behaviors and the development of sustainability-oriented cultures, particularly when supported by green human resource management (GHRM), green reward systems, and sustainability training programs (Amjad et al., 2023).

GTL mechanisms typically operate through:

1. Articulating a compelling green vision (e.g., net-zero, zero-waste, regenerative business) that aligns organizational identity with environmental stewardship.
2. Role modelling pro-environmental behavior, signalling commitment and demonstrating expected actions through personal conduct.
3. Encouraging experimentation with sustainable technologies, eco-innovative processes, and green business models, allowing organizations to test and refine environmental strategies.
4. Empowering employees to propose and implement green initiatives, often supported by green HRM practices such as participatory decision-making, green rewards, and employee-driven sustainability programs.

In many organizations, GTL also interacts with external drivers such as environmental regulations, stakeholder pressure, and market expectations for sustainability. Thus, GTL serves as both an internal motivator and a strategic response to external environmental demands, making it a central leadership model for organizational transitions toward sustainability.

4.2 Adaptive Leadership

Heifetz's theory of adaptive leadership emphasizes how leaders mobilize people to tackle adaptive challenges—problems that do not have clear technical solutions and require changes in values, beliefs, and behaviors. Adaptive challenges require organizations to rethink assumptions, renegotiate priorities, and learn through iterative experimentation, making adaptive leadership particularly relevant for complex environmental issues.

Environmental problems are archetypal adaptive challenges: they involve scientific uncertainty, conflicting interests, and long timeframes. Climate change, biodiversity loss, and pollution cannot be solved solely by applying existing tools or expertise; they require learning across stakeholders, negotiation of trade-offs, and institutional transformation. Adaptive leadership therefore helps organizations confront environmental realities rather than avoid or simplify them.

In environmental governance, adaptive leadership has been linked to climate change adaptation, community resilience, and ecosystem management. Recent work shows that leadership which is adaptive, enabling, and politically savvy is crucial for climate adaptation in sensitive ecosystems such as arid and semi-arid wetlands, where trade-offs between livelihoods, biodiversity, and infrastructure are acute (Karamidehkordi et al., 2024). Other studies illustrate how adaptive leadership supports collaborative environmental governance networks, enabling communities to coordinate cross-scale responses to ecological disturbances (Folke et al., 2021).

Adaptive leaders:

1. Distinguish between technical problems (solvable with current expertise) and adaptive challenges (requiring learning and behavioral change).
2. "Hold the tension" among stakeholders with divergent values, interests, and knowledge systems without prematurely resolving conflict.
3. Create safe spaces for experimentation, dialogue, and collective learning, enabling stakeholders to test new strategies and reflect on failures constructively.
4. Adjust strategies iteratively as conditions, knowledge, and power dynamics evolve, ensuring that responses remain relevant in dynamic environmental contexts.

Adaptive leadership thus provides a framework for navigating the uncertainty, conflict, and long-term horizon inherent in sustainability transitions. It complements GTL by focusing on facilitation, learning, and the enabling of stakeholder-driven solutions rather than solely on inspiration and vision.

4.3 Other Complementary Perspectives

Other leadership perspectives supplement GTL and adaptive leadership in environmental contexts and enrich the theoretical foundation for sustainability leadership research.

1. Ethical leadership – emphasizes fairness, integrity, and responsibility; crucial for avoiding greenwashing and ensuring procedural justice in environmental decisions. Ethical leadership enhances trust, legitimacy, and compliance with sustainability initiatives, especially when environmental policies may impose costs or require behavioral change. It helps organizations balance economic goals with social and ecological responsibility.

2. Servant leadership for sustainability – stresses serving stakeholders and the planet, often predicting employees' proactive environmental behavior (Saleem and Mateou, 2025). Leaders who prioritize serving others demonstrate humility, empathy, and stewardship—qualities that enhance environmental citizenship behaviors and strengthen organizational commitment to sustainability.
3. Collaborative and network leadership – focuses on convening cross-sector coalitions for climate and sustainability initiatives. Such leadership is essential for addressing environmental challenges that span geographic and institutional boundaries, enabling governments, NGOs, businesses, and communities to share information, pool resources, and coordinate strategies.

Although these perspectives contribute valuable insights, green transformational and adaptive leadership are especially useful for explaining how leaders ignite environmental vision and navigate systemic complexity—two central concerns of this chapter. Together, they provide a robust theoretical foundation for understanding how leadership drives sustainability transformations in diverse organizational and ecological contexts.

5. GREEN TRANSFORMATIONAL LEADERSHIP IN ENVIRONMENTAL CONTEXTS

5.1 Mechanisms and Outcomes

Recent empirical studies across sectors and countries show consistent links between GTL and environmental outcomes. For example, research in tourism and hospitality demonstrates that GTL can enhance environmental sustainability through mechanisms such as green team resilience and environmental citizenship behavior (Çop et al., 2021).

Common mechanisms include:

1. Vision framing

   Leaders articulate environmental goals (e.g., "zero waste," "net-zero emissions," "climate-positive operations") not as compliance burdens but as sources of innovation, reputation, and long-term competitiveness.
2. Symbolic actions

   Leaders personally practice low-carbon behavior—such as reducing travel emissions, supporting waste segregation, or championing eco-labels—sending a powerful signal that sustainability is not merely rhetorical.
3. Green HRM and capability building

GTL is often supported by green human resource management: integrating environmental criteria into recruitment, performance appraisal, training, and rewards. This alignment reinforces pro-environmental norms and skills.

4. Empowerment and participatory decision-making

   Employees are invited to co-create green initiatives, for instance by proposing energy-saving ideas or designing eco-innovation projects. GTL encourages experimentation and tolerates "smart failures" in pursuit of environmental learning.

5. Stakeholder engagement

   GTL extends beyond internal followers to customers, suppliers, and local communities, integrating environmental expectations into value chain partnerships and corporate social responsibility (CSR) programs.

5.2 International Case Snapshots

Several international cases illustrate GTL in practice:

1. Telecommunications sector in Sierra Leone – A recent study found that green transformational leadership in the telecom sector improved environmental performance by motivating staff to adopt eco-efficiency practices and by aligning organizational processes with environmental standards (Kebe et al., 2025).
2. Tourism and hospitality – Work in tourism organizations shows GTL can drive green innovation (e.g., waste minimization, energy-efficient operations) when leaders intentionally cultivate green team resilience and provide resources for experimentation (Çop et al., 2020).

These cases highlight that GTL is not limited to high-income countries; it is equally relevant in developing and emerging economies where environmental risks and development needs intersect.

6. ADAPTIVE LEADERSHIP FOR CLIMATE AND ENVIRONMENTAL GOVERNANCE

6.1 Climate Change as an Adaptive Challenge

Climate change is characterized by deep uncertainty, long time horizons, and cross-scale interactions. Leaders cannot rely solely on existing technical expertise or linear planning. Instead, they must facilitate collective learning, experimentation, and institutional flexibility—hallmarks of adaptive leadership.

Studies of climate adaptation in wetlands and water governance show that adaptive and enabling leadership is critical for building coalitions, navigating power asymmetries, and managing trade-offs between agriculture, biodiversity, and infrastructure development (Karamidehkordi et al., 2024).

6.2 Adaptive Leadership and Net-Zero Transitions

Recently, scholarship on "leadership in the pursuit of net-zero emissions" emphasizes that achieving net-zero requires not only technological and financial innovation but also leadership that can overcome institutional inertia, vested interests, and social resistance. Transformational and adaptive leadership together are seen as crucial for overcoming barriers and promoting strategic innovation in the net-zero transition (Awashreh, 2025).

In practice, adaptive climate leaders:

1. Encourage scenario thinking and stress testing for climate and policy uncertainty,
2. Pilot low-regret measures and scale what works,
3. Adjust policies based on feedback from climate indicators, markets, and communities,
4. and design inclusive processes where vulnerable groups participate in shaping climate strategies.

These capabilities become especially important in countries like Indonesia, where climate policies must reconcile rapid development, energy security, and environmental protection.

7. LEADERSHIP IN ENVIRONMENTAL CONTEXTS: THE INDONESIAN EXPERIENCE

7.1 National Climate Commitments as Macro-Level Leadership Signals

Indonesia has articulated ambitious climate goals, including the commitment to achieve Net Zero Emissions (NZE) by 2060 or earlier and to realize the Forestry and Other Land Use (FOLU) Net Sink 2030, indicating that the land use sector should absorb more emissions than it releases by 2030 (Ministry of Environment and Forestry Republic of Indonesia, 2023). These national commitments are not merely policy statements; they function as macro-level leadership signals that cascade across ministries, state-owned enterprises (SOEs), local governments, private corporations, and civil society organizations.

These commitments demonstrate Indonesia's recognition that environmental sustainability is intertwined with national competitiveness, energy security, and socio-economic resilience. NZE goals signal a long-term strategic orientation that requires structural transformation across energy, industry, land use, and transportation sectors. For instance, the FOLU Net Sink 2030 agenda highlights the critical role of forest governance, peatland restoration, mangrove rehabilitation, and sustainable land-use planning as the backbone of Indonesia's climate strategy (Bappenas, 2024).

Within this broader framework, national climate targets require leaders at multiple levels to:

1. Redesign energy and land-use policies

    Government agencies must align national energy planning (RUPTL) with renewable energy targets, expand carbon pricing mechanisms, and integrate climate risk into spatial planning processes.

2. Enforce environmental regulations

    Leaders in ministries and provincial governments must strengthen monitoring systems, resolve regulatory overlaps (especially between forestry, mining, and regional development), and build institutional capacity for compliance.

3. Guide investment into low-carbon pathways

    Leaders in finance, state enterprises, and the private sector must mobilize climate financing, encourage renewable energy deployment, support sustainable forestry, and promote climate-smart agriculture.

Indonesia's climate goals also create normative and symbolic expectations that influence organizational identity. For example, SOEs and multinational companies increasingly view alignment with national sustainability agendas as essential to maintaining legitimacy, stakeholder trust, and international competitiveness (Wijaya & Ernst, 2023). Thus, Indonesia's macro-leadership signals shape the broader ecosystem in which environmental leadership practices emerge.

7.2 Corporate And Organizational Environmental Leadership

Indonesia offers several illustrative examples of environmental leadership in practice, demonstrating how macro-level commitments translate into strategic and operational initiatives at organizational levels.

1. State-Owned Enterprises and Net-Zero Roadmaps

The national electricity company (PLN) has published climate-related disclosure reports outlining strategies to align with Indonesia's NDC and NZE targets, integrating climate risk into corporate governance and signaling leadership in the energy transition (PLN, 2024). PLN's roadmap involves coal retirement scenarios, expansion of solar and hydropower, and grid modernization—illustrating how SOEs can function as catalysts for sector-wide emissions reductions.

Similarly, large energy companies such as PT Pertamina (Persero) are increasingly using CSR innovations and sustainability programs to address environmental challenges and strengthen long-term business sustainability. These initiatives show how corporate leaders reinterpret CSR from philanthropy toward strategic sustainability (Amalia et al., 2025). They incorporate renewable energy adoption, waste-to-energy projects, and green refinery development, positioning Pertamina to transition gradually toward cleaner energy portfolios.

2. Multinational Corporations Operating in Indonesia

A qualitative study of Allianz SE and Azbil Corporation highlights how multinational corporations implement environmental initiatives in Indonesia, motivated by global sustainability commitments and local regulatory expectations. Their leaders champion renewable energy use, eco-efficiency projects, and environmental education activities in local communities (Vandini & Ala'aldin, 2025). These initiatives demonstrate how global companies translate international sustainability frameworks into localized leadership practices.

MNCs often bring advanced sustainability standards, which then interact with Indonesia's regulatory environment, creating opportunities for technology transfer, capacity building, and demonstration of best practices.

3. Tourism Villages and Green Transformational Leadership

Research on tourism villages in the Special Region of Yogyakarta (DIY) shows that green transformational leadership and green knowledge can improve environmental performance, with green creativity mediating the effect. Village leaders who communicate a clear sustainability vision, encourage innovation in waste management and energy use, and involve local residents in decision-making are more successful in achieving eco-tourism outcomes (Hidayati et al., 2024).

These community-level leaders exemplify how GTL can operate in grassroots settings, where relational trust, cultural identity, and local knowledge play significant roles. Tourism villages also demonstrate how sustainability leadership intersects with cultural preservation, livelihood diversification, and community empowerment.

4. Ecopesantren and Environmental Leadership in Education

Studies of ecopesantren (environmentally oriented Islamic boarding schools) show how environmental leadership in education can support local climate action and the achievement of sustainable development indicators. Leaders integrate environmental values into religious teachings, practice waste segregation and organic farming on campus, and involve students in environmental campaigns, thereby cultivating future green leaders (Kejora et al., 2025).

As educational institutions, ecopesantren represent values-based environmental leadership, where moral and religious frameworks reinforce environmental ethics. They also illustrate the potential of education to mobilize youth-driven climate action and embed sustainability across generations.

5. Creative Tourism and Zero-Waste Hospitality in Bali

The Bali-based "Potato Head" hospitality group has become known for its zero-waste and circular design approach, partnering with designers to transform waste into furniture and lifestyle products. The founder's leadership vision—"being sustainable in a beautiful way"—demonstrates how a compelling aesthetic and experiential narrative can mobilize staff, guests, and local artisans around zero-waste practices and in-house recycling systems that reportedly recycle over 99% of waste (wallpaper.com).

This case illustrates design-led sustainability leadership, where creativity, storytelling, and cultural capital serve as mechanisms for behavioral change and environmental innovation. It highlights how hospitality leaders can reposition sustainability as an integral part of brand identity and guest experience. Collectively, these cases exemplify green transformational leadership at different scales—from village chiefs and school principals to CEOs of SOEs and multinational subsidiaries—where leaders frame environmental goals as integral to identity, reputation, and long-term viability.

In addition to examples from the energy sector, tourism, and non-formal education institutions, recent evidence from Indonesian higher education provides a compelling illustration of how *responsible leadership* shapes sustainability outcomes through *green orchestration innovation* (GOI). This case is drawn from a quantitative study involving 227 faculty members across Indonesian universities, examining the relationships among responsible leadership, green innovation processes, and sustainable performance using SEM-PLS analysis (Masyhuri et al., 2025).

6. Responsible Leadership in Indonesian Universities

The study reveals several important insights:

1. Responsible leadership significantly improves sustainable performance

University leaders who demonstrate ethical commitment, stakeholder engagement, environmental awareness, and transparency show higher institutional performance in areas such as waste reduction, resource efficiency, energy management, and sustainability-driven research. Responsible leadership thus acts as a driving force behind environmental governance in higher education institutions.

2. Green orchestration innovation (GOI) is a critical mechanism linking leadership to sustainability outcomes

GOI refers to the institution's ability to orchestrate its diverse resources—faculty expertise, student involvement, curriculum integration, facility management, and external networks—to generate collaborative green innovations. Universities with strong responsible leadership demonstrate:

- more active environmental campaigns,
- cross-faculty collaboration on green initiatives,
- integration of sustainability values into academic programs,
- and reduced bureaucratic barriers to environmental experimentation.

3. GOI mediates the relationship between leadership and sustainable performance

The study finds that responsible leadership does not only have a direct effect on institutional sustainability—it becomes much stronger when combined with coordinated innovation practices. Leaders create enabling structures, mobilize stakeholders, and provide strategic direction, while GOI ensures that the institution's collective capabilities translate into environmental improvements.

This case illustrates how university leaders mobilize academic communities, open spaces for experimentation, integrate sustainability values into daily operations, and leverage partnerships with industry and government to advance environmental outcomes. It demonstrates a powerful form of green transformational leadership in the education sector—one that blends ethical responsibility with strategic innovation orchestration.

7.3 Challenges and Tensions

Despite promising examples, several challenges persist, revealing structural, economic, and social barriers that complicate the exercise of environmental leadership in Indonesia.

1. Policy–implementation gaps
   Ambitious national targets sometimes clash with weak enforcement, fragmented coordination, or conflicting sectoral regulations (e.g., between forestry, mining, and regional development). As highlighted by Santoso and Jaya (2022), overlapping jurisdictional authority often leads to inconsistent land-use practices.
2. Economic pressures and energy security concerns
   Leaders face tension between phasing out fossil fuels and meeting growing energy demand, especially in regions dependent on coal industry employment. Transition policies must therefore incorporate just transition principles to avoid exacerbating unemployment and regional inequalities.
3. Capacity and resource constraints
   Local governments and community organizations may lack technical expertise, funding, and data to design and monitor environmental programs. Limited access to climate financing and weak institutional capacity often delay the realization of sustainability initiatives.
4. Social equity and environmental justice
   Decisions about land use, infrastructure, and resource extraction can exacerbate inequalities if affected communities are not meaningfully involved. Research by Prabowo and Kim (2024) underscores that marginalized groups often bear disproportionate environmental burdens.

These challenges underscore the need for adaptive leadership that can work with uncertainty, negotiate trade-offs, and foster inclusive dialogue, alongside green transformational leadership that keeps the environmental vision visible and inspiring.

The study of Masyhuri et al., (2025) also highlights several leadership challenges unique to higher education institutions in Indonesia. These challenges mirror structural and operational barriers observed in government and corporate sectors, underscoring the systemic nature of sustainability transitions.

1. Gaps Between Sustainability Visions and Organizational Implementation

Although university leaders frequently articulate sustainability commitments, the study reveals several implementation challenges:

- unclear role distribution among faculty and administrative units,
- inconsistencies between rectorate-level policies and faculty-level execution,
- and difficulties in sustaining cross-departmental communication.

This reflects a broader issue of policy–implementation gaps that affects environmental governance in many Indonesian institutions.

2. Limited Resources and Innovation Capacity

Higher education institutions, despite having strong intellectual potential, often face:

- limited funding for green innovation projects,
- insufficient infrastructure to support sustainability measurement or reporting,
- and heavy academic workloads that constrain faculty participation in environmental initiatives.

Resource scarcity and capacity limitations reduce the institution's ability to sustain long-term ecological improvements.

3. Resistance to Change and Unsupportive Organizational Culture

The study notes that environmental programs are sometimes perceived as "additional tasks" rather than core academic responsibilities. This resistance is exacerbated by:

- bureaucratic rigidity,
- limited autonomy in proposing innovative campus initiatives,
- and the absence of strong sustainability values embedded in institutional culture.

Transformational change in universities therefore requires leaders who can shift mindsets, cultivate shared purpose, and motivate collective action.

4. Fragmented Coordination Across Faculties and Administrative Units

GOI demands strong cross-unit collaboration, but the study finds that:

- faculties often operate in silos,
- knowledge-sharing mechanisms are weak,
- and coordination largely depends on individual champions rather than institutional systems.

This fragmentation limits the potential for systemic green innovation and reduces overall campus environmental performance.

Implications for Leadership

These findings underscore the need for adaptive leadership in higher education, complementing green transformational leadership. Adaptive leadership enables university leaders to:

- work with uncertainty,
- negotiate competing priorities,
- encourage organizational learning,
- and create safe spaces for experimentation.

Thus, integrating responsible leadership with GOI, GTL, and adaptive leadership provides a comprehensive model for strengthening environmental leadership within Indonesian universities.

8. INTEGRATING GREEN TRANSFORMATIONAL AND ADAPTIVE LEADERSHIP IN INDONESIA

Environmental sustainability leadership in Indonesia requires a blended framework that integrates green transformational leadership (GTL) and adaptive leadership. GTL provides the compelling future-oriented vision necessary to inspire cultural shifts, while adaptive leadership enables organizations and communities to navigate uncertainties, reconcile competing interests, and engage in continuous learning. Environmental challenges—such as climate change, deforestation, biodiversity loss, and water insecurity—are inherently complex and socially embedded, making neither GTL nor adaptive leadership alone sufficient.

Indonesia's dynamic socio-political context further amplifies the need for a dual approach. The country faces tensions between economic growth and ecological protection, central government directives and local autonomy, and immediate welfare concerns and long-term sustainability. Therefore, integrating GTL and adaptive leadership offers a holistic pathway capable of addressing both strategic visioning and context-specific problem solving.

1. National and Sectoral Leadership

At the national level, GTL appears through Indonesia's high-profile climate commitments, such as the Net Zero Emissions (NZE) 2060 pledge, the Forestry and Other Land Use (FOLU) Net Sink 2030, and the institutionalization of ESG frameworks for State-Owned Enterprises (SOEs). These commitments function as powerful strategic narratives signaling that sustainability is not merely aspirational but foundational for national development. High-level leaders—ministerial officials, agency directors, and SOE commissioners—demonstrate GTL when they align national identity with decarbonization, environmental stewardship, and global climate responsibility.

Yet, GTL alone cannot address policy fragmentation or implementation challenges across Indonesia's diverse geographical and institutional landscapes. Adaptive leadership becomes essential to bridge the gap between ambition and implementation. Policymakers must iteratively refine regulations by learning from pilots—carbon pricing simulations, peatland restoration trials, renewable energy auctions, and mangrove rehabilitation programs. This reflects Heifetz et al.'s (2009) argument that adaptive leadership is crucial when solutions require changes in beliefs, values, and relationships rather than technical adjustments.

Adaptive leadership also helps navigate political economy constraints, such as local resistance to coal phase-out policies, competing interests in forest resource management, and the push-and-pull between central government planning and regional autonomy. Scholars emphasize that climate governance in developing economies requires leaders who can "hold tensions" across stakeholder groups while promoting long-term ecological priorities (Folke et al., 2021).

2. Corporate and Organizational Leadership

In the corporate sector, CEOs and executive leaders embody GTL when they embed sustainability into corporate identity, strategy, and performance evaluations. This includes integrating climate risk into enterprise risk management, adopting science-based decarbonization targets, and championing green innovation as part of organizational culture. GTL also influences how companies communicate sustainability narratives—repositioning environmental goals as core to competitiveness, reputation, and long-term resilience.

However, operationalizing sustainability requires adaptive leadership at middle management and departmental levels. Sustainability officers and managers coordinate cross-functional teams, navigate institutional resistance, translate abstract sustainability mandates into actionable plans, and experiment with green technologies. They operationalize the learning loops that enable organizations to adjust strategies based on real-time feedback and contextual constraints. As Çop et al. (2021) note, GTL produces higher environmental performance when paired with mechanisms that foster experimentation and collaborative learning.

This resonates strongly with findings from the Indonesian university case (from the JARHE study), where responsible leadership improved sustainable performance only when accompanied by Green Orchestration Innovation (GOI)—a process requiring adaptive coordination across academic units, student bodies, and external partners. This reinforces that sustainability transformation is both vision-led and systemically orchestrated.

3. Community and Civil Society Leadership

Community-level leaders—village chiefs, religious scholars, youth activists, customary leaders, and NGO facilitators—play a critical role in shaping grassroots environmental action. GTL appears when these leaders articulate compelling visions of healthy rivers, protected forests, zero-waste villages, or climate-resilient livelihoods. Their moral authority anchors sustainability in cultural identity, local wisdom, and community aspirations.

Adaptive leadership becomes visible in participatory problem solving, where leaders co-design and iterate solutions with farmers, fisherfolk, tourism operators, and neighborhood groups. They test climate-smart farming practices, develop community-based waste management systems, mediate conflicts between livelihood needs and conservation goals, and adjust interventions based on feedback. This form of adaptive leadership is crucial in Indonesian rural and peri-urban settings, where environmental decisions are deeply intertwined with social norms and economic vulnerabilities.

Research on ecopesantren and tourism villages demonstrates that community leaders who combine GTL with adaptive problem solving are more effective in achieving sustainable outcomes (Hidayati et al., 2024; Kejora et al., 2025). Their leadership operates not only through directives but through cultivation of shared purpose and iterative learning.

4. Integrated Framework: Vision + Learning

From a theoretical perspective, Green Transformational Leadership (GTL) and Adaptive Leadership represent two complementary but distinct approaches to understanding how leadership can facilitate environmental sustainability transitions. Although both perspectives share a common concern regarding organizational change and long-term sustainability outcomes, they operate through fundamentally different leadership mechanisms and address different dimensions of organizational transformation.

Green Transformational Leadership primarily focuses on value-driven organizational transformation by shaping environmental vision, inspiring collective commitment, and embedding sustainability principles into organizational culture. Leaders operating within this framework mobilize employees by articulating shared environmental goals, fostering pro-environmental behavior, stimulating green innovation, and aligning organizational identity with long-term ecological responsibility. Existing studies consistently demonstrate that GTL contributes significantly to organizational environmental performance by strengthening employee environmental commitment, encouraging sustainable innovation, and improving the implementation of green organizational practices (Robertson & Barling, 2013; Çop et al., 2021).

However, the GTL framework alone may be insufficient when organizations confront highly uncertain environmental conditions characterized by institutional complexity, rapidly evolving climate risks, conflicting stakeholder interests, and unpredictable policy environments. In such contexts, strong environmental vision alone may produce what can be described as symbolic sustainability commitment or green rhetoric, where organizations formally adopt sustainability commitments but struggle to translate these commitments into flexible and context-responsive implementation strategies.

Adaptive Leadership addresses this limitation by emphasizing organizational learning, experimentation, distributed problem-solving, and the capacity to respond dynamically to complex and uncertain environmental challenges. Rather than relying primarily on visionary influence, Adaptive Leadership enables organizations to continuously adjust strategies, manage conflict among diverse stakeholders, facilitate institutional learning, and maintain organizational flexibility under conditions of environmental uncertainty (Heifetz et al., 2009). This perspective becomes particularly important in environmental governance contexts where sustainability transitions rarely follow predictable implementation pathways.

Nevertheless, Adaptive Leadership operating independently may also encounter important limitations. Organizations may become highly responsive and flexible in managing environmental uncertainty but lack a stable sustainability vision capable of aligning organizational values and generating long-term collective commitment.

Without the normative orientation provided by sustainability-centered leadership frameworks such as GTL, adaptive responses may become fragmented, reactive, and disconnected from broader long-term environmental transformation objectives.

Integrating GTL and Adaptive Leadership therefore provides a more comprehensive framework for sustainability leadership. GTL contributes the normative and motivational dimension by shaping environmental values, organizational commitment, and strategic sustainability orientation, while Adaptive Leadership contributes the operational and dynamic dimension by enabling organizations to navigate uncertainty, facilitate institutional learning, and continuously adjust strategies in response to changing environmental conditions. Effective sustainability leadership therefore requires leaders who are simultaneously capable of inspiring long-term environmental commitment while maintaining adaptive capacity necessary to manage the complexity, uncertainty, and multi-stakeholder dynamics inherent in contemporary environmental governance systems. The integration of both perspectives offers a stronger theoretical foundation for understanding how leadership can effectively support long-term sustainability transitions across increasingly complex organizational and socio-ecological environments.

Bringing GTL and adaptive leadership together highlights an important conceptual insight:

- GTL without adaptive leadership may produce inspiring visions that fail to translate into meaningful change—leading to symbolic sustainability or "green rhetoric."
- Adaptive leadership without GTL may generate small, disconnected improvements without addressing systemic drivers of unsustainability or shifting long-term trajectories.

Therefore, effective sustainability leadership in Indonesia requires synergy:

- GTL provides direction, identity, and inspiration,
- Adaptive leadership provides responsiveness, learning, and resilience.

This integrated model supports multi-level environmental governance in a country where ecological diversity, socio-economic pressures, and institutional complexity intersect.

9. IMPLICATIONS AND DIRECTIONS FOR FUTURE RESEARCH

9.1 Implications for Leadership Development and Practice

For Indonesia and similar emerging economies, leadership development must evolve beyond generic managerial competencies to incorporate specialized sustainability-oriented capabilities. Environmental issues in Indonesia—ranging from land-use conflicts, peatland fires, and biodiversity loss to rapid urbanization and water insecurity—require leaders who can operate across ecological, social, and political systems. Consequently, leadership development should focus on at least four interrelated competencies.

1. Environmental Literacy and Systems Thinking

Leaders must possess a foundational understanding of climate science, biodiversity dynamics, resource cycles, and circular economy principles. This does not require leaders to be technical experts; rather, they must be capable of interpreting scientific information, understanding ecological feedback loops, and anticipating how decisions in one sector (e.g., agriculture or energy) generate consequences in others (e.g., water availability or emissions). Systems thinking equips leaders to recognize non-linear impacts, tipping points, and the interconnected nature of environmental challenges, enabling more informed and strategic decision-making.

2. Ethical and Sustainability-Oriented Competence

Leaders frequently face difficult trade-offs between short-term economic expansion, long-term environmental protection, and the needs of vulnerable communities. Developing ethical competence involves cultivating frameworks to assess fairness, intergenerational responsibility, and ecological integrity. Leaders must be prepared to evaluate distributional impacts of policies and corporate actions and anticipate unintended consequences. In diverse Indonesian contexts—where indigenous rights, land tenure, and resource access are contested—ethical leadership becomes essential for legitimacy and social acceptance.

3. Change Management and Communication Skills

Sustainability transitions require narrative-building that resonates with employees, communities, investors, and policymakers. Leaders must be able to communicate complex environmental ideas in compelling, culturally grounded ways. This includes transforming sustainability from an abstract global agenda into a meaningful local identity—for instance, framing climate resilience as community well-being or linking circular economy practices to cultural values of frugality and mutual care. Effective communication also helps mitigate resistance, build shared purpose, and accelerate behavioral change within organizations and communities.

4. Collaborative and Multi-Stakeholder Facilitation

Environmental issues rarely fall neatly within organizational boundaries. Leaders must be equipped to convene actors across sectors—government agencies, businesses, NGOs, local communities, and international partners—to co-create solutions. Skills in negotiation, conflict mediation, and collaborative governance are especially critical in Indonesia, where environmental decisions frequently involve overlapping jurisdictions, competing interests, and asymmetrical power relations. Leaders who can facilitate multi-stakeholder processes can unlock resources, generate innovation, and build durable coalitions for sustainability.

Given these evolving requirements, leadership programs in universities, business schools, public administration institutes, and civil-service academies should integrate explicit modules on green transformational leadership (GTL) and adaptive leadership. Using Indonesian and ASEAN-based case studies—such as community forestry, eco-tourism, renewable energy transitions, and climate adaptation programs—would ensure contextual relevance. Embedding experiential learning, field immersion, and collaborative problem-solving exercises can further strengthen leaders' ability to apply theory to real-world challenges.

Beyond theoretical relevance, this chapter offers several practical implications for organizations and institutions seeking to accelerate sustainability transitions in increasingly complex environmental contexts. For policymakers, the integrated framework suggests that environmental governance strategies should move beyond regulatory compliance approaches by establishing adaptive policy systems that encourage experimentation, cross-sector collaboration, and institutional learning mechanisms capable of responding to rapidly evolving climate and sustainability challenges. Governments should therefore design sustainability policies that not only impose environmental standards but also strengthen organizational capacity for continuous adaptation under uncertain regulatory and ecological conditions.

For business organizations, the findings indicate that sustainability initiatives should be operationalized through leadership development systems that simultaneously cultivate long-term environmental commitment and adaptive decision-making capability. This requires organizations to institutionalize sustainability performance indicators, integrate environmental targets into corporate strategic planning, establish cross-functional sustainability teams, and create internal innovation mechanisms capable of responding to changing stakeholder expectations and market dynamics.

For universities and professional training institutions, the chapter highlights the need to redesign leadership development programs by integrating sustainability competencies, systems thinking, stakeholder negotiation, interdisciplinary problem-solving, and adaptive leadership capabilities. Preparing future leaders requires educational approaches that combine environmental awareness with practical decision-making skills necessary to manage complex sustainability transitions across diverse institutional environments.

9.2 Directions for Future Research

Despite growing interest, research on leadership in environmental contexts—especially in Indonesia and the broader Global South—remains limited. Future scholarship could advance the field in at least four promising directions:

1. Multi-Level and Cross-Scale Analyses

Researchers should examine how leadership at national, provincial, corporate, and community levels interacts to shape environmental outcomes. For example, studies could investigate how national NZE targets cascade into corporate strategies, provincial climate action plans, or village-level eco-tourism initiatives. Understanding cross-scale dynamics would help clarify how coherence (or fragmentation) in leadership shapes sustainability performance across institutional layers.

2. Comparative Studies Across Sectors and Regions

Indonesia's environmental challenges vary significantly across sectors (energy, forestry, agriculture, manufacturing, tourism) and regions (resource-rich provinces versus less resource-dependent ones). Comparative research could reveal how GTL and adaptive leadership manifest differently depending on economic structures, political contexts, and cultural values. Such studies would strengthen the external validity of leadership theories in sustainability.

3. Longitudinal and Process-Oriented Research

Most existing studies rely on cross-sectional designs. Longitudinal research—through process tracing, ethnography, or multi-year case studies—could capture how leaders navigate political economy constraints, how coalitions evolve, and how sustainability narratives gain or lose traction over time. This approach would illuminate the dynamic and often non-linear nature of sustainability leadership.

4. Leadership, Justice, and Inclusion

Environmental degradation disproportionately affects marginalized communities, including women, indigenous peoples, rural poor, and youth. Future research could explore how leadership intersects with gender equity, indigenous knowledge, land rights, and intergenerational justice. This would help integrate principles of environmental justice into leadership frameworks and decision-making processes.

By deepening empirical understanding of how GTL and adaptive leadership function across diverse Indonesian settings, scholars can contribute to advancing leadership theory and generating actionable guidance for policymakers, educators, and practitioners striving for sustainable development.

9.3 Theoretical Contribution of This Chapter

This chapter contributes to the growing body of scholarship on sustainability leadership by offering a more integrated conceptual understanding of how leadership can facilitate environmental sustainability transitions under increasingly complex organizational and socio-ecological conditions. While existing leadership studies have extensively examined sustainability-oriented leadership from multiple perspectives, prior scholarship often treats Green Transformational Leadership (GTL) and Adaptive Leadership as separate theoretical approaches addressing distinct organizational challenges. This chapter advances existing literature by positioning both frameworks as complementary rather than independent perspectives, thereby providing a more comprehensive analytical foundation for understanding environmental leadership in contemporary sustainability governance contexts.

The first contribution of this chapter lies in its theoretical integration of Green Transformational Leadership and Adaptive Leadership into a unified sustainability leadership framework. Existing GTL literature largely emphasizes the importance of environmental vision, organizational culture transformation, employee environmental commitment, and sustainability-oriented innovation. In contrast, Adaptive Leadership literature focuses more strongly on organizational learning, institutional flexibility, stakeholder negotiation, and adaptive responses under conditions of uncertainty. This chapter demonstrates that effective sustainability leadership requires simultaneous development of both normative environmental commitment and dynamic organizational adaptability, particularly when organizations face increasingly uncertain environmental conditions and complex sustainability transitions.

The second contribution lies in extending sustainability leadership scholarship toward emerging economy contexts. Much of the existing sustainability leadership literature remains heavily concentrated within developed-country organizational environments where institutional systems, regulatory frameworks, and sustainability governance structures are relatively mature. By incorporating Indonesian environmental leadership experiences, this chapter expands existing theoretical discussions by demonstrating how sustainability leadership operates within institutional environments characterized by developmental pressures, fragmented governance structures, resource constraints, and socio-political complexity commonly found across emerging economies.

The third contribution lies in proposing a broader conceptual perspective that sustainability leadership effectiveness should not be understood solely as visionary leadership capacity or organizational environmental commitment, but rather as the ability to continuously align long-term environmental objectives with adaptive organizational learning processes under conditions of uncertainty. This integrated perspective provides a stronger conceptual foundation for future scholarship examining leadership effectiveness within increasingly complex environmental governance systems and long-term sustainability transformation processes.

10. CONCLUSION

Leadership in environmental contexts is fundamentally complex, deeply value-laden, and embedded within dynamic socio-ecological systems. Unlike traditional leadership domains, environmental leadership must confront long-term horizons, scientific uncertainty, and moral dilemmas associated with intergenerational responsibility. This chapter has argued that two complementary lenses—green transformational leadership (GTL) and adaptive leadership—offer a robust theoretical and practical foundation for understanding how leaders can simultaneously inspire environmental vision and navigate systemic uncertainty.

International and interdisciplinary evidence consistently demonstrates that GTL contributes significantly to environmental performance. Leaders with a green transformational orientation articulate compelling ecological visions, cultivate pro-environmental cultural norms, and encourage creativity and experimentation with sustainable technologies. Through mechanisms such as idealized influence, inspirational motivation, and intellectual stimulation, GTL mobilizes employees and stakeholders to transcend short-term interests and commit to sustainability-oriented goals. Studies across manufacturing, tourism, education, and public-sector contexts show that GTL strengthens environmental engagement, enhances green creativity, and fosters organizational readiness for sustainability transitions.

At the same time, adaptive leadership is indispensable for addressing the inherently uncertain and contested nature of environmental problems. Climate change, land degradation, resource depletion, and biodiversity loss are adaptive challenges—problems that cannot be solved solely through technical expertise or fixed policy instruments. Adaptive leadership emphasizes learning, dialogue, conflict mediation, experimentation, and iterative adjustment, enabling systems to respond to emergent information and shifting political or ecological conditions. In climate governance, adaptive leadership helps institutions work through competing interests, reconcile divergent knowledge systems, and engage communities in co-creating solutions that are contextually appropriate and socially legitimate.

In the Indonesian context, the need for integrated leadership is particularly pronounced. Indonesia's environmental landscape is characterized by vast ecological diversity, rapid economic development, and complex governance arrangements that span national ministries, provincial authorities, village governments, state-owned enterprises, multinational subsidiaries, indigenous communities, and civil-society organizations. Examples highlighted in this chapter—from PLN's energy transition planning, to Pertamina's sustainability initiatives, to Allianz and Azbil's multinational environmental programs, to ecopesantren and tourism villages in Yogyakarta, to the creative zero-waste hospitality movement in Bali—demonstrate that environmental leadership is emerging across multiple scales and sectors. These cases show that Indonesian leaders are increasingly embracing sustainability as part of organizational identity, strategic risk management, and cultural transformation.

Yet, despite encouraging progress, persistent challenges continue to hinder Indonesia's environmental transition. Policy–implementation gaps remain pervasive due to fragmented regulatory frameworks and uneven enforcement capacity. Economic dependence on fossil fuels and extractive industries creates structural constraints that complicate the shift toward low-carbon pathways. Local governments and community organizations often face technical, financial, and human resource limitations. Questions of environmental justice—including the disproportionate burden of pollution, land degradation, and climate impacts on marginalized groups—remain unresolved. These structural issues underscore the need for leadership that can balance environmental integrity with economic and social considerations, while navigating complex political-economic realities.

To address these challenges, leadership development and research must intentionally incorporate environmental ethics, systems thinking, adaptive problem-solving, and cross-sector collaboration. Universities, business schools, public administration institutes, and civil-service academies must equip emerging leaders with the knowledge and skills required to manage ecological complexity and social diversity. Future research should deepen understanding of how GTL and adaptive leadership operate across Indonesia's regions and sectors, how they interact across scales, and how they influence environmental outcomes over time.

Ultimately, Indonesia's ability to move from climate commitments on paper to real sustainability transformations will depend on cultivating leaders who can envision, enable, and adapt. Only with such integrated leadership can Indonesia—and the broader global community—reconfigure how we produce, consume, govern, and coexist with the natural world.

REFERENCES

Amalia, R., Fadhilah, N., & Putri, A. (2025). Strategic sustainability and CSR transformation in Indonesian energy companies. *Journal of Asian Business and Sustainability*, *14*(1), 22–38.

Amjad, M., Abbas, M., & Saeed, R. (2023). Linking green transformational leadership and green creativity: The role of green HRM practices. *Environmental Science and Pollution Research International*, *30*(12), 14572–14589.

Avery, G. C. (2005). *Leadership for sustainable futures: Achieving success in a competitive world*. Edward Elgar Publishing. DOI: 10.4337/9781845425494

Awashreh, R. (2025). Leadership in the pursuit of net-zero emissions: Overcoming barriers and embracing strategic innovation. *Arab Gulf Journal of Scientific Research*, *43*(1), 18–36. DOI: 10.1108/AGJSR-02-2025-0030

Bappenas. (2024). *Indonesia Low Carbon Development Update 2024*. Ministry of National Development Planning.

Bass, B. M. (1985). *Leadership and performance beyond expectations*. Free Press.

Bass, B. M., & Riggio, R. E. (2006). *Transformational leadership* (2nd ed.). Psychology Press. DOI: 10.4324/9781410617095

Brundtland Commission. (1987). *Our common future*. Oxford University Press.

Burns, J. M. (1978). *Leadership*. Harper & Row.

Chen, Y., Tang, G., Jin, J., Li, J., & Paillé, P. (2020). Green human resource management and employee green behavior: The role of green self-efficacy and individual green values. *Journal of Business Ethics*, *164*(2), 1–16. DOI: 10.1007/s10551-019-04300-1

Çop, S., Olorunsola, V. O., & Alola, U. V. (2021). Achieving environmental sustainability through green transformational leadership policy: Can green team resilience help? *Business Strategy and the Environment*, *30*(1), 671–682. DOI: 10.1002/bse.2646

Elkington, J. (1997). *Cannibals with forks: The triple bottom line of 21st century business*. Capstone Publishing.

Ferdig, M. A. (2007). Sustainability leadership: Co-creating a sustainable future. *Journal of Change Management*, *7*(1), 25–35. DOI: 10.1080/14697010701233809

Folke, C., Österblom, H., Schultz, L., Olsson, P., Galaz, V., Hentati-Sundberg, J., & de Zeeuw, A. (2021). Governing complex commons: Social–ecological innovations and transformations. *Annual Review of Environment and Resources*, *46*, 319–349. DOI: 10.1146/annurev-environ-012220-011104

Freeman, R. E. (1984). *Strategic management: A stakeholder approach.* Pitman Publishing.

Hasan, M. M., Ali, M., Alam, M. M., & Jan, A. (2024). Sustainability leadership and strategic planning for SDGs. *Journal of Cleaner Production*, *412*, 137405.

Heifetz, R., Grashow, A., & Linsky, M. (2009). *The practice of adaptive leadership.* Harvard Business Press.

Hidayati, N. U., Effendi, M. I., & Kristanto, R. H. (2024). Effect of Green Transformational Leadership and Green Knowledge on Environmental Performance with Green Creativity as Intervening Variable in Tourism Villages in DIY. *West Science Journal Economic and Entreprenurship*, *2*(4), 581–593. Advance online publication. DOI: 10.58812/wsjee.v2i04.1425

Karamidehkordi, E., Karimi, V., Hallaj, Z., Karimi, M., & Naderi, L. (2024). Adaptable leadership for arid/semi-arid wetlands conservation under climate change: Using analytical hierarchy process (AHP) approach. *Journal of Environmental Management*, *351*, 119860. DOI: 10.1016/j.jenvman.2023.119860 PMID: 38128210

Kebe, I. A., Liu, Y., & Kahl, C. (2025). Greener Solutions: Green Transformational Leadership and Sustainable Environmental Outcomes within Sierra Leone's Telecom Sector. *Journal of Cleaner Production*, *517*, 145811. DOI: 10.1016/j.jclepro.2025.145811

Kejora, M. T. B., Komariah, A., Herawan, E., & Sudarsyah, A. (2025). Environmental Leadership of Ecopesantren Supports Sustainable Development Goals. *Didaktika: Jurnal Kependidikan, 14*(2 Mei), 2065-2076.

Masyhuri, M., Muttaqien, F., Rizki, V. L., Fitria, N., Sabrina, C., & Jawo, M. (2025). Empowering leadership and employee agility: The mediating role of knowledge sharing and perceived organizational support. *International Journal of Business Ecosystem & Strategy (2687-2293), 7*(6), 160-170.

Ministry of Environment and Forestry, Republic of Indonesia. (2023). *Indonesia's climate actions towards 2030: FOLU Net Sink.*

PLN (Persero). (2024). *Climate-related disclosure report 2024.*

Prabowo, T., & Kim, S. (2024). Environmental justice and community vulnerability in Southeast Asia: A case study of extractive regions in Indonesia. *Environmental Policy and Governance*, *34*(2), 101–118.

Robertson, J. L., & Barling, J. (2013). Greening organizations through leaders' influence on employees' pro-environmental behaviors. *Journal of Organizational Behavior*, *34*(2), 176–194. DOI: 10.1002/job.1820

Saleem, F., & Mateou, S. (2025). Servant leadership for sustainability: A serial mediation model. *Administrative Sciences*, *15*(10), 380. DOI: 10.3390/admsci15100380

Saleem, M., & Mateou, A. (2025). Servant leadership for sustainability and proactive environmental behavior: A multilevel perspective. *Sustainability*, *17*(2), 509.

Santoso, A., & Jaya, E. (2022). Policy coherence challenges in Indonesia's land-use governance. *Journal of Environmental Management*, *318*, 115–128.

Shriberg, M., & MacDonald, L. (2013). Sustainability leadership programs in higher education: Alumni outcomes and impacts. *Journal of Sustainability Education*, *5*, 1–21.

Vandini, V. A. D., & Ala'aldin, J. (2025). Strategic Initiatives of Multinational Corporations in Environmental Sustainability in Indonesia: Case Study of Allianz and PT Azbil Corporation. *PROIROFONIC*, *1*(1), 627–633.

Visser, W., & Courtice, P. (2011). *Sustainability leadership: Linking theory and practice. Cambridge Programme for Sustainability Leadership*. University of Cambridge.

Wijaya, A., & Ernst, N. (2023). Corporate climate transition strategies in emerging economies: Lessons from Indonesia. *Asia Pacific Business Review*, *29*(5), 765–784.

Chapter 8
When Being Seen Green Matters More Than Being Green:
What Corporate Event Practices Reveal About the Leadership Gap in Sustainability

Thanuja Rathakrishnan
http://orcid.org/0000-0002-5470-8931
School of Management and Marketing, Taylor's University, Subang Jaya, Malaysia

ABSTRACT

This chapter investigates how corporations bring their sustainability commitments to life at events. It argues that companies often focus on visibility than authentic environmental action, revealing both opportunities and gaps in sustainable leadership. While most research focuses on corporate reports, this chapter brings attention to the lived experiences of employees - highlighting the staff preparations in conveying sustainability values effectively. By adopting an interpretivist approach, this chapter introduces the concept of representational readiness, emphasizing that leadership is central to ensuring employees embody organizational sustainability authentically. The findings reveal that booths serve as visible markers of environmental commitment, yet many representatives appeared uncertain, hesitant, or unprepared when discussing their company's practices. The findings also exposed that employees' participation in events appeared more driven by organizational requirements or stakeholder expectations than by direct involvement in sustainability initiatives.

DOI: 10.4018/979-8-3373-7877-0.ch008

1. INTRODUCTION

Sustainability has become a key topic around the globe. Many conferences, workshops and events have surround themselves using the theme "sustainability" to attract event goers and visitors. The purpose of such events is to provide knowledge exchange, demonstrating sustainability or green processes and practices, showcasing their future sustainability plans of an organization and strengthen engagement among stakeholders who seek to understand how organisations respond to evolving sustainability expectations. As such, sustainability conferences and green-focused events have become central platforms for communicating organisational environmental strategies. However, observations from multiple events reveal a recurring gap between event intentions and the quality of their exhibitors' on-site representation (Liu & Cheng, 2022; Wee et al., 2022; Wee et al., 2021). Wee et al (2021) express that it is not suffice if the participating companies are implementing green practices, but other staff that are participating in the green events such as caterers, booth representatives, or event planners should undergone training for sustainability measures. As such, Wee et al (2022) suggests that the frontline communicators must be trained to represent and explain them to attendees; whilst ensuring that sustainability goals are met in practice.

Strong leadership should also openly communicate about their sustainability practices and process to the organization for awareness. By doing so, this would enable more representatives to represent their organization well (Bowdin et al. 2023). Past studies found that representatives who are tasked with presenting their organisations' sustainability initiatives were unable to describe how their companies' activities aligned with environmental goals (Liu & Cheng, 2022; Mair & Smith, 2021; Boström, Uggla, & Hansson, 2018). Some expressed uncertainty about their organisations' initiatives, while others indicated limited awareness of the purpose behind their participation in sustainability events. Although such responses do not necessarily reflect actual organisational performance, they influence how stakeholders interpret the sincerity and substance of sustainability efforts.

This pattern exposes what can be described as a leadership gap. Even when organisations have substantive environmental initiatives, their leadership may not ensure that employees tasked with public engagement are equipped to articulate these initiatives effectively as the exhibitor (Mikhailova, 2025; Bowdin et al. 2023). In some cases, these tasks were assigned at short notice to individuals with lower seniority or responsibilities unrelated to sustainability (e.g., junior level, administrators or human resources staff). As a result, stakeholders would have a misimpression that the organisation is merely wanting to "being seen green" than having genuine sustainability commitments (Cavallin Toscani et al., 2024). Stakeholders such as potential investors, researchers, academicians, potential candidates, or visitors who are interested; expect transparent, consistent, and informed communication, particularly when evaluating environmental claims (Marrucci, Iovino, & Iraldo, 2025; Mair and Smith, 2021). When representatives are not well informed or prepared, the organisation's sustainability message becomes unclear. The alignment between organisational practice and representative knowledge therefore serves as a key indicator of authenticity, accountability, and strategic commitment.

While many organisations invest in formal reports, marketing materials, green events and strategic frameworks, less emphasis is placed on ensuring that exhibitors who engage directly with stakeholders are equipped to communicate these efforts accurately. Although organisations have put in tremendous efforts into their processes and green practices, there seems to be a disconnect between what is being done in the organisation and how it is communicated to the public during events (Zanella et al., 2026; Mehra et al., 2025; Sherif et al., 2025; Zafari et al., 2025.). This shows that the awareness or the need to study on the booth representatives is still in its infancy. Situations like these can be detrimental to organizations globally if left unaddressed. Zhang et al. (2024) studied on the event exhibitors' strategies in shaping the visitors overall experience in the United States. It is found that, when an event makes individuals empowered and involved, they would feel a sense of belonging and would likely recommend and revisit the event. Another study in China by Jiang et al. (2025) also found that visitors would likely to enjoy events when there are guidance and support from the event organisers. Their study stated that "prior to the event, tourists typically seek information from various channels to enhance the rationality of their decisions. These include both formal platforms, such as government websites and event organizers' social media, as well as informal sources, like promotional materials" (p.6). They also added that upon getting initial information before attending the event, visitors would like more personal interaction with the representatives during the event. However, when representatives do not add on to new information or appear disengaged with the stakeholders, this would weaken stakeholder trust, reduced confidence in organisational green narratives, and missed opportunities for meaningful engagement. Over time, these effects can contribute to perceptions of inconsistency and may undermine the broader objectives of sustainability initiatives.

This chapter then asks, "How is organisational sustainability is performed through representative readiness at green-focused events?". To answer this, this chapter explores the representative's readiness to convey their organisations' sustainability legitimacy. The study draws attention to the ways in which leadership decisions or lack thereof, influence not just visibility but the perceived authenticity of corporate sustainability efforts. This chapter is important because it draws attention to a frequently overlooked but influential factor in sustainability communication - the role of organizational representative.

1.1 Key Terms

The key terms that guide this chapter include human representation, institutional pressures, organisational credibility, symbolic participation, and stakeholder interpretation, defined as follows:

Table 1. Interview questions

Key terms	Definitions	Source
Representative	The person who is in charge with communicating ideas, concepts, sharing information and answering questions.	Phan and Ninh (2026); Marshall and Sandberg (2011)
Greenwashing	When an organisation communicates misleading or unsubstantiated environmental claims that create a false impression of environmental performance or benefits.	Eriyawan and Utama (2026); Spaniol et al. (2024)
Organisational Credibility	The extent to which the organization involved, as information source, is perceived as being trustworthy and reliable and to whether the organization could possibly live up to its claims and professional credentials	Rosa, Romeo, and Durst (2026); Botha (2023, p.3)
Symbolic Participation	Individuals that are formally included but exert minimal influence over important matters such as processes, decision-making, operations.	Yusuf, Saliha and Ikbal (2026); Jhunior, de Araujo Góes, and Valentinov (2026); Yang et al., (2026)
Stakeholder Interpretation	Where stakeholders understand or give a meaning to certain messages and actions.	Davidescu et al., 2026; Bajani and Rotimi (2024); Alvarez and Sachs (2023)

2. LITERATURE REVIEW

2.1 Greenwashing in Corporate Contexts

Greenwashing is largely defined as an act of creating an impression that an organisation is more environmentally responsible than it is (Huang, Shi, & Jia, 2025). Existing research has advanced understanding of how organisations communicate their environmental image; however, these studies have focused on static, highly curated artefacts such as sustainability reports, websites, or press releases. Greenwashing is commonly operationalised through practices including selective disclosure, vague or ambiguous environmental claims, symbolic initiatives, and the selective omission of unfavourable information. These studies largely conceptualise greenwashing as a deliberate organisational strategy embedded in formal documentation and carefully crafted messaging.

In such events - exhibitions, sustainability-focused conferences, or stakeholder meetings, representatives would require explaining, defending, or exemplifying the organisation's environmental initiatives. When these individuals possess limited knowledge of sustainability issues, they may attempt to project an environmentally responsible image through carefully rehearsed or improvised statements. In doing so, they risk creating a discrepancy between organisational claims and the knowledge demonstrated, which stakeholders can perceive as a form of greenwashing (Janik & Ryszko, 2025; Forliano et al., 2025; Xu et al., 2025; and Persakis et al., 2025).

Emerging studies highlight the importance of examining how sustainability claims are enacted and interpreted in real-time interactions. Unlike written disclosures, face-to-face encounters allow stakeholders to probe for detail, assess the coherence of claims, and evaluate whether representatives can substantiate the organisation's purported commitments. A misalignment between corporate sustainability narratives and representatives' ability to articulate them - even if unintentional, may signal perceived greenwashing (Marinakou & Mathew, 2025; Montgomery, Lyon, & Barg, 2024; Sajid et al., 2024). Eventually, this would deteriorate credibility and trust.

2.2 Events as Stages for Corporate Image

Corporate events and exhibitions are one of the ways to get hold of public attention and awareness. Its role is to convey messages as part of a general communication process resulting in a presentation of images, texts and symbols that a prospective audience can extract meaning (Boudhane, 2025; Dounavi et al., 2022). Participation in these events is motivated by multiple factors, including the desire to enhance brand/organisation visibility, develop stakeholder trust, and align with institutional expectations (Mitchell et al., 2023; Jung & Lee, 2022). Previously, events were done in public parks or dedicated space. These days, it is also done in large shopping malls which attract not only event goers but mall goers who decided to join the event (Martín-Rojo & Gaspar-González, 2025; Wee et al., 2022). Due to this, the number of event goers have been increasing.

Leadership with green experience believes in equipping their representatives with clearer guidelines, stronger environmental narratives, and consistent messages. They understand the importance of having knowledgeable representatives who can communicate and deliver the message across. According to Qin et al. (2024), representatives experience with green initiatives significantly influences how confidently they communicate sustainability practices. Hence, stakeholders are able to differentiate or might misinterpret organisations that are "greenwashing" or are not being genuine in their participation of the event (Le et al., 2024; Rickly, 2022).

This is because some representatives can clearly explain green practices of their organisation (e.g., waste reduction, digitalisation, green processes, energy efficiency), while others provide superficial or symbolic information. In other words, representatives with limited exposure to sustainability initiatives are more likely to focus on compliance or minimal actions, whereas experienced representatives can translate sustainability into practical, understandable messages for visitors. Qin et al. (2024) also found that organisations adopt and communicate green practices partly to maintain legitimacy in the eyes of stakeholders. In exhibitions, this means representatives often communicate sustainability in ways that align with industry norms and public expectations, rather than deep environmental education.

Oklevik et al. (2022) found that human interaction and information exchange during the event influence stakeholders experience. Stakeholders perceive the event's value and credibility as well as their participation depends on the representatives as they are often a key point of contact for visitors to supplement meaningful interactions and clear information (Troy, Eng, & Overton, 2025). Consequently, the way representatives communicate sustainability-related information may influence not only visitors' understanding of environmental initiatives but also their overall exhibition experience.

All in all, the organisation's representatives become the primary agents responsible for translating corporate sustainability narratives into dialogue with external audiences (Dorfleitner & Utz, 2024; Snowden et al., 2022). Their ability to respond to questions, provide concrete examples, and demonstrate understanding plays a central role in shaping stakeholder perceptions of credibility. When representatives are unable to articulate environmental initiatives beyond surface-level claims, events may expose gaps between organisational messaging and internal understanding. Despite this, existing event literature has paid limited attention to representatives' preparedness; those who enact sustainability claims in real time. As such, this chapter addresses the identified gap by examining the preparedness of representatives at sustainability-focused events. It aims to contribute to literature by offering a deeper understanding of the roles, awareness, and practices of event and organisation's representatives, while generating both theoretical and practical insights. In doing so, this chapter also provides a foundation for future researchers to extend and refine knowledge in this area.

3. METHODOLOGY

This study adopts a qualitative research design using an interpretivist paradigm. The interpretivist approach is particularly suited for examining sustainability communication because it enables the exploration of how organisational messages are constructed, interpreted, and enacted in real-world contexts. Understanding this dynamic requires an approach that captures how organisation's representatives explain, interpret, and enact sustainability initiatives in context. By adopting an interpretivist lens, the study captures the nuances of these interactions, including how representatives explain initiatives, respond to unexpected questions, and rely on visual or symbolic cues to convey meaning.

The researcher took a reflective approach throughout the study, recognising that personal perspectives, the way questions were asked, and what was focused on during observation could influence how participants' behaviours were interpreted. Reflexive notes were kept to record impressions, decisions, and potential biases, helping to make the research process transparent and grounded in what was observed. Participants were aware of the researcher's presence, but care was taken to minimise any influence on their responses, so that their behaviour and interactions remained as natural as possible.

Data collection occured during live sustainability exhibitions and corporate green events in Malaysia, allowing for the capture of real-time, context-rich interactions. Conversations with representatives typically lasted between 20 to 25 minutes per booth, depending on the availability of the staff and the flow of the event. This duration was sufficient to cover core topics, observe engagement practices, and allow participants to elaborate on sustainability initiatives or challenges in explaining them. Spending sustained time at each booth also enabled the researcher to notice behavioural cues, reliance on materials, and interaction styles, adding depth to the understanding of event-based sustainability communication.

This methodology aligns directly with the chapter's objective of understanding how representatives' knowledge, preparedness, and communication practices influence stakeholder perceptions of organisational sustainability initiatives. By examining these interactions in context and over a meaningful duration, the study uncovers patterns of understanding, challenges, and opportunities that shape both the credibility of sustainability claims and the effectiveness of corporate engagement strategies.

3.1 Data Collection Approach

Data were collected through informal conversational interviews conducted at sustainability conferences and green-focused exhibitions in Malaysia. Conversations occurred naturally as representatives engaged in their official roles at event booths, providing opportunities to capture interactions in authentic, real-world contexts.

Representatives were approached politely and invited to discuss their organisations' sustainability initiatives on their own terms. Detailed field notes were taken immediately after each interaction, recording key points, behavioural cues, and engagement styles. No personal identifiers were collected to protect their identities.

This approach provided context-rich insights into how sustainability narratives are communicated in practice, how representatives handle questions, and how information is conveyed to stakeholders in live settings.

3.2 Sampling

A total of nine representatives from different organisations in Malaysia were engaged across the events. The sample was based on convenience sampling, a non-probability sampling approach commonly used in qualitative field research. Representatives were chosen because they were directly responsible for interacting with event attendees and were therefore positioned to communicate organisational sustainability messages. The number of respondents is appropriate for interpretivist inquiry, as the goal is depth rather than numerical representation. Meaningful patterns and recurring gaps in sustainability knowledge emerged across the interactions.

Malaysia was selected due to its growing emphasis on environmental governance, sustainability reporting, and participation in green-related forums. National policies, corporate governance codes, and institutional pressures have contributed to increased visibility of sustainability themes in corporate communication. Data were collected across several sustainability-focused events, including industry conferences, green technology exhibitions, and corporate sustainability forums, which attracted a mix of local and international organisations. The events typically included company booths, poster displays, and interactive sessions, allowing for natural observation of representatives. This variety of event types provided an opportunity to explore how representatives articulate organisational green initiatives across different formats, audiences, and levels of engagement. The Malaysian context, combined with the diversity of events, makes it an ideal setting for studying the practical enactment of sustainability communication within a rapidly developing sustainability landscape

3.3 Data Analysis

The data were analysed using a Braun & Clarke (2006) 6-phrases of iterative thematic analysis approach (refer to Figure 1). First phase, field notes were used in documenting initial concepts, as well as observations to directly record sample behaviours to see what information would emerge. In the second phase, these codes were then examined for recurring patterns across participants, and related codes were grouped into broader thematic categories. Themes were then identified through constant comparison, examining similarities and differences across organisational types and representative roles which are in the third phase. In forth phase, reflexive notes and the themes were revisited to ensure that interpretations were categorised in observed interactions rather than assumptions. Upon satisfaction, the themes will be named in the fifth phase which then leads to writing the report.

Figure 1. 6-phrases of iterative thematic analysis

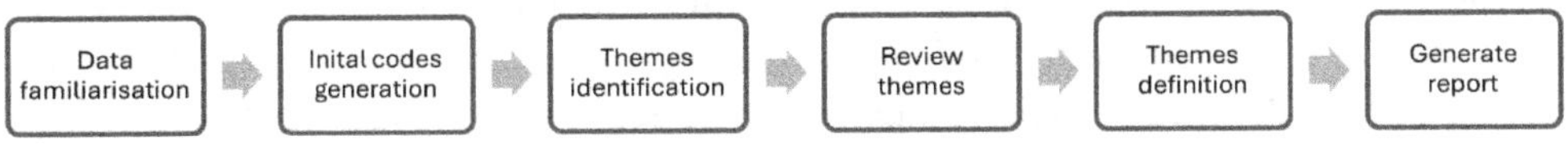

(Braun & Clarke, 2006)

This chapter employs an established criterion from Lincoln and Guba (1986) - credibility, dependability, confirmability, and transferability, to strengthen the trustworthiness of this qualitative study (Kocaman, 2025; Ahmed, 2024).

Credibility was enhanced by comparing patterns and themes across multiple participants. This is done by applying triangulation by cross-referencing conversational accounts with direct observations of representatives' behaviours, including how they interacted with visitors, responded to questions, and used visual materials or promotional content. Reflexive notes taken during data collection allowed the researcher to identify potential biases and maintain awareness of their influence on data interpretation. This iterative process helped confirm that the findings accurately reflected participants' knowledge, communication practices, and perceptions.

Dependability was achieved by using a consistent set of guiding questions across all informal interviews, while allowing follow up questions for natural conversation. This ensured that each interaction addressed core topics without imposing rigid structures that could limit authentic responses. It is important to achieve dependability to ensure the findings are consistent and stable over time.

Confirmability was addressed by recording field notes immediately after each interaction, capturing participants' statements as accurately as possible with minimal interpretation. Reflexive journaling enabled the researcher to acknowledge personal perspectives and their potential influence on data interpretation. Results such as analytical decisions, coding and theme choices were documented systematically to ensure that findings were not based on the researcher's biasness. Direct quotes were included in the findings to allow verification of thematic coding; subsequently, conclusions were drawn from the interpretations of themes.

This chapter provides detailed descriptions of the research context such as participant roles, organisational types, and interaction settings. This allows readers to determine if the findings are transferable and generalizable across other organisations. All in all, the criterion by Lincoln and Guba (1986) ensure that the chapter provides a trustworthy basis for understanding how human representation shapes sustainability communication in real-world event settings.

3.4 Ethical Considerations

Ethical conduct was ensured, especially since the data collected was informal and public. Data was collected through discreet observation of representatives at public sustainability events. Due to the nature of such data collection, care was taken to engage respectfully and minimise any disruption to representatives' responsibilities. To protect participants' privacy and organisational confidentiality, personal identification or corporate information were recorded. The questions asked focused on general insights regarding sustainability communication practices, and observations on the body language and the preparedness of the representatives in entertaining stakeholders into their booth. Field notes were anonymised and data access limited to the researcher.

The approach was done via face-to-face interactions. The researcher encouraged the representatives to share information that they were comfortable with. All representatives were aware of the researcher's role, whereby the researcher partook the event as an event visitor – making sure the representatives feel at easy and not being judged. The representatives were also informed that pseudo names will be used in the chapter's finding.

By implementing these measures, the study upheld ethical principles of privacy, confidentiality, respectful engagement, and minimal risk to participants, ensuring trustworthiness and credibility in the collection and analysis of qualitative data.

3.5 Interview Questions

Data collection relied on informal conversational interviews, which were guided by a concise field guide specifically developed for this study. The field guide consisted of short, open-ended, and flexible questions designed to encourage representatives to discuss their experiences in their own words, without feeling constrained by rigid or overly structured instruments. This approach allowed representatives to share insights naturally, reflecting how they actually communicate sustainability initiatives during real-time interactions with stakeholders at events.

The questions were organised around four key areas: representatives' roles and reasons for participating in events, their understanding of organisational sustainability initiatives, their communication practices and strategies, and their perceptions of how stakeholders interpret these initiatives. The field guide functioned primarily as a conversation facilitator, ensuring that all interviews covered core topics while allowing the researcher to probe deeper when participants offered interesting or unexpected insights.

Through this approach, the chapter captured authentic, context-rich responses, reflecting not only what representatives knew, but also how they explained, framed, or struggled with their organisation's sustainability messages. The conversational format also enabled the researcher to observe non-verbal cues, engagement styles, and interaction patterns, providing additional layers of understanding that extended beyond verbal responses.

Overall, the field guide ensured consistency across interviews in terms of core topics while maintaining the flexibility needed for natural, responsive, and context-sensitive data collection, which is central to interpretivist qualitative research.

Table 2. Interview questions

Topic Area	Questions
Role & Event Participation	1. Could you tell me a bit about your role in the company? 2. What made your company decide to take part in this event?
Understanding of Sustainability Initiatives	3. What kind of green or sustainability work is your company involved in? 4. How did you/your company prepare for being at this event?
Communication Practices	5. Is it easy for you to explain your company's green work to visitors? 6. What kinds of questions do people usually ask you here?
Perceived Impact / Stakeholder Impressions	7. What do you think visitors take away from your company's green efforts after talking to you? 8. What do you think would make it easier for you to explain these things at events like this?

3.6 Profile

The exploration was conducted by interviewing respondents from different industries and roles. Table 3 summarises the organisation type, roles, involvement and preparation for event as well as the observed communication.

Table 3. Respondents profile

ID	Organisation Type	Role / Event Task	Involvement in Green or Sustainability	Preparation for Event	Observed Communication
R1	Oil & Gas	Marketing Officer	Limited; mainly event representation	Brief email; no formal training	Relied on posters; unsure on detailed questions
R2	Manufacturing	Human Resources	Minimal direct involvement	General document	General answers; referred visitors to printed material
R3	FMCG	Administrator	None; only event support	None; simply assigned	Often distracted; appeared hesitant to engage; apprehension toward inquiry. Answers non-suitability questions

continued on following page

Table 3. Continued

ID	Organisation Type	Role / Event Task	Involvement in Green or Sustainability	Preparation for Event	Observed Communication
R4	Technology	Sustainability Assistant	Moderate; assisted in CSR reporting	Familiar with key initiatives	Confident on main points, provide more information on future green initiatives
R5	Banking	Assistant Branch Manager	low; not part of green team	Email instructions; event objectives	Basic answers; mostly referred to visual displays
R6	Energy	CSR support	Moderate; involved in environmental projects	Briefing call; overview of initiatives	Able to explain main projects
R7	Manufacturing	Administrator	Minimal; no direct sustainability work	Assigned by HR; no briefing	Gave vague responses; relied on posters
R8	Technology	Marketing Officer	Moderate; supported CSR campaigns	Briefing session; slide deck	Could explain initiatives
R9	FMCG	Logistic supervisor	Low; mainly logistic support	Briefing only	Basic responses; unsure when probed

4. FINDINGS AND DISCUSSION

4.1 Representative Knowledge and Understanding of Sustainability Initiatives

Analysis of the interactions with the nine event representatives revealed significant variation in their knowledge and understanding of organisational sustainability initiatives. Some participants, such as R4, R6, and R8 demonstrated moderate familiarity with their company's environmental projects and were able to articulate key initiatives clearly. While they occasionally struggled with technical details, their explanations reflected genuine engagement with organisational goals and provided visitors with meaningful insight. For instance, R4, a sustainability assistant at a technology company, confidently described her company's efforts in reducing carbon emissions and supporting community-based environmental projects. She was able to explain not only what initiatives were being implemented but also why they were important for the organisation and its stakeholders. In contrast, other representatives exhibited limited understanding and often relied on visual displays, brochures, or prepared materials to communicate. Some participants openly expressed uncertainty when probed specific initiatives, highlighting gaps between organisational strategy and the representatives' knowledge. This variation underscores the performative nature of sustainability communication - even well-designed initiatives risk being perceived as superficial if the people representing them cannot convey their purpose and outcomes effectively. It also aligns with literature on greenwashing and symbolic legitimacy through communication rather than solely through actual performance; demonstrating that the credibility of corporate sustainability is closely tied to the knowledge and engagement of frontline communicators.

> **R4:** "I am a sustainability assistant in this company. I support CSR reporting and internal sustainability projects. Main initiatives include energy efficiency in our data centres, reducing electronic waste, and responsible sourcing."

R6, who supports CSR works in the company involved in renewable energy projects, also conveyed substantive information while acknowledging the limits of her expertise:

> **R6:** "Mainly renewable energy projects and environmental management at our sites… I can explain the main projects but not technical metrics."

Similarly, R8, a marketing officer supporting CSR campaigns, leveraged her preparation materials to communicate effectively:

> **R8:** "We focus on energy-efficient solutions, reducing operational emissions, and responsible sourcing… The slide deck helps me structure explanations."

In contrast, several representatives demonstrated limited understanding of their organisations' initiatives and often relied on visual materials to convey information. R1, a marketing officer at an oil and gas company, deflected detailed questions on posters, eventually leading to scanning the QR to know more details of their green initiatives:

> **R1:** "Hmm… generally energy transition, reducing emissions… most of the details are on the posters here."
> "You may also scan the QR code, which you can see the videos of the green processes."

R3, an administrative staff member supporting FMCG logistics, similarly relied on displays to convey information. The representative initially appeared apprehensive about interaction, avoiding eye contact until prompted by the researcher's questions.

R3: “Mostly related to packaging… It is all mentioned there.” Both the representative and the researcher were trying to look for the green/sustainability information in the posters.

R7 and R9 also exhibited low knowledge and hesitancy:

R7: “I mostly let the posters explain.”
R9: “Most of the details are there… I can answer basic questions but deeper ones (complex questions) I’m not sure.”

R3, R7 and R9 ended the interaction by pointing the QR code to their company’s website.

These findings illustrate that knowledge gaps among representatives are a critical factor in how sustainability messages are conveyed and perceived by stakeholders. While the low-knowledge respondents were not intentionally misleading, their reliance on posters and prepared materials reduced the richness and credibility of communication as well as the green efforts of their organisations. This reflects the performative dimension of sustainability communication in events; in which, representatives act as the bridge between organisational practice and stakeholder perception. In cases where knowledge is limited, the authenticity of the corporate message is weakened.

In summary, the analysis highlights a key qualitative insight where representative knowledge and understanding plays a central role towards perceived legitimacy of sustainability initiatives. When representatives are moderately informed, they can facilitate meaningful or fair narratives. Likewise, when they are uninformed, genuine initiatives are at risk of being interpreted as superficial.

4.2 Preparation and Support for Representatives

The conversations revealed notable differences in the preparation and support provided to representatives prior to participating in sustainability events. Participants who received structured briefings, orientation sessions, or slide decks generally demonstrated greater confidence and ability to respond to visitor inquiries. For example, R8, a marketing officer in a technology company, described her preparation as including a briefing session and access to a slide deck summarising key sustainability initiatives. This preparation allowed her to answer questions with clarity and to contextualise initiatives within broader organisational goals, illustrating how adequate support enhances representatives’ effectiveness in public-facing communication.

Conversely, several participants reported minimal or no preparation, often receiving only brief emails or simple assignment instructions. These representatives tended to rely on posters or brochures during conversations and struggled to provide detailed answers. Organisations may have comprehensive sustainability initiatives, but without adequate support, representatives cannot translate these into coherent and credible messaging.

> **R8:** "We had a briefing session and were given a slide deck with key messages… the materials help structure the conversation."

R4, who was directly involved in sustainability projects, noted that familiarity with the initiatives through her work also helped her communicate effectively:

> **R4:** "I was already familiar with most of the initiatives because I work with them… it is easier to explain them naturally to visitors."

R6, the energy company event coordinator, mentioned receiving an overview briefing before the event, which enabled her to discuss ongoing projects:

> **R6:** "We had a briefing call and an overview of the initiatives to highlight. I can explain the main projects, though not all the technical details."

In contrast, other representatives reported minimal or no preparation. R3, an administrative staff supporting FMCG logistics, was simply assigned to the booth without guidance:

> **R3:** "I was just told last minute to help at the booth… I mostly let the posters explain."

R7, another administrative staff member, confirmed this lack of preparation:

> **R7:** "HR just assigned me to help here. Since there are posters and video clips to show, I just let the visitors read or hear the clips themselves."

R9, involved mainly in logistics, received only a short briefings:

> **R9:** "We attended the briefings … For basic questions, yes, but deeper ones I am not sure."

R1, a marketing officer at an oil and gas company, also indicated limited preparation:

> **R1:** "To be honest, I do not know why I am here. Since we are in an oil and gas company, I am not sure what is 'green' about our operations. I am quite shocked that the host approved our proposal to participate in the event."

These examples suggest that preparation or the lack thereof directly influences representatives' ability to communicate initiatives with confidence and clarity. Structured preparation, whether through briefings, slide decks, or hands-on involvement in sustainability projects, equips representatives to engage meaningfully with stakeholders. Conversely, minimal guidance leads to reliance on symbolic materials, superficial answers, and uncertainty, which can negatively affect stakeholder perceptions.

From a qualitative perspective, these findings highlight the interplay between organisational support and the performative quality of communication. Even representatives with some knowledge of sustainability projects can be hindered by inadequate preparation, while structured support can compensate for moderate knowledge gaps, allowing representatives to convey initiatives more effectively. The data emphasize that preparation is not just procedural; it is a strategic factor in shaping credibility, stakeholder trust, and perceived authenticity of corporate sustainability messages.

4.3 Communication Practices and Challenges

The interactions highlighted a strong connection between representatives' knowledge, preparation, and communication practices. Participants with moderate involvement in sustainability (R4, R6, R8) were able to engage visitors substantively, describing initiatives in depth and providing context about their relevance and impact. For example, R4 described her role in monitoring environmental projects and could navigate questions regarding project scope, goals, and outcomes with relative ease. This reflects a higher level of performative competence, where representatives act as bridges between organisational strategy and stakeholder understanding.

By contrast, representatives with limited knowledge or preparation often defaulted to symbolic communication, pointing to visual displays or reading from brochures without elaboration. They found it challenging to respond to unexpected or technical questions, highlighting the gap between formal organisational initiatives and the capacity of individuals to communicate with them effectively. This demonstrates that event-based interactions are not merely informational - they are dynamic performances through which credibility, authenticity, and organisational commitment are judged.

> **R4:** "...visitors usually ask about our energy efficiency projects and waste management. I can explain the process and how we track results."

R6, despite some limits on technical metrics, could convey ongoing environmental projects:

> **R6:** "I can explain our renewable energy projects and environmental management at our sites. When visitors; especially those exploring potential future collaborations ask detailed technical questions, I usually refer them to our project teams."

R8 leveraged structured materials to communicate initiatives clearly:

> **R8:** "The slide deck helps structure the conversation… visitors can see what's ongoing and what's planned. It makes answering questions easier."

By contrast, low-knowledge representatives often defaulted to symbolic communication. Many relied heavily on posters, brochures, or slides and were hesitant to answer in-depth questions. R1, R3, R7, and R9 provide illustrative examples:

> **R1:** "Most of the details are on the posters here… Maybe the green initiative is because of the operational part? To be really honest, I do not know… haha"
> **R3:** "It's all mentioned there… (pointing at the brochure)."
> **R7:** R7 pointed at the posters and appeared to be reading them for the first time.
> **R9:** "For basic questions, yes… but deeper ones I cannot be 100% sure. Most of the information is in the materials."

These communication practices highlight the interplay between knowledge, preparation, and performative delivery. Moderate knowledge combined with structured preparation allows representatives to navigate inquiries, contextualize projects, and engage meaningfully. Low knowledge or inadequate preparation, however, limits dialogue and reinforces symbolic representation, potentially weakening stakeholder trust. As R7 noted,

> **R7:** "I give vague responses or rely on posters… I wish I knew more to explain properly."

In sum, the findings suggest that communication practices are not just procedural but constitute a core aspect of perceived organisational credibility. For qualitative research, these observed patterns illuminate how frontline representatives enact the sustainability narrative, revealing both strengths and vulnerabilities in event-based engagement.

4.4 Perceived Impact on Stakeholder Impressions

The conversations further indicated that the quality of representative communication directly shapes stakeholder perceptions of organisational sustainability efforts. Knowledgeable and well-prepared representatives (R4, R6, R8) reported that their ability to explain initiatives contributed to a coherent and credible image of their company. For example, R4 observed that visitors appeared engaged and often expressed appreciation for the clarity of the explanations provided, suggesting that the representative's competence reinforced trust and legitimacy.

In contrast, representatives who were unsure or relied heavily on materials risked creating impressions of superficiality or limited organisational commitment. Stakeholders may interpret gaps in knowledge as misalignment between strategy and practice, potentially undermining the perceived authenticity of sustainability initiatives. These findings reinforce the theoretical argument that sustainability communication is not only about what organisations do, but also about how representatives perform and articulate these initiatives in real-world interactions, bridging internal organisational efforts with external stakeholder perceptions.

> **R4:** "I think visitors see that sustainability is integrated into operations, not just a marketing thing. They understand we are doing something meaningful."

R8 similarly emphasized how structured communication shaped visitors' understanding:

> **R8:** "They understand that sustainability is part of our business strategy, not just a side activity. The slide deck helps me show what's happening now and what's planned."

R6 noted that even when technical questions arose, conveying key initiatives clearly helped visitors appreciate the company's commitment:

> **R6:** "The visitors seem to understand our efforts. I try to explain the best I can and hopefully they see there is a real program behind it."

In contrast, representatives with limited knowledge or preparation (R1, R3, R7, R9) expressed concerns that their interactions might leave visitors with a weaker impression. R1 admitted:

> **R1:** "I am quite worried if more people ask questions because I am unsure what to tell them. But it is a good thing you asked because now I can call my boss

and ask about the green initiatives of our company… before more visitors ask me questions. I am equally curious about it too. But please do visit our website and from there you can see more pictures of our initiatives". R3 reflected:
R3: "Usually when we go to events, we just stick the posters. We do not really want people to ask us questions to be honest haha. Because I do not know how to answer them and worried this will give a bad impression on the company. But I also hope the company would send someone who is directly involved with these initiatives as I am only the administrator"

R7 and R9 echoed this sentiment, noting that limited personal knowledge could hinder meaningful engagement:

R7: "I wish I knew more to answer their questions properly. Usually, they see us struggling to answer, they would just smile and walk away"
R9: "For basic questions, I can respond… but when they ask about deeper details, I am not always sure. I will just get them to visit our website. Then they would just walk away"

These reflections indicate that representative knowledge and confidence serve as mediators between organisational initiatives and stakeholder perception. Even genuine sustainability efforts can be perceived as superficial or symbolic when representatives are unsure or overly reliant on materials. Conversely, knowledgeable representatives enhance credibility, build trust, and create a more coherent organisational narrative.

The patterns highlight the interpretive nature of stakeholder impressions. Visitors do not merely observe visual displays; they interpret interactions, responses, and confidence levels as signals of authenticity. The findings reinforce the idea that effective human representation is central to sustaining organisational legitimacy, and that alignment between strategy, knowledge, and communication is crucial for shaping positive perceptions.

In essence, stakeholder impressions at events are constructed through interaction. Representatives act as active interpreters of an organisation's sustainability narratives, and the depth of their knowledge affects whether initiatives are seen as meaningful or symbolic.

5. CONCEPTUAL FRAMEWORK AND THEORY

Figure 2. Conceptual framework

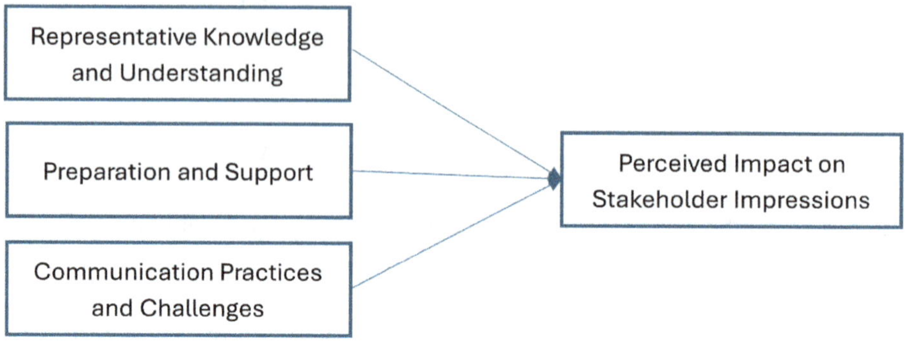

Based on empirical findings, this chapter conceptualises an organisation's sustainability operations at green-focused events as an interactional process; which is influenced more by the individuals that represent the organisation and how they engage with the stakeholders. While participation in sustainability conferences and exhibitions reflects broader institutional pressures for organisations to appear aligned with environmental norms, this chapter's findings revealed that visibility alone is not sufficient to how sustainability commitments are interpreted. Instead, the framework highlights representative role and their preparation as key conditions in providing impact to stakeholders. From the findings, representatives who were directly involved in sustainability work or who received structured briefings demonstrated greater confidence and clarity during interactions, while those with limited involvement relied heavily on posters, brochures, or surface-level explanations. The level of representative preparedness also shaped how sustainability was communicated in real time, including the depth of explanation offered, the ability to respond to questions, and the capacity to sustain dialogue with visitors. For instance, highly prepared representatives were encouraging and happy to welcome stakeholders to their booth, as opposed to those with low level or preparedness that appeared to be distracted.

As such, conceptual framework in Figure 2 is justified through legitimacy theory, whereby stakeholder's confidence and trust are shaped through human interaction. Organisational representatives thus function as the critical link between organisational sustainability efforts and stakeholder evaluation. This illustrates how legitimacy is constructed through representative's knowledge and understanding, preparedness and communication.

6. PATTERNS AND EMERGING THEMES

Analysis of the nine representatives' conversations and observations revealed several recurring patterns that illuminate how sustainability communication is enacted in event settings. These patterns provide insights into the broader implications for organisational credibility and stakeholder perception.

6.1 Knowledge Gaps

A consistent theme was the variation in representatives' understanding of sustainability initiatives. Several respondents demonstrated limited knowledge, relying heavily on visual materials or providing vague responses. R1 noted:

> **R1:** "Most of the details are on the posters here… I am not sure if I can explain more."

Similarly, R3 admitted:

> **R3:** "Mostly related to packaging… It is all mentioned in the poster."

These gaps highlight a disconnect between the organisation's sustainability strategy and the capacity of its representatives to communicate initiatives meaningfully. Knowledge gaps were particularly pronounced among administrative or logistics staff who were assigned to booths without direct involvement in sustainability projects.

6.2 Reliance on Visuals and Materials

Many representatives defaulted to posters, brochures, or slide decks to convey information. This reliance often served as a proxy for personal knowledge but limited engagement depth. R7 explained:

> **R7:** "I mostly let the posters explain"

R9 echoed this approach:

> **R9:** "Most of the details are there… For basic questions, yes, but deeper ones I am not always sure."

While visuals can support communication, over-reliance signals symbolic participation rather than informed dialogue. Stakeholders interpreting these interactions may perceive initiatives as superficial, even when genuine.

6.3 Inconsistent Preparation

Preparation varied widely across organisations, affecting representatives' confidence and effectiveness. Structured briefings, slide decks, and prior involvement in sustainability initiatives enabled representatives to communicate more clearly. For instance, R8 stated:

> **R8:** "We had a briefing session and were given a slide deck with key messages… the materials help structure the conversation."

In contrast, minimal preparation limited engagement. R3 and R7 described being assigned to booths without guidance:

> **R3:** "I was just told to help at the booth… I mostly let the posters explain."
> **R7:** "HR just assigned me to help here. I give vague responses or point to the posters."

This variation demonstrates that organisational support plays a crucial role in enabling representatives to act as effective communicators. Additionally, all booths have a QR to scan which directly leads to their company's website to know more details. From the visit, all 9 representatives requested to scan the QR to know more about their green initiatives.

6.4 Impact on Credibility

Representative knowledge and preparedness were closely tied to perceived organisational credibility. Confident and informed communication enhanced stakeholder trust, while uncertainty or reliance on materials risked skepticism. R4 noted:

> **R4:** "I think visitors see that sustainability is integrated into operations, not just a marketing thing."

R1, by contrast, reflected on potential limitations:

> **R1:** "I am quite worried if more people ask questions because I am unsure what to tell them."

These contrasting perspectives illustrate how frontline communication mediates the perceived authenticity of sustainability efforts.

6.5 Opportunities for Improvement

Many representatives identified a need for additional guidance, structured briefings, or accessible talking points to communicate effectively. R9 emphasized:

> **R9:** "A short briefing or key talking points would help a lot."

R7 also suggested:

> **R7:** "… even a quick overview of the main projects would help."
> **R1:** "…I also hope the company would send someone who is directly involved with these initiatives as I am only the administrator"

These reflections indicate that targeted preparation, selecting the right representative and resources can strengthen representatives' capacity to convey organisational sustainability narratives, improving both stakeholder understanding and credibility.

The emerging patterns collectively underscore the chapter's central argument: human representation is a critical determinant of perceived organisational sustainability credibility. Knowledge gaps, reliance on symbolic materials, and inconsistent preparation weaken communication, even when initiatives are genuine. Conversely, moderate knowledge combined with structured preparation enables representatives to facilitate meaningful engagement, bridging the gap between strategy and public perception. From a qualitative perspective, these findings reveal the performative, interpretive, and relational nature of sustainability communication, highlighting the strategic importance of aligning representative knowledge, organisational support, and public interaction.

7. IMPLICATIONS

7.1 Theoretical Implications

The findings of this chapter offer theoretical insights that deepen understanding of how organisational credibility and legitimacy are constructed through human interactions. Prior research focused on tangible outputs such as formal strategies, reports, or corporate social responsibility campaigns. This chapter, however, reinforces key principles of legitimacy theory by illustrating that the real-time performance of employees is mainly on how they explain, interpret, and respond to stakeholder inquiries. Representatives who relied primarily on posters or printed materials, or who were unsure how to answer specific questions, inadvertently signaled ambiguity or superficial engagement. Even when organisational initiatives were substantive and well-intentioned, the limited knowledge or preparation of representatives reduced perceived credibility. Conversely, employees who had direct experience with projects, such as a sustainability assistant explaining concrete initiatives, were able to provide contextual detail and nuanced explanations, actively reinforcing organisational legitimacy in the eyes of stakeholders.

These observations provide a nuance theoretical standpoint, which is legitimacy is performative and relational. It emerges in the interaction between organisations and their stakeholders, shaped by knowledge, confidence, and the ability to articulate experience. The study also extends the discourse on greenwashing by suggesting that credibility gaps do not always result from deliberate misrepresentation. Instead, they may arise from inconsistencies in human enactment, highlighting that the translation of strategy into action requires alignment between organisational design, employee competence, and public-facing roles. In this sense, the study bridges legitimacy theory with practical enactment, showing that legitimacy is not only signaled symbolically but also co-constructed through interpersonal encounters.

7.2 Practical Implications

Practically, the study suggests several steps managers can take to strengthen sustainability communication. First, leadership should prioritise structured training for event representatives. This involves not only providing detailed or sufficient information about initiatives but also preparing employees to handle potential questions confidently. For example, short briefing sessions, role-play exercises, or concise "cheat sheets" with key talking points can help representatives internalize information and respond naturally, rather than relying solely on posters or brochures. The reason is clear; representatives that are prepared convey authority and authenticity, increasing stakeholder trust and reducing the risk of perceived superficiality.

Second, selecting representatives who are directly involved in sustainability initiatives can enhance credibility. Employees who participate in project design, implementation, or monitoring are better able to discuss objectives, challenges, and outcomes in concrete terms. This how-to can be achieved by coordinating with internal sustainability teams and identifying staff whose experience aligns with public-facing roles. The why is that first-hand experience provides richer, more persuasive communication, bridging the gap between organisational strategy and stakeholder perception. In contrast, sending representatives who lack relevant knowledge can be damaging. For example, one of the staff expressed uncertainty about why they were participating in sustainability events or struggled to identify 'green' aspects of their operations. Such scenarios risk undermining the organisation's credibility, as uninformed representatives may struggle to explain initiatives convincingly, leaving stakeholders with doubts about the authenticity of sustainability claims.

Third, internal alignment and ongoing communication are essential. Organisations should ensure that all employees who might represent the company in public-facing contexts understand the organisation's sustainability goals and messaging. Regular updates, newsletters, or internal briefings can keep staff informed about new initiatives, milestones, or targets. This approach ensures that knowledge gaps are minimized and that representatives' messages remain consistent across different events and audiences.

Finally, evaluation and feedback loops are important. Managers can gather feedback from both visitors and representatives after events to identify knowledge gaps, common questions, and areas where communication could be improved. Insights from these evaluations can guide the development of future training, briefing materials, and selection of representatives. The why is that iterative learning strengthens the organisation's capacity to communicate sustainability initiatives effectively, reinforces credibility, and reduces the risk of miscommunication or perceived greenwashing.

In sum, these implications illustrate that effective sustainability communication requires deliberate alignment between organisational strategy, representative knowledge, and stakeholder interaction. By focusing on who communicates, how they are prepared, and how interactions are structured, organisations can enhance credibility, foster trust, and translate sustainability initiatives into meaningful engagement rather than symbolic gestures.

8. LIMITATIONS AND FUTURE RESEARCH SUGGESTIONS

While this study provides valuable insights into the role of human representation in sustainability communication at events, several limitations should be acknowledged. First, the study relied on a small sample of nine representatives from various organisations in Malaysia. Although this sample allowed for in-depth interpretive exploration of communication practices, it limits the generalizability of the findings to other contexts, industries, or countries. Future research could expand the sample size and include diverse geographic and sectoral contexts to examine whether similar patterns emerge in different institutional environments. Additionally, the sample and selection of interviewees may not initially be assigned to represent sustainability events. This could be due to last-minute changes or unavailability of the representative; the organisation may assign staff who might not have sufficient knowledge or time to prepare. Since the objective of the chapter is to explore real time experiences of sustainability events, perhaps future studies could consider representatives who are directly involved in sustainability or green efforts.

Second, the data was collected using informal conversational interviews at events. This style of data collection offers a rich, naturalistic understanding of representative practices but may be influenced by the presence of the researcher and the dynamic environment of events. Some representatives may have felt uncertain or self-conscious, potentially affecting the depth and accuracy of their responses. Future studies could focus on a formal or traditional route of interviewing their subject respondents (e.g., to get formal approval for interview and providing sample questions to prepare prior to interview).

Additionally, all respondents in this chapter had low to moderate involvement in sustainability initiatives, which may have influenced their ability to communicate effectively about organisational practices. Future research should include representatives with high levels of involvement to explore how deeper knowledge and direct responsibility affect credibility and stakeholder perception. Lastly, the data collected was at free, public sustainability events, where organisations may be more likely to send staff who are not directly involved in sustainability. Future studies could examine paid or invitation-only events to assess whether the type of event influences the selection of representatives and the quality of sustainability communication.

Overall, these limitations highlight opportunities for future research to build on the present study by expanding methodological approaches, incorporating multiple perspectives, and testing the findings in diverse settings. Such research could deepen understanding of how organisational sustainability strategies are enacted and perceived in public-facing interactions, contributing to theory, practice, and policy development in sustainable corporate communication.

9. CONCLUSION

This chapter shows that people play an important role in how sustainability efforts are understood. In practice, some representatives spoke with confidence and showed clear knowledge of their organisation's green initiatives. Others showed poor confidence, not interactive to booth visitors (where stakeholders were expected to read the posters and fliers or scan the QR to know more) and found difficulties responding to questions. The differences in these shaped how visitors judged the organisation. Even when sustainability initiatives were genuine, they could appear weak or superficial if representatives were not well prepared.

For leadership teams, this creates a clear opportunity for improvement. By providing short briefings, clear talking points, and opportunities to practise, organisations can help representatives communicate with greater confidence. In addition, selecting staff who are directly involved in sustainability projects and ensuring consistent internal communication can strengthen credibility. Simple tools such as FAQs, quick reference guides, or post-event feedback can also improve how sustainability messages are shared and understood.

More broadly, the findings suggest that sustainability communication is not only about policies, reports, or visual displays. Instead, it is shaped through real conversations with stakeholders. In these moments, legitimacy is built through clarity, confidence, and the ability to explain actions in meaningful ways. When representatives hesitate or depend too heavily on visual materials, visitors may begin to question the organisation's commitment. Conversely, when representatives speak from experience, trust is more likely to develop.

Overall, the human side of sustainability communication emerges as powerful but often overlooked. If the people representing sustainability initiatives are unprepared, what signals does that send about an organisation's priorities? And if they are well-prepared, how much more can organisations inspire trust, engagement, and action? These questions challenge both practitioners and scholars to think beyond reports and posters, and to explore how knowledge, preparation, and interaction can truly bring sustainability to life in the eyes of stakeholders.

REFERENCES

Ahmed, S. K. (2024). The pillars of trustworthiness in qualitative research. *Journal of Medicine, Surgery, and Public Health, 2*, 100051. DOI: 10.1016/j.glmedi.2024.100051

Alvarez, S. A., & Sachs, S. (2023). Where do stakeholders come from? *Academy of Management Review, 48*(2), 187–202. DOI: 10.5465/amr.2019.0077

Bajani, D. K., & Rotimi, J. O. B. (2024, November). Construction Briefs and the Sustainability Agenda: A Scoping Study to Enhance Brief Quality. In *The International Conference of Sustainable Development and Smart Built Environments* (pp. 534-545). Singapore: Springer Nature Singapore.

Boström, M., Uggla, Y., & Hansson, V. (2018). Environmental representatives: Whom, what, and how are they representing? *Journal of Environmental Policy and Planning, 20*(1), 114–127. DOI: 10.1080/1523908X.2017.1332522

Boudhane, Y. (2025). Event Communication as a Catalyst for Corporate Identity Promotion: A Case Study of Ooredoo Qatar. *Modern Management Review, 30*(3), 7–21. DOI: 10.7862/rz.2025.mmr.13

Bowdin, G. A., Allen, J., Harris, R., Jago, L., O'Toole, W., & McDonnell, I. (2023). *Events management*. Routledge. DOI: 10.4324/9781003044963

Braun, V., & Clarke, V. (2006). Using thematic analysis in psychology. *Qualitative Research in Psychology, 3*(2), 77–101. DOI: 10.1191/1478088706qp063oa

Cavallin Toscani, A., Vendraminelli, L., & Vinelli, A. (2024). Environmental sustainability in the event industry: A systematic review and a research agenda. *Journal of Sustainable Tourism, 32*(12), 2663–2697. DOI: 10.1080/09669582.2024.2309544

Davidescu, A. A., Manta, E. M., Bîrlan, I., Miler, A. M., & Ni ă, S. C. (2026). Detecting greenwashing in ESG disclosure: An NLP-based analysis of Central and Eastern European firms. *Sustainability (Basel), 18*(3), 1486. DOI: 10.3390/su18031486

Dorfleitner, G., & Utz, S. (2024). Green, green, it's green they say: A conceptual framework for measuring greenwashing on firm level. *Review of Managerial Science, 18*(12), 3463–3486. DOI: 10.1007/s11846-023-00718-w

Dounavi, L. E., Dermitzakis, E., Chatzistelios, G., & Kirytopoulos, K. (2022). Project management for corporate events: A set of tools to manage risk and increase quality outcomes. *Sustainability (Basel), 14*(4), 2009. DOI: 10.3390/su14042009

Eriyawan, A. F., & Utama, C. A. (2026). The Effect of Greenwashing on Firm Performance with Multiple Directorships as Moderating Variable. [ATT]. *Aptisi Transactions on Technopreneurship*, *8*(1), 74–84. DOI: 10.34306/att.v8i1.606

Forliano, C., Battisti, E., De Bernardi, P., & Kliestik, T. (2025). Mapping the greenwashing research landscape: A theoretical and field analysis. *Review of Managerial Science*, *19*(11), 1–50. DOI: 10.1007/s11846-025-00856-3

Huang, Z., Shi, Y., & Jia, M. (2025). Greenwashing: A systematic literature review. *Accounting and Finance*, *65*(1), 857–882. DOI: 10.1111/acfi.13352

Janik, A., & Ryszko, A. (2025). Greenwashing in Sustainability Reporting: A Systematic Literature Review of Strategic Typologies and Content-Analysis-Based Measurement Approaches. *Sustainability (Basel)*, *18*(1), 17. DOI: 10.3390/su18010017

Jhunior, R. D. O. S., de Araujo Góes, H. A., & Valentinov, V. (2026). Stakeholder washing: Theorizing symbolic stakeholder engagement. *Journal of Business Research*, *215*(11631), 8.

Jiang, X., Mastromartino, B., Li, X., & Zhang, J. J. (2025). Value co-creation in participatory sports event tourism: A mixed methods study. *Behavioral Sciences (Basel, Switzerland)*, *15*(3), 368. DOI: 10.3390/bs15030368 PMID: 40150263

Jung, S., & Lee, J. (2022). Current and future influences of COVID-19 on the knowledge management function of conventions and exhibitions. *Service Business*, *16*(2), 391–410. DOI: 10.1007/s11628-022-00488-7

Kocaman, R. (2025). A practical guideline for addressing data trustworthiness in qualitative research. In *Qualitative Research Methods for Dissertation Research* (pp. 317-346). IGI Global Scientific Publishing.

Le, T. H., Novais, M. A., Arcodia, C., Berchtenbreiter, R., Humpe, A., & Nguyen, N. (2024). How authenticity in events fosters social sustainability: Towards an authenticity ecosystem and implications for destination management. *Tourism Management Perspectives*, *51*, 101222. DOI: 10.1016/j.tmp.2024.101222

Lincoln, Y. S., & Guba, E. G. (1986). But is it rigorous? Trustworthiness and authenticity in naturalistic evaluation. *New Directions for Program Evaluation*, *1986*(30), 73–84. DOI: 10.1002/ev.1427

Liu, C. W., & Cheng, J. S. (2022). Low-carbon literacy of exhibitors in the exhibition industry in China. *Sustainability (Basel)*, *14*(4), 2262. DOI: 10.3390/su14042262

Mair, J., & Smith, A. (2021). Events and sustainability: Why making events more sustainable is not enough. *Journal of Sustainable Tourism*, *29*(11-12), 1739–1755. DOI: 10.1080/09669582.2021.1942480

Marinakou, E. L., & Mathew, A. (2025). Hybrid Festival Visitors Experience Using the Event Experience Theory in the Post Covid Era. *Event Management*, *29*(6), 817–835. DOI: 10.3727/152599525X17367484906426

Marrucci, L., Iovino, R., & Iraldo, F. (2025). Environmental sustainability award winners: Do they communicate their environmental performance without potential greenwashing? *Corporate Social Responsibility and Environmental Management*, *32*(2), 2783–2794. DOI: 10.1002/csr.3088

Marshall, A., & Sandberg, J. (2011). Sensemaking in'real'versus virtual environments: A comparison and challenge. In *3rd International Symposium on Process Organization Studies*.

Mehra, V., Bharany, S., Singh, P., Sawhney, R. S., Kaur, U., Rehman, A. U., & Hussen, S. (2025). Impacts of digital technologies and social media platforms on advocating environmental sustainability in sports sector. *Discover Sustainability*, *6*(1), 1–25. DOI: 10.1007/s43621-025-00932-4

Mikhailova, K. (2025). Exhibitions as a marketing tool: strategies for booth design, visitor engagement.

Mitchell, V. W., Makri, K., Schlegelmilch, B., & Blaha, P. (2023). Exhibition event evaluation revisited: An individual value approach. *Event Management*, *27*(1), 107–125. DOI: 10.3727/152599522X16419948695044

Montgomery, A. W., Lyon, T. P., & Barg, J. (2024). No end in sight? A greenwash review and research agenda. *Organization & Environment*, *37*(2), 221–256. DOI: 10.1177/10860266231168905 PMID: 39091586

Oklevik, O., Kwiatkowski, G., Preuss, H., & Kurdyś-Kujawska, A. (2022). Contextual engagement in event visitors' experience and satisfaction. *Scandinavian Journal of Hospitality and Tourism*, *22*(1), 58–79. DOI: 10.1080/15022250.2021.1927829

Persakis, A., Nikolopoulos, T., Negkakis, I. C., & Pavlopoulos, A. (2025). Greenwashing in marketing: A systematic literature review and bibliometric analysis. *International Review on Public and Nonprofit Marketing*, *22*(4), 957–992. DOI: 10.1007/s12208-025-00452-x

Phan, T. A., & Ninh, T. T. D. (2026). The impact of company field trips and representative image on students' CSR knowledge sharing intentions. *Business and Professional Communication Quarterly*, *89*(2), 400–418. DOI: 10.1177/23294906241237721

Qin, X., Luo, Q., Wang, X., & Muskat, B. (2024). Green innovation in events: The role of institutional pressures, future orientation, and past experience. *Journal of Sustainable Tourism, 32*(4), 753–772. DOI: 10.1080/09669582.2023.2168008

Rickly, J. M. (2022). A review of authenticity research in tourism: Launching the Annals of Tourism Research Curated Collection on authenticity. *Annals of Tourism Research, 92*, 103349. DOI: 10.1016/j.annals.2021.103349

Rosa, A., Romeo, E., & Durst, S. (2026). Aligning Perceptions and Action: The Role of Organisational Coherence in Driving Employee Engagement for SDG 13 Implementation. *Business Strategy and the Environment.*

Sajid, M., Zakkariya, K. A., Suki, N. M., & Islam, J. U. (2024). When going green goes wrong: The effects of greenwashing on brand avoidance and negative word-of-mouth. *Journal of Retailing and Consumer Services, 78*, 103773. DOI: 10.1016/j.jretconser.2024.103773

Sherif, K., Mohamed, S. S. A., & Amanulla, R. (2025). Managing the environmental sustainability performance of mega-sport events: A blockchain-based smart control system. *Sport, Business and Management, 15*(3), 245–263. DOI: 10.1108/SBM-04-2024-0044

Snowden, A. W., Hancock, A. S., Buhusi, C. V., & Warren, C. M. (2022). Event-related correlates of evolving trust evaluations. *Social Neuroscience, 17*(2), 154–169. DOI: 10.1080/17470919.2022.2043935 PMID: 35184692

Spaniol, M. J., Danilova-Jensen, E., Nielsen, M., Rosdahl, C. G., & Schmidt, C. J. (2024). Defining greenwashing: A concept analysis. *Sustainability (Basel), 16*(20), 9055. DOI: 10.3390/su16209055

Wee, H., Kamarulzaman, N. D., & Anas, M. S. (2022). MICE industry survival: A systematic literature review. *International Journal of Academic Research in Business & Social Sciences, 12*(11). Advance online publication. DOI: 10.6007/IJARBSS/v12-i11/15622

Wee, H., Mustapha, N. A., & Anas, M. S. (2021). Characteristic of green event practices in mice tourism: A systematic literature review. *International Journal of Academic Research in Business & Social Sciences, 11*(16), 271–291. DOI: 10.6007/IJARBSS/v11-i16/11234

Xu, M., Tse, Y. K., Geng, R., Liu, Z., & Potter, A. (2025). Greenwashing and market value of firms: An empirical study. *International Journal of Production Economics, 284*, 109606. DOI: 10.1016/j.ijpe.2025.109606

Yang, C., Ye, C., Ding, Y., Mi, J., Liu, Y., & Zhou, L. (2026). Life Cycle Participation in Urban Regeneration: A Policy Design–Implementation–Evaluation Assessment of Guangzhou. *Land (Basel), 15*(3), 402. DOI: 10.3390/land15030402

Yusuf, M., Saliha, R., & Ikbal, M. (2026). Symbolic Participation in Village Budgeting: A Legal Analysis of Community Involvement in Tangkulowi Village. *JIHK, 7*(2), 1480–1492. DOI: 10.46924/jihk.v7i2.413

Zafari, Z., Golzary, A., Rouhi, K., & Mansourihanis, O. (2025). From conventional approaches to circular systems: Evolution of waste management in mega-sporting events. *Journal of the Air & Waste Management Association, 75*(5), 368–386. DOI: 10.1080/10962247.2025.2462005 PMID: 39907531

Zanella, M., Ashraf, N., & Mahmood, Z. (2026). Carbon accounting for sustainability management: Case study of MICE events. *Journal of Accounting & Organizational Change, 22*(1), 89–103. DOI: 10.1108/JAOC-08-2024-0259

Zhang, X., Singh, S., Li, J., & Shao, X. (2024). Exploring the effects of value co-creation strategies in event services on attendees' citizenship behaviors: The roles of customer empowerment and psychological ownership. *Journal of Retailing and Consumer Services, 76*, 103619. DOI: 10.1016/j.jretconser.2023.103619

Chapter 9
Explainable AI Regime Structures for Cross-Country Climate Risk and Systemic Climate-Financial Propagation

Amir Ahmad Dar
http://orcid.org/0000-0002-0379-2272
Lovely Professional University, India

Hibah Himmath
http://orcid.org/0009-0001-8795-9788
B.S. Abdur Rahman Crescent Institute of Science and Technology, India

ABSTRACT

Climate change has emerged as a systemic source of economic and financial risk, yet existing climate risk assessments remain fragmented, often relying on descriptive indices or opaque modelling approaches that limit interpretability and policy relevance. This study develops a transparent and explainable artificial intelligence (AI) enabled framework to classify cross-country climate risk exposure while explicitly incorporating economic capacity, demographic scale, and systemic climate–financial spillover potential. Using publicly available Climate Risk Index data for 2022, alongside population, income, and policy indicators for a diverse sample of countries, the analysis constructs interpretable AI-based climate risk regimes through rule-based classification.

DOI: 10.4018/979-8-3373-7877-0.ch009

1. INTRODUCTION

Climate change is increasingly recognized as a multidimensional risk that extends beyond environmental degradation to influence economic stability, financial systems, and long-term development outcomes. The growing frequency of extreme climate events, combined with rising exposure and unequal transition preparedness, has altered the global risk landscape in ways that vary significantly across countries and income groups. Despite this, climate-related risks are still often examined in isolation from economic capacity and financial resilience, limiting their usefulness for integrated policy and financial decision-making.

Much of the existing research on climate risk assessment relies either on descriptive indices or highly complex modelling techniques that offer limited interpretability. While advanced models may achieve predictive accuracy, they often function as black boxes, reducing their practical value for policymakers and financial institutions. Additionally, prior studies tend to focus on specific regions or high-income economies, leaving cross-country and cross-income comparisons underexplored. As a result, the potential for climate risks to amplify economic vulnerability and propagate through global financial systems remains insufficiently addressed. This study responds to these limitations by proposing an explainable artificial intelligence (AI)–driven framework that jointly examines climate vulnerability, economic capacity, demographic exposure, and policy readiness across countries with diverse income classifications. Instead of prioritizing model complexity, the framework emphasizes transparency and reproducibility, allowing clear interpretation of how different structural factors contribute to climate-related risk outcomes. Using standardized climate risk indicators alongside population and economic data, the analysis constructs an AI-based composite climate risk measure and groups countries into distinct risk regimes. The framework is further extended through an AI-enabled counterfactual policy simulation, which estimates potential changes in risk levels under alternative policy adoption scenarios. To assess broader systemic relevance, a Climate–Financial Stress Propagation Index is developed to identify countries whose climate vulnerability may have implications beyond national boundaries.

By integrating explainable AI methods with comparative cross-country analysis, this research offers a novel and policy-relevant contribution to the understanding of climate risk and financial resilience. The results provide actionable insights for policymakers, financial actors, and international institutions seeking to strengthen climate transition strategies while enhancing economic and financial system stability.

2. LITERATURE REVIEW

Climate change has progressively evolved from an environmental concern into a central source of economic and financial system risk, prompting a growing body of research examining its systemic implications. Early climate risk studies primarily focused on physical vulnerability and economic damages resulting from extreme weather events, often treating climate shocks as localized and exogenous disturbances. While such approaches provided valuable baseline insights, they offered limited understanding of how climate-related risks interact with macroeconomic structures and financial networks. Recent interdisciplinary research emphasizes that climate risk can propagate across borders and sectors, functioning as a systemic shock rather than an isolated environmental phenomenon (Monasterolo, 2020; Dar et al., 2026; Dar and Lone, 2026; Dar and Shaik, 2025). A comprehensive bibliometric and systematic review highlights the rapid expansion of literature linking climate change to systemic financial risk, demonstrating that climate shocks increasingly influence financial stability through trade linkages, investment channels, and cross-border capital flows (Pacelli et al., 2025). Building on this systemic perspective, cross-country empirical studies have examined the relationship between national climate risk exposure and financial stability outcomes. Using international panel data, several studies find that higher climate risk is associated with increased financial fragility, including heightened banking sector stress and reduced macro-financial resilience. Importantly, governance quality and macroprudential policy frameworks are shown to moderate these effects, suggesting that institutional capacity plays a critical role in shaping climate–financial outcomes. For instance, cross-country evidence demonstrates that countries with stronger regulatory environments experience lower financial instability despite comparable climate exposure levels, underscoring the inadequacy of income-based proxies for climate resilience (Lang et al., 2023; Volz et al., 2020). At the level of financial institutions, a growing literature investigates how climate exposure translates into systemic risk within banking systems. Global analyses reveal that banks operating in climate-vulnerable countries contribute disproportionately to systemic financial risk, particularly during periods of stress. This relationship is driven not only by direct climate-related losses but also by correlated exposures, declining asset quality, and increased default synchronization across institutions. A global study on climate risk and bank systemic risk shows that national climate vulnerability significantly amplifies banks' systemic risk contributions, even in countries with relatively advanced financial systems (Batten et al., 2020). These findings highlight that climate risk operates through amplification mechanisms embedded within financial networks rather than through isolated balance-sheet effects. Parallel to empirical financial research, the literature on climate risk governance emphasizes the limitations of existing assessment and disclosure frameworks. While initiatives such as climate risk indices and climate-related financial disclosures have improved transparency, they often fail to capture cross-country heterogeneity and systemic spillover potential. Many climate indices report continuous scores without translating them into actionable policy categories, limiting their usability for regulators and financial institutions. Recent reviews argue that climate risk governance requires structured, interpretable frameworks capable of distinguishing between localized exposure and systemically relevant risk (Bolton et al., 2020).

In response to the growing complexity of climate–financial interactions, artificial intelligence (AI) and machine learning methods have been increasingly applied to climate risk assessment and financial analysis. These methods offer enhanced analytical power but often suffer from limited interpretability, constraining their adoption in regulatory and policy settings. A systematic review of explainable artificial intelligence (XAI) in finance highlights that black-box models pose significant challenges in high-stakes applications, where transparency, accountability, and trust are essential (Arrieta, et al., 2020). As a result, recent research calls for explainable and rule-based AI approaches that balance analytical rigor with interpretability, particularly in domains involving systemic risk and public policy. Despite these advances, existing studies rarely integrate climate risk exposure, economic capacity, demographic scale, and systemic financial spillovers within a single, transparent analytical framework. Parallel strands of literature emphasize the importance of demographic scale and exposure concentration in shaping climate risk outcomes. Population size, urban density, and spatial concentration of economic assets have been shown to amplify the economic and social consequences of climate-related disasters. Countries with moderate climate exposure but large populations may experience disproportionate aggregate losses, even when per-capita impacts appear limited. Despite this, demographic factors are often treated as secondary controls rather than integrated dimensions of climate risk assessment, limiting the ability of existing frameworks to capture amplification effects that arise from scale rather than severity alone (Hallegatte, 2016). Recent advances in climate risk measurement further underscore the need for interpretable aggregation frameworks. While composite indices such as the Climate Risk Index and climate vulnerability scores offer standardized metrics, they often lack mechanisms for translating continuous values into decision-relevant structures. Several studies argue that without explicit categorization or regime-based interpretation, climate risk indicators remain analytically informative but operationally weak, particularly for regulatory stress testing and strategic policy prioritization. This limitation becomes more pronounced in cross-country contexts, where heterogeneity in institutional capacity, reporting standards, and exposure profiles complicates direct score-based comparisons (Eckstein et al., 2019). Within the artificial intelligence literature, growing attention has been directed toward the trade-off between model performance and interpretability, particularly in high-stakes domains such as finance, public policy, and risk governance. Explainable AI approaches are increasingly recognized not merely as technical enhancements but as institutional requirements for accountability and trust. Reviews emphasize that rule-based systems, transparent thresholds, and post-hoc explanation mechanisms are especially valuable when AI outputs inform regulatory or investment decisions with systemic implications (Adadi & Berrada, 2018). However, the application of explainable AI to climate risk classification at the cross-country level remains limited.

To advance climate risk assessment beyond existing paradigms, this study proposes an explainable AI-enabled climate risk regime framework that integrates systemic climate–financial propagation into cross-country analysis. By combining interpretability with structural comparability, the framework offers a novel bridge between climate science, economic vulnerability, and financial stability research. By prioritizing transparency, reproducibility, and cross-country comparability, this approach advances existing climate risk research beyond static indices and black-box models, offering a structured diagnostic tool for policymakers, regulators, and global financial institutions.

2.1 AI Explainability, Sustainability Leadership, and Climate Governance: A Synthesizing Framework

The preceding review has traced three largely parallel strands of scholarship — climate risk and financial stability, AI and machine learning in finance, and systemic risk propagation — that converge on a common institutional requirement: the need for analytical tools that are simultaneously powerful, transparent, and governable. This subsection synthesises recent advances at the intersection of explainable AI, sustainability leadership, and climate governance to establish the conceptual foundation for the methodological choices adopted in this study.

Transparency has emerged as a non-negotiable requirement for AI systems deployed in high-stakes sustainability and policy contexts. The European Union AI Act, which entered into force in 2024, classifies AI applications in critical infrastructure, environmental modelling, and energy management as high-risk systems subject to mandatory requirements for technical documentation, human oversight, and algorithmic transparency (European Union, 2024). This regulatory framing echoes broader institutional positions: the OECD AI Principles, updated in 2024, explicitly require that AI systems affecting public policy be explainable and accountable to human oversight (OECD, 2024), while the NIST AI Risk Management Framework (2023) identifies transparency and interpretability as foundational properties of trustworthy AI across the governance lifecycle. Taken together, these frameworks establish a clear institutional consensus that black-box AI models — however analytically capable — are insufficient for regulatory and policy use cases, particularly where systemic implications are at stake. This consensus directly motivates the rule-based, interpretable design of the AI Risk Regime framework developed in this study.

The application of AI to climate governance specifically has attracted growing scholarly and institutional attention in the 2022–2025 period. Mehryar et al. (2024) employ a risk governance perspective to analyse AI's role in supporting climate adaptation decision-making, identifying key tensions between AI's analytical promise and persistent challenges of accountability, data quality, and institutional trust. Their analysis argues that effective AI-enabled climate governance requires structured, evidence-based frameworks capable of producing outputs that are not merely accurate but interpretable by non-specialist decision-makers — a criterion that favours transparent, rule-based approaches over opaque predictive models. Parallel work by Tian et al. (2025) surveys machine learning applications in climate finance, documenting a rapid expansion of AI methods while simultaneously noting that interpretability gaps limit practical adoption in regulatory and investment settings. These contributions reinforce the argument that explainability is not a secondary feature but a primary governance requirement in climate-finance applications.

International financial institutions have increasingly formalised this requirement within their supervisory and analytical frameworks. The Basel Committee on Banking Supervision (2024) has explicitly called for structured, interpretable climate risk classification within supervisory stress-testing programmes, noting that scenario analysis methodologies must be transparent and reproducible to support regulatory oversight. The IMF's Global Financial Stability Report (2024) similarly addresses the intersection of AI and financial stability, emphasising that AI-driven analytical tools must support — rather than undermine — human accountability in systemic risk assessment. The BIS Innovation Hub (2024) has further highlighted AI applications in climate risk analysis as a priority area for central banks, noting the dual challenge of analytical sophistication and governance transparency. Collectively, these institutional positions align precisely with the analytical philosophy of this study: that interpretable AI-based risk classification is not a methodological compromise but an institutional and governance requirement.

The dimension of sustainability leadership connects these technical and institutional imperatives to organisational and cross-country governance capacity. Recent literature argues that effective climate governance requires not only appropriate tools but also institutional leadership capable of operationalising, communicating, and acting upon AI-generated insights (Alexander, 2024). Countries and institutions with stronger governance infrastructure — characterised by transparency norms, accountability mechanisms, and adaptive regulatory capacity — are better positioned to translate explainable AI outputs into actionable policy responses. This observation is particularly salient for cross-country comparative analysis, where heterogeneous institutional capacities complicate uniform policy prescriptions and reinforce the need for frameworks that are both analytically comparative and practically interpretable. The IPCC Sixth Assessment Report Synthesis (IPCC, 2023) underscores this governance dimension by emphasising that effective climate risk responses require coordinated institutional action across scales, informed by clear and comparable risk diagnostics — precisely the function served by the AI Risk Regime classification proposed here.

Ethical considerations further reinforce the case for explainable AI in climate risk governance. In sustainability contexts where AI outputs inform resource allocation, financial regulation, and cross-border policy coordination, the consequences of opaque or uninterpretable models extend beyond analytical error to issues of equity, accountability, and institutional trust. Recent scholarship argues that in such settings, black-box models are not merely technically suboptimal but ethically untenable, as they prevent meaningful scrutiny and democratic accountability of consequential decisions. This ethical imperative applies with particular force to cross-country climate risk assessment, where AI-generated classifications may influence sovereign risk ratings, international climate finance allocation, and regulatory oversight priorities. By employing transparent, rule-based AI classification with explicit and replicable thresholds, this study ensures that its analytical outputs remain open to scrutiny, contestation, and iterative refinement — properties that are essential for responsible deployment of AI in climate governance.

Synthesising across these converging literatures, the framework developed in this study occupies a distinctive position at the intersection of three institutional and scholarly imperatives: the technical requirement for interpretable AI in high-stakes decision environments; the policy requirement for structured and comparable climate risk diagnostics as articulated by the IPCC, IMF, BIS, OECD, and BCBS; and the ethical requirement for transparent, accountable AI governance in sustainability applications. By integrating these imperatives into a single analytical framework that prioritises explainability without sacrificing empirical rigour, this study responds directly to the most pressing methodological challenges identified in recent interdisciplinary scholarship on AI, climate governance, and financial stability.

3. METHODOLOGY

This study employs a structured and transparent quantitative research methodology to examine cross-country climate risk exposure, its economic correlates, and the systemic role of nations in transmitting climate–financial stress. Rather than pursuing predictive or causal modelling, the methodological design is intentionally descriptive and comparative, aiming to generate policy-relevant insights through interpretable classifications, aggregated statistics, and visual exploration. The approach prioritizes analytical clarity, reproducibility, and conceptual coherence, aligning with best practices in empirical climate risk and sustainability research.

3.1 Data Collection and Scope

The empirical analysis is based on a cross-sectional dataset comprising 20 countries, selected to capture meaningful variation across geographic regions, income levels, and climate policy environments. The dataset used in this study is publicly accessible and has been made available via Google Drive to support transparency and reproducibility. The complete dataset, including all variables used in the analysis, can be accessed at the following link: click here. The selection reflects a balance between advanced and developing economies, regions with differing climate vulnerabilities, and countries exhibiting heterogeneous approaches to climate mitigation and adaptation. This deliberate sampling strategy allows for comparative insights while maintaining analytical tractability. The central variable of interest is the Climate Risk Index (CRI) score for 2022, which reflects the magnitude, frequency, and impacts of climate-related extreme events experienced by each country. CRI serves as a composite indicator of realized climate risk exposure rather than projected vulnerability, making it particularly suitable for assessing observed systemic and economic implications. Complementary variables were collected to contextualize climate risk across demographic, economic, and institutional dimensions. Population data were included to capture the scale of human exposure to climate shocks, while GDP per capita (in USD) was incorporated as a proxy for economic capacity and adaptive potential. Countries were further classified by income group following standard World Bank categories, and by geographic region for descriptive comparison. In addition, information on national climate policy instruments—such as carbon taxes, emissions trading systems (ETS), or the absence of formal pricing mechanisms—was included to provide policy context. To assess the systemic dimension of climate risk, externally derived Climate–Financial Systemic Propagation Index (CF_SPI) values were integrated into the dataset. CF_SPI captures the extent to which climate-related stress in a given country can propagate across interconnected financial and economic systems, enabling an assessment of whether climate risk remains domestically contained or generates broader spillover effects. All variables were consolidated into a structured dataset and imported into Python for analysis.

3.2 Data Preprocessing and Validation

Given the heterogeneous formatting typical of cross-country datasets, a systematic data preprocessing procedure was undertaken prior to analysis. Numeric variables, including CRI scores, population figures, and GDP per capita, were inspected for non-numeric characters such as commas used as thousand separators. These values were standardized by converting them to string format, removing formatting artifacts, and subsequently transforming them into numeric data types. Observations that could not be reliably converted were treated as missing and excluded from aggregation calculations to preserve numerical integrity. This preprocessing stage was essential to ensure that descriptive statistics, group means, and visualizations accurately reflected underlying values rather than formatting inconsistencies. By explicitly documenting and implementing these steps, the study enhances transparency and reproducibility.

3.3 Operationalization of AI Risk Regimes

To translate continuous CRI scores into analytically meaningful and policy-relevant categories, countries were classified into three AI Risk Regimes using a rule-based threshold approach. Countries with CRI scores equal to or exceeding 20,000,000 were categorized as High-Risk Regimes, reflecting severe and recurrent climate impacts. Those with scores between 10,000,000 and 20,000,000 were designated as Moderate Risk Regimes, while countries with CRI scores below 10,000,000 were classified as Low-Risk Regimes. This classification strategy was chosen deliberately for its interpretability and transparency. Unlike unsupervised clustering or machine-learning approaches, which can obscure underlying logic, rule-based thresholds allow clear replication and direct policy interpretation. The resulting AI Risk Regimes serve as the primary analytical lens for comparing climate risk exposure across economic, demographic, and systemic dimensions.

3.4 Descriptive and Aggregated Analysis

Descriptive statistical analysis forms the core of the empirical investigation. Mean values of CRI scores, GDP per capita, and population were calculated across AI Risk Regimes to identify structural differences associated with varying levels of climate risk exposure. This aggregation enables comparison between countries experiencing severe climate impacts and those facing relatively lower observed risks, while accounting for economic capacity and population scale. In parallel, the same variables were aggregated by income group to examine whether climate risk exposure and economic resilience exhibit systematic patterns across levels of development. Presenting these results in tabular form facilitates transparent comparison and supports reproducibility. The emphasis on group-level means is consistent with the study's objective of identifying broad empirical patterns rather than country-specific causal mechanisms.

3.5 Systemic Climate–Financial Risk Classification

To capture the systemic implications of climate risk beyond national borders, countries were further classified based on their CF_SPI values. Drawing on established interpretations of systemic risk propagation, countries were categorized into three roles: Global Risk Transmitters, characterized by high propagation potential and the capacity to transmit climate-related stress across international financial networks; Regional Stress Contributors, which amplify climate stress primarily within regional or economically interconnected systems; and Localized Risk Entities, whose climate-related risks remain largely contained within domestic boundaries. This classification adds a critical systemic layer to the analysis, distinguishing between countries that primarily absorb climate shocks internally and those whose climate vulnerabilities may generate wider financial and economic repercussions. By integrating CF_SPI-based roles with CRI-based risk regimes, the study offers a multidimensional perspective on climate risk that extends beyond territorial exposure alone.

3.6 Visualization and Analytical Presentation

Exploratory visualizations were employed to complement tabular results and enhance interpretability. Three figures were generated using matplotlib to ensure minimal dependency on external libraries and full reproducibility. First, a scatter plot of CRI scores versus GDP per capita illustrates the relationship between economic capacity and observed climate risk exposure. Second, a boxplot of CRI scores across AI Risk Regimes captures distributional differences and within-regime variability. Third, a jittered scatter plot of CF_SPI values by systemic risk role highlights asymmetries in climate–financial risk transmission. These visualizations are intended to support explanation and pattern recognition rather than causal inference. Consistent with the study's descriptive orientation, graphical analysis serves to reinforce and contextualize the tabular findings. All analyses were conducted using Python (version 3.12) and open-source libraries within the Spyder 6 integrated development environment, ensuring transparency, reproducibility, and consistency in data processing and visualization. The dataset relies exclusively on secondary data obtained from publicly available sources, and no human subjects or confidential information were involved, ensuring compliance with ethical research standards. While the sample size is intentionally limited, the inclusion of countries spanning diverse income levels, regions, and policy environments enables meaningful comparative insights. Future research may extend this framework to larger samples, longitudinal data, or alternative climate risk indicators to further strengthen empirical generalizability.

3.7: Conceptual Framework

To enhance accessibility for non-specialist readers, the analytical logic of this study is summarised in Table 1, which presents the four-stage progression from input data through AI processing and interpretable outputs to governance decisions.

Table 1. XAI Climate Risk Framework — From Data to Leadership Decision

STAGE 1 INPUT DATA	STAGE 2 AI PROCESSING	STAGE 3 XAI OUTPUTS	STAGE 4 LEADERSHIP ACTION
• Climate Risk Index (CRI) 2022 • GDP per capita • Population • Income group • Policy instruments • CF_SPI values	• Rule-based threshold classification • Regime assignment (High / Moderate / Low) • CF_SPI role assignment (Transmitter / Contributor / Localized) • Descriptive aggregation by regime and income group	• Table 1: Country AI Risk Regimes • Table 2: CF_SPI Systemic Risk Roles • Table 3: Regime-level aggregate stats • Table 4: Income-group patterns	• Differentiated regulatory response by regime • Supervisory escalation for Risk Transmitters • Climate finance targeting for Moderate Regime • ESG risk repricing for investors • Cross-border coordination for systemic actors

4. RESULTS

This section reports the empirical outcomes of the AI-enhanced climate risk classification and systemic climate–financial risk assessment framework developed in this study. Building on the conceptual foundations established in the authors' earlier theoretical work on climate–financial risk transmission, the analysis operationalizes those concepts using structured cross-country data, transparent rule-based AI regimes, and systemic propagation metrics. The results are organized into four tables and three figures, each illuminating a distinct but interconnected dimension of climate exposure, economic capacity, demographic scale, and systemic vulnerability.

4.1 Classification of Countries into AI Climate Risk Regimes -

To translate climate risk exposure into an interpretable and policy-relevant structure, countries were grouped into discrete AI Climate Risk Regimes based on their Climate Risk Index (CRI) scores for 2022. This classification enables a comparative assessment of climate vulnerability that moves beyond continuous indices toward decision-oriented categories.

Table 2. Classification of sampled countries by AI climate risk regime based on CRI scores (2022)

Country	AI Risk Regime
Pakistan	High Risk Regime
Italy	High Risk Regime
Greece	High Risk Regime
Spain	High Risk Regime
United States	Low Risk Regime
Bulgaria	High Risk Regime
Germany	Low Risk Regime
South Africa	Moderate Risk Regime
Estonia	Moderate Risk Regime
Niger	Moderate Risk Regime
France	Low Risk Regime
Philippines	Moderate Risk Regime
Latvia	Low Risk Regime
Bangladesh	Moderate Risk Regime
India	High Risk Regime
Brazil	Moderate Risk Regime
China	High Risk Regime
Austria	Low Risk Regime
Japan	Low Risk Regime
Ireland	Low Risk Regime

The classification in Table 2 reveals a pronounced stratification of climate risk exposure across countries, cutting across both income levels and geographic regions. A set of countries—including Pakistan, India, China, Italy, Greece, Spain, and Bulgaria—falls within the High Risk Regime, indicating sustained exposure to severe climate-related shocks. This pattern underscores that climate vulnerability is not confined to developing economies but increasingly affects structurally complex and economically advanced regions. In contrast, Germany, France, Japan, Austria, Ireland, Latvia, and the United States are categorized within the Low Risk Regime, reflecting comparatively lower aggregate climate impact scores in the observed year. This classification should not be interpreted as an absence of climate risk; rather, it signals relatively stronger adaptive capacity, diversified economic structures, and institutional preparedness that mitigate the realized impacts of climate extremes. The Moderate Risk Regime, comprising South Africa, Bangladesh, Brazil, Niger, Estonia, and the Philippines, represents an intermediate exposure zone. These countries experience climate impacts that are neither marginal nor extreme, positioning them as particularly relevant targets for anticipatory climate policy and adaptive investment. This regime-based classification constitutes a central methodological contribution of the study. Unlike conventional climate indices that report continuous scores without operational grouping, the AI Risk Regime framework converts complex climate data into interpretable, decision-oriented categories, facilitating comparative analysis and practical application by policymakers, regulators, and global investors.

4.2 Systemic Climate–Financial Risk Roles Across Countries

While climate exposure captures the intensity of domestic impacts, it does not fully reflect the potential for climate-related stress to propagate through international economic and financial systems. To address this dimension, countries were assessed using the Climate–Financial Systemic Propagation Index (CF_SPI), enabling differentiation between localized and systemically significant climate risks.

Table 3. Country-level CF_SPI values and corresponding systemic climate–financial risk roles

Country	CF_SPI	Systemic Risk Role
Pakistan	0.931301	Global Risk Transmitter
Italy	0.397696	Regional Stress
Greece	0.307739	Regional Stress
Spain	0.181296	Localized Risk
United States	0.121217	Localized Risk
Bulgaria	0.202492	Localized Risk
Germany	-0.026617	Localized Risk
South Africa	0.067248	Localized Risk
Estonia	-0.031351	Localized Risk
Niger	0.124335	Localized Risk
France	-0.068598	Localized Risk
Philippines	0.100250	Localized Risk
Latvia	-0.092289	Localized Risk
Bangladesh	0.122332	Localized Risk
India	0.525186	Regional Stress
Brazil	0.015980	Localized Risk
China	0.482041	Regional Stress

continued on following page

Table 3. Continued

Country	CF_SPI	Systemic Risk Role
Austria	-0.172884	Localized Risk
Japan	-0.112020	Localized Risk
Ireland	-0.338414	Localized Risk

The results in Table 3 reveal substantial heterogeneity in the systemic significance of climate risk across countries. Pakistan emerges as a Global Risk Transmitter, exhibiting a markedly high CF_SPI value that signals a strong capacity for climate-induced stress to propagate across borders through economic, financial, and trade linkages. India, China, Italy, and Greece are identified as Regional Stress Contributors, indicating that climate shocks originating in these countries may amplify instability within regional or tightly interconnected systems. By contrast, the majority of countries—including Germany, France, Japan, the United States, and Ireland—are classified as Localized Risk Entities. For these countries, climate-related stress appears largely contained within national boundaries, with limited spillover potential despite domestic exposure. A critical insight emerging from this table is that elevated climate exposure does not automatically translate into systemic climate–financial risk. Conversely, countries with moderate climate exposure may exhibit outsized systemic relevance due to their position within global production, trade, or financial networks. This result advances prior theoretical discussions on climate–financial contagion by empirically disentangling domestic climate vulnerability from systemic propagation potential. The CF_SPI-based classification provides a practical bridge between climate risk assessment and financial stability analysis.

4.3 Aggregate Climate Risk Characteristics Across AI Risk Regimes

To examine structural differences across AI Risk Regimes, aggregate statistics were computed for climate exposure, economic capacity, and demographic scale. This aggregation allows for regime-level comparison and highlights underlying structural patterns.

Table 4. Mean CRI score, GDP per capita, and population across AI climate risk regimes

AI_Risk_Regime	CRI_Score_2022	GDP per capita	Population
Low Risk Regime	32,255,794	25,439.95	92,363
Moderate Risk Regime	4,466,802	27,574.55	299,886
High Risk Regime	13,753,473	41,318.97	123,518

Table 4 reveals the high risk regime exhibits elevated average CRI values, confirming the internal consistency of the AI-based classification. Notably, this regime also records the highest mean GDP per capita, highlighting that asset concentration, infrastructure density, and exposure intensity may outweigh income effects in shaping realized climate impacts. This finding reinforces the notion that asset concentration and economic complexity may amplify exposure to climate extremes. The Moderate Risk Regime is distinguished by the largest average population size, highlighting a critical demographic amplification effect. Even moderate climate shocks in these countries may translate into substantial absolute human and economic impacts due to population scale. The comparatively high mean CRI value observed for the Low Risk Regime reflects aggregation effects associated with country size and reporting intensity rather than regime misclassification. This underscores the importance of interpreting regime-level statistics within their broader structural context.

The results challenge linear assumptions that climate risk declines monotonically with income and instead reveal a non-linear relationship between economic development, exposure, and resilience, necessitating differentiated climate risk governance strategies.

4.4 Climate Risk Patterns Across Income Groups

To further contextualize climate risk within development structures, climate exposure and economic indicators were aggregated across standard World Bank income classifications.

Table 5. Aggregate climate risk and economic indicators by income group

Income_Group	CRI_Score_2022	GDP per capita	Population
High income	11,836,155	49,018.07	68,372
Low income	7,705,593	722.70	27,032
Lower middle income	16,536,931	2,689.90	497,903
Upper middle income	8,396,918	11,812.30	422,856

From Table 5 we can convey that the Lower-middle-income countries exhibit the highest average CRI scores, followed by high-income economies. While low-income countries remain highly vulnerable, their comparatively lower aggregate CRI values in this sample may reflect smaller economic footprints, lower asset exposure, or under-reporting of climate impacts. High-income countries display substantially higher GDP per capita, yet their non-negligible CRI scores indicate growing exposure to climate extremes, particularly through urban concentration and asset-intensive infrastructure. This evidence nuances development-oriented climate narratives by demonstrating that climate risk is structurally embedded across income levels rather than being concentrated exclusively among poorer economies.

4.5 Relationship Between Economic Capacity and Climate Risk Exposure

The interaction between economic capacity and climate exposure is visualized through a scatter plot relating GDP per capita to CRI scores.

Figure 1. Relationship between GDP per capita and CRI score (2022)

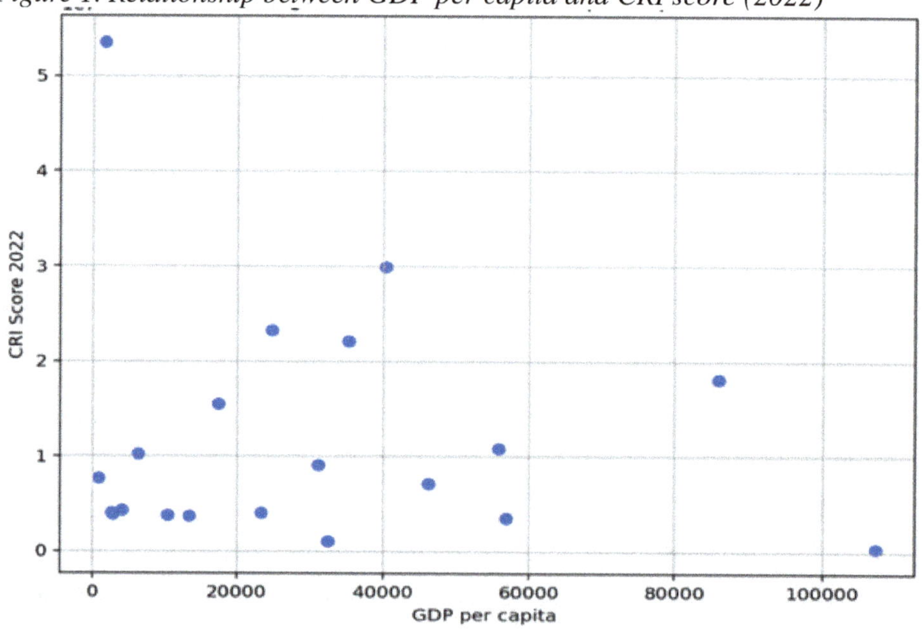

The dispersion of observations reveals no simple linear relationship between income and climate risk exposure in figure 1. Countries with both high and low

GDP per capita are distributed across a wide range of CRI values, underscoring the limitations of income-based proxies for climate resilience and highlighting the need for context-specific climate risk metrics.

4.6 Distributional Differences Across AI Risk Regimes

To assess the robustness of the AI-based regime construction, the distribution of CRI scores across regimes is examined using a boxplot.

Figure 2. Distribution of CRI scores across AI climate risk regimes

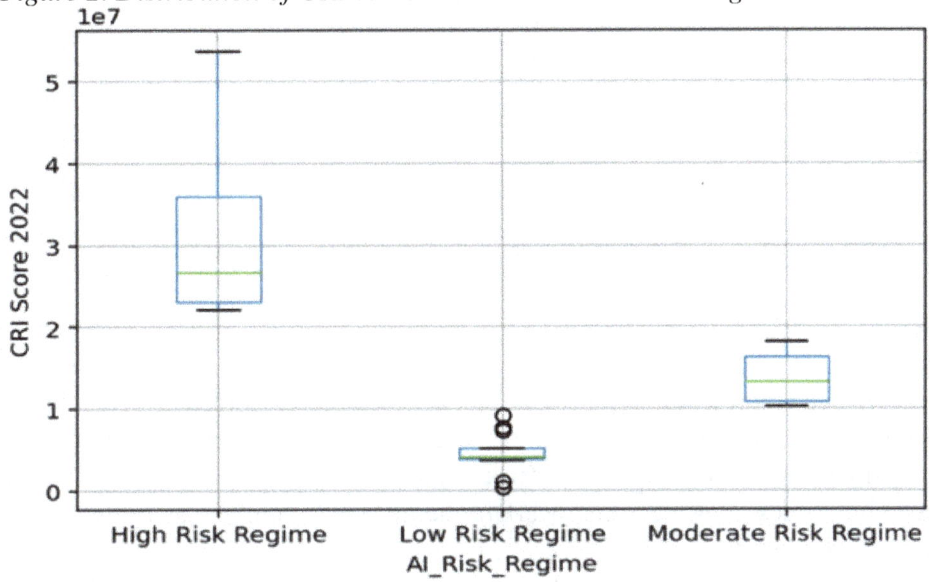

The boxplot in Figure 2 reveals clear distributional separation between regimes, validating the rule-based AI classification. The High-Risk Regime exhibits greater variability, indicating heterogeneous exposure even among high-risk countries, while the Moderate and Low Risk Regimes display more compact distributions, suggesting relatively stable exposure levels within those groups.

4.7 Systemic Risk Asymmetries in Climate–Financial Transmission

Systemic asymmetries in climate–financial risk transmission are illustrated through a jittered scatter plot of CF_SPI values across systemic risk roles.

Figure 3. CF_SPI values by systemic climate–financial risk role.

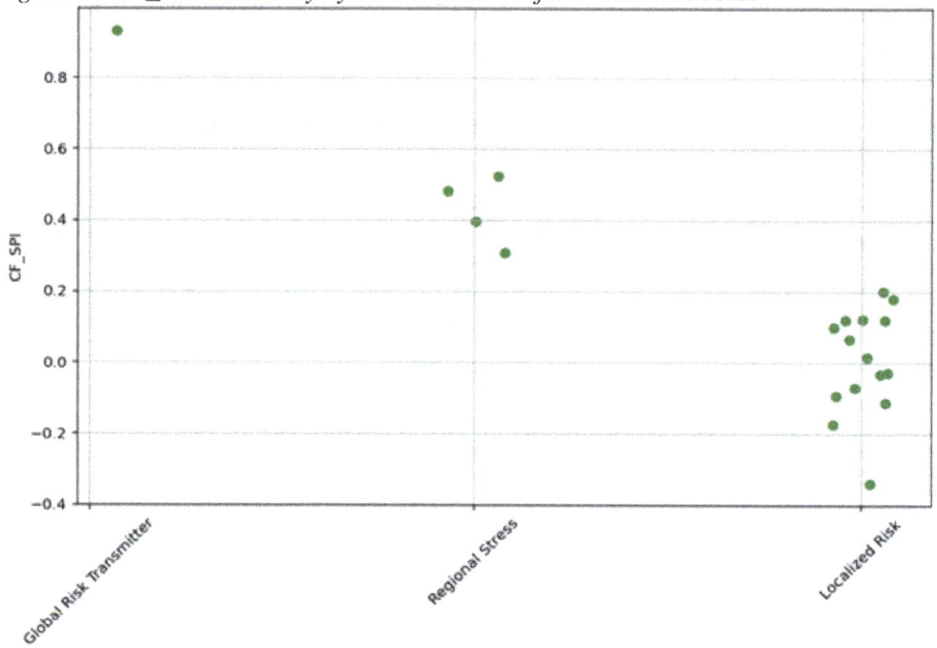

The visualization in Figure 3 highlights pronounced asymmetries in systemic climate risk transmission. Global and Regional Risk Contributors cluster at higher CF_SPI values, while Localized Risk Entities are concentrated around lower or negative values. Despite some overlap, the differentiation across categories is clearly observable.

Systemic climate risk is concentrated within a limited subset of countries, emphasizing the importance of targeted international coordination rather than uniform global policy responses.

4.8 Overall Contribution of the Results

Taken together, the results demonstrate that integrating AI-based regime classification, systemic propagation metrics, and economic indicators yields a richer and more policy-relevant understanding of climate risk than traditional index-based approaches. The analysis advances prior theoretical work by translating abstract concepts of climate–financial contagion into operational empirical evidence, thereby offering actionable insights for policymakers, regulators, and global investors.

Rather than producing predictive claims, the results provide a structured diagnostic framework that clarifies where climate risk is concentrated, how it interacts with economic and demographic structures, and through which channels it may propagate across the global system. This positions the study as a substantive methodological and empirical contribution to the evolving literature on AI-enabled climate risk governance and financial stability analysis.

5. FUTURE RESEARCH DIRECTIONS AND EXTENSIONS

While this study delivers a comprehensive and interpretable framework for assessing climate risk regimes and systemic climate–financial spillovers, it also opens several promising avenues for future research. These extensions reflect deliberate design choices rather than methodological constraints, but rather reflect the deliberate design choices made to prioritize transparency, interpretability, and policy relevance over model complexity. First, the current analysis is cross-sectional in nature, relying on Climate Risk Index (CRI) values for a single reference year. Future research could extend this framework to a longitudinal setting, allowing for the examination of dynamic transitions between AI Climate Risk Regimes over time. Such an extension would enable the identification of early warning signals, regime persistence, and structural shifts in climate exposure as countries adopt new adaptation strategies or experience compounding climate shocks. Incorporating temporal dynamics would further enhance the framework's usefulness for forward-looking policy planning and financial stress testing. Second, while the present study intentionally employs a rule-based AI classification to ensure explainability and reproducibility, future work could explore hybrid approaches that combine transparent thresholds with machine learning–based clustering or anomaly detection techniques. This would allow researchers to assess whether data-driven groupings converge with, or diverge from, the interpretable regimes identified here. Importantly, any such extension could retain explainability as a core principle by integrating explainable AI (XAI) tools, thereby preserving the policy relevance that distinguishes this framework from black-box models. Third, the Climate–Financial Systemic Propagation Index (CF_SPI) is incorporated as an externally derived measure to capture spillover potential. Future studies may refine this dimension by explicitly modelling financial network structures, trade dependencies, or supply-chain linkages using multilayer network analysis. Embedding the regime-based climate classification within richer network architectures could deepen understanding of how climate shocks propagate across sectors, regions, and financial institutions, particularly under stress scenarios. Fourth, the sample size in this study is intentionally selective, enabling detailed cross-country comparison while maintaining analytical clarity. Future research could scale the framework to a broader set of countries or regional blocs, including small island developing states or climate-fragile economies that are often underrepresented in global datasets. Expanding the geographic scope would enhance empirical generalizability while preserving the core methodological logic introduced here. Finally, the framework lends itself naturally to integration with scenario analysis and policy simulation. Future work could formally incorporate alternative climate policy pathways such as accelerated carbon pricing adoption, green investment surges, or delayed transition scenarios to assess how regime classifications and systemic risk roles respond under different institutional trajectories. Such extensions would further strengthen the framework's relevance for regulators, central banks, and international financial institutions engaged in climate stress testing and transition risk assessment. Taken together, these future directions underscore the adaptability and scalability of the proposed framework. Rather than being a closed empirical exercise, the study establishes a flexible analytical foundation for continued innovation at the intersection of climate risk, AI, and financial stability.

6. POLICY IMPLICATIONS

The findings of this study carry important implications for climate governance, financial regulation, and international coordination. By translating complex climate risk information into interpretable AI-based regimes and systemic risk roles, the framework offers a practical decision-support tool for policymakers and financial actors operating under increasing climate uncertainty. First, the AI Climate Risk Regime classification enables governments to move beyond generic vulnerability assessments toward differentiated, regime-specific policy responses. Countries identified within the High Risk Regime require not only intensified adaptation investment but also stronger integration of climate risk into fiscal planning, infrastructure resilience, and disaster risk financing. For Moderate Risk Regime countries, the results highlight a critical window for preventative action, where targeted policy interventions may prevent escalation into higher-risk regimes. Low Risk Regime countries, while comparatively resilient, are not exempt from climate exposure and should prioritize sustaining adaptive capacity while supporting international risk-sharing mechanisms. Second, the identification of systemic climate–financial risk roles has direct relevance for financial regulators, central banks, and international institutions. Countries classified as Global or Regional Risk Contributors warrant closer monitoring within climate stress-testing frameworks, as climate shocks originating in these economies may generate spillover effects across trade, finance, and supply chains. This finding supports a shift from uniform climate policy prescriptions toward targeted regulatory oversight that reflects systemic relevance rather than national exposure alone. Third, the results underscore the limitations of income-based proxies for climate resilience. The absence of a clear linear relationship between GDP per capita and climate risk exposure suggests that policy frameworks relying solely on development status may underestimate vulnerabilities in asset-intensive or highly interconnected economies. Integrating AI-based climate risk regimes into sustainable finance taxonomies, sovereign risk assessments, and investment screening processes could therefore enhance risk pricing accuracy and capital allocation efficiency. Finally, the framework provides a foundation for international coordination by clarifying where climate risks are most likely to propagate across borders. Rather than dispersing resources uniformly, global climate finance and adaptation funding can be strategically aligned with systemic risk profiles, improving the effectiveness of collective action. In this sense, the study supports a transition from reactive climate risk management toward anticipatory and system-aware governance.

6.1 Applied Contexts, Policy Translation, and Ethical Considerations

The AI Risk Regime and CF_SPI classifications developed in this study find direct institutional parallels in ongoing policy efforts. Within the European Union, the European Central Bank's climate stress-testing framework requires banks to assess climate risk exposures across geographies and translate them into supervisory capital and liquidity considerations — a process that depends critically on structured, comparable, and interpretable risk categories of the kind produced here. Similarly, the ASEAN Taxonomy for Sustainable Finance, adopted progressively since 2021, requires member states to classify economic activities and country-level exposures according to climate transition readiness, highlighting the regional demand for cross-country AI-enabled risk diagnostics that are both granular and policy-legible. These institutional contexts illustrate that the framework's outputs are not merely analytical outputs but are directly applicable to regulatory classification, sovereign risk screening, and climate finance prioritisation exercises currently underway in major financial systems.

Translating AI Risk Regime outputs into regulatory and leadership action requires a structured policy logic. Countries classified within the High Risk Regime warrant immediate integration of climate risk into fiscal stress-testing, infrastructure resilience planning, and disaster risk financing mechanisms. Moderate Risk Regime countries represent a critical preventative policy window, where targeted adaptation investment and early-warning systems can forestall regime escalation. Global and Regional Risk Transmitters identified through CF_SPI classification require heightened supervisory attention within international financial architecture — including monitoring by the Financial Stability Board, IMF Article IV consultations, and NGFS scenario analyses — given their potential to amplify climate-financial stress beyond national borders. This output-to-action mapping provides a practical bridge between the study's technical classifications and the leadership decisions of regulators, central banks, and multilateral institutions.

Implementation of such frameworks at scale faces three practical constraints that future research and institutional standardisation should address. First, cross-country data heterogeneity — particularly under-reporting of climate impacts in lower-income and data-scarce economies — may compress CRI scores in ways that understate true exposure, calling for investment in reporting infrastructure and methodological harmonisation. Second, political economy dynamics may limit cross-border coordination even where systemic risk roles are clearly identified, as national governments may resist risk classifications with implications for sovereign borrowing costs or capital flows. Third, differences in supervisory mandates and regulatory philosophies across jurisdictions complicate uniform adoption, requiring flexible implementation guidance rather than prescriptive standards.

Finally, the ethical and socio-political dimensions of deploying AI-based climate risk tools globally warrant explicit acknowledgement. Risk regime classifications, if mechanically embedded in sovereign credit assessments or climate finance eligibility criteria, risk compounding capital disadvantage for countries already bearing disproportionate climate burdens — a concern that aligns with broader debates on climate justice and equitable transition finance. The interpretability of AI outputs is therefore not only a technical virtue but an equity imperative: classifications must remain legible, contestable, and contextually qualified by development constraints, so that affected governments and communities retain meaningful agency in how their risk profiles are understood and acted upon by international actors.

6.2 XAI Outputs and Sustainability Leadership: An Operationalization Framework

The analytical outputs of this study — the AI Risk Regime classification and the CF_SPI systemic risk role — are designed to be directly actionable across multiple layers of climate governance. For national policymakers, the AI Risk Regime classification provides a structured basis for differentiating fiscal and adaptation responses: High Risk Regime countries face an immediate imperative to integrate climate exposure into sovereign stress-testing, infrastructure resilience budgets, and disaster risk financing, while Moderate Risk Regime countries occupy a preventative policy window in which targeted investment can forestall regime escalation. For financial regulators and central banks, the CF_SPI classification identifies which jurisdictions warrant elevated supervisory scrutiny within climate scenario analysis frameworks — particularly those designated as Global or Regional Risk Transmitters, whose climate vulnerabilities carry cross-border financial implications that demand proportionate prudential oversight. For multilateral institutions, the cross-country comparability of both outputs enables evidence-based prioritisation of climate finance and surveillance resources toward countries where domestic climate vulnerability and systemic propagation potential intersect. Institutional investors can overlay AI Risk Regime designations with economic capacity indicators (GDP per capita) to refine sovereign ESG risk assessments and portfolio climate risk weights in ways that go beyond conventional rating-agency proxies. For the benefit of executive and non-specialist readers, Table 6 synthesises the study's principal findings and corresponding leadership actions across five strategic dimensions.

For the benefit of executive and non-specialist readers, Table 6 synthesises the study's principal findings and corresponding leadership actions across five strategic dimensions."

Table 6. Key Findings and Leadership Takeaways

Strategic Dimension	What the Study Shows	What Leaders Should Do
CLIMATE RISK EXPOSURE	7 of 20 countries fall in the High Risk Regime — cutting across both developing and high-income economies. Climate vulnerability is not confined to poorer nations.	Governments in High Risk Regime countries must integrate CRI-based exposure into national adaptation budgets, infrastructure resilience plans, and sovereign stress-test scenarios immediately.
SYSTEMIC FINANCIAL RISK	Pakistan is the sole Global Risk Transmitter; India, China, Italy, and Greece are Regional Stress Contributors — meaning their climate shocks can destabilize connected financial systems.	Financial regulators and central banks should apply elevated supervisory scrutiny to Global and Regional Risk Transmitters within climate scenario analysis (BCBS/NGFS frameworks).
ECONOMIC CAPACITY vs. EXPOSURE	High-income countries are not insulated: the High Risk Regime records the highest average GDP per capita, showing that asset concentration amplifies — not reduces — climate impact.	Investment screening and sovereign ESG ratings must move beyond income-based proxies. AI Risk Regime classifications offer a more accurate basis for climate risk pricing.
DEMOGRAPHIC AMPLIFICATION	Moderate Risk Regime countries have the largest average populations. Even moderate shocks translate into massive aggregate human and economic losses at scale.	Multilateral institutions should prioritise adaptation finance for densely populated Moderate Risk Regime countries — particularly those with low GDP per capita — as preventative investments.
AI EXPLAINABILITY AS GOVERNANCE TOOL	Rule-based AI classification produces transparent, replicable risk regimes that are directly interpretable without specialist modelling expertise.	Leaders and regulators can embed these classifications directly into policy workflows, supervisory mandates, and investment guidelines without requiring bespoke technical capacity.

7. CONCLUSION

This study presents an original and interpretable AI-enabled framework for assessing climate risk exposure and systemic climate–financial vulnerability across countries with diverse economic and demographic characteristics. By integrating Climate Risk Index data with economic capacity indicators, population scale, and systemic propagation metrics, the analysis moves beyond conventional climate risk assessments toward a multidimensional and policy-relevant understanding of climate vulnerability. The results demonstrate that climate risk is neither geographically nor economically uniform. High climate exposure is observed across income levels, while systemic climate–financial risk is concentrated among a limited subset of countries whose economic and financial interconnectedness amplifies spillover potential. These findings challenge simplified narratives that equate development with resilience or climate exposure with systemic risk, revealing instead a complex and non-linear risk landscape. A key contribution of this work lies in its emphasis on explainability. Rather than relying on opaque predictive models, the study employs transparent, rule-based AI classification to generate interpretable climate risk regimes that can be readily understood and applied by policymakers, regulators, and financial practitioners. The incorporation of systemic risk roles further extends the analysis beyond territorial exposure, offering empirical insight into how climate shocks may transmit through global systems. Importantly, this research does not seek to replace existing climate indices or forecasting tools. Instead, it complements them by providing a structured diagnostic framework that clarifies *where* climate risk is concentrated, *how* it interacts with economic and demographic structures, and *through whom* it may propagate internationally. To the best of current knowledge, this integrated combination of explainable AI-based regime classification and systemic climate–financial risk assessment has not been jointly operationalized in prior empirical studies. As climate change increasingly intersects with financial stability, development planning, and global governance, the need for transparent and system-aware analytical tools becomes more urgent. The framework developed in this study offers a scalable and adaptable foundation for future research and policy application, contributing meaningfully to the evolving literature on AI-enabled climate risk governance and financial resilience. By jointly operationalizing explainable AI-based climate risk regimes and systemic climate–financial spillover roles within a single empirical framework, this study establishes a methodological approach that is distinct from, and not replicated in, existing climate risk or financial stability literatures.

Beyond its methodological contributions, this study carries direct implications for the practice of sustainability leadership. Policymakers, financial regulators, and institutional decision-makers increasingly operate in an environment where climate risk is systemic, cross-border, and structurally embedded in financial networks — yet where the analytical tools available to them often lack the transparency required for accountable governance. The AI Risk Regime classification and CF_SPI-based systemic risk roles developed here are designed to close this gap: they convert complex, multi-dimensional climate data into interpretable, decision-ready outputs that leaders can act upon without requiring specialist modelling expertise. In this sense, explainability is not merely a technical property of the framework — it is a leadership enabler, equipping those responsible for climate and financial governance with the clarity, comparability, and institutional legitimacy needed to make and defend consequential decisions under uncertainty.

REFERENCES

Adadi, A., & Berrada, M. (2018). Peeking inside the black-box: A survey on explainable artificial intelligence (XAI). *IEEE Access : Practical Innovations, Open Solutions*, *6*, 52138–52160. DOI: 10.1109/ACCESS.2018.2870052

Alexander, C. (2024). *Relationship Between Performance Feedback and Job Satisfaction Environment of Accountants at Accounting Firms* (Doctoral dissertation, Capella University).

Arrieta, A. B., Díaz-Rodríguez, N., Del Ser, J., Bennetot, A., Tabik, S., Barbado, A., & Herrera, F. (2020). Explainable Artificial Intelligence (XAI): Concepts, taxonomies, opportunities and challenges toward responsible AI. *Information Fusion*, *58*, 82–115. DOI: 10.1016/j.inffus.2019.12.012

Basel Committee on Banking Supervision (BCBS). (2024). The role of climate scenario analysis in strengthening the management and supervision of climate-related financial risks. Bank for International Settlements. https://www.bis.org/bcbs/publ/d572.pdf

Batten, S., Sowerbutts, R., & Tanaka, M. (2020). Climate change: Macroeconomic impact and implications for monetary policy. *Ecological, societal, and technological risks and the financial sector*, 13-38.

Bolton, P., Despres, M., Da Silva, L. A. P., Samama, F., & Svartzman, R. (2020). *The green swan*. BIS Books.

Dar, A. A., Arora, V., & Yas, H. (2026). Sustainable Intelligence: How AI Powers a Greener, Smarter Planet. In *Predicting and Monitoring Climate Risks Through Green AI: Weather Forecasting, Disaster Prediction, and Biodiversity Monitoring* (pp. 147-178). IGI Global Scientific Publishing.

Dar, A. A., & Lone, W. Z. (Eds.). (2026). *Integrative Approaches to Quality, Data Analysis, and Interdisciplinary Research*. CRC Press. DOI: 10.1201/9781779641014

Dar, A. A., & Shaik, A. J. (2025). Spatiotemporal clustering and multivariate forecasting of air quality index across Indian cities using machine learning and deep learning models. *Franklin Open*, *13*, 100435. DOI: 10.1016/j.fraope.2025.100435

Eckstein, D., Künzel, V., Schäfer, L., & Winges, M. (2019). Global climate risk index 2020. *Bonn. Germanwatch*, *1*, 1–50.

European Union. (2024). Regulation (EU) 2024/1689 of the European Parliament and of the Council laying down harmonised rules on artificial intelligence (Artificial Intelligence Act). Official Journal of the European Union.

Hallegatte, S. (2016). *Shock waves: managing the impacts of climate change on poverty*. World Bank Publications.

Innovation Hub, B. I. S. (2024). The use of artificial intelligence for policy purposes. Bank for International Settlements. https://www.bis.org/publ/othp100.pdf

International Monetary Fund (IMF). (2024). *Global Financial Stability Report, October 2024: Steadying the Course: Uncertainty, Artificial Intelligence, and Financial Stability*. IMF.

IPCC. (2023). [Synthesis Report. Contribution of Working Groups I, II and III to the Sixth Assessment Report of the Intergovernmental Panel on Climate Change. IPCC.]. *Climatic Change*, 2023.

Lang, Q., Ma, F., Mirza, N., & Umar, M. (2023). The interaction of climate risk and bank liquidity: An emerging market perspective for transitions to low carbon energy. *Technological Forecasting and Social Change*, *191*, 122480. DOI: 10.1016/j.techfore.2023.122480

Mehryar, S., Yazdanpanah, V., & Tong, J. (2024). AI and climate resilience governance. *iScience*, *27*(5), 109812. DOI: 10.1016/j.isci.2024.109812 PMID: 38784017

Monasterolo, I. (2020). Climate change and the financial system. *Annual Review of Resource Economics*, *12*(1), 299–320. DOI: 10.1146/annurev-resource-110119-031134

OECD. (2024). OECD Principles on Artificial Intelligence (updated). Organisation for Economic Co-operation and Development. https://oecd.ai/en/ai-principles

Pacelli, V., Foglia, M., & Mariano, D. (2025). The link between climate and systemic risk: A bibliometric and systematic literature review. *Research in International Business and Finance*, *79*, 103072. DOI: 10.1016/j.ribaf.2025.103072

Tian, Y., Wen, H., & Guo, K. (2025). Machine learning applications in climate finance: An overview. *Research in International Business and Finance*, *79*, 103063. DOI: 10.1016/j.ribaf.2025.103063

Volz, U., Beirne, J., Ambrosio Preudhomme, N., Fenton, A., Mazzacurati, E., Renzhi, N., & Stampe, J. (2020). Climate change and sovereign risk.

Chapter 10
Leadership for Sustainable Internationalization:
A Survey of Senior International Officers in U.S. Higher Education

Holly Alderman
http://orcid.org/0009-0008-4653-8240
University of North Georgia, USA

ABSTRACT

The chapter explores leadership for sustainability transformations in U.S. higher education with a focus on senior international officers (SIOs) and environmental sustainability in education abroad. Student mobility for cross-border learning has been increasing, often without accounting of its environmental impacts. This research surveyed U.S. SIOs and found that many had the personal conviction and competencies for leading sustainable internationalization, but few had promoted institution-wide strategies addressing environmental impacts of education abroad. Many believed their institutions had the will to advance environmental sustainability, but few saw institutional action to reconcile the international agenda with sustainability goals. The chapter concludes by suggesting that U.S. SIOs seek training, communicate with sustainability experts, and assess their institutions' readiness for sustainable practices in education abroad. Furthermore, by engaging in sustainability-oriented networks, SIOs may strengthen their position for leading environmentally responsible global engagement.

DOI: 10.4018/979-8-3373-7877-0.ch010

INTRODUCITON

Many U.S. higher education institutions have committed to comprehensive internationalization while prioritizing sustainability and climate action. Although international education indirectly promotes sustainability values in some ways, academic mobility for education abroad has been problematized by polluting the environment, impacting sensitive cultural and environmental sites, and contributing to global warming. Education abroad entails students physically crossing international borders for academic study. The environmental impacts of education abroad are an important issue because academic mobility worldwide involves millions of people, and many U.S. institutions have been intent on growing their student participation in such programs. Environmental sustainability practices in education abroad aimed at resolving these tensions have often been described as ad hoc and bottom up, suggesting the need for a closer examination of organizational leadership for the challenge. Therefore, in this chapter, I explore an emerging dialogue in U.S. international education research and practice around sustainable internationalization by emphasizing the senior international officer's (SIO) role in advancing environmentally sustainable education abroad. SIOs are positioned to shape future development of institution-wide internationalization agendas including for education abroad. I describe the leadership capacity for promoting sustainability transformations through the lens of regenerative leadership theory, using survey data gathered from SIOs in U.S. higher education institutions (HEIs). The observations and analysis, though small and exploratory, further the understanding about SIO leadership at the nexus of internationalization and sustainability in U.S. HEIs and provide a basis for future research and practice in international education. Importantly, it contributes lesser-known SIO leadership perspectives to the broader conversation about the critical role of international higher education in achieving global sustainability goals. Administrators outside of higher education with an interest in leadership for promoting environmentally responsible global engagement will also find this study relevant.

Tension at the Nexus of Internationalization and Sustainability in Higher Education

Many institutions have responded to the macro-level phenomena of globalization and climate change with policies that foster both internationalization and sustainability agendas in higher education (Cortese, 2010; deWit & Merkx, 2021). Globalization has often been considered positively due to the economic and technological development resulting from greater cross-border trade and market competitiveness, but climate change has been seen negatively, given the environmental impacts that threaten human societies. In this context, HEIs are positioned to impact both phenomena, with programs aimed at resilience, innovation, and cooperation.

The main goal of internationalization is "to infuse international, global, and comparative content and perspectives throughout the teaching, research, and service missions of higher education" (Hudzik, 2018, p. 6). Accordingly, many presidents, chancellors, deans, and governing bodies of HEIs have instituted significant policy changes toward comprehensive internationalization (Freidheim, 2012), and this move has had several benefits. Comprehensive internationalization promoted international mobility, research, and networking opportunities for the faculty (deWit & Merkx, 2021). Comprehensive internationalization fostered a global mindset, foreign language skills, and appreciation of cross-cultural engagement among students (Soler et al., 2022). Learning outcomes from education abroad included "attitudes and capabilities such as openness, critical thinking, perspective taking, dealing with complexity, and intercultural dialogue" (Gregersen-Hermans, 2021, p. 468). Additionally, comprehensive internationalization has enhanced institutional competitiveness for talent, external funds, and global rankings (Stromquist, 2007). Moreover, international capacity in U.S. higher education has been found across various institutional types and sizes, public and private, but mainly in research universities (Koehn et al., 2011; Kwai, 2020; McBride & Greenberg, 2021; Ruther, 2002). Some scholars have disparaged internationalization, noting that it sometimes promoted market behaviors in HEIs and valued economic ends over intellectual and cooperative goals (deWit & Merkx, 2021; Stromquist, 2007). Nevertheless, the trend has been to increase international education by, for example, expanding global engagement through cross-border student mobility.

Education abroad is a significant facet of the internationalization agenda. Simply defined, such programs give students the opportunity to complete curricular coursework while becoming immersed in another country and culture for the purpose of their study (Forum on Education Abroad, 2020). The predominant form of education abroad entails students' physical mobility to go to another country for an academic degree or short-term program (e.g., exchange semester or faculty-led trip). Other types of mobility include cross-border study through branch campuses or study abroad agencies as well as virtual global studies, which connect students with global learning online (Woodfield, 2013). According to the Association of American Colleges & Universities (AAC&U, 2014), global learning values include global self-awareness, perspective taking, cultural diversity, personal and social responsibility, understanding global systems, and applying knowledge to contemporary global contexts. In the USA, education abroad has helped students achieve academic and personal development to become global citizens capable of cooperation on civic issues and to be competitive in a global workforce (Gullahorn & Gullahorn, 1958; Stromquist, 2007). Additionally, education abroad was reported to be a high-impact educational practice as it increased the rates of student retention and academic engagement (Kuh, 2008). Furthermore, after the 9/11 terrorist attacks in the USA, many policy makers and international educators resolved to increase students' understanding of others and themselves through self-reflection in the context of education abroad programs (NAFSA Strategic Task Force on Education Abroad, 2003). Consequently, the number and variety of scholarships enabling U.S. students to study abroad has proliferated in recent decades (Freidheim, 2012). For example, scholarship programs such as Gilman, Fulbright, and Boren targeted diverse U.S. students with substantial funding for different global learning and development experiences supporting U.S. foreign policy interests (Freidheim, 2012). Migration data from the United Nations Educational, Scientific, and Cultural Organization (UNESCO) in 2020 indicate that academic mobility worldwide, both outward and inward flow, was in the millions (cited in Ramaswamy et al., 2021). In 2019 alone, of the 17.5 million undergraduates enrolled in degree-granting postsecondary institutions (National Center for Educational Statistics, 2023), outbound U.S. students participating in education abroad programs numbered 347,099, but this number faced a 53% decline in 2020 due to the COVID-19 global pandemic (Institute of International Education [IIE], 2021). However, the post-pandemic rebound was dramatic, with outbound U.S. students numbering 188,753 and rising (IIE, 2023). Of the U.S. institutions that reported setting education abroad participation targets, the average target was approximately one-third of the undergraduate and graduate student populations (Soler et al, 2022). Thus, many U.S. HEIs identified global engagement as a growth area, but the acceleration of education abroad entailed some less desirable externalities which were counterproductive to institutional sustainability goals.

Notably, education abroad was forecasted as having a growing share of the overall greenhouse gas (GHG) emissions (Shields, 2019), which contribute to global climate change and negatively impact natural and human systems (Intergovernmental Panel on Climate Change, 2014; Khalili, 2011). Education abroad has also fostered exploration into sites and areas that were environmentally and culturally sensitive to visitor incursions, especially when concurrent with regular tourist seasons (Hale, 2019). Host communities benefitted economically from educational tourism, but in some places, education abroad programs inadvertently promoted income inequality and dependency (Schroeder et al., 2009). Additionally, overreliance on purchasing credits from carbon-reducing projects to offset the GHG emissions of institutional activities has raised environmental justice issues for marginalized populations (Barron et al., 2021). For example, communities of color have experienced higher air pollution from fossil fuel use and may not directly benefit from offsetting projects. Several scholars have described the tension between education abroad and sustainability goals as a moral dilemma and concluded that further investigation is needed into the benefits of international academic mobility and its costs to society to advance international higher education policy responsibly (Dvorak et al., 2011; Rumbley, 2020; Shields, 2019). These environmental, social, and economic impacts of internationalization activities are indicative of a sustainability problem within the field.

Environmental Sustainability Drivers and Barriers in International Education

Sustainability, broadly defined, is a state of the world in which the needs of the present can be met without compromising the ability of future generations to meet their needs (Mebratu, 1998; Khalili, 2011). The idea of sustainability encompasses environmental, social, and economic dimensions, and it has been and continues to be the subject of multinational dialogue on climate change. Of particular interest to this study was that environmental sustainability has crystalized the notions that natural resources are finite and must be conserved to meet present and future needs (Mebratu, 1998) and that the integrity of natural systems must be protected (U.S. Environmental Protection Agency, 2021). The United Nations' (UN) *Agenda 2030* provides 17 Sustainable Development Goals (SDGs) across an array of human interests, including environmental targets to mitigate or avoid the negative effects of global climate change (Jackson, 2007; UN Department of Economic & Social Affairs, 2021). As noted, HEIs are in a unique position to accelerate the progress on climate change and sustainable development through innovation, knowledge transfer, and engagement with diverse local and global constituents (Barron et al., 2021; Ramaswamy et al., 2021).

There are multiple motivators for HEIs to prioritize environmental sustainability. From 2005 to 2014, which was dubbed the "Decade of Education for Sustainable Development" (ESD), the UN urged educators to integrate sustainability values into all levels of education (Huckle & Wals, 2015) and support sustainable development research and cooperation (Kohl et al., 2022) to create a carbon-literate population equipped to advance a sustainable future. The U.S. Higher Education Opportunity Act (2008) authorized university sustainability programs (USPs) to boost capacity-building for resilience. These programs fostered activities such as reforming the curriculum for greater climate and sustainability literacy and designing "green" degree and certificate programs (Cortese, 2010; Stober et al., 2021); engaging students, faculty, and staff members in actions that address climate change (Stober et al., 2021); and engaging community members in local greening initiatives (Stober et al., 2021). The outcomes of such institutional sustainability interventions in some cases resulted in a culture-shift and increased climate awareness among HEI leaders (Barron, et al., 2021). Across the USA, over 700 higher education leaders have adopted Presidents' Climate Leadership Commitments since 2008 to prioritize decarbonization and resilience in institutional missions and campus operations (Barron et al., 2021;

Cortese, 2010). The 2015 UN *Global Action Program on ESD* further encouraged university leaders to transform HEIs with "whole-institution approaches" of incorporating ESD (Kohl et al., 2022, p. 225). These signals from global and institutional leaders together with the actions of a diverse base of interested organizational and community members have gained the attention of international educators.

Increasingly, many scholars have called upon higher education leaders to align internationalization strategies with sustainability goals (Davies, 2016; Greenberg, 2011; Hale et al., 2013; Long et al., 2014; Ramaswamy et al., 2021) and develop sustainability standards in education abroad programs (Dvorak et al., 2011; McBride, 2022). Leggett (2012) noted that "No one was talking about sustainability at study abroad conferences, and no one was talking about study abroad at sustainability conferences" (p. 42). Furthermore, cross-disciplinary literature on environmental sustainability and education abroad, though limited, has centered around four themes: environmental impacts of education abroad, education for sustainable development, administration of sustainability-oriented programs, and assessment of ESD and other sustainability interventions in education abroad. While in-person global engagement has remained a cornerstone of international education, mitigating environmental impacts and contributing to sustainable development has been the order of the day (Bowman, 2022).

Notably, approaches to environmentally sustainable study abroad have been varied. For instance, students have been provided with critical models for thinking about educational travel, focusing on balancing personal enrichment with improving the lives of the people and places visited (Simblach, 2008, as cited in McClelland, 2015; Gregersen-Hermans, 2021). Pre-departure orientation meetings have been used to inform students of the environmental impacts of study abroad, encouraging them to measure their carbon footprint and discuss pro-environmental behaviors for travel (Dvorak et al., 2011; Greenberg, 2011; Long et al., 2014). Hybrid program designs that comprise part online learning with the host nationals and part in-person engagement have been used to help mitigate environmental impacts without diminishing the intercultural experiences (Leggett, 2012). Some programs and/or their participants have purchased carbon offsets for air travel (Davies, 2016; Dvorak et al., 2011; Long et al., 2014; Rumbley, 2020), though this practice has detractors. For example, carbon offsetting has been sometimes perceived as a license to pollute, it has lacked a standard measurement or verification, and its benefits have often been far removed in time and place from those impacted by pollution (Davies, 2016; Long et al., 2014). Returning students have shared their sustainability insights and connected with eco-interest groups on campus to continue developing pro-environmental mindsets and behaviors that they experienced abroad (Greenberg, 2011). For measuring ESD outcomes, there has been an overreliance on student self-assessments of learning, but typical assessment methods such as oral defense, exams, critical self-reflection, projects, and feedback from group work have been observed (Gregersen-Hermans, 2021). Generally, however, institution-wide sustainability policies have often lagged behind program-level policies and individual actions to travel responsibly (Hale et al., 2013).

The literature review also revealed a variety of external influences on U.S. HEIs and international education leaders to encourage a shift toward sustainable internationalization. The American Council on Education (ACE), which shaped postsecondary education policy in the USA (ACE, 2023), reframed model comprehensive internationalization to foster global engagement that is "sustainable and just" (Internationalization Lab, 2022, p. 1). Particularly in education abroad, ACE espoused that "special consideration is given to environmental sustainability and the social, economic, and cultural impacts of off-campus mobility" (p. 5). The NAFSA Association of International Educators (NAFSA, 2019) added sustainability to its "Statement of Ethical Principles," emphasizing environmental responsibility in internationalization. Additionally, the Forum on Education Abroad (2021) developed an administrative framework to align the programs with the SDGs and guidance for establishing ESD in education abroad. Furthermore, special interest groups such as the Climate Action Network for International Educators (CANIE) fostered pro-environmental discussions and responsible education abroad (McBride, 2022). The Association of International Education Administrators (AIEA, 2023) signed the *CANIE Accord* acknowledging climate change and the need for international education leaders to actively work towards net-zero carbon emissions. Overall, the development of a sustainability mandate and practical tools for international education professionals have laid important foundations for advancing environmentally sustainable education abroad, but organization-level cues are key for activating wider change.

On this, some studies noted HEIs' societal obligation and institutional commitments as factors promoting engagement with sustainability goals (Dvorak et al., 2011; Davies, 2016). Long et al. (2014) observed that program mission statements that foster an ethos of sustainability could compel pro-environmental behaviors. An institution's financial position also affects its environmental stance. For instance, an organization's ability to purchase carbon offsetting or incentivize the use of alternative transportation modes were found to affect institutional policies on sustainable travel (Davies, 2016). In some cases, an institution's reputation also led to climate action and support for green programs (Davies, 2016). More importantly, institutions that proactively engaged with the climate crisis were often influenced by an environmental champion and leadership commitment to sustainability (Hale et al., 2013; Davies, 2016). Paradoxically, individuals not in a leadership position are often at the forefront of sustainable practices in HEIs (Hale et al., 2013). For example, engaged faculty develop sustainability-themed courses and activities

(Dvorak et al., 2011; Maragakis & Van den Dobbelsteen, 2013), students demand green programs (Hale et al., 2013; Maragakis & Van den Dobbelsteen, 2013), and staff members are a "driving force" toward greening initiatives (Maragakis & Van den Dobbelsteen, 2013, p. 4). Structural and normative contexts have been considered important factors of local environmental stewardship (Bennett et al., 2018). For example, Bennett et al. (2018) found that 'local assets' such as human and financial capital as well as extrinsic rewards or sanctions enable environmental stewardship. Organizational aspects such as these provide useful context for leading sustainable internationalization.

Despite the various drivers, barriers to environmental sustainability in education abroad have persisted. Internationalization in higher education has indirectly promoted some of the socio-economic goals of sustainability (Gregersen-Hermans, 2021), yet education abroad programs have been largely disconnected from university climate commitments (Long et al., 2014). This weakness is particularly salient because education abroad is a top priority for future growth in many U.S. HEIs' internationalization strategies (Soler et al., 2022). Notwithstanding the academic aims of education abroad, many HEIs become reliant on international student tuition to compensate for higher operating costs and stagnant state or donor support locally (Davies, 2016). Many of them also develop offshore branch campuses and online programs to further attract international students and faculty (deWit & Merkx, 2021). However, university sustainability programs and carbon management schemes often do not account for the environmental impacts of education abroad programs and other university-sponsored travel (Hale et al., 2013). An independent study spanning 1999–2014 suggested that travel for education abroad constituted a substantial portion global GHG emissions (Shields, 2019). Moreover, the internationalization and sustainability agendas in many HEIs operate in silos, stifling necessary cross-communication and collaboration (Davies, 2016; Hale et al., 2013; Leggett, 2012). Other barriers to reform include resistance to change, procrastination, and power struggles, which necessitate resources and incentives for sustainable internationalization (Ramaswamy et al, 2021). Overall, international educators and administrators face many challenges to intensify education abroad while addressing its adverse environmental impacts. The literature is clear that individual- and program-level actions alone are not enough to drive meaningful environmental sustainability in education abroad without organizational-level leadership and structural support for practical change (Davies, 2016; Hale et al., 2013; Hale, 2019; Long et al., 2014).

Meanwhile, there have been scant observations on the role of international education administrators and SIOs in promoting sustainable internationalization in U.S. HEIs. SIOs are senior-level administrators devoted to overseeing international strategies, programs, and services and to leading comprehensive internationalization in such institutes (AIEA, 2022). They typically have multiple responsibilities within the scope of international education, such as the institution's comprehensive internationalization strategy, education abroad administration, international research and partnerships administration, international student and scholar services, and global campus operations, though their expertise may lie in only one of these subfields (NAFSA, 2019). SIOs most often work in public, four-year, doctoral- and masters-granting institutions as well as other types and levels of HEIs (Kwai, 2020). The AIEA coined the term SIO, defined the standards of professional practice, and conducted annual conferences and meetings of SIOs (AIEA, 2016). Notably, these high-level administrators did not exclusively use the title SIO, making it difficult to identify them for the purpose of research (AIEA, 2016 and 2022). There is a lack of research on to what extent U.S. SIOs engage with sustainability and what factors enable or constrain leadership for promoting environmental sustainability in education abroad. Therefore, I investigated executive international education administrators in U.S. HEIs and focused on their roles in promoting environmental sustainability and education abroad (Alderman, 2024). As previously stated, the concepts of sustainability and internationalization are multi-faceted, and this study had a narrow scope of addressing an emerging issue in the international education field.

SIOs have been described as transformational leaders for their role in moving HEIs toward comprehensive internationalization (Heyl & Hunter, 2019). However, transformational leadership style has been criticized for being neither focused on influencing pro-environmental behaviors (Robertson & Barling, 2012) nor engaged with the wider society for activating structural equity and social justice (Shields, 2010). As the literature lacks sufficient conceptualization of leadership for bringing successful practices of environmental sustainability in education abroad up to scale, regenerative leadership theory (Hardman, 2012), which has been used to study sustainability mindsets and behaviors in educational leaders (Hardman, 2012, 2017), provided a framework for this study. Hardman (2012) described regenerative leaders as those who improve individuals' and organizations' level of engagement with sustainability. Building upon transformational leadership's foundation of moral values and trust within organizations, regenerative leadership adapts models of thinking and behaving responsibly to encompass more diverse stakeholders (e.g., including external stakeholders both locally and abroad) and much longer planning horizons than previous models (Hardman, 2012; Huron Consulting Group et al., 2019). Regenerative educational leaders have further been described as purpose-driven, systems thinkers who inspire and engage stakeholders in co-created goals that transform organizational cultures and behaviors (Hardman, 2017). Additionally, this leadership model is influenced by organizational and personal aspects that enable or constrain regenerative practices in certain contexts, such as organizational structure and culture as well as members' personal beliefs and habits (Roosen, 2022). Moreover, the regenerative concept pushes past sustainable development and carbon neutrality toward renewable and carbon negative (net-positive) practices (Hardman, 2012; McBride & Ponce-Taylor, 2022). Thus, the regenerative leadership model responds to calls from the international education field for a leadership style that moves sustainable internationalization toward regenerative practices (Lee & Lundemo, 2021; Tremblay, 2021).

Research has noted that international education leaders feel "stuck, frustrated, and overwhelmed" trying to reconcile the internationalization and sustainability agendas (Campbell et al., 2022, p. 13). Many environmental sustainability practices of U.S. international educators are ad hoc (Greenburg, 2011; Gregersen-Hermans, 2021) and bottom up (Bell, 2010; Fischer, 2022; Hale et al., 2013; Ramaswamy et al., 2021), suggesting the need to examine leadership for driving institutional-level and system-wide reforms. Thus, from a leadership lens, I examined U.S. SIOs regarding the extent to which they develop mindsets and behaviors for promoting environmental sustainability in education abroad as well as their perceptions of their institutional contexts for leading sustainable internationalization. The next section discusses the survey findings.

U.S. SIOs on Environmental Sustainability in Education Abroad

Little is known about SIOs' roles in the emerging ethos of sustainability within international higher education, particularly in terms of bringing successful environmental sustainability practices in education abroad up to scale for meaningful progress. The limited literature on SIOs has ascribed the role with consensus-building (West, 2014), coordination (Merkx, 2018), and institutional change agency for leading comprehensive internationalization in HEIs (Deardorff et al., 2018; Heyl & Hunter, 2019). SIOs are positioned to influence their institution's internationalization strategy including education abroad policy and practice (AIEA, 2022). Therefore, I chose this population as they would provide a valuable snapshot of the international education leadership landscape for addressing the problem. From October 2023 through January 2024, I conducted a targeted survey of SIOs in U.S. HEIs to describe the current state of international education leadership and to close a gap in the international education literature. The participants were screened based on their primary responsibilities rather than by title, to account for anticipated differences in position titles across U.S. HEIs (AIEA, 2022). The survey examined U.S. SIOs' sustainability mindsets and behaviors and how they perceived their institutions' levels of engagement with environmental sustainability in education abroad. The survey questions were based on the developmental and contextual constructs of regenerative leadership theory (Hardman, 2012) and on relevant recommendations from the available literature.

SIOs have been described as "time pressured" and a challenging population to recruit (Koehn et al., 2011, p. 337). In alignment with this finding, the participation in the current survey was lower than expected ($n = 41$) and therefore, more than one sampling frame was used. The first approach yielded a 3% response rate and the second yielded a 13% response rate to the survey. Prior online surveys of SIOs have yielded a range of response rates (Cotton et al., 2019; IIE, n.d.; Koehn et al., 2011; Kwai, 2020). A notable difference in most of the prior studies with higher response rates was the recruiter's executive position or high name recognition in the international education field, and this authoritative approach may be hard for respondents to disregard (Daikeler et al., 2020). In general, for descriptive research, a response rate of 10–20% is acceptable (Gay & Diehl, 1992, as cited in Hill, 1998). After analyzing the sample demographics for nonresponse bias, responses from 35 participants were included in the study. In exploratory research such as this, small samples of 10–30 people are sufficient (Gay & Diehl, 1992, as cited in Hill, 1998). Moreover, as SIOs, the qualified participants provided the kind of descriptive data needed for the purpose of this study. These data were quantified and analyzed for frequency, and descriptive statistics were compiled. Responses from a few open-ended questions were used to further explain the results.

SURVEY FINDINGS AND DISCUSSION

Using regenerative leadership as the theoretical framework to survey SIOs, I described leadership and organizational conditions for sustainable internationalization in U.S. higher education. I emphasized leadership for the promotion of environmental sustainability because education abroad is a growth area in many internationalization strategies that has been problematized by adverse environmental impacts and tensions with sustainability agendas. According to extant studies on various pro-environmental interventions in education abroad, implementation appears to be ad hoc and bottom up, with scant observations of the role of SIOs in leading change. Thus, little is known about U.S. SIOs' engagement with environmental sustainability personally or within the contexts of their academic institutions. The current investigation posed questions on personal mindset, personal behaviors, organizational mindset, organizational behaviors, and system leadership in relation to sustainability. An online survey of the SIOs from the AIEA membership listserv and select IPEDS institutions yielded low response rates (3% and 13%, respectively), as discussed above, but provided adequate data for a descriptive analysis and discussion of an emerging issue in the international education field (Gay & Diehl, 1992 as cited in Hill, 1998). The findings are not expected to describe individuals other than SIOs who are engaged in internationalization and sustainability initiatives, but they provide additional insights on institutional and system dimensions regarding regenerative leadership in U.S. higher education in general.

Demographically, the descriptions of SIOs found in past literature reasonably applied to the current sample. They were seasoned executives working across all institutional types and sizes in the U.S. higher education sector. Most of them reported being 46–55 years old and having 20 or more years' experience in international education. Their cumulative expertise and longevity in the field meant that they likely had institutional knowledge and professional networks that could enable regenerative practices (Hardman, 2012). However, the presence of more late-career SIOs in the population increased the urgent need for them to act on their potential advantages while still in office. Most of the respondents were in medium to large doctoral-granting research universities. When asked to describe their institutions' education abroad programs, respondents reported a combined total of 17,084 student-participants (AY 2022-23) and their top destinations were Europe, Latin America, and Asia. The current data describing their institutional education abroad programs aligned with the wider U.S. study abroad trends in terms of the host regions and participation (IIE, 2021, 2023). Since the active education abroad programs were found to impact every region worldwide, and inbound and outbound student mobility are top priorities for the strategic growth of internationalization in many HEIs (Soler et al., 2022), the urgency of effective U.S. SIO leadership in sustainability initiatives has been reinforced. Moreover, because the environmental impacts of education abroad are often unevenly felt across the globe (Davies, 2016; McClelland, 2015), sustainability-oriented leaders need to be aware of the places where issues of environmental justice are particularly salient and adapt their programs accordingly. The Forum on Education Abroad (2021) has provided a resource for international educators to address the ethical issues in educational tourism including how to engage host communities in mutually beneficial ways.

The findings on the SIOs' mindsets and behaviors suggested their strengths and weaknesses in advancing environmentally responsible internationalization. First, when asked to describe their personal drive and capacity to approach the problem, most of the SIOs reported believing environmental sustainability to be important to them personally, which was unsurprising since professional organizations of international educators have adopted sustainability as an ethical principle (NAFSA, 2019) and committed to climate action leadership (AIEA, 2023). Environmental stewardship is predicated on the notion of an agent's personal needs aligning with the interests of their organization (Davis et al., 1997). Davies (2016) found that for all the postsecondary schools studied, "the benefits of the internationalization agenda in terms of income generation and enhancing the student experience were perceived to outweigh the additional induced emissions" from student mobility (p. 293). In relation to the cost-benefit concern and the importance placed on ensuring that the benefits of education abroad outweigh the environmental costs to society, the current study's SIOs had varied beliefs. Most indicated that the benefits of education abroad in terms of students' learning outcomes, cross-cultural understanding, and global citizenship, for example, outweighed any adverse environmental impacts. Many believed that education abroad was a transformative experience that fostered mindsets which complimented sustainability goals and could have a mitigating effect on the problem. Paradoxically, Morais and Ogden (2011), who examined the benefits of education abroad among U.S. students, found more gains in global workforce competencies than in social justice learning. Meanwhile, many of the current respondents were uncertain about the cost-benefit tradeoff of environmental sustainability initiatives in education abroad and favored a balanced or staged approach to reform while developing comprehensive internationalization strategies. Some of them shared that their institutional and state contexts played a key role in prioritizing internationalization over sustainability. Davies (2016) recommended that HEIs and SIOs should not simply believe in internationalization's indirect contributions to a low-carbon society but take steps to substantiate the beneficial claims of their global programs to justify further environmental costs to society. Many SIOs also expressed a willingness to take diverse approaches to address the issue, but few had formal knowledge about environmental sustainability or the UN SDGs through coursework or training. This aligned with Bell (2010), who observed that many international educators felt ill-equipped to understand the problem or how to resolve it. Moreover, the SIOs' lack of requisite knowledge made the subsequent discussion about personal capacity to connect, observe, and listen relevant.

Next, the SIOs were asked to describe their personal orientation to act on the problem. The survey findings revealed both capacity and deficiency among international education leaders in reconciling their global programs with sustainability. Regenerative leaders are relational, are inclusive, and exercise humility by being open to listen and learn from others when resolving differences and working toward a common objective (Hardman, 2012). Relatedly, environmental stewards derive satisfaction in taking action, such as working to reduce pollution or redress environmental impacts (Bennett et al., 2018). The respondents felt competent in their abilities to connect a diverse range of stakeholders and effectively listen to others' ideas, which were key skills for promoting environmental sustainability in education abroad. This response was expected because the role of an SIO often entails facilitating global partnerships and bridging disciplinary silos for activating comprehensive internationalization (Heyl & Tullbane, 2012; IIE, n.d.). Additionally, the role requires cross-cultural intelligence for representing institutions across borders (Heyl & Tullbane, 2012; Kwai, 2020) and perspective-taking for leading the various responsibilities under an SIO's purview such as international student and scholar services (Mestenhouser, 2011, as cited in Deardorff et al., 2018). Notably, the respondents were much less engaged in working with sustainability experts (internal or external to the organization), learning best practices, or seeking funding to support environmental sustainability initiatives in education abroad. This confirmed what Greenberg (2011) and others (Davies, 2016; Forum on Education Abroad, 2021) previously acknowledged, that there is a disconnect between university sustainability programs and international student mobility. Extant literature has called for financial support to promote environmentally responsible travel (Hale, 2019; Ramaswamy et al., 2021) and highlighted the capacity to support a collective goal regardless of personal agendas (which is understood to include collaboration and financial support in this study) as an important leadership behavior for regenerative practice (Hardman, 2012). Overall, since both capacity and deficiency were found among international education leaders in relation to sustainability, it was important to contextualize the participants' responses.

The findings on institutional mindsets and behaviors from the SIOs' viewpoints indicated challenges and opportunities for promoting environmental sustainability in education abroad. The first set of survey questions asked the SIOs to describe their institutions' motivations and capacity to take a collective approach to the problem. Gauging an organization's position on sustainability is important for assessing whether the institutional environment is conducive to leading with a shared purpose (Hardman, 2012). Most of the respondents viewed their institutions as valuing environmental sustainability. This finding was expected given the federal support for university sustainability programs (Higher Education Opportunity Act, 2008), high number of U.S. university presidents' commitments to carbon reduction (Barron et al., 2021; Cortese, 2010), and the reputation-seeking behaviors of U.S. institutions including participating in rankings such as the *Times Higher Education (THE) Impact Rankings* (Burmann et al., 2021; NAFSA, 2015). Clugston and Calder (2014) found that a majority of faculty members wanted their institutions to prioritize sustainability and empower them to help lead the change. However, a few SIOs did not perceive that their institutions valued sustainability while some were unsure. Additionally, engaging a broad range of internal and external stakeholders to the organization as well as being cognizant of future generations is important for sustainability transformation and regenerative practice (Hardman, 2017; Huron Consulting Group et al., 2019). Collective approaches require high levels of trust that group members' ideas will be heard and valued even if they diverge from dominant organizational norms and assumptions (Hardman, 2012). Most respondents believed that their institutions valued diverse and inclusive approaches to sustainability initiatives. Others were uncertain whether collectively generated solutions should conform to the institutional culture, traditions, or the institutional expectations of strictly empirical ideas. Kezar and Eckel (2002) observed that structural and socio-cultural factors are important to change strategies in higher education. Whether or not an institution has explicit value statements or commitments to sustainability, for example, could influence a leader's ability to address emerging issues with new guidelines and standards of professional practice across an organization (Kezar & Eckel, 2002). These observations of the respondents' organizational contexts in terms of mindset was followed by their perceptions of organizational behavior, namely the capacity to envision and promote regenerative practices.

The next set of questions asked the SIOs to describe their institutions' orientation toward collectively acting on the problem. On institutional visioning and planning, most respondents reported conventional 3–5-year planning horizons. A short planning horizon was criticized as being "not fast enough to be agile or future-facing enough to be transformational" (Huron Consulting Group et al., 2019 p.12). Sustainability planning is meant to consider the needs of future generations (Mebratu, 1998; Khalili, 2011), and regenerative visions should include renewable and carbon-negative futures (Hardman, 2012, McBride & Ponce-Taylor, 2022). Likely related to a lack of deep future-casting, most of the SIOs in the current study reported a lack of environmental sustainability goals for institution-sponsored global programs. While sustainability's emergence as a priority in the international education field is relatively recent, the call to action has been clear (Forum on Education Abroad, 2021; McBride & Greenberg, 2011). A sustainable internationalization strategy includes a common purpose with environmental, social, and economic goals (Ramaswamy et al., 2021). Additionally, regenerative solutions aspire to raise people's living standards, celebrate community, and improve the health of all species for all time (i.e., *triple top line*) (Hardman, 2012). Meanwhile, many of the current respondents reported that their institutions lacked a sustainable international travel policy for university-sponsored travel. They also indicated that their institutions did not provide training about environmentally responsible travel. However, sustainable educational travel should strive to "minimize the negative economic, socio-cultural, and environmental effects of travel and to maximize the potential beneficial effects" (Long et al., 2014, p. 422). For example, a sustainable international travel policy can provide guidance for choosing program destinations and travel services (e.g., transportation, housing, and food) that align with the institutional sustainability values and climate commitments (Long et al., 2014). Next, most of the SIOs reported a general lack of measurement of the adverse environmental impacts of university-sponsored travel, including for education abroad. Many scholars have previously examined this problem (Dvorak et al., 2011; Hale et al., 2013; Hale, 2019; Long et al., 2014; Shields, 2019; Soler et al., 2022), with some developing methods for estimating environmental impacts such as the GHG emissions of student air travel (Davies, 2016; Hale, 2019; Shields, 2019). The organizational deficiencies uncovered in the current study further demonstrate what other scholars found: The higher education sector is poorly motivated to mitigate the impacts of international student mobility because of a lack of reliable data, competing priorities, and low sense of responsibility for the GHG emissions of student travel (Davies, 2016; Hale et al., 2013).

In response to the final set of questions pertaining to system leadership for promoting environmental sustainability in education abroad, the SIOs described the general emerging consciousness and collective engagement with the problem. Most of them perceived institutional vision/values/priorities as the top driver of sustainability in their institutions. Many identified collegial engagement with environmentalism, followed by institutional tracking or measuring sustainability performance, as additional drivers. According to Roosen (2022), organizations with a collective sustainability vision and goals and those which practice sustainability principles enable regenerative leadership. Conversely, the top barriers to sustainability initiatives in education abroad are a lack of institutional funding/incentives, lack of awareness, and competing priorities. On this, Ramaswamy et al. (2021) observed that sustainability initiatives "may be challenging for an institution due to lack of awareness, resistance toward change, procrastination, and power struggle issues," requiring a policy-level approach (p. 395). Reflecting on the barriers and drivers may have helped some of the respondents identify the push-pull levers for promoting organizational change. Even though the barriers and drivers were likely unique to individual institutions, there was practical value in aggregating the SIOs' observations to consider system-level agendas at scale (i.e., statewide and global scales). Systems that support regenerative leadership have been found to have the capacity for internal collaborations that extend through institutional consortiums, coalitions, and global networks (Hardman, 2012). Accordingly, HEIs are essential players in facilitating partnerships for implementing the SDGs, including cooperating with and learning from each other regarding sustainable development, though many North American HEIs tend to connect at the local and regional levels (Mallow et al., 2020). Approximately half of the SIO respondents reported participating in sustainability-themed groups and networks, primarily in internal/institution-wide or national networks. As the SIOs faced common challenges when advocating for sustainability in their work, the importance of their participation in networks that support capacity- and coalition-building on this issue became apparent.

In summary, the survey results describe a hard-to-reach population of SIOs in U.S. HEIs who are responsible for institution-wide internationalization strategy and services. The findings provide a multidimensional description of the SIOs and their institutional contexts in terms of leadership suitable for addressing environmentally responsible internationalization, with a special focus on education abroad. The sample participants were diverse in their responses across all dimensional measures. There was individual and collective willingness to prioritize sustainability and regenerative leadership competencies among SIOs and to reform the education abroad programs; however, institution-level awareness of and commitment to sustainable global engagement were perceived as largely missing. This likely relates to what previous scholars have observed in U.S. education abroad programs, that environmental sustainability practices take place in an ad hoc (Greenburg, 2011; Gregersen-Hermans, 2021) and bottom-up (Bell, 2010; Fischer, 2022) manner. Even though most of the respondents recognized the importance of environmental sustainability personally, many believed that the benefits of education abroad outweighed the environmental costs to society. This view was likely because most of the respondents had spent the bulk of their careers advancing internationalization in higher education, and they believed in the potential of a global mindset resulting from education abroad to stimulate sustainability learning and action. Similarly, most of them perceived that sustainability was important to their institutions generally, but the institutional capacity to achieve sustainability goals collectively and with consideration of future generations was less obvious. Thus, there is a need for broader participation, including SIOs, in sustainability-oriented networks for collective learning and engagement. These findings and analyses, though limited and exploratory, extend the discussion of international education leadership and practice, particularly in relation to environmental sustainability in education abroad. The findings provide a basis for future research and practice to advance the international education field. The following section discusses the practical implications of the findings for U.S. SIOs who recognize the emergent sustainability priority in international education.

Lessons for Sustainability Leadership in Global Education

This study aimed to describe leadership capacity for sustainability transformations in U.S. international higher education, with a special focus on SIOs and environmental sustainability in education abroad. Since sustainability is an emerging priority in U.S. international education, it was encouraging to find that many of the surveyed SIOs had the personal conviction and competencies to lead sustainable internationalization, and many perceived the institutional will to advance environmental sustainability in principle. However, only a few of the SIOs had taken action to address the adverse environmental impacts of education abroad programs and most of them perceived little organizational movement to address the tension at the nexus of sustainability and internationalization agendas. Hardman (2012) theorized that regenerative leadership capacity, capable of creating the conditions in which sustainability transformations move toward regenerative practices, arises from an internal-to-external journey, from mindsets to behaviors and to systems. The following ideas span personal, institutional, and system domains to help international education leaders and SIOs implement sustainable global engagement.

The surveyed SIOs reported various dispositions for addressing the environmental impacts of education abroad. Stewardship entails a collectivist orientation, long term outlook, and intrinsic rewards (Davis et al., 1997). Therefore, the recommendations for practice may need to be tailored to individual capacities and institutional contexts. Overall, based on the findings, the following four key lessons are suggested for leaders in global education:

1. Complete training on sustainability, environmental stewardship, and the UN SDGs. Understanding environmental stewardship is a pathway to involvement in promoting sustainable policies and programs (Bennett et al., 2018). Previous studies have emphasized the importance of awareness among leaders regarding the science behind global climate change and how to minimize the impacts of educational travel on the environment (Bell, 2010; Hale, 2019). A key lever for leading change and engagement with regenerative practice entails critically reflecting on the dominant mindset within one's organization (Doppelt, 2009; Roosen, 2022). Additionally, according to Gregersen-Hermans (2021), international educators need to reflect on their own implicit, normative, and cultural understandings of sustainability when designing responsible education abroad programs. Hardman (2012) theorized a sustainability mindset developed in five stages: resistance, skepticism, compliance, commitment, and engagement. Others have recommended inner development, such as the ability to relate to people, for taking collective approaches to sustainability policymaking and projects (Hardman, 2012; Roosen, 2022). On this, the Inner Development Goals (IDGs) created in 2020 to complement the UN SDGs provides a framework for cultivating the skills and abilities necessary to support long-term systemic transformations (Cooper & Gibson, 2022; Thomas, 2021). For example, the IDG dimension *relating* stimulates the inner capacity needed for SDG Goal 17: global partnerships, which has direct applications for HEIs (Cooper & Gibson, 2022).
2. Connect with sustainability experts and environmental stewards. Regenerative leaders are connectors, are inclusive, and exercise humility by listening and learning from others and by resolving differences when working toward a common objective (Hardman, 2012). Various scholars and practitioners have suggested that study abroad administrators should collaborate with the campus sustainability officer or eco-minded faculty and interest groups on campus (Greenberg, 2011; Forum on Education Abroad, 2021). Notably, many U.S. HEIs have established university sustainability programs that have spurred capacity-building for resilience. U.S. SIOs without the internal resources of a USP can connect with external sustainability organizations such as the Association for the Advancement of Sustainability in Higher Education (AASHE) for guidance

and benchmarking. Thus, they can begin to bridge the silos in which internationalization and sustainability agendas are developed and operated and address possible points of tension collectively.

3. Assess institutional values around social and environmental justice issues. Values may be explicit (e.g., mission statement or strategic plan) or implicit (e.g., social norms or customs), but they are key motivations for environmental stewardship (Bennett et al., 2018). Analyzing change readiness in one's organization is important to leading sustainability transformation (Doppelt, 2009). Organizational contexts that support regenerative leadership are conducive to collective strategies for desired futures, from developing a vision to implementation and scaling up sustainable and regenerative practices (Hardman, 2012). Therefore, U.S. SIOs need to ascertain the organizational will to take collective approaches to sustainable internationalization, engage a broader range of stakeholders (internal, external, and future generations), and consider diverse solutions to sustainability problems even if some ideas may challenge the organizational culture, traditions, or norms.
4. Engage in sustainability-oriented networks. The SIOs in this study reported participating in groups at the institutional, state, and regional levels that comprised an emerging system for promoting environmental sustainability in education abroad. The optimal form of regenerative practice is when power and decision-making are distributed throughout an organization and across local, national, and global systems through inclusivity and interdependence (Hardman, 2012). Therefore, SIOs should utilize their networks to reflect upon environmental sustainability drivers and barriers across institutions and to consider multilateral agendas and priorities for implementing sustainable internationalization. Participation in such networks can also support capacity and coalition building for tackling this issue.

In addition, while laying the groundwork for reform with the above suggestions, SIOs can help prepare for institutional-level action. For example, they can map the programmatic mission and goals regarding the SDGs (Forum on Education Abroad, 2021), collectively define a unified concept of sustainability with diverse stakeholders (Singh & Segatto, 2020), reflect on the impacts (positive and negative) of education abroad for the participating students and the destinations visited (Bowman, 2022; Hale, 2019), collectively identify which sustainability metrics should be employed to track the adverse environmental impacts and mitigation efforts (Barron et al., 2021), and develop training and policy guidance for sustainable international travel (Long et al., 2014). In a recent case study of University of Minnesota's SDG Initiative, Allen et al. (2024) found model practices in comprehensive sustainability transformation founded in global-to-local connections. For example, the initiative utilized *THE*

Impact Ranking metrics as a framework for mapping institutional activities to the UN SDGs and collecting data. Successful environmental stewardship efforts rely on careful alignment of outside prescriptions to local realities (Bennett et al., 2018). It is possible that as international education leaders and SIOs develop sustainability mindsets and behaviors while promoting these ideas in their institutions and networks, they will begin to address the conflict observed between internationalization and sustainability agendas in higher education, thus advancing international higher education in relation to global sustainability goals.

CONCLUSION

This study extended an emerging dialogue in U.S. international education research and practice on sustainable internationalization by emphasizing the SIO's role in promoting environmental sustainability in education abroad. Institutional climate commitments and the UN SDGs have challenged international education leaders to achieve organizational goals for global engagement while being mindful of the societal impacts of success. Guided by the regenerative leadership theoretical framework (Hardman, 2012), this study addressed the need for more inclusive, longer-term, and systems-based constructs of leadership that encompass social, economic, and environmental outcomes. This research is significant because it addresses the tension between internationalization and sustainability agendas in U.S. higher education, particularly the environmental impacts of education abroad programs, and fills the gap in literature by describing leadership capacity for promoting sustainable internationalization. The findings illustrated the individual and collective aspects of the leadership landscape in U.S. international education and suggested immediate actions that SIOs can take to help advance education abroad responsibly. More than improving only policy, the recommendations contribute to inspiring environmental stewardship development among leaders who influence global engagement and more broadly. These lessons will be of interest to international educators and administrators as well as higher education consultants, institutional partners, study abroad faculty and students, and educational travel service providers. Administrators outside of higher education, such as in Tourism or Transportation, who are challenged with prioritizing sustainability and promoting environmental stewardship in their organizations' global strategies and operations may also find this study useful. Moreover, there is significant scope for future research on this topic, in terms of improving the research design as well as conducting longitudinal studies, qualitative research, case studies, and related leadership studies.

REFERENCES

Allen, I., Brown, K., & Pelican, K. (2024). Connecting higher education to the SDGs: Strategies from the University of Minnesota SDG Initiative. In Shoemaker, J., & Regulska, J. (Eds.), *Global Goals, Global Education: Advancing the United Nations Sustainable Development Goals* (pp. 55–62). NAFSA Association of International Educators.

Alderman, H. (2024). *Leadership for environmental sustainability in education abroad* [Doctoral dissertation, University of North Georgia]. Nighthawks Open Institutional Repository. https://ir.ung.edu/work/sc/569236aa-6121-4d0f-a8ff-2ba42eba8861

American Council on Education. (2023, February 18). *About us*. https://www.acenet.edu/About/Pages/default.aspx

Association of American Colleges and Universities. (2014). *Global learning VALUE rubric*. https://www.aacu.org/initiatives/value-initiative/value-rubrics/value-rubrics-global-learning

Association of International Education Administrators. (2016, November). *Standards of professional practice for international education leaders and senior international officers*. https://www.aieaworld.org/standards-of-professional-practice.html#gsc.tab=0

Association of International Education Administrators. (2022, February 1). *What is an SIO?* https://www.aieaworld.org/sio

Association of International Education Administrators. (2023). *2022-23 AIEA annual report*. https://aieaworld.org/wp-content/uploads/2026/03/2022-23_aiea_annual_report-LR.pdf

Barron, A.R., Domeshek M., Metz, L.E., Draucker, L.C, & Strong, A.L. (2021). Carbon neutrality should not be the end goal: Lessons for institutional climate action from U.S. higher education. *One Earth, 4*, 1248-1258. Elsevier Inc. https://doi.org/ DOI: 10.1016/j.oneear.2021.08.014

Bell, K. (2010, December 4). *Where travel abroad meets sustainability: The green passport*. https://melibeeglobal.com/blog/2010/12/where-travel-abroad-meets-sustainability-the-green-passport/

Bennett, N. J., Whitty, T. S., Finkbeiner, E., Pittman, J., Bassett, H., Gelcich, S., & Allison, E. H. (2018). Environmental steweardship: A conceptual review and analytical framework. *Environmental Management, 61*(4), 597–614. DOI: 10.1007/s00267-017-0993-2 PMID: 29387947

Bowman, K. D. (2022, July 12). Future-proofing: How international education leaders foster resilience and fuel innovation. *International Educator.* https://www.nafsa.org/ie-magazine/2022/7/12/future-proofing-international-education

Burmann, C., Garcia, F., Guijarro, F., & Oliver, J. (2021). Ranking the performance of universities: The role of sustainability. *Sustainability (Basel), 13*(23), 13286. Advance online publication. DOI: 10.3390/su132313286

Campbell, A. C., Nguyen, T., & Steward, M. (2022). *Promoting international student mobility for sustainability? Navigating conflicting realities and emotions of international educators.* Association for Studies in International Education., 6.DOI: 10.1177/1028315322112138

Clugston, R., & Calder, W. (2014). Higher education for strong sustainability. In Corcoran, P. B., & Hollingshead, B. P. (Eds.), *Intergenerational learning and transformative leadership for sustainable futures.* Wageningen Academic Publishers. DOI: 10.3920/978-90-8686-802-5_6

Cooper, K. J., & Gibson, R. B. (2022). A novel framework for inner-outer sustainability assessment. *Challenges, 13*(64), 64. Advance online publication. DOI: 10.3390/challe13020064

Cortese, A. (2010, January 28). Higher education's purpose: a healthy, just, and sustainable society. *Fast Company.* https://www.fastcompany.com/1527689/higher-educations-purpose-healthy-just-and-sustainable-society

Cotton, D. R. E., Morrison, D., Magne, P., Payne, S., & Heffernan, T. (2019). Global citizenship and cross-cultural competency: Student and expert understandings of internationalization terminology. *Journal of Studies in International Education, 23*(3), 346–364. DOI: 10.1177/1028315318789337

Daikeler, J., Bosnjak, M., & Manfeda, K. L. (2020). Web versus other survey modes: An updated and extended meta-analysis comparing response rates. *Journal of Survey Statistics and Methodology, 9*(3), 513–539. DOI: 10.1093/jssam/smz008

Davies, J. C. (2016). *Exploring the inherent conflict between the internationalization and carbon management agendas in the UK higher education sector.* [Doctoral thesis, Manchester Metropolitan University]. e-space Manchester Metropolitan University's Research Repository. https://e-space.mmu.ac.uk/id/eprint/619558/

Davis, J. H., Schoorman, F. D., & Donaldson, L. (1997). Toward a stewardship theory of management. *Academy of Management Review, 22*(1), 20–47. https://www.jstor.org/stable/259223. DOI: 10.2307/259223

Deardorff, D. K., Diabate, D. B., & Rosenbaum, K. L. (2018). Foundations in leading internationalization: Lessons from Mestenhouser. In Deardorff, D. K., & Charles, H. (Eds.), *Leading internationalization: A handbook for international education leaders* (pp. 16–22). Stylus Publishing.

deWit, H., & Merkx, G. (2021). The history of the internationalization of higher education. In Leask, et al. (Eds.), pp. 23-52. *The Handbook of International Higher Education*. Stylus Publishing. https://ebookcentral.proquest.com/lib/northgeorgia/detail.action?docID=6822677

Doppelt, B. (2009). *Leading change toward sustainability: A change-management guide for business, government and civil society*. Taylor & Francis Group.

Dvorak, A. M. W., Christiansen, L. D., Fischer, N. L., & Underhill, J. B. (2011). A necessary partnership: Study abroad and sustainability in higher education. *Frontiers: The Interdisciplinary Journal of Study Abroad, 21*(1), 143–166. DOI: 10.36366/frontiers.v21i1.307

Fischer, K. (2022, April 27). An accord aims to make international education more sustainable. *Latitudes, a newsletter of The Chronicle of Higher Education.*

Forum on Education Abroad. (2020). *Standards of good practice for education abroad*. https://www.forumea.org/standards-of-good-practice.html

Forum on Education Abroad. (2021). *Advancing the UN sustainable development goals through education abroad*. https://forumea.org/resources/guidelines/advancing-the-un-sdgs/

Freidheim, S. J. (2012). *Valuing study abroad: the global mandate for higher education* [Policy brief]. Institute of International Education, Center for Academic Mobility Research. https://www.iie.org/publications/valuing-study-abroad/

Greenberg, D. (2011, February 17). *5 ways to design sustainable study abroad*. https://my3.my.umbc.edu/gropus/sustainability/posts/5279

Gregersen-Hermans, J. (2021). Toward a curriculum for the future: Synthesizing education for sustainable development and internationalization of the curriculum. *Journal of Studies in International Education, 4*(25), 461–481. DOI: 10.1177/10283153211031033

Gullahorn, J. T., & Gullahorn, J. E. (1958). American objectives in study abroad. *The Journal of Higher Education, 7*(29), 369–374. https://www.jstor.org/stable/1978752. DOI: 10.1080/00221546.1958.11779560

Hale, B. W. (2019). Understanding potential impacts from university-led educational travel. *International Journal of Sustainability in Higher Education, 20*(2), 245–263. DOI: 10.1108/IJSHE-09-2018-0163

Hale, B. W., Vogelaar, A., & Long, J. (2013). A-broad spectrum: Sustainability in educational travel. *International Journal of Sustainability in Higher Education, 4*(14), 349–366. DOI: 10.1108/IJSHE-07-2011-0049

Hardman, J. (2012). *Leading for regeneration: Going beyond sustainability in business, education, and community.* Routledge; Earthscan.

Hardman, J. (2017). Educational leadership and environmental justice in a climate-challenged world. In White, D., & Bogotch, I. (Eds.), *The Wiley international handbook of educational leadership* (1st ed., pp. 139–156). John Wiley & Sons, Inc. DOI: 10.1002/9781118956717.ch8

Heyl, J. D., & Hunter, F. J. H. (2019). *The senior international officer as change agent* (2nd ed.). Association of International Education Administrators.

Heyl, J. D., & Tullbane, J. (2012). Leadership in international higher education. In Deardorff, D. K. (Eds.), *The sage handbook of international higher education* (pp. 113–130). SAGE Publications., DOI: 10.4135/9781452218397.n7

Higher Education Opportunity Act, 110 U.S.C.§4137 (2008). https://www.congress.gov/bill/110th-congress/house-bill/4137

Hill, R. (1998). What sample size is "enough" in internet survey research? *Interpersonal Computing and Technology, 6*(3-4), 1-10. https://cadcommunity.pbworks.com/f/what%20sample%20size.pdf

Huckle, J., & Wals, A. E. J. (2015). The UN decade of education for sustainable development: Business as usual in the end. *Environmental Education Research, 21*(3), 491–505. DOI: 10.1080/13504622.2015.1011084

Hudzik, J. K. (2018). *Comprehensive and strategic internationalization: lessons learned and prospects*. NAFSA Association of International Educators.

Huron Consulting Group. American Council on Education, & Georgia Institution of Technology. (2019). *The transformation-ready higher education institution: How leaders can prepare for and promote change.* https://www.huronconsultinggroup.com/insights/transformation-ready-higher-education-institution

Institute of International Education. (2021). *Open doors report: 2021 fast facts.* https://opendoorsdata.org/wp-content/uploads/2021/11/OD21_Fast-Facts-2021.pdf

Institute of International Education. (2023). *U.S. study abroad for academic credit trends, 1989/90-2021/22*. https://opendoorsdata.org/data/us-study-abroad/u-s-study-abroad-for-academic-credit-trends/

Institute of International Education. (n.d.). *A brief history* of *senior international officers on U.S. higher education campuses*. https://www.iie.org/publications/a-brief-history-of-senior-international-officers-on-u-s-higher-education-campuses/

Intergovernmental Panel on Climate Change. (2014). *Climate change 2014 synthesis report summary for policymakers*. United Nations Framework Convention on Climate Change. https://www.ipcc.ch/site/assets/uploads/2018/02/AR5_SYR_FINAL_SPM.pdf

Internationalization Lab. (2022). *Model for comprehensive internationalization*. American Council on Education. https://www.acenet.edu/Research-Insights/Pages/Internationalization/CIGE-Model-for-Comprehensive-Internationalization.aspx

Jackson, P. (2007). From Stockholm to Kyoto: A brief history of climate change. *UN Chronicle*, *44*(2), 1. https://www.un.org/en/chronicle/article/stockholm-kyoto-brief-history-climate-change

Kezar, A., & Eckel, P. D. (2002). The effect of institutional culture on change strategies in higher education: Universal principles or culturally responsive concepts? *The Journal of Higher Education*, *73*(4), 435–460. https://www.jstor.org/stable/1558422. DOI: 10.1080/00221546.2002.11777159

Khalili, N.R. (2011). *Practical sustainability: From grounded theory to emerging strategies*. Palgrave Macmillan; ProQuest Ebook Central.

Koehn, P. H., Deardorff, D. K., & Bolognese, K. D. (2011). Enhancing international research and development project activity on university campuses: Insights from U.S. senior international officers. *Journal of Studies in International Education*, *15*(4), 332–350. DOI: 10.1177/1028315310365540

Kohl, K., Hopkins, C., Barth, M., Michelsen, G., Dlouha, J., Razak, D. A., Bin Sanusi, Z. A., & Toman, I. (2022). A whole-institution approach towards sustainability: A crucial aspect of higher education's individual and collective engagement with the SDGs and beyond. *International Journal of Sustainability in Higher Education*, *23*(2), 218–236. DOI: 10.1108/IJSHE-10-2020-0398

Kuh, G. D. (2008). *High impact educational practices: What they are, who has access to them, and why they matter.* Association of American Colleges and Universities. https://dgmg81phhvh63.cloudfront.net/content/user-photos/Publications/PDFs/E-HIGHIMP.pdf

Kwai, C. K. (2020). *The SIO profile: A preliminary analysis of the survey on senior international education officers, their institutions and offices.* Association of International Education Administrators. https://www.aieaworld.org/uploads/1/4/5/2/145269424/2020_aiea_sio_profile_survey_executive_summary.pdf

Lee, J. L., & Lundemo, O. A. (2021, June 5). *Why sustainability is not enough in international education.* University World News. https://www.universityworldnews.com/post.php?story=20210531123436636

Leggett, K. (2012). Leaving light footprints. [NAFSA Association of International Educators.]. *International Educator*, (May-June), 41–52.

Long, J., Vogelaar, A., & Hale, B. (2014). Toward sustainable educational travel. *Journal of Sustainable Tourism*, *22*(3), 421–439. DOI: 10.1080/09669582.2013.819877

Mallow, S., Toman, I., & van't Land, H. (2020). *Higher education and the 2030 Agenda: Moving into the 'Decade of Action and Delivery for the SDGs'.* International Association of Universities. https://static1.squarespace.com/static/64ca66e5f086975ce8783d53/t/653bcfbcaf643c254532ba19/1698418623774/iau_hesd_survey_report_final_jan2020.pdf

Maragakis, A., & Van den Dobbelsteen, A. (2013). Higher education: Features, trends, and needs in relation to sustainability. *Journal of Sustainability Education*, *4*, 1–20. https://repository.tudelft.nl/islandora/object/uuid%3A4b307380-2960-48e1-a6cc-bb186d518095

McBride, K. (2022, July 25). *Scope 3 emissions: The big exclusion.* Bound International. https://bound-intl.com/scope-3-emissions-the-big-exclusion/

McBride, K., & Greenberg, D. (2021). *Survey of sustainability in education abroad.* Bound International. https://bound-intl.com/wp-content/uploads/2021/12/Survey-of-Sustainability-in-Education-Abroad-2.pdf

McBride, K., & Ponce-Taylor, D. (2022). *Environmental sustainability and internationalization: challenges and opportunities* (Trends & Insights, May). NAFSA Association of International Educators. https://www.nafsa.org/professional-resources/research-and-trends/trends-insights/environmental-sustainability-and-internationalization

McClelland, K. (2015). *Cultivating social responsibility in study abroad: A phenomenological exploration of study abroad program design* [Doctoral dissertation, Plymouth State University]. ProQuest Dissertations and Theses Global.

Mebratu, D. (1998). Sustainability and sustainable development: Historical and conceptual review. *Environmental Impact Assessment Review*, *18*(6), 493–520. DOI: 10.1016/S0195-9255(98)00019-5

Merkx, G. W. (2018). Phases of internationalization and the senior international officer role. In Deardorff, D. K., & Charles, H. (Eds.), *Leading internationalization: A handbook for international education leaders* (pp. 9–15). Stylus Publishing.

Morais, D. B., & Ogden, A. C. (2011). Initial development and validation of the Global Citizenship Scale. *Journal of Studies in International Education*, *5*(15), 445–466. DOI: 10.1177/1028315310375308

NAFSA Association of International Educators. (2019, March). *NAFSA's statement of ethical principles.* https://www.nafsa.org/about-us/about-nafsa/nafsas-statement-ethical-principles

NAFSA Strategic Task Force on Education Abroad. (2003). *Securing America's future: global education for a global age.* NAFSA Association of International Educators. https://www.nafsa.org/policy-and-advocacy/policy-resources/securing-americas-future

National Center for Education Statistics. (2023). *Fast facts: enrollment.* U.S. Department of Education. https://nces.ed.gov/fastfacts/display.asp?id=98

Northouse, P. G. (2010). *Introduction. Leadership theory and practice* (5th ed.). SAGE Publications.

Ramaswamy, M., Marciniuk, D. D., Csonka, V., Colo, L., & Saso, L. (2021). Reimagining internationalization in higher education through the United Nations Sustainable Development Goals for the betterment of society. *Journal of Studies in International Education*, *4*(25), 388–406. DOI: 10.1177/10283153211031046

Robertson, J. L., & Barling, J. (2012). Greening organizations through leaders' influence on employees' pro-environmental behaviors. *Journal of Organizational Behavior*, *34*(2), 176–194. DOI: 10.1002/job.1820

Roosen, N. (2022). *Regenerative leadership: Understanding the possibilities of a new leadership paradigm for the corporate sector* [Master's thesis, Swedish University of Agricultural Sciences]. SLU Epsilon Archive for Student Projects. https://stud.epsilon.slu.se/18030/

Rumbley, L. E. (2020, Winter). Internationalization of higher education and the future of the planet. *International Higher Education*, (100), 32–34. https://ihe.bc.edu/pub/usg1w08q

Ruther, N. L. (2002). *Barely there, powerfully present: Years of U.S. policy on international higher education.* Taylor & Francis Group. https://ebookcentral.proquest.com/lib/northgeorgia/detail.action?docID=1602099

Schroeder, K., Wood, C., Galiardi, S., & Koehn, J. (2009). First, do no harm: Ideas for mitigating negative community impacts of short-term study abroad. *The Journal of Geography*, *108*(3), 141–147. DOI: 10.1080/00221340903120866

Shields, C. M. (2010). Transformative leadership: Working for equity in diverse contexts. *Educational Administration Quarterly*, *46*(4), 558–589. https://DOI.org/10.1177/0013161X10375609. DOI: 10.1177/0013161X10375609

Shields, R. (2019). The sustainability of international higher education: Student mobility and global climate change. *Journal of Cleaner Production*, *217*, 594–602. DOI: 10.1016/j.jclepro.2019.01.291

Singh, A. S., & Segatto, A. P. (2020). When relational capabilities walk in education for sustainability scenario. *Journal of Cleaner Production*, *263*, 1–10. DOI: 10.1016/j.jclepro.2020.121478

Soler, M. C., Kim, J. H., & Cecil, B. G. (2022). *Mapping internationalization on US campuses: 2022 edition.* American Council on Education. https://www.acenet.edu/Documents/Mapping-Internationalization-2022.pdf

Stober, H., Gaebel, M., & Morrisroe, A. (2021). *Greening in European higher education institutions.* European University Association. https://www.eua.eu/publications/reports/greening-in-european-higher-education-institutions2.html

Stromquist, N. P. (2007). Internationalization as a response to globalization: Radical shifts in university environments. *Higher Education*, *53*(1), 81–105. https://link.springer.com/article/10.1007/s10734-005-1975-5. DOI: 10.1007/s10734-005-1975-5

Thomas, J. (2021). Inner Development Goals: Background, method, and the IDG framework. https://drive.google.com/file/d/13fcf9xmYrX9wrsh3PC3aeRDs0rWsWCpA/

Tremblay, C. (Host). (2021, November 03). *Sustainability is not enough, enter regenerativity* (No. 13) [Audio podcast episode]. In CANIE Climate Dialogues. https://canie.org/learn/podcast/13-sustainability-is-not-enough-enter-regenerativity

U.S. Environmental Protection Agency. (2021, November 22). *Sustainability Primer.* https://www.epa.gov/sites/default/files/2015-05/documents/sustainability_primer_v9.pdf

West, C. (2014). Puzzle masters. *International Educator,* (January-February), 26-31. https://www.nafsa.org/sites/default/files/ektron/files/underscore/ie_janfeb14_puzzlemasters.pdf

Woodfield, S. (2013). Key trends and emerging issues in international student mobility (ISM). In Maringe, F., & Foskett, N. (Eds.), *Globalization and internationalization in higher education: Theoretical, strategic, and management perspectives.* Bloomsbury Publishing.

KEY TERMS AND DEFINITIONS

Comprehensive Internationalization: This is a strategy to expand international education and globalize teaching, research, and service across all disciplines in higher education (Hudzik, 2018).

Education Abroad: This is education that occurs outside a student's home country that may result in their progress toward an academic degree at the student's home institution (Forum on Education Abroad, 2020).

Education for Sustainable Development: Sustainable development is at the nexus of natural, economic, and social challenges and is defined as a state in which the needs of the present can be met without compromising the ability of future generations to meet their own needs (Mebratu, 1998; Khalili, 2011). Education for sustainable development (ESD) is a pilar of the United Nations' Sustainable Development Goals (SDGs) in which educators are urged to integrate sustainability values into all levels of education (Huckle & Wals, 2015) and to support sustainable development research and cooperation (Kohl et al., 2022).

Environmental Stewardship: This entails actors and actions motivated by biophilia to protect, care for, and responsibly use natural systems (Bennett et al., 2018).

Environmental Sustainability: This is the idea that natural resources are finite and must be conserved to meet present and future needs (Mebratu, 1998) and that the integrity of the environment such as air, water, and ecosystems must be protected (U.S. Environmental Protection Agency, 2021). Therefore, environmental sustainability is also an ethical principle and practice to reduce adverse environmental impacts and improve natural systems for sustaining life (Association for the Advancement of Sustainability in Higher Education, 2023; National Association of International Educators, 2019).

Leadership: This embodies an individual's traits, behaviors, power and position, knowledge, and skills for influencing subordinates to achieve organizational goals or affect change (Northouse, 2010).

Regenerative Leadership: This leadership theory comprises personal and collective mindset, behaviors, and strategies for engaging with sustainability goals and moving toward regenerative practices that restore and secure natural and human systems (Hardman, 2012).

Senior International Officer (SIO): This is a higher education administrative role devoted to leading and facilitating institution-wide internationalization strategy, overseeing multiple functions of international education such as education abroad, international student recruitment, and global partnerships (Association of International Education Administrators, 2022).

Chapter 11
Repositioning Leaders in Malaysia's Construction Industry Through Bourdieusian Field Analysis

Mohd Ruzaini Che Zahari
http://orcid.org/0000-0002-7750-0957
Universiti Teknologi Malaysia, Malaysia

Mohd Zahierruddin Ismail
Universiti Teknologi Malaysia, Malaysia

Ramly Hasan
Universiti Malaysia Kelantan, Malaysia

Mohammad Syahir Aiman Abd Khair
http://orcid.org/0009-0003-6965-3229
Universiti Malaya, Malaysia

Shazwan Mohamed Shaari
Universiti Malaysia Sarawak, Malaysia

Mohamad Haider Abu Yazid
http://orcid.org/0000-0003-4301-9994
Universiti Teknologi Malaysia, Malaysia

ABSTRACT

The Malaysian construction industry has contributed significantly to national GDP growth since 2022, while industry liberalisation and the Construction Industry Transformation Programme (CITP) have increased the presence of foreign firms. This development underscores the need for local leaders to strengthen their competitive advantage in addressing economic, social, and environmental challenges. This paper repositions leadership within this context using Pierre Bourdieu's field theory. It examines the industry as a social arena structured by power relations, habitus,

DOI: 10.4018/979-8-3373-7877-0.ch011

and capital accumulation through the exploration of how local leaders navigate and consolidate their influence amidst intensifying globalisation and foreign competition. Findings indicate that local leaders can enhance their competitiveness by strategically accumulating and deploying multiple forms of capital. Ultimately, this study advances the understanding of socio-structural mechanisms that shape leadership and offers field-specific strategies to reinforce local competitiveness in Malaysia

INTRODUCTION

The construction industry is vital to Malaysia's economic development. According to Dehdasht et al. (2022), the construction industry is one of the main sectors for the Malaysian economy. Its significance is highlighted by its growth and contribution to the national GDP; for instance, it grew at 7.1% per annum in 2016-2017 and recently reached a 17% growth rate in the second quarter of 2024. The Construction Industry Development Board (CIDB) is the governing body of the Malaysian construction industry, tasked with providing leadership, strategic coordination, and regulation to industry players.

The emergence of rapid technologies, such as the Internet of Things (IoT), computer automation, and Artificial Intelligence (AI) in recent years, underlines the need for local practitioners—specifically leaders—to equip themselves with updated skills, higher thinking, and modern knowledge. These technological shifts fundamentally alter the nature of competition and affect how companies operate. Through trade agreements like the Malaysia-Australia Free Trade Agreement (MAFTA) and the ASEAN Framework Agreement on Services (AFAS), the Malaysian government aims to increase local players' involvement in the global market. This liberalisation has simultaneously increased the presence of foreign players in the local Malaysian market, many of whom are equipped with superior technology and higher competitive advantages.

However, the performance of Malaysian firms in the global market has not been consistently impressive despite these initiatives. This underscores the urgent need for the local Malaysian construction industry to enhance its competitiveness across economic, social, and environmental dimensions. Economics and society determine a firm's competitive edge. The emergence of climate change has created opportunities for firms to compete by providing assurance of environmental stewardship. In the environmental context, leaders have a role in inspiring change, novelty, and innovation to encourage team members to preserve the environment (Yukl et al., 2022). Green construction is one of the initiatives to achieve a low-carbon future, aiming to reduce carbon emissions in construction projects (Azizi and Abidin, 2025). Past studies (Nawaz Khan, 2023; Pham et al., 2023; Ren et al., 2024) have shown that leadership is associated with green construction and environmental stewardship. It is also one of the competitive edges for construction companies.

Modern organisational theory suggests that sources of competitive edge have shifted from purely financial resources to intangible resources such as individual attitudes, competencies, skills, trust, and communication, which constitute social capital. This shift emphasises the role of social capital as a critical intangible resource that can confer a competitive advantage on an organisation. Social capital is a central concept highlighted by Pierre Bourdieu. In his theory, The Forms of Capital (1986), he posited that social capital, along with cultural and economic capital, plays a decisive role in shaping and mobilising an individual's position within a competitive field. This idea is deeply rooted in his broader conceptualisation of Field Theory. Grenfell (2014) emphasised that Bourdieusian narratives of field, capital, and habitus are interconnected, together structuring the social and professional context of his ideas.

Scholars globally have adopted or adapted Bourdieu's theories into various disciplines, yet there have been relatively few researchers who have applied Bourdieu's frame of reasoning specifically to the construction industry. Sang and Powell (2012) delved into gender inequality within the industry using Bourdieu's concept of symbolic violence, discussing how women must learn the masculine habitus or dispositions of the construction field to assimilate. Askland et al. (2013) investigated project organisations through the lens of habitus, capital, and field to understand sector processes and interactions. In the Malaysian context, Ariffin et al. (2019a; 2019b) and Zahari et al. (2018; 2019a; 2019b) explored the influence of leaders' capitals—such as human, social, and emotional capitals—on subordinates' creative behaviours and motivation.

These scholars laid a foundation for exploring Bourdieusian concepts in the industry. However, there remains a lack of comprehensive analysis, grounded in Bourdieu's field theory, that specifically guides leaders in the construction industry in strategically repositioning themselves. Such analysis is necessary for leaders to navigate the increasingly competitive local and global markets shaped by liberalisation and rapid technological adoption. By understanding the industry as a social field structured by power relations and the interplay of various forms of capital, local leaders can better adapt to shifting field dynamics and enhance their competitive edge against foreign practitioners.

This study adopts a conceptual analytical approach grounded in Bourdieusian relational sociology. Instead of advancing new empirical findings, it offers a theoretically driven reinterpretation of leadership by synthesising existing literature through Bourdieu's field theory. The aim is to enrich the conceptual understanding of how leadership can be repositioned, shaped by the value of capitals, habitus-influenced and the logics of the field (Bourdieu's theoretical framework). By mobilising Bourdieusian constructs as interpretive tools, this study could deepen understanding of leadership structures within professional environments in the Malaysian construction industry.

This chapter contributes to leadership scholarship by advancing Bourdieusian field theory through an examination of leadership structures within professional fields in the Malaysian construction industry. First, it deepens theoretical conversations by mobilising Bourdieu's relational sociology of the field into leadership paradigms, demonstrating the influence of capital, habitus, and doxa on leadership decisions. Second, it provides insights into leadership discourse by analysing the situation in the Malaysian construction industry, especially in withstanding global competitive pressures. Third, it builds a conceptual bridge between sociological theory and applied professional domains, illustrating how Bourdieusian concepts such as capital conversion, habitus, and doxa can benefit the repositioning of leaders in the construction industry.

BOURDIEU KEY CONCEPT

Pierre Bourdieu (1930–2002) was a French sociologist recognised as one of the 20th century's most influential thinkers. His work introduced transformative conceptual tools, including habitus, field, capital, and symbolic violence, which provide a comprehensive framework for understanding social structures through the lens of individual practices. The profound impact of his scholarship led to his appointment as the Chair of Sociology at the College de France, a position he held until his passing in 2002. It was during this period that he refined his conceptual triad, which is habitus, field, and capital, to analyse the operation of power across various social domains, including education (Homo Academicus, 1988), art and taste (Distinction, 1984), and the state (The State Nobility, 1996).

Habitus

Bourdieu's foundational theory of habitus first appeared in his work *Outline of a Theory of Practice* (1977). He defined habitus as a system of "durable, transposable dispositions". These dispositions are deeply ingrained attitudes, beliefs, and perceptions shaped by an individual's past experiences and social conditions. Habitus operates as an internal generative schema that guides how an individual perceives, thinks, and acts within the social world, often at an unconscious or pre-reflexive level.

Field

Building upon the concept of habitus, Bourdieu proposed the theory of the field in *The Logic of Practice* (1990). He conceptualised society as being composed of multiple, distinctive yet interconnected "fields"—such as education, art, politics, or the construction industry. A field is a structured social space governed by its own specific regulations, "rules of the game," and internal logic. Within these spaces, social actors engage in a constant struggle to maintain or improve their positions.

Capital

The resources that actors compete for and utilise within a field are defined as capital. In *The Forms of Capital* (1986), Bourdieu identified three primary forms:

- Economic Capital: Tangible assets, including financial reserves, equipment, and property.
- Social Capital: Resources derived from a durable network of institutionalised relationships and mutual recognition.
- Cultural Capital: Knowledge, professional skills, and educational credentials that signal competence and status.
- Symbolic Capital: This represents prestige, social honour, or reputation that validates and legitimises a practitioner's other assets. Unlike other forms, symbolic capital acts as a form of "credit" that can often be recognised and utilised across different professional fields.

Collectively, the total amount of these resources (volume) and how these capitals are convertible (conversion) constitute the "capital mobilisation". This concept is the primary determinant of a leader's relative status, organisational power, and overall competitive advantage within the field.

Symbolic Interaction

The dynamic interaction between these three elements is fundamental to Bourdieu's framework. Habitus provides the internalised understanding required to navigate the field's rules without deliberate reasoning, while capitals provide the necessary assets to execute strategies effectively. How an individual utilises their capital and leans into their ingrained habitus ultimately determines their trajectory and influence within the field's social hierarchy.

UNDERSTANDING BOURDIEU'S FIELD THEORY

Bourdieu's field theory offers a relational framework for understanding how individuals act, compete, and position themselves within structured social environments. Rather than focusing on individual agency alone, it situates behaviour within the interconnected triad of field, habitus, and capital. Capital is viewed as accumulated labour that enables access to social power in material or embodied forms, functioning as the "energy" that sustains and transforms social structures over time (Grenfell, 2014). Consequently, everyday actions are embedded within objective networks of relations known as fields (Nolan, 2012).

A field is a structured space governed by its own rules and power relations, where actors compete for resources and recognition. Habitus complements this by representing durable, internalised dispositions shaped by past experiences, enabling individuals to navigate the doxa of the field—the taken-for-granted norms guiding behaviour—without constant reflection (Bourdieu, 1986; Grenfell, 2014). Capital, in economic, social, cultural, and symbolic forms, determines positions and influence. These dynamics are often illustrated through analogies such as the science-fiction force-field, football field, and force-field models (Grenfell, 2014).

Science-Fiction Force Field

The science-fiction force-field analogy portrays the field as a structured and hierarchical space where actors occupy unequal positions based on the volume and composition of their capital. Power is unevenly distributed, allowing dominant actors to influence norms, outcomes, and trajectories within the field.

Central to this analogy is doxa, the taken-for-granted understanding of how the field operates. Myles (2004) describes doxa as a socially instilled "feel for the game," while Jones, Bradbury, and Boutillier (2011) view it as shared commonsense knowledge. Grenfell (2014) further explains doxa as pre-reflexive, embodied knowledge developed through experience, enabling actors to recognise opportunities and constraints without conscious calculation.

Bourdieu (1992) argues that individuals participate in multiple fields, each governed by distinct logics and expectations. Consequently, influence varies according to the alignment between habitus and field conditions. Within professional settings, accumulated experience and capital become strategic resources that enable actors to navigate complexities, reposition themselves within power structures, and promote sustainable leadership balancing economic, social, and environmental goals (Liao, 2022).

Football Field

The football field analogy illustrates the structured and competitive nature of social life, where actors occupy positions and compete strategically within the boundaries and rules of a field. Social action is both enabled and constrained by the field's structure.

Each actor enters the field with a habitus, a system of dispositions shaped by personal history and socialisation. Bourdieu (1986) describes habitus as a "structured and structuring structure," while Lizardo (2004) views it as a set of generative schemes guiding perception and behaviour. Habitus develops through primary and secondary socialisation, including family, education, and professional experiences (Grenfell, 2014).

Competition within the field is driven by economic, cultural, social, and symbolic capital, which influence actors' ability to secure advantageous positions. Success is therefore shaped not only by merit but also by unequal access to capital. The analogy highlights the dynamic interplay between structural constraints and strategic agency in shaping social trajectories.

Forces-Field

The field analogy presents a structural perspective in which actors occupy positions along a continuum between cultural and economic capital. Similar to a magnetic field, an actor's position is shaped by the composition and volume of their capital rather than fixed status.

This perspective highlights the concept of distinction, where social value is constructed through comparisons between cultural and economic resources. Examples include elite versus mass-market products, prestigious versus non-prestigious educational institutions, and architect-designed versus prefabricated housing. Such distinctions demonstrate that legitimacy and prestige arise from capital configurations rather than inherent qualities.

According to Grenfell (2014), capital comprises factors such as educational attainment, institutional prestige, professional networks, and social origin, which collectively determine status and influence. Actors who effectively combine economic and cultural capital often gain strategic advantages. This analogy underscores fields as relational systems where positions, competition, hierarchy, and social mobility are shaped by the dynamic interplay of different forms of capital.

Symbolic Interaction and Field Dynamics

Across all three analogies, the interaction between field, habitus, and capital remains central to understanding Bourdieu's theoretical framework. Habitus provides the internalised logic that guides behaviour, capital supplies the resources required for strategic action, and the field structures the conditions under which struggles unfold. Together, these elements form a dynamic system in which actors continuously negotiate their positions within hierarchical social environments.

Bourdieu suggests that actors operate using two forms of knowledge. The first is the logic of practice, which refers to the intuitive, practical mastery of the field's rules. This form of knowledge reflects the tacit understanding gained through immersion and experience. The second is reflexive knowledge, which involves a conscious awareness of the field's structures and one's own positioning within it. Reflexivity enables actors to critically evaluate and potentially transform existing power relations within the field.

Importantly, fields do not operate in isolation. Instead, they are interconnected systems where shifts in one domain can influence dynamics in another. For example, developments in the economic field may reshape hierarchies within educational or professional fields. This interconnectedness highlights the relational nature of social structures and reinforces Bourdieu's view of society as a network of overlapping and interacting fields.

Summary of Bourdieu's Field Analogies

Collectively, the three analogies provide a comprehensive understanding of Bourdieu's field theory. The science-fiction force-field highlights the role of doxa and hierarchical power distributions within structured social spaces. The football field emphasises the competitive and rule-bound nature of social interaction, illustrating how actors deploy capital and habitus within bounded arenas. The field analogy introduces a structural lens, demonstrating how positions are shaped by the relational balance between cultural and economic capital.

Across these perspectives, Bourdieu conceptualises social fields as arenas of struggle where actors with varying strengths compete according to implicit norms and rules. The distribution and conversion of capital determine actors' positions, while habitus shapes their strategic responses within the field. By integrating structural constraints with individual agency, Bourdieu's framework provides a powerful analytical lens for examining power, hierarchy, and mobility across social contexts.

Ultimately, field theory explains how individuals accumulate and mobilise capital to maintain or enhance their positions within complex social environments. Through the interaction of doxa, habitus, and capital, actors continuously navigate structured arenas of competition and influence. This relational perspective offers a robust foundation for analysing leadership, competition, and social positioning within professional domains characterised by dynamic power relations and evolving structural conditions. This knowledge is useful for leaders to be sustainable through various economic, social, and environmental conditions.

REFRAMING LEADERSHIP THROUGH BOURDIEUSIAN CAPITALS

While Bourdieu's field theory provides a robust socio-structural lens for analysing power relations, its integration with leadership scholarship offers a deeper understanding of how authority and influence are enacted within professional domains. Traditional leadership theories, such as trait theory and implicit leadership theory, have historically focused on individual characteristics, personality attributes, or socially constructed perceptions of leadership effectiveness. These perspectives often assume that leadership influence is rooted in stable personal traits or cognitive prototypes rather than relational socio-structural dynamics. This arguably tends to overlook the social context domain, especially the influence of values (capitals, habitus, doxa) of the leaders to influence decisions and behaviours within structured fields.

Leadership influence is deeply embedded within social networks and institutional norms rather than exercised in isolation. Consequently, attributes like relational trust and emotional intelligence are better understood as forms of capital rather than static personality traits (Bourdieu, 1986). Effective leadership, therefore, relies less on inherent characteristics and more on successfully mobilising capital configurations within a specific field.

Building on Bourdieu's original conceptualisation of economic, social, and cultural capital, scholars have expanded the capital framework to introduce additional forms of capital. For example, gender capital (Huppatz and Goodwin, 2013; Ross-Smith and Huppatz, 2010; Bridges, 2009), emotional capital (Zahari et al., 2019a; Zahari et al., 2019b; Feeney & Lemay Jr,, 2012), political capital (French, 2011; Nee and Opper, 2010), human capital (Goldin, 2024; Pasban and Nojedeh, 2016). These capitals were studied in relation to their field of study. According to Zahari et al. (2019a) and Ariffin et al. (2019a), five interrelated forms of capital are particularly relevant in understanding leadership influence in the Malaysian construction industry. These capitals are social capital, cultural capital, human capital, emotional capital, and political capital. Social capital reflects the diversity and strength of relational networks that enable access to opportunities and resources. Cultural capital encompasses embodied dispositions, institutional credentials, and objectified assets that signal legitimacy and professional authority. Human capital represents accumulated knowledge, skills, and experiential expertise often validated through institutional recognition. Emotional capital captures affective dispositions such as empathy and interpersonal sensitivity, while political capital reflects symbolic authority derived from positional legitimacy (Bourdieu, 1986).

This expanded capital typology reframes leadership as a relational phenomenon embedded within social fields, rather than a static individual attribute. Beyond formal authority, leaders influence outcomes by strategically accumulating, converting, and deploying various capitals—such as utilising social capital for successful environmental stewardship (Waters, 2022). The ability to convert these capitals to meet contextual demands reinforces leadership influence. This aligns with Bourdieu's (1986) proposition that capital configurations shape positional advantages within structured social spaces through the successful navigation of economic, societal, and environmental factors.

Leadership also plays a critical role in shaping relational capital structures within organisations. Strömgren et al. (2016) highlight leadership as a key driver of social capital formation, reinforcing the argument that leadership influence is relationally constructed rather than individually determined. Integrating leadership into Bourdieusian capital analysis, therefore, introduces a richer understanding of how leaders shape not only organisational outcomes but also behavioural dynamics through capital-mediated interactions.

Contextual factors further shape how capitals are accumulated and legitimised across professional environments. Cultural values and institutional norms influence how individuals interpret authority and legitimacy within organisational settings. Hofstede (1984) demonstrated that national culture shapes work values and organisational practices, while Meyer (2014) highlighted the role of cultural environments in shaping social interpretation. These perspectives suggest that capital deployment effectiveness is not universal but depends on context.

By reframing leadership through Bourdieusian capitals, leadership emerges as a socio-relational construct shaped by capital configurations and field dynamics. This perspective strengthens the theoretical foundation of Bourdieusian field analysis and provides a more nuanced understanding of how leadership operates within complex professional fields such as the construction industry. Rather than treating leaders as isolated agents, the Bourdieusian lens reveals leadership as a dynamic interplay between individual dispositions, accumulated capitals, and structured social environments.

The analytical value of this Bourdieusian perspective becomes more pronounced when situated within industries marked by complex interdependencies and institutional stratification, such as the construction industry. The construction sector, with its layered professional hierarchies and capital-intensive interactions, provides an interesting ground for examining how capital is accumulated, contested, and legitimised.

Examining capital configurations also extends to ecological outcomes and environmental stewardship in the construction sector. Bourdieu's framework aligns with emerging climate change challenges (Thapa, 2024), demonstrating how firms can maintain sustainability and competitiveness within dynamic environments (Chang et al., 2017). Leveraging these multilayered capital structures drives sustainable leadership, which requires integrating environmental, social, and economic goals (Piwowar-Sulej and Iqbal, 2023). Crucially, leadership intervention is a key leverage point for facilitating stewardship (Bennett et al., 2018), as leaders can instil a sense of environmental responsibility and pride among employees (Aisha et al., 2026). Particularly in the Malaysian context, Bourdieusian concepts offer valuable insights into how institutional and cultural norms shape these sustainable industry practices.

BOURDIEU AND THE MALAYSIAN CONSTRUCTION INDUSTRY

According to Sang and Powell (2012), there is a paucity of publications and research that specifically apply Bourdieu's theory to the construction industry. In Malaysia, the theoretical exploration of Bourdieu's concepts in the field of construction has appeared primarily in the works of Ariffin et al. (2019a; 2019b) and Zahari et al. (2018; 2019a; 2019b). The lack of research on Bourdieu's ideas does not indicate irrelevance, as these scholars have provided empirical evidence of the theory's usefulness in a specific industry.

The construction industry is a complex meta-field comprising various sub-fields, such as architectural, engineering, contracting, and development. It encompasses the entire lifecycle of the built environment, including planning, design, execution, and maintenance. As a determinant of a country's technological and technical advancement, the construction industry regulates the growth of infrastructural development. However, the industry faces numerous systemic challenges, including:

- Shortages of skilled craftworkers.
- Health, safety, environmental, and sustainability challenges.
- Procurement and supply chain complexities.
- Lagging technological adoption and economic fluctuations.
- Regulatory changes and ineffective collaboration and communication between parties.
- Lack of measures on environmental management

In Malaysia, these challenges are compounded by the increasing presence of foreign players. This influx was facilitated by the Internationalisation Thrust of the Construction Industry Master Plan (CITP) 2016–2020 and the broader liberalisation of the industry through various Multilateral, Regional, and Bilateral Trade Agreements. Consequently, local practitioners face intense competition to secure projects from foreign firms that often possess superior speciality, technical competency, and larger market shares.

The impact of this globalised "football field" is evident: according to Veerasanai and Alias (2022), an increasing number of Malaysian local firms are going dormant due to their inability to secure projects, raising serious concerns about the future sustainability of local practitioners. Within this competitive field, leaders play a pivotal role in determining the success of projects and the survival of their firms. There is a growing need for leadership development that moves beyond technical expertise to embrace soft skills and strong character. Strong leadership qualities can influence employees through role modelling and effective management of intangible resources.

Beyond technical competencies, leadership effectiveness in Malaysian construction firms is increasingly shaped by the diversity and configuration of capital possessed by organisational actors. While economic capital remains a foundational resource, relational and symbolic capitals often determine access to opportunities and legitimacy within project networks. Social capital, manifested through professional networks and long-term industry relationships, plays a crucial role in facilitating collaborations, securing tenders, and navigating institutional complexities. Cultural capital, including professional credentials and tacit industry knowledge, further enhances legitimacy and reinforces professional authority within multi-disciplinary project environments (Bourdieu, 1986).

Human capital, driven by experiential knowledge and technical mastery, remains a critical differentiator. Alongside this, emotional capital—expressed through empathy and relational awareness—is vital in people-intensive sectors like construction for building trust and managing project uncertainties. Ultimately, these capitals do not operate in isolation; rather, their combined configuration dictates a leader's positioning within the broader professional field.

In the Malaysian context, capital diversity is further shaped by socio-cultural and institutional factors that influence how legitimacy is constructed and recognised. Cultural expectations, hierarchical norms, and relational orientations affect how authority is exercised and interpreted within organisations. Hofstede (1984) emphasised that national culture significantly shapes work values and organisational behaviours, while Meyer (2014) highlighted the role of cultural environments in shaping social interpretation. These perspectives suggest that the accumulation and deployment of capital are not universally uniform but are mediated by local cultural logics that influence professional interactions and leadership dynamics.

Empirical evidence from Ariffin et al. (2019a) and Zahari et al. (2019a) has demonstrated the significant impact of leaders' Bourdieu capitals—specifically, social, human, and emotional capitals—on their subordinates' creative behaviours and motivation. In landscape architecture firms, for instance, a leader's emotional and social capital are primary predictors of subordinate creativity.

Understanding leadership influence through a Bourdieusian lens shifts attention from purely individual attributes to field-structured dynamics that shape decisions. In construction-related design domains such as architecture and landscape architecture, creativity emerges from the interaction between accumulated capital and field expectations. Leaders who possess strong social and emotional capital are better positioned to cultivate psychologically supportive environments that encourage experimentation and knowledge exchange. At the same time, cultural capital in the form of professional expertise and aesthetic sensibility contributes to the symbolic valuation of creative outputs. These dynamics illustrate how creativity functions not merely as an individual outcome but as a relational product shaped by capital configurations within structured professional fields.

Leadership acts as a catalyst, transforming capital configurations into creative outcomes by mobilising relational and symbolic resources to shape idea generation. Crucially, leadership drives the development of social capital, fostering the collaborative environments necessary for innovation (Strömgren et al., 2016). In the complex, multi-stakeholder Malaysian construction ecosystem, this interplay between capital accumulation and creativity becomes essential for driving organisational adaptability and competitive positioning.

However, existing research has focused primarily on leaders' possession of these capitals, with minimal discussion of how Bourdieu's broader field theory shapes their strategic *position* within intra- and inter-organisational structures. This leaves a significant research gap, particularly in the Malaysian context with its unique racial and cultural profile. By foregrounding the interrelations between habitus, doxa, and capital conversion, the analysis shifts leadership from an individualistic construct to a field-embedded phenomenon shaped by structured power relations and symbolic hierarchies. This framing not only clarifies the positional realities faced by Malaysian leaders but also highlights how strategic capital mobilisation may serve as a mechanism for sustaining relevance and competitiveness within an increasingly globalised construction field.

Taken together, this repositioning foreground the importance of examining leadership within its broader field conditions rather than as an isolated organisational phenomenon. If leadership is fundamentally shaped by relational positioning, then understanding its dynamics requires closer attention to the structured environments in which professional actors operate. This entails moving beyond abstract theorisation to consider how Bourdieusian constructs unfold within the lived realities of industry-specific contexts.

THE FIELD OF THE MALAYSIAN CONSTRUCTION INDUSTRY

In the Malaysian construction industry, agents are identified through various individuals who work on a project across different scopes of built environment professions. Within this complex social space, agents include landscape architects, architects, engineers, contractors, and developers. Bourdieu states that a field is profoundly tied to hierarchical order and that dominant agents have the decisive power to determine the dynamic interaction within it. This hierarchy is not static but is rather a site of constant struggle where agents leverage their resources to maintain or improve their standing. Within these agents, the leaders are critical, as they have the power to influence decisions and set the field's direction.

To discuss and thus reposition the local leaders' position in the construction field, three analytical steps were suggested by Bourdieu:

- **Analyse the position of the field vis-à-vis the field of power**: This involves understanding how the construction industry as a whole relates to the broader meta-field of power, where economic, political, and social capital are exchanged and dominated.
- **Mapping the objective structure of relations**: This step requires identifying the positions occupied by social agents who compete for the legitimate forms of specific authority—such as project approvals, design ownership, or technical credibility—within the field.
- **Analyse the possession of habitus**: This focuses on the varying systems of dispositions social agents have acquired by internalising a determinate type of social and economic condition, which guides their perceptions and actions within the industry.

By expanding on these steps, local leaders can better navigate the "arena of struggle" that defines the industry. For example, repositioning involves recognising that the construction field is highly regulated and politically entangled, where capital accumulation—especially economic and social capital—is vital for shaping authority. By understanding the objective relations between regulatory bodies and private consultants, leaders can strategically convert their symbolic capital, such as professional recognition, into institutional influence. Ultimately, this comprehensive field analysis allows local practitioners to recognise the *doxa*—the unwritten rules of the game—and leverage their *habitus* to compete more effectively against the foreign players in the Malaysian market.

Local Leader's Position vis-à-vis the Field of Power

The field of power serves as a meta-field in Bourdieu's framework, acting as a space where holders of different forms of capital—economic, cultural, political, and social—compete to impose the legitimate principle of domination over other specific fields. It is not a field like others; rather, it is structuring and overarching, exerting a profound influence over all other social fields. Local leaders in any field who accumulate enough capital, especially convertible symbolic capital, can ascend into this field of power.

In the Malaysian context, the construction industry is characterised by high regulation, hierarchy, and political entanglement. Within this arena, capital accumulation—particularly economic and social capital—plays a critical role in shaping authority and influence. The following Table 1 maps the relationship between various construction leaders, the capitals they hold, and their resulting position relative to the field of power:

Table 1. Construction leaders, capitals, and position in the field

Local Leaders/ Entities	Capitals Held	Position vis-à-vis Field of Power
Top-tier Developers	High economic + high political/social capital	Close alignment; often integrated via GLCs
Politically-connected Contractors	High social capital (political parties link etc.), moderate cultural	Embedded in power networks; influential in public procurement
Regulatory Leaders	High symbolic + institutional capital	Agents of translation between field and state power
Renowned Architects	High cultural + symbolic capital	Peripheral to power field; symbolic influence only
Academics/Policy Advisors	Cultural capital, some symbolic	Advisors to power, rarely direct agents of power
Small Local Builders	Low capital across all forms	Distant from power field; depend on subsidies/quotas

Local leaders in the Malaysian construction industry navigate and challenge the field of power through the conversion of capital, often in patterns of Economic Capital to Political Capital. This pattern utilises the company's assets to support political events, thereby maintaining its position in the field of power. The second pattern is Symbolic to Institutional Capital, which leverages design awards or professional recognition as status or an asset to secure seats on institutional boards. The acquisition of institutional capital may allow leaders to gain influence over prestige and visibility. Next, the conversion of Social Capital to Economic Capital, which is described as the networking assets for local leaders to win tenders, gain land access, or licenses. This conversion is common and is highlighted by Bourdieu (1986) as " *The convertibility of the different types of capital is the basis of the strategies aimed at ensuring the reproduction of capital (and the position occupied in social space*" This underlines the proposition that, by Bourdieu's view of the field of power, local leaders shall align themselves with the field of power, not resisting its existence, and the dynamic interaction and conversion of capitals within the field reinforce the dominance of actors with higher political and economic capitals and marginalise agents with higher cultural capital (less influence).

While capital conversion offers pathways for repositioning, it is important to recognise that not all forms of capital are equally convertible across contexts. Bourdieu emphasised that the convertibility of capital is always conditioned by field-specific logics, meaning that the value and exchangeability of a given capital depend on how it is recognised within the prevailing symbolic order. In the Malaysian construction field, this uneven convertibility shapes the strategic possibilities available to different categories of actors.

Economic capital, for instance, often demonstrates high convertibility because it aligns with the institutional and political structures that govern large-scale development projects. Financial resources can be translated into influence through partnerships, project sponsorships, or strategic alliances that facilitate entry into influential networks. By contrast, cultural capital rooted in professional expertise or design excellence may face limitations in convertibility, particularly in segments of the industry where procurement decisions are driven more strongly by economic and relational considerations than by aesthetic or intellectual merit.

Symbolic capital occupies an especially complex position within this conversion landscape. While professional recognition, awards, or reputational prestige can enhance visibility and legitimacy, their conversion into institutional power is not automatic. Symbolic capital often requires mediation through relational channels before it can be effectively translated into tangible influence. This explains why actors who possess strong symbolic recognition but limited relational embeddedness may struggle to exert broader field-level impact despite high professional standing.

The uneven convertibility of capital also reveals structural asymmetries within the field. Actors positioned at the intersection of multiple capital forms, such as developers with strong economic resources and embedded political networks, enjoy greater flexibility in navigating the field of power. Conversely, practitioners whose capital portfolios are concentrated in less convertible forms may encounter structural constraints that limit their upward mobility. These asymmetries underscore how field dynamics shape not only capital accumulation but also the strategic horizons available to different actors.

Understanding capital conversion reveals that effective leadership repositioning depends on recognising which forms of capital translate into influence, rather than simply accumulating more of it. For leaders in the globalised construction sector, this awareness allows them to strategically align capital mobilisation with the field's institutional logics. Empirically, mobilising social capital is a proven strategy for leaders to drive competitiveness, build trust, and successfully reposition themselves in the business landscape (Mishchuk et al., 2023).

Objective Structures of Relations in the Field

The Malaysian construction field is conceptualised as a multidimensional space of structured positions in which various agents, including contractors, consultants, regulatory bodies, and academics, compete for legitimate forms of authority. This competition centres on securing project approvals, design ownership, technical credibility, contracts, and regulatory influence. The field is strictly governed by Malaysian regulatory bodies and ingrained industry norms, including Bumiputera quotas and procurement laws. Furthermore, market pressure, reinforced by liberalisation, has added a layer of globalised competition to these existing structures.

The typical agents occupying these various positions, as exhibited through Bourdieu's field theory, can be categorised by their specific capital configurations and the authority they seek, as visualised in Table 2.

Table 2. Position, agent, capital and authority

Position	Typical Agent	Capital Configuration	Authority Sought
Regulatory Dominant	Regulatory bodies	High symbolic, high institutional capital	Control of standards, licensing
Cultural Elite	Award-winning architects, senior academics	High cultural, symbolic, low/moderate economic	Thought leadership, design legitimacy
Economic Dominant	G7 Contractors, Developer Corporations (e.g., GLC/MNC coorporations)	High economic, moderate cultural/social	Project control, procurement power
Hybrid Actors	Design-and-Build Firms	Balanced cultural-economic-social	Project streamlining, niche dominance
Marginal Practitioners	Small G1–G3 contractors, fresh graduates	Low in all capitals	Entry, survival, local relevance

In this complex field, the dynamic of struggles to reposition revolves around how leaders deploy their unique capital portfolios. Leaders in dominant positions use their influence to leverage high symbolic and institutional capital to streamline the bureaucratic processes of licensing and standard control. Meanwhile, the cultural elites—such as established architects or landscape architects—often struggle to preserve design integrity and professional autonomy, relying on their high cultural and symbolic status to counterbalance their sometimes-lower economic capital.

The field also contains marginal actors, primarily newcomers and fresh graduates, who often struggle just to gain entry or initial recognition. For these agents, the possession and strategic use of social capital is vital. They must engage in capital conversion—turning social connections into economic or cultural opportunities—to shape a viable competitive advantage.

Local leaders equipped with context-specific social, cultural, and symbolic capital hold a significant home-field advantage. While foreign firms may introduce superior technology, they often face disadvantages in unfamiliar environments due to a lack of local knowledge (Fang et al., 2016). Conversely, local leaders intuitively understand the Malaysian market's doxa—the implicit, unwritten rules and taken-for-granted assumptions that shape behaviours, procurement practices, and decision-making hierarchies. This localised knowledge enables them to effectively navigate regulatory and cultural complexities, strategising for long-term survivability. Ultimately, this deeper grasp of local needs and social structures empowers these leaders to successfully transform government initiatives (Rami et al., 2021).

Unlike formal regulations, doxa functions through implicit, shared understandings embedded in everyday professional routines, such as culturally informed communication styles and implicit hierarchies. Locally socialised actors intuitively internalise these norms, developing a practical mastery for navigating complex institutional landscapes that cannot be replicated by technical competence alone.

Competitively, doxa acts as an invisible boundary between insiders and outsiders. While foreign firms entering the Malaysian construction sector may possess strong economic or technological capital, they often struggle with the implicit cultural codes governing negotiations. Consequently, these external actors risk misaligning with local expectations. In contrast, embedded local leaders can better interpret these symbolic nuances, strategically leveraging their relational trust and contextual fluency.

In addition, doxa can be embedded into the management of environmental stewardship by leaders in the construction industry. In the Malaysian construction industry, the Environmental, Social and Governance (ESG) report is important to measure the impact of developed projects on the environment. This measurement ensures the sustainability of environmental stewardship and has become a priority for every player in the Malaysian construction industry. Malaysian leaders can leverage environmental stewardship to their advantage by understanding and incorporating ESG requirements, informed by familiarity with local environmental conditions, natural resources, social impact, governance systems, and so on. This doxa enables leaders to compete more effectively for projects than foreign leaders.

Understanding doxa, therefore, reframes competitive advantage beyond measurable capital accumulation. It highlights how field-embedded knowledge functions as a subtle yet influential form of symbolic competence that shapes legitimacy and acceptance within professional networks. By recognising the role of doxa, Bourdieusian analysis provides a richer explanation of why locally grounded leaders may sustain relevance even within increasingly globalised construction environments, where formal indicators of capability alone do not fully determine success.

Local Leader's Habitus Possession

Habitus is conceptualised as a system of durable, transposable dispositions—specifically the ways of seeing, acting, and reacting—that individuals acquire through life experiences, primarily during early socialisation. It is structured by past conditions, such as social class, family background, and education, and serves as a generative schema that structures future perceptions, practices, and strategies. Crucially, habitus is embodied and often operates at a pre-reflexive level, meaning it is felt intuitively rather than consciously reasoned.

In the Malaysian construction industry, different parties, including developers, architects, and contractors, exhibit distinct habitus shaped by their professional training and social origins. For design professionals, such as architects and landscape architects, the key disposition is often aesthetic-oriented and culturally refined. These agents are typically protective of their design integrity, which significantly influences their professional practice and preferences for design work.

The internalisation of these dispositions creates a sharp contrast in how leadership and strategy are enacted within the field:

- **Elite Backgrounds**: Agents from elite educational or social backgrounds tend to possess the confidence required to navigate complex bureaucracies. They hold specific "symbolic codes" in terms of language and taste (cultural capital) and established networks (social capital) that afford them a naturalised advantage in assuming leadership roles.
- **Working-Class Backgrounds**: In contrast, agents from working-class or rural backgrounds may exhibit deep practical, embodied knowledge but often lack institutional fluency. They may tend to show more deference to regulators and operate reactively rather than strategically within the field's power structures.

This socio-structural contrast can induce a scenario Bourdieu described as symbolic violence. This occurs when the culturally recognised dispositions shaped by elite institutions—often exemplified by large foreign firms—are misrecognized as "natural talent" or superior "professionalism". This misrecognition creates a form of dominance over local practitioners, who may be perceived as less professional simply because their habits differ from the elite standard.

These dynamics highlight the subtle operation of symbolic violence within professional fields. Rather than overt exclusion, it functions through misrecognition, naturalising historically contingent standards as universal benchmarks of competence. In the construction and design domains, professional legitimacy privileges specific forms of cultural capital, such as institutional affiliations and accreditations. Consequently, actors aligning with dominant norms are readily recognised as authoritative, while those with alternative habitus configurations are often marginalised despite their practical expertise.

In the Malaysian context, symbolic violence often emerges through the implicit privileging of globally recognised professional trajectories, such as Western education and transnational affiliations. These markers carry disproportionate symbolic weight, unfairly overshadowing the equally valuable contextual competence of local practitioners and reinforcing hierarchies of credibility. Furthermore, institutionalised evaluative mechanisms—including award systems and accreditations—reproduce these dominant standards. As these ideals become normalised, alternative expressions of professional identity are undervalued, gradually reproducing symbolic asymmetries across generations.

Recognising these dynamics is critical for understanding leadership positioning within the Malaysian construction industry. Leaders who are attuned to the workings of symbolic violence are better equipped to navigate legitimacy structures that extend beyond formal performance metrics. By strategically mobilising diverse forms of capital and reframing local competencies as sources of symbolic strength, such actors can cultivate more contextually grounded pathways to influence. In this sense, Bourdieusian analysis not only exposes hidden structures of domination but also illuminates the reflexive awareness required for repositioning within complex professional fields.

Despite initial habitus disadvantages, local leaders can successfully reposition themselves and boost their competitiveness by consciously acquiring diverse capitals and mastering the field's doxa. This highlights the vital role of reflexivity: a Bourdieusian perspective encourages leaders to recognise how their behaviours are conditioned by historically accumulated habitus rather than being purely intentional. Integrating this theory with leadership scholarship demonstrates that effective influence stems from the strategic mobilisation of these elements within professional networks, rather than from static traits. In the Malaysian construction sector, this lens exposes how implicit norms and symbolic violence govern legitimacy. Ultimately, reflexive leadership empowers locally embedded actors to interrogate inherited assumptions, recalibrate their capital, and seize moments of structural flexibility to maintain contextually grounded authority against foreign competition.

STUDY LIMITATIONS

This study has a few limitations. First, the discussion of Bourdieu's theory of the field is based on secondary data, which may be outdated. This risks its applicability to generalise to a larger population. However, this study's scope is the Malaysian construction industry. The discussion's derivation may be less likely to have an impact given this specific scope. However, a future study is recommended to advance this discussion into empirical research, which can test its relevance statistically. Secondly, this study limits its discussion of the field's theory to a leadership perspective. This may introduce sample bias, raising concerns about the study's internal and external validity. This dataset may yield results that do not reflect broader industry perspectives, which are comprised of several other variables, such as subordinate-leader relationships, policy impact, and others.

CONCLUSION

Bourdieu conceptualised the social field as a dynamic site of struggle where agents must constantly negotiate their positions using their accumulated resources. Within the Malaysian construction industry, a leader's ability to maintain or advance their position is profoundly dependent on the interplay among their ingrained habitus, their understanding of the field's doxa, and the strategic mobilisation of various capitals. This paper has discussed the repositioning of local practitioners through an analytical Bourdieusian lens, expanding his theoretical frameworks into the specific socio-structural mechanisms of the Malaysian construction scene. The analysis confirms that the industry is not merely a technical domain but a structured field of power relations where local leaders must strategically navigate regulatory hurdles and globalised competition.

Key Conclusions

The following conclusions are drawn from the application of field theory to leadership repositioning:

- The Significance of Intangible Resources: In a liberalised market, traditional financial resources are no longer sufficient to maintain a competitive edge; leaders must instead focus on cultivating intangible resources such as the capitals proposed by Bourdieu: economic capital, social capital, and cultural capital to navigate their position towards surviving in the presence of foreign competition.
- Mastering the Field of Power: To compete effectively against foreign firms with superior technological infrastructure, local leaders must align themselves with the field of power by converting economic or symbolic assets into political and institutional influence. This provides them with the benefits of implicit influence, which could help them navigate regulations in the construction field.
- Overcoming Habitus Constraints: While a leader's professional trajectory is often shaped by their background and early socialisation, the strategic acquisition of capital and a reflexive understanding of the industry's "rules of the game" allow for successful repositioning and the mitigation of symbolic violence.

Recommendations for Local Practitioners and Industry Bodies

To enhance the competitiveness of the leaders in the Malaysian construction industry, the following strategies are recommended:

1. Strategic Capital Accumulation: Leaders should focus on increasing their social capital within intra- and inter-organisational frames to facilitate knowledge acquisition and problem-solving, as well as to secure future projects and navigate regulations.
2. Leadership Development through Emotional Capital: Governing bodies should implement training focused on empathy, trust, and collaborative efforts, as these components could promote trust and creative domain. In addition, leaders shall enrol in courses to strengthen their emotional capital and better reposition themselves in the construction field.
3. Institutional Repositioning: Local leaders should seek to convert their professional recognition into institutional capital by assuming roles in regulatory boards, thereby gaining the authority sought to control standards and licensing.
4. Sustainability Leadership for Environment: To ensure environmental sustainability, local leaders shall also leverage capital advantages and their social positioning to exert influence on improved environmental assessment and conditions in construction projects.

In conclusion, leaders in the Malaysian construction industry should understand the implicit impact of Bourdieu's capital, habitus, and doxa, and the interplay of knowledge of the field, which could help them reposition themselves intra- and inter-organisationally. This study provides an in-depth discussion of how to relate Bourdieu's lens to leadership perspectives in the Malaysian construction industry through secondary data. This secondary data derivation might be more influenced by opinions than by empirical evidence of the current situation. Therefore, future studies are encouraged to advance this theoretical idea into empirical research across all sectors of the Malaysian construction industry, including landscape architecture, architecture, engineering, contractors, developers, and local authorities. Such a study would further contribute to understanding this relation, grounded in evidence rather than conceptual discussion.

REFERENCES

Aisha, R., Channa, N. A., Casali, L., Bhutto, M. H., & Qureshi, N. A. (2026). Servant leadership for a sustainable future: transforming organisations through environmental stewardship. Journal of Knowledge, Risk, and Sustainable Management, 1-21.

Ariffin, M. H., Abdul Rahman, A. R., Arshad, R. A., & Zahari, R. (2019b). Bourdieu Capitals And Leadership: The Case Of Design Consultant Firms' Managers. In *The Malaysian Construction Industry*. EDP Sciences., DOI: 10.1051/matecconf/201926603005

Ariffin, M. H., Arshad, R. A., Rahim, A. R. A., & Zahari, R. (2019a). Leadership-Relevant Bourdieu Capitals Of Design Consultant Firms' Managers in The Malaysian Construction Industry. [MCRJ]. *Malaysian Construction Research Journal*, *8*(3), 60–76.

Askland, H. H., Gajendran, T., & Brewer, G. (2013). Project organisations as organisational fields: Expanding the level of analysis through Pierre Bourdieu's Theory of Practice. *Engineering Project Organisation Journal*, *3*(2), 116–126. DOI: 10.1080/21573727.2013.768986

Azizi, Z. M., & Abidin, N. Z. (2025). Green Construction Project Design Decision-making: The Role of Design Leaders. In Developing a Body of Knowledge for Green Construction Project Management (pp. 275-298).

Bennett, N. J., Whitty, T. S., Finkbeiner, E., Pittman, J., Bassett, H., Gelcich, S., & Allison, E. H. (2018). Environmental stewardship: A conceptual review and analytical framework. *Environmental Management*, *61*(4), 597–614. DOI: 10.1007/s00267-017-0993-2 PMID: 29387947

Bourdieu, P. (1977). Outline of a theory of practice (R. Nice, Trans.). Cambridge University Press. (Original work published in 1972)

Bourdieu, P. (1984). Distinction: A social critique of the judgement of taste (R. Nice, Trans.). Harvard University Press. (Original work published 1979)

Bourdieu, P. (1986). The forms of capital. In J. Richardson (Ed.), Handbook of theory and research for the sociology of education (pp. 241–258). Greenwood Press.

Bourdieu, P. (1988). *Homo academicus* (Collier, P., Trans.). Stanford University Press. (Original work published 1984)

Bourdieu, P. (1990). The logic of practice (R. Nice, Trans.). Stanford University Press. (Original work published 1980)

Bourdieu, P. (1992). The logic of practice (R. Nice, Trans.). Stanford University Press. (Original work published 1975)

Bourdieu, P., & Passeron, J.-C. (1970). *La reproduction: Éléments pour une théorie du système d'enseignement*. Paris: Éditions de Minuit. (Translated as: Bourdieu, P., & Passeron, J.-C. (1990). Reproduction in education, society and culture (2nd ed., R. Nice, Trans.). Sage.)

Bridges, T. S. (2009). Gender capital and male bodybuilders. *Body & Society*, *15*(1), 83–107. DOI: 10.1177/1357034X08100148

Chang, R. D., Zuo, J., Soebarto, V., Zhao, Z. Y., & Zillante, G. (2017). Dynamic interactions between sustainability and competitiveness in construction firms: A transition perspective. *Engineering, Construction, and Architectural Management*, *24*(5), 842–859. DOI: 10.1108/ECAM-01-2016-0025

CIDB (Construction Industry Development Board). (2024, December 12)Part 1: Malaysia's Construction Industry Roars Back to Life: A Boom Fueled by Investment and Infrastructure https://www.cidb.gov.my/eng/part-1-malaysias-construction-industry-roars-back-to-life-a-boom-fueled-by-investment-and-infrastructure/

Dehdasht, G., Ferwati, M. S., Abidin, N. Z., & Oyedeji, M. O. (2022). Trends of the construction industry in Malaysia and its emerging challenges. *Journal of Financial Management of Property and Construction*, *27*(2), 161–178. DOI: 10.1108/JFMPC-08-2020-0054

Fang, E., Kotlar, J., Memili, E., Chrisman, J. J., & De Massis, A. (2016). Tapping the power of local knowledge: A local-global interactive perspective. *Journal of World Business*, *51*(4), 641–653. DOI: 10.1016/j.jwb.2016.04.001

Feeney, B. C., & Lemay, E. P.Jr. (2012). Surviving relationship threats: The role of emotional capital. *Personality and Social Psychology Bulletin*, *38*(8), 1004–1017. DOI: 10.1177/0146167212442971 PMID: 22535924

French, R. D. (2011). Political capital. *Representation (McDougall Trust)*, *47*(2), 215–230. DOI: 10.1080/00344893.2011.581086

Goldin, C. (2024). Human capital. In *Handbook of cliometrics* (pp. 353–383). Springer International Publishing. DOI: 10.1007/978-3-031-35583-7_23

Grenfell, M. (2014). *Pierre Bourdieu: Key Concepts* (2nd ed.). Routledge. DOI: 10.4324/9781315729923

Huppatz, K., & Goodwin, S. (2013). Masculinised jobs, feminised jobs and men's 'gender capital' experiences: Understanding occupational segregation in Australia. *Journal of Sociology (Melbourne, Vic.), 49*(2-3), 291–308. DOI: 10.1177/1440783313481743

Jones, L., Bradbury, T., & Boutillier, S. (2011). Understanding doxa: The common sense of field-specific knowledge. *Journal of Organisational Studies, 18*(3), 45–59.

Liao, Y. (2022). Sustainable leadership: A literature review and prospects for future research. *Frontiers in Psychology, 13*, 1045570. DOI: 10.3389/fpsyg.2022.1045570 PMID: 36420377

Lizardo, O. (2004). The cognitive origins of Bourdieu's habitus. *Journal for the Theory of Social Behaviour, 34*(4), 375–401. DOI: 10.1111/j.1468-5914.2004.00255.x

Mishchuk, H., Bilan, Y., Androniceanu, A., & Krol, V. (2023). Social capital: Evaluating its roles in competitiveness and ensuring human development. *Journal of Competitiveness, 15*(2).

Myles, J. (2004). From Doxa to Experience: Issues in Bourdieu's Adoption of Husserlian Phenomenology. *Theory, Culture & Society, 21*(2), 91–107. DOI: 10.1177/0263276404042136

Nawaz Khan, A. (2023). Is green leadership associated with employees' green behaviour? Role of green human resource management. *Journal of Environmental Planning and Management, 66*(9), 1962–1982. DOI: 10.1080/09640568.2022.2049595

Nee, V., & Opper, S. (2010). Political capital in a market economy. *Social Forces, 88*(5), 2105–2132. DOI: 10.1353/sof.2010.0039

Nolan, K. (2012). Dispositions in the field: Viewing mathematics teacher education through the lens of Bourdieu's social field theory. *Educational Studies in Mathematics, 80*(1-2), 201–215. DOI: 10.1007/s10649-011-9355-9

Pasban, M., & Nojedeh, S. H. (2016). A Review of the Role of Human Capital in the Organisation.Procedia-social and behaviou*ral sciences,230*, 249-253.

Pham, H. T., Pham, T., Truong Quang, H., & Dang, C. N. (2023). Impact of transformational leadership on green learning and green innovation in construction supply chains. *Engineering, Construction, and Architectural Management, 30*(5), 1883–1901. DOI: 10.1108/ECAM-05-2021-0379

Piwowar-Sulej, K., & Iqbal, Q. (2023). Leadership styles and sustainable performance: A systematic literature review. *Journal of Cleaner Production, 382*, 134600. Advance online publication. DOI: 10.1016/j.jclepro.2022.134600

Rami, A. M., Aziz, F., Zaremohzzabieh, Z., & Ibrahim, A. (2021). Assessing the Challenges of Local Leaders in Rural Community Development: A Qualitative Study in Malaysia. *Pertanika Journal of Social Science & Humanities*, •••, 29.

Ren, X., Liu, X., Li, L., & Lin, F. (2024). How perceived environmental leadership influences employee green behaviour in the construction industry: The mediating role of environmental awareness and the moderating role of perceived pro-environmental climate. *Current Psychology (New Brunswick, N.J.)*, *43*(37), 29419–29432. DOI: 10.1007/s12144-024-06565-6

Ross-Smith, A., & Huppatz, K. (2010). Management, women and gender capital. *Gender, Work and Organization*, *17*(5), 547–566. DOI: 10.1111/j.1468-0432.2010.00523.x

Sang, K., & Powell, A. (2012). Gender inequality in the construction industry: Lessons from Pierre Bourdieu. *Management*, *237*, 247.

Strömgren, M., Eriksson, A., Bergman, D., & Dellve, L. (2016). Social capital among healthcare professionals: A prospective study of its importance for job satisfaction, work engagement and engagement in clinical improvements. *International Journal of Nursing Studies*, *53*, 116–125. DOI: 10.1016/j.ijnurstu.2015.07.012 PMID: 26315780

Thapa, R. (2024). An application of Pierre Bourdieu's key concepts to the domain of climate change education. *Journal of Teacher Education for Sustainability*, *26*(2), 105–120. DOI: 10.2478/jtes-2024-0018

Veerasanai, R., & Alias, A. (2022). A review of foreign construction firm presence in the Malaysian construction industry. *International Journal of Construction Management*, *22*(10), 1830–1843. DOI: 10.1080/15623599.2020.1742635

Waters, A. (2022). Activating Cosmic Capital: Evaluating the Role of Social Capital and Sense of Place in Environmental Stewardship and Adaptive Governance.

Yukl, G., Gordon, A., & Taber, T. (2022). A hierarchical taxonomy of leadership behaviour: Integrating a half century of behaviour research. *Journal of Leadership & Organizational Studies*, *9*(1), 15–32. DOI: 10.1177/107179190200900102

Zahari, R., Ariffin, M. H., & Othman, N. (2018, February). Influential aspects of the leader's Bourdieu capital on subordinates' creativity in Malaysian landscape architecture. In IOP *Conference Series: Earth and Environmental Science*(Vol. 117, No. 1, p. 012008). IOP Publishing.

Zahari, R., Ariffin, M. H., & Othman, N. (2019a). Leaders' Bourdieu Capitals and Subordinates' Creative Behaviours: Empirical Study In Malaysian Landscape Architecture Firms. [MRCJ]. *Malaysian Construction Research Journal*, *8*(3), 133–148.

Zahari, R., Ariffin, M. H., & Othman, N. (2019b). Measuring Bourdieu Capitals of Landscape Architects. In *Malaysian Landscape Architecture Firms*. EDP Sciences.

Chapter 12
Saudization and Leadership Development

Yasser Almaloy
http://orcid.org/0009-0009-6407-2517
Al Yamamah University, Saudi Arabia

ABSTRACT

Saudization have traditionally been examined from the perspectives of employment generation and labor market reform. This chapter extends the discussion by exploring Saudization as a mechanism for sustainable leadership development, human capital enhancement, and environmental stewardship in Saudi Arabia. Drawing on sustainable leadership theory and experiential learning perspectives, the chapter examines how workforce localization creates opportunities for Saudi professionals to acquire leadership competencies, managerial experience, and strategic decision-making capabilities. It further argues that Saudization can contribute to environmental leadership capacity by preparing leaders who support sustainability initiatives aligned with Vision 2030, the Saudi Green Initiative, and broader environmental transformation efforts. The chapter discusses key developmental mechanisms, leadership challenges, competency gaps, and policy implications while highlighting the importance of sustainability competencies and environmental stewardship in leadership development.

1. INTRODUCTION

Saudization and the development of national leadership capabilities have created new opportunities for Saudi professionals to expand their managerial, economic, and environmental awareness in support of the Kingdom's long-term development objectives (Nurunnabi, 2017). While much of the existing literature focuses on

DOI: 10.4018/979-8-3373-7877-0.ch012

employment outcomes, compliance costs, and productivity effects, far less attention has been given to how such policies influence the development of leadership skills among national employees and foster a stronger commitment to environmental stewardship (Al-Qutb & Ghaith, 2018; Bennett et al., 2018).

These developments have become increasingly important as Saudi Arabia advances major sustainability initiatives, including the Saudi Green Initiative and its commitment to achieving net-zero emissions by 2060 (Abdelraouf & Nagasawa, 2023; Amran et al., 2020). Within this context, developing leaders who can balance economic growth, social progress, and environmental stewardship has become a national priority (Fry & Egel, 2021; Sajjad et al., 2024).

In the Saudi context, Saudization refers to the systematic replacement of expatriate workers with Saudi nationals in private sector employment (Al-Qutb & Ghaith, 2018). Historically the Kingdom's efforts toward workforce localization were part of a broader national strategy aimed at stimulating economic development and reducing unemployment among Saudi citizens. Over time, Saudization has evolved from a labor market intervention into a central pillar of national economic planning, human capital development, and institutional reform.

Despite these long-standing efforts, Saudi Arabia has historically faced a significant structural challenge related to employment concentration among expatriates, particularly within the private sector. For decades, foreign workers occupied a substantial share of operational, technical, and managerial roles, limiting opportunities for nationals to gain experience, advance professionally, and assume leadership responsibilities (Al-Qutb & Ghaith, 2018). According to data reported by the General Authority for Statistics, the number of expatriate employees in the private sector reached approximately 8.58 million in 2016, highlighting the scale of reliance on foreign labor and the depth of the localization challenge at that time (General Authority for Statistics, 2023). This imbalance not only intensified competition for jobs but also contributed to persistent concerns regarding unemployment rates among Saudi nationals and the sustainability of national workforce development.

Recent labor market indicators suggest gradual progress in addressing these challenges, yet the issue remains multifaceted (Rahman & Qattan, 2021). Official statistics released by the General Authority for Statistics indicate that the unemployment rate among Saudi citizens reached 3.4 percent in the most recent reporting period, reflecting a slight increase of 0.2 percent compared to the second quarter of the same year (General Authority for Statistics, 2023). While such figures may appear modest in comparative terms, they underscore the continued sensitivity of the Saudi labor market and the importance of policies that extend beyond job creation to focus on long-term capability building (Alshuwaikhat & Mohammed, 2017).

Within this context, an important question emerges: Why Saudization? Is it merely a policy designed to reduce unemployment and replace expatriate workers with Saudi nationals, or does it represent a broader strategy aimed at preparing future leaders who will guide the Kingdom and manage its resources? This chapter argues that the latter perspective is more appropriate. The long-term objective of Saudization extends beyond workforce localization to the development of national leaders who can contribute to the economic, social, and environmental advancement of Saudi Arabia (Chaaben et al., 2024; Selim & Alshareef, 2025).

Preparing future leaders requires looking beyond the needs of the present. The future will not be defined solely by the traditional oil economy, but increasingly by sustainability, renewable energy, smart cities, sustainable tourism, and circular economy initiatives (Amran et al., 2020; Rodriguez-Anton et al., 2019). Saudi Arabia is already investing billions of dollars in transformative projects such as NEOM, the Red Sea Project, and Green Riyadh. These initiatives raise an important question: Who will lead these projects over the next ten or twenty years? Will they continue to rely primarily on expatriate expertise, or will they be led by highly trained Saudi professionals who possess a deeper understanding of the local environment, culture, and national priorities? From a long-term strategic perspective, developing capable national leaders represents a more sustainable approach to ensuring the success of these projects (Fry & Egel, 2021; McCann & Holt, 2010).

This chapter proposes that Saudization contributes to leadership development among Saudi professionals. However, leadership development alone does not capture the full impact of workforce localization. Saudization may also contribute to the development of sustainable leadership and environmental stewardship capabilities (Boeske, 2023; Bennett et al., 2018). As Saudi professionals assume leadership roles in major national projects, they become increasingly exposed to sustainability-related practices, environmental policies, and long-term development objectives (Leal Filho et al., 2020; Jayashree et al., 2022). Such experiences can strengthen environmental awareness, sustainability literacy, and resource-efficiency thinking, all of which are becoming essential competencies for future leaders.

In this regard, Saudization should not be viewed solely as a mechanism for developing traditional leadership capabilities. Rather, it can serve as a pathway for developing sustainable leaders who are capable of balancing economic performance with environmental and social responsibilities. Future leaders must not only ask how organizations can increase profitability, but also how economic growth can be achieved while protecting environmental resources and supporting broader societal interests. Since many of Saudi Arabia's major development projects place environmental protection and sustainability at the center of their strategic objectives, Saudization can play an important role in preparing environmentally responsible leaders by providing them with leadership opportunities and direct exposure to sustainability-oriented initiatives.

Strategic Reframing of Saudization for Sustainable Leadership

Saudization can also be understood as a strategic recalibration of authority structures within the national economy. For decades, leadership hierarchies in many private-sector institutions were shaped primarily by expatriate managerial dominance and external technical expertise. This configuration was not accidental; it reflected the historical trajectory of rapid industrialization and economic expansion that required immediate access to global knowledge and specialized competencies. While this model enabled accelerated development and institutional growth, it simultaneously produced a structural asymmetry in leadership ownership. Decision-making authority, strategic planning influence, and executive discretion were often concentrated in non-national leadership tiers (Den Hartog & De Hoogh, 2024).

The gradual reallocation of authority toward Saudi nationals therefore represents far more than employment redistribution. It signifies a shift in decision-making sovereignty and institutional voice. Leadership positions are not merely symbolic markers of seniority; they determine who sets priorities, allocates resources, and defines organizational values **(Connerley & Pedersen, 2005).** When authority begins to reflect national representation, institutional direction becomes increasingly aligned with domestic socio-economic aspirations. In this respect, Saudization contributes to the localization of strategic control.

Leadership, in this sense, should not be reduced to a functional managerial role. It is a form of strategic participation in shaping the trajectory of firms, sectors, and ultimately the national economy. When Saudi professionals assume roles that influence budgeting processes, negotiation frameworks, partnership strategies, and long-term vision development, they move from operational execution to institutional authorship. They become architects of organizational futures rather than executors of predefined plans.

This transition also carries symbolic significance. Leadership visibility affects collective perceptions of opportunity and representation. When emerging professionals observe nationals occupying executive and supervisory roles, it alters expectations regarding career progression and social mobility. Representation within leadership structures signals the accessibility of influence. Over time, such shifts can reshape professional aspirations among younger cohorts entering the workforce (Shriberg & MacDonald, 2013).

Moreover, leadership formation under localization policies introduces a distinctly intergenerational dimension. Emerging Saudi managers are not simply filling vacated roles; they are actively constructing the managerial norms and professional cultures that will define future institutions. Their decisions shape internal governance routines, team dynamics, and performance standards. The cumulative impact of these contributions extends beyond individual careers and gradually embeds national perspectives within corporate governance and strategic discourse.

As localized leadership becomes normalized, institutions may experience increased continuity between organizational objectives and national development goals. Alignment between corporate strategies and broader socio-economic visions strengthens institutional coherence. Saudization, when viewed through this lens, represents not only a labor market instrument but a structural rebalancing of institutional stewardship.

One of the factors that may strengthen the effectiveness of localized leadership in sustainability-related projects is the long-term connection that Saudi leaders have with their country and its future development. Unlike expatriate managers who may work in the Kingdom for a limited period, Saudi leaders are more likely to experience the long-term consequences of their decisions. The environmental outcomes of major projects will affect not only them but also their families and future generations (Bramston et al., 2011). This creates a stronger sense of ownership and responsibility toward environmental stewardship and sustainable development.

From this perspective, Saudization may contribute to sustainability objectives by placing decision-making authority in the hands of leaders who possess a deeper connection to the local environment and society. Saudi leaders have been raised within the cultural, social, and environmental context of the Kingdom and therefore may have a stronger personal commitment to protecting national resources and supporting long-term environmental goals (Ahsan, 2025).

In addition, the involvement of Saudi leaders in major sustainability-oriented projects can strengthen accountability and encourage long-term thinking. When leaders perceive themselves as stakeholders in the future of their communities, they may place greater emphasis on environmental protection, resource conservation, and sustainable development outcomes. In this sense, Saudization may contribute not only to leadership development but also to the creation of sustainable leaders who are committed to advancing environmental stewardship and supporting the Kingdom's long-term sustainability ambitions (Mahran et al., 2025).

National Capacity Building and Sustainable Leadership Development

Saudization should also be interpreted as an investment in national leadership sovereignty. Economies that rely heavily on external managerial capital may experience efficiency gains in the short term, yet overdependence can limit the internal development of strategic capacity. When critical decision-making authority resides predominantly outside the national talent pool, long-term institutional memory and localized strategic insight may remain underdeveloped (Groves, 2007). Leadership localization therefore strengthens endogenous capacity—the ability of the national workforce to generate, adapt, and sustain organizational direction independently.

Over time, such capacity building contributes to economic resilience. Leaders who are socially embedded within the national context possess a deeper awareness of regulatory dynamics, cultural expectations, and demographic shifts. This embeddedness enhances contextual intelligence, enabling more sustainable and culturally aligned strategic decisions (Javidan et al., 2006).

From a sustainability perspective, national leadership capacity plays an increasingly important role in supporting Saudi Arabia's long-term environmental ambitions. Achieving the objectives of Vision 2030, the Saudi Green Initiative, renewable energy programs, and other sustainability-oriented projects requires leaders who possess both contextual understanding and long-term commitment. Localized leadership can strengthen continuity in environmental decision-making because national leaders are more directly connected to the future social, economic, and environmental well-being of the country. As a result, leadership sovereignty extends beyond economic considerations and becomes an important component of environmental stewardship and sustainable development (Liao, 2022). In this sense, Saudization contributes not only to national capacity building but also to the development of sustainable leaders capable of supporting the Kingdom's long-term sustainability goals.

2. SAUDIZATION AND SUSTAINABLE LEADERSHIP DEVELOPMENT: A CONCEPTUAL PERSPECTIVE

Saudization should not be viewed solely as a policy instrument designed to address unemployment challenges. While its direct impact on job creation is evident, its broader contribution lies in expanding opportunities for Saudi nationals to engage in meaningful professional experiences that foster leadership development. By enabling greater participation of Saudi employees in private sector roles, the policy creates pathways for individuals to practice leadership in real organizational settings, learn through responsibility, and gradually build confidence in decision-making (Heslin & Keating, 2017). In this sense, workforce localization contributes not only to economic inclusion but also to the formation of leadership capacity through continuous exposure to real workplace demands.

Saudi Arabia is currently undergoing a major sustainability transformation. Many initiatives associated with Vision 2030, including NEOM, the Saudi Green Initiative, renewable energy projects, and sustainable tourism developments, require a new generation of leaders who understand environmental sustainability and long-term development challenges. Consequently, workforce localization may contribute not only to leadership development but also to the formation of sustainable leaders who are capable of supporting environmental stewardship and advancing national sustainability objectives.

In this regard, Saudization should not be viewed merely as a mechanism for creating employment opportunities. Rather, it can serve as a platform for developing leaders who possess the competencies required in a sustainability-oriented economy. In addition to traditional leadership skills, future Saudi leaders may need to acquire capabilities related to green innovation, resource efficiency, environmental awareness, and long-term strategic thinking (Hu et al., 2022). Such competencies will become increasingly important as Saudi professionals assume leadership roles in projects that place sustainability and environmental protection at the center of their strategic objectives.

2.1 Sustainable Leadership Theory and Workforce Localization

In recent years, interest in sustainable leadership theory and practice has increased significantly among both researchers and practitioners. Sustainable leadership differs from traditional leadership in several important ways. Traditional leadership primarily focuses on organizational performance, profitability, communication skills, and managing relationships with customers and stakeholders. In contrast, sustainable leadership expands this focus by incorporating broader responsibilities related to environmental stewardship, social responsibility, long-term thinking, resource preservation, and green innovation. These competencies are becoming increasingly important as organizations face growing sustainability challenges and are expected to contribute to the wellbeing of both present and future generations (McCann & Holt, 2010).

Within this context, Saudization can be viewed as an important mechanism for developing sustainable leadership capabilities among Saudi professionals. Workforce localization not only provides leadership opportunities and decision-making authority, but also exposes emerging leaders to sustainability-oriented projects and initiatives. Through participation in projects such as NEOM, Green Riyadh, and other sustainability-focused developments, Saudi professionals gain valuable experience and a deeper understanding of environmental policies, sustainable development practices, and long-term strategic planning. Such exposure helps develop competencies associated with sustainable leadership, including environmental awareness, resource efficiency, and sustainability literacy (Leal Filho et al., 2020).

Furthermore, sustainability-oriented projects provide important learning environments in which future Saudi leaders can acquire both managerial and sustainability-related capabilities. This process may be particularly beneficial because Saudi leaders possess a stronger connection to the local environment, society, and long-term development goals of the Kingdom. As they and their families will continue to live within the communities affected by these projects, they may have a greater sense of responsibility toward environmental protection and sustainable development outcomes. Therefore, Saudization should be viewed not only as an employment policy, but also as a sustainable leadership development mechanism capable of supporting environmental stewardship and advancing Saudi Arabia's long-term sustainability ambitions.

2.2 Leadership Development Through Experiential Exposure

The relationship between Saudization and leadership development can be better understood through experiential learning theory, particularly the model proposed by Kolb. This perspective emphasizes that meaningful learning occurs through experience, reflection, conceptual understanding, and active experimentation. In organizational settings, leadership capabilities are rarely formed through classroom instruction alone; instead, they emerge through direct involvement in real tasks, responsibility, and problem-solving situations. When Saudi employees are given opportunities to occupy supervisory or middle-management roles, they move through these stages of experiential learning in a natural and progressive manner (Heslin & Keating, 2017).

For instance, assuming a second- or third-line leadership position exposes individuals to concrete experiences such as managing teams, handling operational challenges, and making decisions under pressure. These experiences are followed by reflection, where individuals evaluate their actions and outcomes. Over time, they begin to develop conceptual understanding about leadership styles, communication approaches, and strategic thinking. Finally, they test these ideas in practice through new decisions and initiatives. This continuous cycle strengthens leadership competence and allows learning to evolve from theory into lived professional capability.

It is important to note that leadership is already emphasized within the Saudi educational system. Universities, professional institutes, and training centers offer leadership courses, and various government entities—including the Ministry of Human Resources and Social Development—support skill development through targeted programs and intensive training initiatives. These efforts play a valuable role in building awareness, knowledge, and initial competence. However, theoretical preparation alone cannot fully translate into effective leadership unless individuals are given the opportunity to apply what they have learned in real organizational contexts (Grocutt et al., 2022).

Without access to leadership roles, knowledge acquired through education and training may remain abstract and underutilized. Practical exposure is what enables individuals to develop judgment, accountability, and resilience. Learning to take responsibility, make decisions, deal with uncertainty, and grow from failure are all processes that require real involvement in leadership environments (Lovelace et al., 2007). In this regard, Saudization creates the conditions necessary for transforming theoretical learning into practical capability. By opening pathways to managerial roles, it supports the long-term development of human capital and contributes to building a generation of leaders capable of supporting both organizational performance and national economic progress.

2.2.1 Experiential Learning and Sustainable Leadership Development

Experiential learning theory and sustainable leadership theory complement one another in explaining how future leaders can acquire the competencies required in modern organizations. While sustainable leadership theory identifies the knowledge, values, and capabilities needed to balance economic, social, and environmental objectives, experiential learning theory explains how such competencies are developed in practice. Sustainable leadership requires more than theoretical understanding; it requires direct exposure to real organizational challenges, decision-making situations, and sustainability-related responsibilities (Sajjad et al., 2024; Heslin & Keating, 2017).

Within this context, Saudization creates opportunities for Saudi professionals to develop sustainable leadership capabilities through practical experience. As individuals assume leadership roles within organizations and participate in major national projects, they gain firsthand exposure to issues related to environmental sustainability, resource management, and long-term strategic planning. Through continuous learning, reflection, and application, Saudi professionals can strengthen competencies associated with sustainable leadership, including environmental awareness, sustainability literacy, resource efficiency, and responsible decision-making. In this sense, workforce localization not only facilitates leadership development but also provides an important pathway for cultivating sustainable leaders who can support Saudi Arabia's long-term sustainability ambitions.

2.3 Structural Shifts in Leadership Opportunities

In recent years, Saudi Arabia has witnessed noticeable structural shifts in the distribution of leadership opportunities, particularly following the announcement of Vision 2030 and the expansion of large-scale national projects across multiple sectors. Notably, many of these large-scale national projects are not only economic transformation initiatives but also platforms for advancing environmental sustainability. Projects associated with Vision 2030 increasingly emphasize renewable energy integration, resource efficiency, sustainable urban development, and environmental protection. Examples include NEOM and the Red Sea Project, where environmental stewardship is embedded within strategic planning and operational practices (Alshuwaikhat & Mohammed, 2017; Amran et al., 2020; Abdelraouf & Nagasawa, 2023; Almatar, 2025). These projects incorporate sustainability principles through renewable energy adoption, resource conservation, waste reduction, recycling initiatives, and long-term environmental protection measures.

Saudization plays an important role in preparing the next generation of leaders within these environments. By placing young Saudi professionals in managerial and leadership positions, organizations provide opportunities for them to participate in planning, implementation, and decision-making processes that directly influence sustainability outcomes. Through such experiences, they develop not only leadership capabilities but also a deeper appreciation of environmental responsibility. Over time, this exposure can contribute to the emergence of a generation of Saudi leaders who understand the importance of environmental stewardship and are capable of transferring these values and practices to future generations. As Saudi professionals participate in and gradually lead such initiatives, they are exposed not only to advanced managerial practices but also to emerging sustainability expectations. This creates a dual developmental pathway in which Saudization contributes to both leadership capability and environmental stewardship awareness.

The involvement of international companies and expertise in major projects has introduced new technologies, systems, and management approaches. Saudi professionals working within these environments are increasingly exposed to advanced operational models and leadership practices, allowing them to observe, learn, and gradually take on more responsibility. This exposure also extends to environmental sustainability and conservation technologies, particularly when Saudi professionals are given opportunities to occupy supervisory and managerial roles. Through direct involvement in major projects, they gain practical experience with innovative approaches to resource management, renewable energy, environmental protection, and sustainable development practices. Such exposure contributes not only to leadership development but also to the cultivation of environmental awareness and stewardship capabilities that are increasingly important in modern organizations.

Moreover, placing Saudi employees in second- and third-tier leadership roles has created a transitional space where individuals can develop their capabilities while being guided by more experienced professionals. This gradual exposure allows emerging leaders to gain confidence, understand organizational dynamics, and build the practical skills needed for higher-level responsibilities. Over time, such structural changes contribute to strengthening the leadership pipeline within organizations and preparing a new generation of Saudi executives who are better equipped to manage complex institutions and contribute to future national development.

2.3.1 Leadership Opportunities and Sustainable Development

The expansion of leadership opportunities through Saudization has implications that extend beyond organizational advancement and career progression. As Saudi professionals gain access to managerial and decision-making positions, they become increasingly involved in shaping strategies, policies, and initiatives that influence long-term development outcomes. This is particularly important in sectors associated with sustainability, renewable energy, environmental protection, and resource management, where leadership decisions can have lasting economic, social, and environmental consequences. Greater participation of Saudi leaders in such roles therefore contributes to building national capacity to support sustainable development objectives.

Furthermore, leadership opportunities created through workforce localization provide a foundation for developing future leaders who are capable of balancing organizational performance with broader national priorities. Exposure to sustainability-oriented projects enables emerging leaders to understand the complex relationship between economic growth, environmental stewardship, and social wellbeing. As more Saudi professionals assume positions of influence within major national initiatives, the potential for integrating sustainability considerations into organizational decision-making increases. In this regard, the structural expansion of leadership opportunities can be viewed as an important mechanism for supporting both sustainable leadership development and the Kingdom's long-term sustainability ambitions.

2.4 Leadership Identity Formation Under Localization

Beyond structural access to managerial positions, Saudization plays a critical role in shaping leadership identity among emerging professionals. Leadership identity refers to the internalized belief that one is both capable of and entitled to exercise influence within organizational systems. It develops gradually through repeated engagement with responsibility, accountability, and recognition.

Merely occupying a managerial title does not automatically produce leadership identity. Instead, identity formation evolves through experiential reinforcement. When Saudi employees are entrusted with supervisory responsibilities—such as overseeing teams, approving budgets, or resolving conflicts—they begin to reinterpret their professional self-concept. They shift from perceiving themselves as task performers to understanding themselves as decision-makers whose actions carry broader implications.

This psychological transition alters their relationship with authority. Rather than deferring to external expertise, they begin to evaluate information independently, weigh competing interests, and assume ownership of outcomes. Over time, repeated exposure to consequential decision-making strengthens confidence and reduces hesitation in exercising leadership discretion.

Leadership identity formation is particularly significant in contexts where historical leadership pathways were limited. When access to managerial influence was previously constrained, professionals may have internalized implicit ceilings regarding advancement. Localization policies disrupt these implicit boundaries by institutionalizing opportunity. As structural barriers diminish, psychological barriers may also weaken.

Furthermore, leadership identity contributes directly to sustainable leadership practice. Competence without internal legitimacy may result in tentative decision-making or excessive reliance on hierarchical approval. In contrast, individuals who internalize their leadership role demonstrate greater initiative, resilience, and strategic engagement. They are more likely to take calculated risks, advocate for innovation, and accept accountability for outcomes (Boeske, 2023; Liao, 2022; Sajjad et al., 2024).

Localization therefore affects not only organizational charts but also internal narratives. It reshapes how emerging professionals interpret their capacity to lead, influence, and shape institutional direction. Sustainable leadership development depends as much on identity consolidation as on skill acquisition.

2.4.1 Leadership Identity and Environmental Stewardship

Leadership identity may also play an important role in shaping environmental stewardship and sustainability-related behavior. Individuals who view themselves as leaders are more likely to take responsibility for the long-term consequences of their decisions and to consider the broader impact of organizational activities on society and the environment. As sustainability becomes an increasingly important organizational and national priority, leadership identity extends beyond traditional managerial responsibilities to include environmental accountability and stewardship **(Alshuwaikhat & Mohammed, 2017; Selim & Alshareef, 2025)**.

In the context of Saudization, the development of leadership identity may contribute to the emergence of leaders who are committed not only to organizational success but also to sustainable development objectives. As Saudi professionals gain confidence in their ability to influence organizational outcomes, they may become more willing to advocate for sustainability initiatives, support environmental protection measures, and promote responsible resource management practices. In this regard, leadership identity formation can serve as an important foundation for the development of sustainable leaders capable of advancing Saudi Arabia's long-term environmental and sustainability ambitions.

2.5 Leadership Legitimacy and Social Trust

Beyond skill acquisition, Saudization influences perceptions of leadership legitimacy within organizations. Employees often assess leaders not only on competence but also on representational alignment. When leadership reflects the broader workforce demographic, symbolic trust may increase. This representational congruence can reduce psychological distance between management and employees, strengthening communication flows and organizational cohesion.

Legitimacy also affects followership dynamics. When employees perceive leaders as legitimate stakeholders in the national development project, compliance transforms into engagement. The shift from hierarchical compliance to voluntary commitment is critical for sustainable leadership effectiveness (Connerley & Pedersen, 2005; Den Hartog & De Hoogh, 2024).

2.5.1 Social Trust and Sustainability Initiatives

Social trust is particularly important when organizations seek to implement sustainability initiatives and environmental policies. Many sustainability-related programs require behavioral change, long-term commitment, and cooperation from employees and stakeholders. When leaders are perceived as legitimate and trustworthy, they are more likely to gain support for sustainability objectives and encourage active participation in environmental initiatives. Trust therefore serves as an important mechanism through which sustainable leadership can influence organizational behavior and long-term outcomes (Fry & Egel, 2021; Nielsen, 2022).

In the context of Saudization, leadership legitimacy may strengthen the implementation of sustainability and environmental stewardship initiatives within Saudi organizations. Employees who trust their leaders are often more willing to support organizational change, adopt new sustainability practices, and contribute to broader environmental goals (Muff et al., 2020; Rahman & Qattan, 2021). Consequently, the development of leadership legitimacy and social trust can play an important role in supporting sustainable leadership and advancing Saudi Arabia's sustainability ambitions under Vision 2030.

3. MECHANISMS THROUGH WHICH SAUDIZATION SHAPES LEADERSHIP AND SUSTAINABLE LEADERSHIP DEVELOPMENT

3.1 Accelerated Responsibility Allocation

One of the most visible consequences of Saudization is the accelerated progression of Saudi nationals into supervisory and middle-management roles. As workforce localization reduces structural barriers that previously limited access to managerial pathways, Saudi employees increasingly find themselves positioned to assume responsibilities that extend beyond routine operational tasks. This shift often requires them to oversee teams, manage resources, and participate in decision-making processes at earlier stages of their careers than was historically common (Al-Qutb & Ghaith, 2018; Groves, 2007).

Such accelerated responsibility inevitably exposes individuals to complex operational and administrative pressures. Managing performance targets, coordinating teams, and responding to unexpected challenges demand the exercise of judgment under conditions of uncertainty. While some decisions may initially prove imperfect, it is precisely through this process of trial, reflection, and adjustment that managerial maturity develops (Fusarelli et al., 2018). Leadership competence is rarely formed through flawless execution; rather, it evolves through experience, corrective learning, and the gradual refinement of professional judgment (Grocutt et al., 2022).

Moreover, occupying managerial roles necessitates the development of complementary skills, including time management, delegation, prioritization, and the ability to balance professional obligations with personal commitments. Exposure to real accountability encourages emerging leaders to align daily decisions with broader organizational objectives, strengthening their strategic awareness. In this way, Saudization contributes not only to positional advancement but also to the cultivation of practical leadership capabilities grounded in lived organizational experience.

3.1.1 Responsibility and Environmental Accountability

The allocation of responsibility through Saudization may also contribute to the development of environmental accountability among emerging leaders. As Saudi professionals assume greater managerial responsibilities, they become increasingly aware that leadership decisions often have consequences that extend beyond immediate operational outcomes. In modern organizations, leaders are expected not only to achieve performance objectives but also to consider the environmental implications of their decisions and actions **(McCann & Holt, 2010; Bennett et al., 2018)**.

This broader sense of accountability is particularly important in an era where sustainability has become a strategic priority for many organizations and national development initiatives. Through direct involvement in planning, resource allocation, and operational management, emerging Saudi leaders gain experience in balancing organizational objectives with environmental considerations (Shriberg & MacDonald, 2013; Cebrián et al., 2019). As a result, responsibility allocation serves not only as a mechanism for leadership development but also as an opportunity to strengthen environmental stewardship and long-term sustainability awareness among future leaders.

3.2 Learning Through Structured Exposure

While accelerated promotion can create pressure, it also accelerates exposure to complexity. Leadership development is often nonlinear; it requires encountering ambiguity, managing incomplete information, and navigating competing stakeholder interests. Early exposure to these realities builds strategic endurance. Leaders learn to tolerate uncertainty without paralysis.

Structured exposure also enhances judgment calibration. Through repeated decision cycles, emerging leaders refine their ability to distinguish between operational noise and strategic signals. This discernment gradually differentiates managerial competence from executive readiness **(Lovelace et al., 2007; Den Hartog & De Hoogh, 2024).**

3.2.1 Sustainability Learning Through Structured Exposure

Through the structured exposure that Saudization programs can provide, Saudi leaders may also develop sustainability-related knowledge and competencies. While training programs and formal education can increase awareness and theoretical knowledge, direct exposure to sustainability-oriented projects can lead to a deeper understanding of environmental issues and the practical challenges associated with addressing them. Such experiences allow emerging leaders to move beyond classroom learning and gain firsthand insight into the realities of sustainable development.

Repeated exposure to sustainability challenges, environmental initiatives, and resource management decisions enables leaders to move beyond theoretical awareness and develop practical insight into the complexities of sustainable development **(Hughes et al., 2025; Groves, 2007)**. Through continuous involvement in sustainability-related activities, Saudi leaders can strengthen their environmental awareness, sustainability literacy, and understanding of the long-term consequences of organizational decisions. As a result, structured exposure serves not only as a mechanism for leadership development but also as an important pathway for developing sustainable leadership capabilities and environmental stewardship competencies.

3.3 Accountability and Decision-Making Under Pressure

The rapid integration of Saudi nationals into second- and third-tier managerial roles also introduces heightened levels of accountability. Leadership positions inherently involve responsibility for outcomes, whether related to financial performance, operational efficiency, or team morale. Newly appointed managers often encounter significant pressure as they navigate expectations from senior leadership while simultaneously guiding their teams through demanding projects and performance benchmarks (O'Donoghue & Van Der Werff, 2022).

Although such environments can be challenging, they also serve as powerful developmental platforms. Confronting setbacks, managing uncertainty, and responding constructively to failure are critical elements in the formation of resilient leaders (Schmidt, 2010). Over time, repeated exposure to these pressures strengthens decision-making capacity, risk evaluation skills, and the ability to connect operational choices to the organization's mission and long-term strategy.

It is realistic to acknowledge that not every individual will adapt immediately to these demands, and some may struggle during the early stages of their leadership journey. However, sustained engagement in real managerial contexts provides invaluable experiential learning. With appropriate institutional support and mentoring, these challenges often translate into enhanced communication skills, improved conflict management, and deeper strategic insight. The opportunity to operate within demanding leadership environments allows individuals to reveal and refine capabilities that might otherwise remain undeveloped in less challenging roles (Dawley et al., 2008; Wals, 2010).

3.4 Adaptive Capacity and Strategic Maturity

Accelerated leadership exposure under Saudization cultivates adaptive capacity—an increasingly essential competency in rapidly transforming economic environments. Leaders operating within emerging markets and reform-driven economies must navigate shifting regulations, evolving sectoral priorities, technological disruption, and fluctuating global dynamics. Early immersion in such environments accelerates the development of strategic maturity (Soni et al., 2023).

Adaptive capacity involves more than reactive problem-solving. It requires anticipatory thinking, systems awareness, and the ability to recalibrate strategy in response to evolving constraints. Saudi managers who assume leadership roles during periods of institutional transition develop familiarity with change as a constant rather than an exception. This familiarity reduces cognitive rigidity and strengthens flexibility in evaluating alternatives.

Operating under uncertainty encourages integrative reasoning. Leaders learn to balance short-term operational performance with long-term strategic positioning. They recognize that immediate efficiency gains must align with sustainable institutional growth. This balancing process enhances strategic depth and prevents narrow decision-making driven solely by immediate metrics (Schroeder et al., 2019).

Moreover, exposure to organizational complexity strengthens cross-functional awareness. Emerging leaders begin to understand how financial decisions affect workforce morale, how operational delays influence stakeholder confidence, and how regulatory adjustments shape competitive positioning. Such integrative insight marks the transition from functional management to systemic leadership.

Adaptive maturity also enhances resilience. Leaders who develop within evolving institutional contexts become more comfortable navigating ambiguity and recalibrating plans when assumptions prove inaccurate. This capacity to adjust without paralysis strengthens institutional stability in volatile environments.

Through repeated engagement with complexity, Saudization indirectly fosters a generation of leaders whose strategic perspective is shaped by exposure to reform, expansion, and modernization. Their leadership orientation reflects not only technical proficiency but also adaptive intelligence.

Adaptive capacity is also closely associated with sustainable leadership. Sustainability challenges are often characterized by uncertainty, evolving regulations, technological change, and shifting stakeholder expectations. As organizations increasingly pursue sustainability objectives, leaders must be capable of adapting to new environmental requirements, emerging green technologies, and changing societal priorities. Consequently, sustainable leadership requires a high degree of flexibility, long-term thinking, and the ability to respond effectively to complex and dynamic environments. In this regard, the adaptive capacity developed through leadership experiences under Saudization may also contribute to the preparation of leaders who are better equipped to support sustainability initiatives and environmental stewardship objectives in the Kingdom.

3.5 Cross-Cultural Leadership Development

The expansion of large-scale national projects in recent years—particularly following the launch of Vision 2030—has transformed the organizational landscape in Saudi Arabia. Mega-projects across sectors such as urban development, tourism, technology, and infrastructure have attracted significant international participation, including global firms, technical experts, and multinational teams (Nurunnabi, 2017;

Almatar, 2025). This influx of expertise has positioned the Kingdom as an active participant in global knowledge exchange and technological advancement.

As a result, organizational environments have become increasingly multicultural, bringing together professionals from diverse linguistic, cultural, and professional backgrounds (Connerley & Pedersen, 2005; Den Hartog & De Hoogh, 2024). For Saudi managers, leading within such settings requires more than technical competence; it demands cultural sensitivity, adaptive communication, and the ability to bridge differing perspectives. Managing diverse teams fosters the development of cultural intelligence, empathy, and inclusive leadership practices.

Cross-cultural collaboration often stimulates innovation and creativity, as varied experiences and viewpoints contribute to problem-solving. At the same time, differences in communication styles, expectations, and professional norms may generate misunderstandings or conflict. Navigating these dynamics provides emerging Saudi leaders with valuable opportunities to strengthen negotiation skills, conflict resolution strategies, and multilingual communication abilities. The necessity of operating in international environments frequently enhances proficiency in global business languages, particularly English, further broadening professional capacity.

Over time, sustained engagement in multicultural leadership contexts contributes to greater openness, adaptability, and mutual understanding. Such competencies are increasingly essential in globally interconnected economies and reinforce the long-term development of leaders capable of operating effectively across cultural boundaries.

It is important to note that collaboration with large and innovative companies from around the world is essential for achieving the goals of major projects such as NEOM, Diriyah Gate, and the Red Sea Project. Such collaboration brings a vast amount of experience, expertise, and technological knowledge to Saudi Arabia. As a result, young Saudi leaders who emerge through Saudization programs are provided with valuable opportunities to learn from these technologies, management practices, and professional experiences.

In addition, Saudi leaders possess a strong understanding of the characteristics of their local environment and the challenges associated with it. This enables them to play a more effective role in achieving the objectives of these projects. Furthermore, they can adapt and refine globally proven environmental policies and practices to better align with the land, culture, and society of Saudi Arabia.

In conclusion, cross-cultural leadership development offers significant benefits for emerging Saudi leaders. It facilitates knowledge transfer, expands professional experience, and exposes leaders to a variety of environmental policies and sustainability practices from different countries. Through this exposure, Saudi leaders can learn how to adapt international best practices to local conditions while becoming more innovative, creative, and effective in addressing environmental challenges.

3.6 Global Integration and Competitive Positioning

As Saudi organizations integrate into global markets, leadership competence increasingly determines competitive positioning. Leaders who can translate national priorities into internationally credible strategies strengthen institutional legitimacy abroad. Cross-cultural competence is therefore not merely a soft skill but a strategic asset (Javidan et al., 2006; Den Hartog & De Hoogh, 2024). Localization within globalized environments creates a hybrid leadership profile—rooted locally yet fluent internationally. Such hybridity enhances adaptability, allowing organizations to compete while maintaining cultural coherence.

3.7 Organizational Cultural Transformation

As leadership representation becomes increasingly localized, organizational cultures undergo gradual yet meaningful transformation. Leadership behavior influences informal norms, communication patterns, and value prioritization. When Saudi nationals assume greater managerial presence, workplace cultures begin to reflect local social dynamics alongside global professional standards (McCann & Holt, 2010).

This process produces hybrid organizational identities. Institutions do not simply replace one cultural template with another; rather, they integrate localized perspectives with established operational practices. Over time, leadership narratives, motivational frameworks, and performance expectations may align more closely with national socio-cultural values.

Cultural transformation also affects employee engagement. Representation within leadership structures can reduce perceived psychological distance between management and workforce. Employees may experience greater relatability and trust when leadership reflects shared social contexts. This alignment can strengthen organizational cohesion and collective commitment (Dawley et al., 2008).

Furthermore, localized leadership may influence decision-making rhythms. Leaders embedded within local cultural contexts often possess nuanced understanding of stakeholder expectations and community sensitivities. Such awareness enhances responsiveness and institutional legitimacy.

Localization therefore extends beyond individual leadership development to influence collective organizational identity. Over time, culturally grounded leadership models may emerge, reflecting a synthesis between national heritage and global management standards (Abiddin et al., 2022).

3.7.1 Sustainability Culture and Environmental Stewardship

Sustainability is not achieved solely through policies, regulations, or technological solutions. It also depends on the development of organizational cultures that encourage responsible behavior, environmental awareness, and long-term thinking. Organizational culture influences how employees perceive sustainability initiatives, how they respond to environmental policies, and the extent to which sustainability becomes integrated into everyday decision-making processes. As a result, culture plays a critical role in supporting environmental stewardship within organizations (Bennett et al., 2018; Boeske, 2023).

Leadership is one of the most important drivers of organizational culture. Leaders influence values, priorities, and behavioral expectations through both formal decisions and daily interactions (Sajjad et al. 2024). As Saudi nationals increasingly occupy leadership positions through Saudization, they gain greater influence over the development of workplace cultures. This influence provides opportunities to promote sustainability-oriented values, encourage responsible resource management, and strengthen environmental awareness among employees and stakeholders.

Over time, the combination of localized leadership and sustainability-oriented organizational cultures may contribute to stronger environmental stewardship practices within Saudi organizations. When sustainability becomes embedded in organizational values rather than treated as a standalone initiative, environmental responsibility is more likely to be sustained over the long term. In this regard, organizational cultural transformation represents an important mechanism through which Saudization can contribute to sustainable leadership development and support the Kingdom's broader sustainability objectives (Schroeder et al., 2019).

3.8 Institutional Learning and Succession Planning

Beyond individual development, Saudization influences organizational behavior at a structural level. As Saudi nationals assume greater representation in managerial roles, organizations are encouraged to adopt more systematic approaches to leadership development and succession planning. Rather than relying primarily on external recruitment for senior roles, firms increasingly invest in identifying, preparing, and mentoring internal Saudi talent for progressive advancement (Fusarelli et al., 2018).

This shift encourages the establishment of leadership pipelines, structured training programs, and mentoring systems designed to prepare second- and third-line managers for future executive responsibilities. Financial and organizational resources are gradually redirected toward capability building, reflecting a longer-term commitment to human capital investment. Such institutional learning processes strengthen organizational stability while fostering continuity in leadership transitions **(Grocutt et al., 2022; Dawley et al., 2008).**

From a broader perspective, sustained investment in leadership development contributes to economic resilience and social advancement. When national talent is systematically prepared for leadership roles, the benefits extend beyond individual careers to encompass organizational performance, workforce stability, and national competitiveness (Fusarelli et al., 2018; Groves, 2007). In this sense, Saudization functions not merely as an employment policy but as a catalyst for structured leadership formation and institutional capacity building within the Saudi economy.

3.9 Negotiating Authority in Multicultural Environments

Leadership within multicultural teams requires the negotiation of authority across divergent cultural expectations. In some professional contexts, authority is expressed through directive clarity and hierarchical distance. In others, it is exercised through participatory dialogue and collaborative consensus-building. Saudi leaders operating within international project environments must navigate these contrasting assumptions while maintaining credibility.

This negotiation process strengthens interpretive sensitivity. Leaders learn to decode behavioral cues, adjust communication tone, and calibrate assertiveness based on audience composition. They become attentive to contextual nuances that influence how authority is perceived and received. Such sensitivity enhances diplomatic competence and reduces friction within diverse teams (Javidan et al., 2006; Den Hartog & De Hoogh, 2024).

Exposure to multicultural leadership contexts also broadens strategic imagination. Interaction with varied professional philosophies introduces alternative problem-solving approaches and governance practices. Leaders who operate within such environments develop pluralistic thinking patterns that enable them to evaluate decisions from multiple vantage points.

Moreover, multicultural leadership demands emotional regulation and patience. Misinterpretations rooted in cultural divergence may generate tension. Effective leaders learn to respond constructively, emphasizing clarification rather than escalation. Over time, this practice strengthens negotiation skills and conflict resolution capacity **(Nielsen, 2022)**.

The cumulative impact of such experiences extends beyond immediate project performance. Leaders who develop competence in multicultural negotiation acquire portable skills that enhance their effectiveness in global partnerships, cross-border collaborations, and multinational negotiations. As Saudi Arabia deepens its integration within the global economy, these competencies assume increasing strategic relevance.

4. CHALLENGES AND COMPETENCY GAPS IN LEADERSHIP AND SUSTAINABLE LEADERSHIP FORMATION UNDER SAUDIZATION

4.1 Rapid Promotion and Preparedness Gaps

While Saudization has the potential to play a significant role in cultivating future leaders and strengthening national human capital, it is important to recognize that accelerated leadership pathways may also generate unintended challenges (Heslin & Keating, 2017). The expansion of opportunities for Saudi nationals to assume managerial roles represents a positive structural shift; however, rapid promotion does not automatically guarantee readiness. Leadership competence requires not only opportunity, but also depth of experience, judgment, and gradual exposure to complex organizational environments.

In some cases, individuals may be appointed to administrative or supervisory positions before they have fully developed the practical expertise necessary to manage projects, navigate organizational dynamics, or lead teams effectively. The ability to make sound decisions, resolve conflicts, and align operational activities with strategic objectives evolves over time and through accumulated experience. When such experience is limited, organizations may encounter performance inconsistencies or operational strain (Hughes et al., 2025; Grocutt et al., 2022).

Addressing this gap requires early and systematic preparation. Leadership development should not begin at the point of promotion; rather, it should be integrated into earlier stages of professional formation. Collaboration between universities, training institutions, and relevant government bodies can facilitate early identification of promising talent and provide structured developmental pathways before individuals assume formal leadership roles. Early exposure to project management, decision-making simulations, and supervised leadership assignments can mitigate preparedness gaps and ensure smoother transitions into managerial responsibilities.

Proactive talent identification and gradual capability building are therefore essential in minimizing the risks associated with rapid promotion. When leadership pipelines are intentionally developed, the benefits of Saudization can be realized more effectively without compromising organizational stability.

4.1.1 Sustainability Preparedness and Green Leadership Skills

The challenge of preparedness extends beyond traditional leadership competencies and increasingly includes sustainability-related capabilities. As Saudi Arabia advances ambitious initiatives under Vision 2030, future leaders are expected to operate in environments where environmental sustainability, resource efficiency, and responsible development play an increasingly important role. Consequently, leadership readiness should not be evaluated solely in terms of managerial and technical competence, but also in terms of preparedness to address sustainability-related challenges (Ghanem & Alamri, 2023; Edgell, 2019).

Rapid promotion may create situations in which emerging leaders possess strong operational or administrative skills but have limited exposure to environmental management, sustainability principles, or green innovation practices. While such competencies were not traditionally considered central leadership requirements, they are becoming increasingly relevant in organizations that operate within sustainability-oriented sectors and national development projects (Berniak-Woźny & Rataj, 2023; Saleem et al., 2025). Without adequate preparation, leaders may find it difficult to integrate environmental considerations into strategic and operational decision-making.

Addressing this challenge requires the incorporation of sustainability competencies into leadership development pathways. Universities, professional training institutions, and organizational development programs can play an important role in preparing future leaders to understand environmental issues and sustainability objectives alongside traditional management responsibilities. In this regard, leadership preparedness should be viewed as a multidimensional concept that includes not only managerial capability but also the knowledge and awareness necessary to support long-term sustainable development (Redman et al., 2021; Wals, 2011).

4.2 Tokenism and Compliance-Driven Appointments

Another challenge arises when localization efforts are reduced to numerical compliance rather than substantive development. In certain instances, organizations may prioritize meeting Saudization targets without ensuring meaningful role integration or structured leadership preparation (Fusarelli et al., 2018). Such practices risk creating symbolic appointments rather than genuine leadership participation.

When positions exist primarily on paper, or when managerial titles are granted without corresponding responsibility or engagement, the long-term developmental benefits of localization are undermined. Beyond limiting individual growth, this approach may erode trust within the labor market and weaken perceptions of transparency and accountability. Sustainable leadership formation cannot occur in environments where advancement is detached from actual performance, responsibility, and contribution (O'Donoghue & Van Der Werff, 2022; Dawley et al., 2008).

Ensuring authenticity in leadership appointments requires robust regulatory oversight and organizational commitment to real developmental outcomes. Monitoring mechanisms, clear performance expectations, and transparent reporting structures can discourage superficial compliance. More importantly, organizations should invest in structured onboarding, training, and mentorship programs that enable Saudi professionals to actively participate in decision-making and operational leadership. Localization achieves its intended purpose only when positions are accompanied by genuine responsibility and measurable contribution (Grocutt et al., 2022; Dawley et al., 2008).

This challenge may be particularly important in the context of sustainable leadership development. Environmental stewardship and sustainability initiatives require leaders who possess genuine authority, responsibility, and the ability to influence organizational decisions. When leadership appointments are primarily symbolic or driven by compliance considerations, opportunities to develop sustainability-related competencies may be limited. Consequently, meaningful leadership participation is essential not only for effective localization but also for preparing future leaders who can contribute to Saudi Arabia's long-term sustainability and environmental objectives (Boeske, 2023; Fry & Egel, 2021).

4.3 Mentorship and Institutional Support Limitations

Leadership development is rarely an individual endeavor; it is shaped by the institutional environment within which leaders emerge. Despite growing emphasis on human capital investment, some organizations may lack structured mentorship systems or comprehensive development programs for newly appointed managers. Without adequate guidance, emerging leaders may find themselves navigating complex responsibilities with limited support (Barkhuizen et al., 2022).

The absence of mentoring relationships can intensify pressure, reduce confidence, and slow the acquisition of managerial competence. Experienced leaders play a crucial role in transferring tacit knowledge, modeling professional conduct, and offering strategic perspective. When such support structures are weak or inconsistent, the learning process becomes more fragmented and reactive.

Strengthening institutional support requires coordinated efforts between private organizations and public entities. Structured leadership development programs, executive coaching, rotational assignments, and continuous professional education can provide emerging leaders with the tools needed to grow sustainably. Early-stage talent development—beginning at secondary and university levels—can further enhance preparedness by cultivating leadership awareness before individuals enter managerial tracks. Systematic investment in mentoring and institutional learning mechanisms ensures that opportunity is complemented by guidance.

4.3.1 Sustainability Mentorship Gaps

The limitations of mentorship and institutional support may also create challenges for the development of sustainability-related leadership competencies. As organizations increasingly face environmental and sustainability expectations, emerging leaders require guidance not only in traditional management practices but also in areas such as environmental stewardship, sustainability strategy, and responsible resource management (Fisher et al., 2012). When mentoring systems are weak or inconsistent, opportunities to transfer such knowledge and experience may be significantly reduced.

This challenge may be particularly important in sectors associated with sustainability initiatives and environmental transformation. Without experienced mentors, structured development programs, or institutional support mechanisms, future leaders may struggle to acquire the sustainability literacy and environmental awareness required to address complex environmental challenges. Consequently, gaps in mentorship and institutional support can slow the development of sustainable leadership capabilities and limit the effectiveness of future environmental stewardship efforts within organizations (Ryan et al., 2001).

4.4 Psychological Pressure and Performance Anxiety

Leadership roles inherently involve pressure, yet the context of workforce localization may amplify this experience. Emerging Saudi leaders may perceive themselves not only as organizational representatives, but also as symbols of broader national policy success. Such expectations, whether explicit or implicit, can create heightened psychological strain (Nielsen, 2022; Schmidt, 2010).

In addition to external expectations, managerial roles bring continuous accountability to senior executives, performance targets, and stakeholder demands. Sustained exposure to these pressures may contribute to stress, decision fatigue, or performance anxiety. Over time, excessive pressure can reduce creativity and encourage risk-averse behavior, as individuals may prioritize caution over innovation in order to avoid failure (Lovelace et al., 2007).

If not managed constructively, such dynamics may limit the developmental potential of leadership roles. Leaders operating under chronic stress may become overly conservative, hesitant to experiment, or reluctant to pursue bold strategic initiatives. This cautious posture, while understandable, may restrict organizational agility and innovation.

Addressing these challenges requires more than technical training; it calls for psychological and organizational support systems. Leadership awareness programs, stress management initiatives, and open discussions about the realities of managerial responsibility can normalize the challenges associated with leadership roles. Creating environments where constructive failure is treated as part of learning can also reduce fear-driven decision-making. When psychological resilience is cultivated alongside technical competence, leadership development becomes more sustainable and balanced (Schmidt, 2010; Nielsen, 2022).

4.5 The Risk of Performance Compression

One overlooked challenge of accelerated leadership integration is performance compression. Emerging leaders may simultaneously face expectations to demonstrate competence, validate localization policies, and outperform predecessors. This layered expectation compresses developmental timelines and heightens scrutiny (Heslin & Keating, 2017; Hughes et al., 2025).

Such compression can generate disproportionate pressure to deliver immediate outcomes. While urgency can stimulate growth, sustained compression risks emotional exhaustion and defensive leadership behavior. Leaders operating under intense evaluative scrutiny may prioritize short-term certainty over long-term innovation.

Performance compression may also inhibit experimentation. When perceived tolerance for error is low, leaders may avoid strategic risks that could otherwise generate competitive advantage. Over time, this caution may restrict creative exploration and institutional agility (Heslin & Keating, 2017).

Balancing accountability with developmental patience is therefore critical. Leadership excellence rarely emerges from acceleration alone. It requires structured reinforcement, constructive feedback, and psychological safety. Institutions that recognize and manage performance compression enhance the sustainability of leadership formation.

4.5.1 Sustainability Under Performance Pressure

Performance compression may create additional challenges for sustainability-oriented leadership. Environmental stewardship and sustainability initiatives often require long-term planning, experimentation, and gradual implementation. However, leaders operating under intense performance pressure may feel compelled to prioritize short-term outcomes that can be measured and reported quickly. As a result, sustainability objectives may receive less attention when immediate operational and financial expectations dominate managerial decision-making.

Furthermore, sustainability initiatives frequently involve uncertainty, innovation, and the possibility of setbacks during implementation. Leaders who perceive little tolerance for failure may become reluctant to support new environmental practices, green technologies, or sustainability-related innovations. Consequently, excessive performance pressure may unintentionally discourage the experimentation and long-term thinking that are often necessary for sustainable leadership (Ahsan, 2025; Saleem et al., 2025). Balancing performance expectations with developmental flexibility can therefore play an important role in supporting both leadership growth and environmental sustainability objectives (McCann & Holt, 2010; Fry & Egel, 2021).

4.6 The Risk of Symbolic Advancement

Another challenge lies in symbolic advancement without substantive authority. Leaders who hold formal titles but lack real decision-making autonomy may experience constrained development. Without exposure to consequential decision-making, leadership growth remains superficial. Organizational structures must therefore align authority with accountability to ensure authentic developmental progression (O'Donoghue & Van Der Werff, 2022; Dawley et al., 2008). Sustainable leadership formation requires meaningful discretion. Developmental depth emerges when leaders experience both the weight and consequence of their strategic choices.

4.7 The Environmental Competence Gap

As mentioned above, the Kingdom of Saudi Arabia is undergoing a significant transformation socially, economically, politically, and environmentally. As a result, a new generation of Saudi leaders is expected to face challenges that were either less prominent or did not exist in previous decades. Among the most important of these challenges are environmental issues. Global warming, air pollution, water scarcity—particularly in arid regions such as Saudi Arabia—and the loss of natural habitats that may lead to species extinction are all serious concerns that require effective leadership and long-term strategic responses (Chaaben et al., 2024; Abdelraouf & Nagasawa, 2023). Therefore, the traditional set of leadership skills may no longer be sufficient on its own. Historically, leadership development focused on managing employees, communicating with customers and stakeholders, implementing strategic change, and developing financial and marketing capabilities. While these skills remain important, the environmental challenges facing modern organizations require an expanded leadership skill set. This raises an important question: Are traditional leadership skills alone sufficient for young Saudi leaders to address emerging environmental challenges? As discussed throughout this chapter, Saudization plays an important role in developing the capabilities of young Saudi professionals and can be viewed as a long-term investment in national human capital. However, future Saudi leaders who emerge through workforce localization initiatives will also need to acquire competencies related to environmental sustainability. These include environmental awareness, sustainability literacy, and an understanding of how organizations can adopt and improve sustainable practices **(Redman et al., 2021; Cebrián et al., 2019)**. Waste management and the adoption of global best practices, particularly within major projects such as NEOM and the Red Sea Project, are essential for protecting wildlife and preserving natural habitats **(Almatar, 2025; Bennett et al., 2018)**. Similarly, renewable energy—especially solar energy—represents a significant opportunity for Saudi Arabia to pursue cleaner and more efficient forms of development. Saudi leaders who are appointed through Saudization initiatives can play an important role in supporting these efforts and enhancing quality of life throughout the Kingdom. Furthermore, environmental impact assessment can serve as an important tool to help leaders evaluate decisions and ensure that organizational activities align with broader environmental objectives. Saudization opens the door to leadership opportunities for young professionals; however, this represents only part of the solution. The other part lies in education, training, and awareness. Universities, training centers, and professional development institutions can play a critical role in equipping future leaders with the knowledge and skills required to address environmental challenges **(Shriberg & MacDonald, 2013; Wals, 2011)**. Government agencies, in collaboration with national and international organizations, can develop programs that integrate environmental awareness and sustainability competencies into leadership development initiatives. In other words, Saudization programs should not focus solely on traditional leadership, managerial, and financial skills, but should also incorporate environmental and sustainability-related competencies. Saudization opens leadership opportunities, but environmental education and sustainability training ensure that future leaders are capable of supporting the Kingdom's environmental ambitions.

5. POLICY AND ORGANIZATIONAL IMPLICATIONS

The preceding discussion demonstrates that Saudization extends beyond labor market regulation to influence leadership formation, organizational capability, and long-term institutional sustainability. However, realizing these benefits requires deliberate alignment between public policy objectives and organizational development practices. Workforce localization alone does not automatically generate leadership competence; rather, its effectiveness depends on the structures, systems, and support mechanisms that accompany it.

From a policy perspective, localization strategies should increasingly emphasize quality alongside quantity. While employment targets remain important, leadership development outcomes must also be considered as part of the broader evaluation framework. Policymakers may consider integrating leadership readiness indicators into national workforce assessments, encouraging organizations not only to hire and promote Saudi nationals but also to demonstrate structured development pathways. Incentive schemes could be linked to mentorship programs, leadership training initiatives, and succession planning frameworks that prepare emerging leaders for progressively complex roles.

Furthermore, early-stage talent development represents a critical area for policy attention. Collaboration between universities, vocational institutions, and government agencies can help identify high-potential individuals and provide them with exposure to leadership simulations, internships, and applied learning opportunities. Embedding leadership development within higher education curricula ensures that theoretical preparation is complemented by practical engagement before graduates enter managerial tracks. Such alignment strengthens the transition from academic learning to organizational leadership (Heslin & Keating, 2017)..

At the organizational level, firms must recognize that accelerated promotion requires parallel investment in support systems. Structured onboarding for new managers, executive coaching, and rotational leadership assignments can reduce preparedness gaps and mitigate performance risks. Clear performance metrics, constructive feedback mechanisms, and mentoring relationships allow emerging leaders to refine their capabilities in a supportive environment rather than through isolated trial and error.

Organizations should also cultivate cultures that normalize learning and responsible risk-taking. When emerging leaders perceive that measured experimentation and reflective learning are encouraged, they are more likely to innovate and adapt strategically. Conversely, environments driven solely by performance pressure may inhibit creativity and long-term development. Leadership formation thrives in settings where accountability is balanced with guidance and institutional trust (Heslin & Keating, 2017; Nielsen, 2022)..

Finally, succession planning must become a central component of organizational strategy. Rather than treating leadership appointments as isolated events, firms should establish continuous pipelines that identify, develop, and evaluate future leaders over time. Such institutionalization of leadership development ensures continuity, enhances organizational resilience, and aligns workforce localization efforts with sustainable economic growth.

In sum, Saudization offers a unique opportunity to reshape leadership ecosystems within Saudi Arabia. Its long-term success depends on coordinated policy frameworks and organizational practices that transform access to leadership roles into sustained capability development. When opportunity is matched with preparation, mentorship, and institutional commitment, workforce localization can evolve into a powerful engine for leadership renewal and national competitiveness.

5.1 Policy Implications for Sustainable Leadership Development

Achieving Saudi Arabia's sustainability ambitions requires more than environmental regulations and sustainability initiatives. It also requires policy frameworks that support the development of leaders capable of advancing environmental stewardship and sustainable development objectives. Therefore, policymakers should establish clear performance measures and evaluation systems that assess both sustainability outcomes and leadership effectiveness. Such indicators can help organizations monitor progress, identify areas for improvement, and strengthen accountability (Sajjad et al., 2024; Bennett et al., 2018).

Effective implementation also requires coordination among multiple government entities, including the Ministry of Education, the Ministry of Environment, Water and Agriculture, and the Ministry of Human Resources and Social Development. Collaborative efforts among these institutions can support the development of common evaluation frameworks, performance indicators, and incentive systems that encourage sustainability-oriented leadership. Linking recognition and rewards to sustainability achievements can further motivate leaders to integrate environmental considerations into organizational decision-making processes.

In addition, sustainability leadership competencies should be embedded within educational curricula, executive training programs, and professional development initiatives across both the public and private sectors. Future leaders will increasingly require knowledge of environmental stewardship, sustainability literacy, resource efficiency, and long-term strategic thinking. Integrating these competencies into leadership development pathways can help prepare a generation of leaders capable of supporting the Kingdom's long-term sustainability and environmental objectives (Cebrián et al., 2019; Redman et al., 2021)..

5.2 Toward a Leadership Development Ecosystem

Saudization contributes to the emergence of a broader leadership development ecosystem that extends beyond individual organizations. This ecosystem encompasses educational institutions, regulatory frameworks, professional associations, and corporate networks. Leadership formation becomes distributed across interconnected actors rather than confined to isolated organizational experiences (Fusarelli et al., 2018; Groves, 2007).

Within such an ecosystem, leadership development is conceptualized as a national investment. Universities prepare students with foundational competencies, organizations provide experiential exposure, and regulatory bodies shape developmental incentives. Collaboration among these actors enhances continuity between academic preparation and executive succession.

Over time, this systemic alignment reduces volatility in leadership transitions and strengthens institutional depth. Instead of relying on sporadic recruitment cycles, organizations cultivate internal pipelines supported by broader national structures. Leadership capability thus becomes cumulative and self-reinforcing.

When workforce localization operates within a coordinated ecosystem, its long-term effects extend beyond employment metrics. It contributes to governance quality, innovation capacity, and institutional resilience (Fry & Egel, 2021; Sajjad et al., 2024). Sustainable leadership development becomes embedded within the structural fabric of the economy rather than treated as a reactive adjustment.

5.3 Long-Term Institutional Memory Formation

As localization deepens, institutions begin to accumulate leadership continuity grounded in national experience. Successive generations of leaders inherit organizational knowledge shaped by shared socio-economic transitions. This continuity strengthens strategic coherence across decades. Institutional memory built through localized leadership reduces volatility associated with leadership turnover. Over time, this continuity supports policy alignment, innovation sustainability, and stable governance structures (Groves, 2007).

Long-term institutional memory may be particularly important for achieving sustainability objectives. Many environmental initiatives and sustainability programs require decades of continuous implementation before their full benefits become visible. As leadership transitions occur, accumulated knowledge regarding environmental policies, sustainability practices, and lessons learned from previous initiatives can be preserved and transferred to future generations of leaders. This continuity reduces the risk of repeating past mistakes, strengthens organizational learning, and supports the long-term implementation of sustainability strategies. In this regard, institutional memory serves as an important mechanism for maintaining momentum toward environmental stewardship and sustainable development goals (Wals, 2011; Bennett et al., 2018).

5.4 From Employment Policy to Leadership Architecture

Ultimately, Saudization should not be viewed solely as a workforce adjustment mechanism. It functions as an architectural redesign of leadership pipelines within the national economy. By expanding access to authority, accelerating experiential exposure, and reshaping identity formation, localization policies gradually construct a domestic leadership infrastructure capable of sustaining long-term development.

The enduring significance of this transformation lies not in numerical employment metrics, but in the quiet maturation of strategic competence across institutions. Leadership depth, once established, becomes self-perpetuating—nurturing innovation, strengthening governance, and reinforcing national confidence.

5.5 Saudization and Environmental Leadership Capacity

As discussed above, the Kingdom of Saudi Arabia is embarking on some of the largest and most ambitious projects in the world. Among these projects, several initiatives are specifically aimed at preserving the environment and protecting natural resources. Examples include Green Riyadh, the Green Hydrogen Project in NEOM, and the Saudi Green Initiative. These projects involve investments worth billions of dollars and reflect the Kingdom's commitment to environmental sustainability. To support the achievement of their objectives, advanced technologies have been introduced into Saudi Arabia, and a wide range of environmental policies and best practices have been studied and, in many cases, implemented. In short, environmental projects in Saudi Arabia represent some of the most significant sustainability initiatives globally, supported by both advanced technologies and progressive environmental policies (Abdelraouf & Nagasawa, 2023; Almatar, 2025; Awan, 2019).

As noted throughout this chapter, Saudi Arabia is investing heavily in environmental and sustainability-related projects. If Saudization continues to place Saudi professionals in leadership and managerial positions, these individuals will increasingly be responsible for managing and leading such initiatives. This represents one of the most important benefits of workforce localization. Saudi leaders possess a deeper understanding of their country, environment, and the unique challenges associated with local development. Consequently, they are often better positioned to provide culturally appropriate environmental stewardship. In addition, they understand local customs, social values, and community expectations. This enables them to adapt and refine international environmental practices in ways that align with the Kingdom's culture and societal needs while minimizing potential conflicts between imported policies and local realities.

Saudization therefore contributes to the development of leaders who understand the land, the culture, and the long-term development goals of their country. Such understanding can improve the implementation of environmental initiatives and increase their acceptance within society. At the same time, future Saudi leaders must be equipped with a range of environmental competencies, including renewable energy awareness, environmental impact assessment, resource efficiency, and green innovation. These skills are becoming increasingly important as organizations face growing environmental responsibilities. Future leaders must be able to integrate environmental considerations into strategic decision-making processes. This is perhaps the most important point: Saudi leaders need not only to understand environmental policies and sustainability strategies, but also to possess the capability to implement them effectively within their organizations and communities.

The future success of Saudi Arabia's sustainability ambitions will depend not only on technological advancement, but also on the availability of capable leaders who can translate environmental goals into organizational practice. These leaders are likely to be even more effective when they possess a deep understanding of the country, its culture, and its social context. Such understanding enables them to apply environmental strategies in ways that are both practical and culturally appropriate, ultimately supporting the Kingdom's long-term sustainability objectives.

6. CONCLUSION

This chapter set out to examine Saudization not merely as a workforce localization policy, but as a structural mechanism with significant implications for leadership development in Saudi Arabia. While much of the existing discourse surrounding localization focuses on employment figures and regulatory compliance, this analysis has argued that the policy also reshapes leadership pathways, managerial exposure, and institutional learning processes within organizations.

Through accelerated responsibility allocation, increased accountability, cross-cultural engagement, and institutional restructuring, Saudization creates conditions that can foster experiential leadership development. By expanding access to supervisory and middle-management roles, it enables Saudi professionals to acquire practical competencies in decision-making, team management, and strategic alignment. In this sense, localization functions as a catalyst for experiential growth, transforming theoretical preparation into lived managerial capability.

At the same time, the chapter has highlighted that such transformation is neither automatic nor without risk. Rapid promotion, compliance-driven appointments, limited mentorship structures, and psychological pressure may constrain leadership formation if not addressed systematically. These challenges underscore the importance of coupling opportunity with preparation, support, and institutional commitment (Heslin & Keating, 2017). Sustainable leadership development requires deliberate coordination between policy frameworks, educational institutions, and organizational systems.

Looking ahead, further empirical research is needed to assess the long-term leadership outcomes associated with workforce localization. Future studies may explore comparative sectoral differences, the effectiveness of mentorship programs, and the relationship between localization intensity and leadership performance indicators. Longitudinal research designs could provide deeper insight into how early managerial exposure shapes executive competence over time.

Ultimately, Saudization represents more than a labor market adjustment; it reflects a broader national effort to strengthen human capital and institutional capacity. When supported by coherent development strategies, workforce localization has the potential to contribute meaningfully to the formation of a new generation of leaders capable of navigating complex economic and organizational environments. Its long-term success will depend on the continued integration of opportunity, accountability, and structured development within both public policy and private sector practice.

6.1 Expanding the Leadership Horizon

The long-term value of leadership localization also lies in its capacity to reshape professional ambition across generations. When leadership becomes visibly attainable within national institutions, career trajectories expand in both scope and imagination. Emerging professionals begin to perceive executive influence not as a distant or externalized function, but as a realistic extension of their own development pathway. This shift in aspiration alters educational choices, professional investments, and long-term career planning across the workforce.

In this context, Saudization contributes to a subtle but powerful cultural transformation: it normalizes national authority within strategic spaces. Leadership ceases to be associated with external expertise and becomes embedded within domestic competence. Over time, this normalization reduces dependency narratives and strengthens institutional self-confidence. Organizations begin to rely on internally cultivated insight rather than defaulting to imported strategic direction.

Another enduring implication concerns decision-making continuity. Leaders who grow within the same institutional ecosystem often develop deeper alignment with long-term organizational objectives. Their professional identity is shaped not only by quarterly performance expectations but also by awareness of national transformation goals and generational development. This continuity enhances strategic patience. Rather than prioritizing immediate gains at the expense of structural sustainability, localized leadership may be more inclined to pursue gradual, cumulative progress.

Furthermore, leadership localization enhances accountability ecosystems. When decision-makers are socially and economically embedded within the national context, the consequences of their strategic choices extend beyond abstract metrics. Their professional networks, reputational standing, and societal engagement are directly linked to institutional outcomes. This proximity can strengthen ethical awareness and reinforce responsible governance (O'Donoghue & Van Der Werff, 2022; Muff et al., 2020).

It is also important to recognize that leadership development is not a linear progression but a layered process of refinement. Early exposure may introduce complexity and pressure, yet sustained experience gradually stabilizes judgment. As leaders accumulate exposure to negotiation, crisis management, and long-term planning, their strategic intuition deepens. What initially requires deliberate analysis eventually becomes structured discernment. Over time, this maturation process strengthens institutional foresight.

Ultimately, the maturation of localized leadership contributes to national strategic autonomy. Economic independence is not defined solely by production capacity or capital investment; it is equally determined by decision-making capability. When strategic direction is shaped by domestically developed leaders, institutions gain confidence in their ability to navigate uncertainty without excessive reliance on external stewardship.

The broader narrative, therefore, is not merely about employment distribution but about cultivating enduring leadership capital. Leadership capital consists of accumulated experience, institutional memory, adaptive intelligence, and strategic coherence. Unlike financial capital, it cannot be imported instantly; it must be developed through structured opportunity and sustained responsibility. Saudization, when effectively implemented, accelerates this accumulation.

As the leadership architecture continues to evolve, the most significant outcome may be the gradual alignment between institutional ambition and national aspiration. When leaders emerge from within the society they serve, strategic direction reflects lived realities rather than abstract projections. This alignment strengthens governance legitimacy, enhances organizational trust, and supports long-term economic resilience.

In this sense, workforce localization transcends its immediate objectives. It becomes a formative mechanism through which leadership capability is cultivated, institutional depth is strengthened, and strategic independence is secured for future generations.

Finally, the future success of Saudization should not be measured solely by employment figures, but also by its ability to develop a generation of leaders capable of balancing economic growth, social development, and environmental stewardship. Saudization can serve as an effective mechanism for developing well-rounded leaders who are able to integrate economic, social, and environmental objectives into their decision-making processes. Furthermore, workforce localization can contribute to the development of culturally appropriate environmental stewardship, as leaders who possess a deep understanding of their country's culture, society, and environmental context are often better positioned to design and implement solutions that align with local needs and expectations. Such understanding can enhance both the effectiveness of environmental initiatives and the long-term sustainability of development outcomes (Sajjad et al., 2024; Bennett et al., 2018).

6.2 Saudization and the Future of Sustainable Leadership

As Saudi Arabia continues to pursue its Vision 2030 agenda, the role of leadership will increasingly extend beyond traditional managerial responsibilities. Future leaders will be expected not only to achieve economic objectives, but also to balance social development and environmental sustainability considerations. In this context, Saudization can contribute to the development of sustainable leaders who possess both traditional leadership capabilities and a broader understanding of environmental stewardship, long-term planning, and responsible resource management. Such competencies are becoming increasingly important as organizations operate within a rapidly changing environment characterized by sustainability challenges and growing expectations for responsible leadership (Liao, 2022; Boeske, 2023).

The future success of Saudi Arabia's sustainability ambitions will depend not only on technological advancement and financial investment, but also on the availability of capable leaders who can translate sustainability objectives into organizational practice. By expanding leadership opportunities for Saudi nationals, workforce localization can contribute to the development of leaders who are deeply familiar with the Kingdom's culture, society, and long-term development priorities. This combination of leadership capability, local knowledge, and sustainability awareness may strengthen environmental stewardship and support the achievement of the Kingdom's long-term economic, social, and environmental goals.

AI DISCLOSURE STATEMENT

The author used OpenAI's ChatGPT as a language support tool during the preparation and revision of this chapter. The tool was utilized to assist with language editing, grammar improvement, readability enhancement, and organizational refinement of the manuscript. All conceptual development, content decisions, interpretation, analysis, and final manuscript preparation were conducted and verified by the author. The author assumes full responsibility for the accuracy, originality, and integrity of the content presented in this chapter.

REFERENCES

Abdelraouf, M., & Nagasawa, A. (2023). Saudi green initiatives and their role in achieving environmental goals in the Middle East. Gulf Research Center, 19.

Abiddin, N. Z., Ibrahim, I., & Abdul Aziz, S. A. (2022). Non-governmental organisations (NGOs) and their part towards sustainable community development. *Sustainability (Basel)*, *14*(8), 4386. DOI: 10.3390/su14084386

Ahsan, M. J. (2025). Green leadership and innovation: Catalysts for environmental performance in Italian manufacturing. *The International Journal of Organizational Analysis*, *33*(7), 1646–1669. DOI: 10.1108/IJOA-04-2024-4450

Al-Qutb, H. R., & Ghaith, N. A. (2018). The impact of job localization on organizational commitment: A field study of employees at the Saudi Geological Survey Authority in Jeddah. Journal of Economic. *Administrative and Legal Sciences*, *2*(6), 81–96.

Almatar, K. M. (2025). Sustainable Urban Futures: Transportation and Development in Riyadh, Jeddah, and Neom. *Sustainability (Basel)*, *17*(22), 10133. DOI: 10.3390/su172210133

Alshuwaikhat, H. M., & Mohammed, I. (2017). Sustainability matters in national development visions—Evidence from Saudi Arabia's Vision for 2030. *Sustainability (Basel)*, *9*(3), 408. DOI: 10.3390/su9030408

Amran, Y. A., Amran, Y. M., Alyousef, R., & Alabduljabbar, H. (2020). Renewable and sustainable energy production in Saudi Arabia according to Saudi Vision 2030; Current status and future prospects. *Journal of Cleaner Production*, *247*, 119602. DOI: 10.1016/j.jclepro.2019.119602

Awan, A. B. (2019). Performance analysis and optimization of a hybrid renewable energy system for sustainable NEOM city in Saudi Arabia. *Journal of Renewable and Sustainable Energy*, *11*(2), 025905. DOI: 10.1063/1.5071449

Barkhuizen, E. N., Masakane, G., & van der Sluis, L. (2022). In search of factors that hinder the career advancement of women to senior leadership positions. *SA Journal of Industrial Psychology*, *48*, 1986. DOI: 10.4102/sajip.v48i0.1986

Bennett, N. J., Whitty, T. S., Finkbeiner, E., Pittman, J., Bassett, H., Gelcich, S., & Allison, E. H. (2018). Environmental stewardship: A conceptual review and analytical framework. *Environmental Management*, *61*(4), 597–614. DOI: 10.1007/s00267-017-0993-2 PMID: 29387947

Berniak-Woźny, J., & Rataj, M. (2023). Towards green and sustainable healthcare: A literature review and research agenda for green leadership in the healthcare sector. *International Journal of Environmental Research and Public Health, 20*(2), 908. DOI: 10.3390/ijerph20020908 PMID: 36673663

Boeske, J. (2023). Leadership towards sustainability: A review of sustainable, sustainability, and environmental leadership. *Sustainability (Basel), 15*(16), 12626. DOI: 10.3390/su151612626

Bramston, P., Pretty, G., & Zammit, C. (2011). Assessing environmental stewardship motivation. *Environment and Behavior, 43*(6), 776–788. DOI: 10.1177/0013916510382875

Cebrián, G., Segalàs, J., & Hernández, À. (2019). Assessment of sustainability competencies: A literature review and pathways for future research and practice. *Central European Review of Economics and Management, 3*(3), 19–44.

Chaaben, N., Elleuch, Z., Hamdi, B., & Kahouli, B. (2024). Green economy performance and sustainable development achievement: Empirical evidence from Saudi Arabia. *Environment, Development and Sustainability, 26*(1), 549–564. DOI: 10.1007/s10668-022-02722-8 PMID: 36320556

Connerley, M. L., & Pedersen, P. B. (2005). *Leadership in a diverse and multicultural environment: Developing awareness, knowledge, and skills*. Sage Publications.

Dawley, D. D., Andrews, M. C., & Bucklew, N. S. (2008). Mentoring, supervisor support, and perceived organizational support: What matters most? *Leadership and Organization Development Journal, 29*(3), 235–247. DOI: 10.1108/01437730810861290

Den Hartog, D. N., & De Hoogh, A. H. (2024). Cross-cultural leadership: What we know, what we need to know, and where we need to go. *Annual Review of Organizational Psychology and Organizational Behavior, 11*(1), 535–566. DOI: 10.1146/annurev-orgpsych-110721-033711

Edgell, D. L.Sr. (2019). *Managing sustainable tourism: A legacy for the future*. Routledge.

Fisher, D. R., Campbell, L. K., & Svendsen, E. S. (2012). The organisational structure of urban environmental stewardship. *Environmental Politics, 21*(1), 26–48. DOI: 10.1080/09644016.2011.643367

Fry, L. W., & Egel, E. (2021). Global leadership for sustainability. *Sustainability (Basel), 13*(11), 6360. DOI: 10.3390/su13116360

Fusarelli, B. C., Fusarelli, L. D., & Riddick, F. (2018). Planning for the future: Leadership development and succession planning in education. *Journal of Research on Leadership Education, 13*(3), 286–313. DOI: 10.1177/1942775118771671

General Authority for Statistics. (2023). *Labor market statistics: Unemployment rate among Saudi nationals*. General Authority for Statistics. Retrieved from the official website of the General Authority for Statistics, Saudi Arabia.

Ghanem, A. M., & Alamri, Y. A. (2023). The impact of the green Middle East initiative on sustainable development in the Kingdom of Saudi Arabia. *Journal of the Saudi Society of Agricultural Sciences, 22*(1), 35–46. DOI: 10.1016/j.jssas.2022.06.001

Grocutt, A., Gulseren, D., Weatherhead, J. G., & Turner, N. (2022). Can mentoring programmes develop leadership? *Human Resource Development International, 25*(4), 404–414. DOI: 10.1080/13678868.2020.1850090

Groves, K. S. (2007). Integrating leadership development and succession planning best practices. *Journal of Management Development, 26*(3), 239–260. DOI: 10.1108/02621710710732146

Heslin, P. A., & Keating, L. A. (2017). In learning mode? The role of mindsets in derailing and enabling experiential leadership development. *The Leadership Quarterly, 28*(3), 367–384. DOI: 10.1016/j.leaqua.2016.10.010

Hu, X., Li, R. Y. M., Kumari, K., Ben Belgacem, S., Fu, Q., Khan, M. A., & Alkhuraydili, A. A. (2022). Relationship between green leaders' emotional intelligence and employees' green behavior: A PLS-SEM approach. *Behavioral Sciences (Basel, Switzerland), 13*(1), 25. DOI: 10.3390/bs13010025 PMID: 36661597

Hughes, H., Hanneman, K., & Patlas, M. N. (2025). First year in a new leadership role: Lessons learned. *Canadian Association of Radiologists Journal, 76*(4), 659–663. DOI: 10.1177/08465371251332504 PMID: 40245170

Javidan, M., Dorfman, P. W., De Luque, M. S., & House, R. J. (2006). In the eye of the beholder: Cross cultural lessons in leadership from project GLOBE. *The Academy of Management Perspectives, 20*(1), 67–90. DOI: 10.5465/amp.2006.19873410

Jayashree, P., El Barachi, M., & Hamza, F. (2022). Practice of sustainability leadership: A multi-stakeholder inclusive framework. *Sustainability (Basel), 14*(10), 6346. DOI: 10.3390/su14106346

Leal Filho, W., Eustachio, J. H. P. P., Caldana, A. C. F., Will, M., Lange Salvia, A., Rampasso, I. S., Anholon, R., Platje, J., & Kovaleva, M. (2020). Sustainability leadership in higher education institutions: An overview of challenges. *Sustainability (Basel), 12*(9), 3761. DOI: 10.3390/su12093761

Liao, Y. (2022). Sustainable leadership: A literature review and prospects for future research. *Frontiers in Psychology*, *13*, 1045570. DOI: 10.3389/fpsyg.2022.1045570 PMID: 36420377

Lovelace, K. J., Manz, C. C., & Alves, J. C. (2007). Work stress and leadership development: The role of self-leadership, shared leadership, physical fitness and flow in managing demands and increasing job control. *Human Resource Management Review*, *17*(4), 374–387. DOI: 10.1016/j.hrmr.2007.08.001

Mahran, K., Albarrak, H., Ibrahim, B. A., & Elamer, A. A. (2025). Leadership and sustainability in tourism and hospitality: A systematic review and future research agenda. *International Journal of Contemporary Hospitality Management*, *37*(7), 2219–2242. DOI: 10.1108/IJCHM-07-2024-0985

McCann, J. T., & Holt, R. A. (2010). Defining sustainable leadership. *International Journal of Sustainable Strategic Management*, *2*(2), 204–210. DOI: 10.1504/IJSSM.2010.032561

Muff, K., Liechti, A., & Dyllick, T. (2020). How to apply responsible leadership theory in practice: A competency tool to collaborate on the sustainable development goals. *Corporate Social Responsibility and Environmental Management*, *27*(5), 2254–2274. DOI: 10.1002/csr.1962

Nielsen, K. (2022). Anxiety, trust, leadership, communication and stress. *Nordic Psychology*, *74*(3), 150–151. DOI: 10.1080/19012276.2022.2124079

Nurunnabi, M. (2017). Transformation from an oil-based economy to a knowledge-based economy in Saudi Arabia: The direction of Saudi vision 2030. *Journal of the Knowledge Economy*, *8*(2), 536–564. DOI: 10.1007/s13132-017-0479-8

O'Donoghue, D., & Van Der Werff, L. (2022). Empowering leadership: Balancing self-determination and accountability for motivation. *Personnel Review*, *51*(4), 1205–1220. DOI: 10.1108/PR-11-2019-0619

Rahman, R., & Qattan, A. (2021). Vision 2030 and sustainable development: state capacity to revitalize the healthcare system in Saudi Arabia. INQUIRY: The Journal of Health Care Organization, Provision, and Financing, 58.

Redman, A., Wiek, A., & Barth, M. (2021). Current practice of assessing students' sustainability competencies: A review of tools. *Sustainability Science*, *16*(1), 117–135. DOI: 10.1007/s11625-020-00855-1

Rodriguez-Anton, J. M., Rubio-Andrada, L., Celemín-Pedroche, M. S., & Alonso-Almeida, M. D. M. (2019). Analysis of the relations between circular economy and sustainable development goals. *International Journal of Sustainable Development and World Ecology*, *26*(8), 708–720. DOI: 10.1080/13504509.2019.1666754

Ryan, R. L., Kaplan, R., & Grese, R. E. (2001). Predicting volunteer commitment in environmental stewardship programmes. *Journal of Environmental Planning and Management*, *44*(5), 629–648. DOI: 10.1080/09640560120079948

Sajjad, A., Eweje, G., & Raziq, M. M. (2024). Sustainability leadership: An integrative review and conceptual synthesis. *Business Strategy and the Environment*, *33*(4), 2849–2867. DOI: 10.1002/bse.3631

Saleem, F., Sundarasen, S., & Malik, M. I. (2025). Green leadership and environmental performance in hospitals: A multi-mediator study. *Sustainability (Basel)*, *17*(12), 5376. DOI: 10.3390/su17125376

Schmidt, M. J. (2010). Is there a place for emotions within leadership preparation programmes? *Journal of Educational Administration*, *48*(5), 626–641. DOI: 10.1108/09578231011067776

Schroeder, P., Anggraeni, K., & Weber, U. (2019). The relevance of circular economy practices to the sustainable development goals. *Journal of Industrial Ecology*, *23*(1), 77–95. DOI: 10.1111/jiec.12732

Selim, M. M., & Alshareef, N. (2025). Trends and opportunities in renewable energy investment in Saudi Arabia: Insights for achieving vision 2030 and enhancing environmental sustainability. *Alexandria Engineering Journal*, *112*, 224–234. DOI: 10.1016/j.aej.2024.10.107

Shriberg, M., & MacDonald, L. (2013). Sustainability leadership programs: Emerging goals, methods & best practices. The Journal of Sustainability Education.

Soni, V., Gnekpe, C., Roux, M., Anand, R., Yaroson, E. V., & Banwet, D. K. (2023). Adaptive distributed leadership and circular economy adoption by emerging SMEs. *Journal of Business Research*, *156*, 113488. DOI: 10.1016/j.jbusres.2022.113488

Wals, A. E. (2010). Mirroring, Gestaltswitching and transformative social learning: Stepping stones for developing sustainability competence. *International Journal of Sustainability in Higher Education*, *11*(4), 380–390. DOI: 10.1108/14676371011077595

Wals, A. E. (2011). Learning our way to sustainability. *Journal of Education for Sustainable Development*, *5*(2), 177–186. DOI: 10.1177/097340821100500208

Chapter 13
The Role of Education in Cultivating Future Stewardship Leaders

Deepak Gupta
http://orcid.org/0000-0003-3929-1362
Institute of Technology and Management, Gwalior, India

ABSTRACT

Whether stewardship leaders are products of educational design or biographical accident has never been straightforward, though literature increasingly favours the former. This chapter takes that view seriously. Drawing on peer-reviewed scholarship from 2021 to 2026, it examines how formal schooling, transformative pedagogy, outdoor and nature-based learning, service learning, indigenous knowledge integration, and digital modalities contribute—independently or in combination—to the knowledge, values, and competencies required for environmental stewardship leadership. The analysis is framed by Education for Sustainable Development and transformative learning theory, proposing a five-stage model mapping development from ecological awareness to sustained leadership. A comparative table and effectiveness graph support the discussion. The chapter concludes with recommendations for educators, curriculum designers, and researchers.

1. INTRODUCTION

Ask almost any senior conservation practitioner or environmental policy officer how they arrived at their vocation and the account will invariably involve a mix of formative experience—a teacher who lit something, a place that mattered, a community that modelled care for the land—alongside degrees, fieldwork, and profes-

DOI: 10.4018/979-8-3373-7877-0.ch013

sional training that gave that initial spark both shape and traction. The biographical particularity of these accounts can make stewardship leadership look like something that happens to certain people rather than something that educational systems can reliably produce. That appearance, this chapter argues, is misleading. The conditions under which stewardship leaders develop are not random; they are educationally reproducible, and there is now a sufficient research base to say so with reasonable confidence (Boeske, 2023; McLeod et al., 2024).

This matters more than it might seem. Global projections from biodiversity science, climate research, and land-use modelling consistently describe ecological trajectories that require not marginal adjustment but deep systemic change—in governance, in land management, in economic organisation, and in the cultural frameworks through which societies relate to natural systems. Achieving such change requires leaders: people with the knowledge to diagnose what is wrong, the values to care about fixing it, the skills to mobilise others toward collective action, and the staying power to persist when institutions resist. These capacities do not appear spontaneously. They are cultivated. And since cultivating them at scale is ultimately an educational challenge, the design of educational systems bears directly on whether transformational stewardship leadership becomes widespread enough to matter (Al-khamaiseh et al., 2025; Chaudhry & Noureen, 2023).

The phrase 'stewardship leader' as used here describes someone who combines functional ecological literacy with a values-driven orientation toward environmental care, sufficient civic and relational skill to work with and through communities, and the adaptive capacity to navigate socio-ecological complexity without foreclosing on longer-term sustainability goals. These qualities are, notably, all present in the scholarly literature on effective environmental leadership—but they are rarely theorised as a developmental package, and the educational pathways most reliably associated with their combined development remain underspecified. This chapter attempts to address that gap.

The organisation is as follows. Section 2 develops the conceptual framework linking education, leadership, and stewardship, introducing Figure 1. Section 3 examines what formal schooling and higher education institutions have actually contributed to this developmental pipeline. Section 4 turns to transformative and experiential pedagogies—outdoor learning, critical reflection programmes, and service learning. Section 5 considers the particular contribution of indigenous and local knowledge integration. Section 6 addresses newer and more experimental approaches: digital and mobile platforms, marine biodiversity curricula, and competency-based conservation genetics programmes. Section 7 draws the threads together through a five-stage developmental model (Figure 2), a comparative table (Table 1), and a synthesised effectiveness graph (Graph 1). Section 8 derives policy and institutional implications. Section 9 concludes.

2. CONCEPTUAL FRAMEWORK: EDUCATION, LEADERSHIP, AND STEWARDSHIP

2.1 Three Domains in Conversation

The scholarly literature on sustainability, stewardship, and educational leadership has grown considerably over the past decade, but it has grown somewhat separately. Researchers in environmental education, leadership studies, and sustainability science have each produced substantial bodies of work—yet the points of contact among these traditions have been relatively few, and theoretical frameworks that genuinely span all three remain rare. Boeske's (2023) review of the sustainability leadership literature is one of the more useful recent attempts at synthesis. He identifies three partially overlapping constructs. Sustainable leadership refers to organisational and interpersonal leadership practices designed with long-term institutional viability in mind; sustainability leadership describes those who actively push institutions and systems toward ecological and social sustainability; and environmental leadership focuses specifically on advocacy for and protection of natural systems. All three forms share a developmental prerequisite, though Boeske does not fully spell it out: each requires individuals whose values, knowledge base, and action repertoire have been substantively shaped by educationally mediated encounters with ecological reality.

What McLeod et al. (2024) add to this picture is empirical traction on the mechanisms involved. Their systematic scoping review of environmental stewardship research—spanning psychology, sociology, ecology, and education—identifies a cluster of antecedents that predict stewardship behaviour: ecological knowledge, sense of place, responsibility attributions, social norms, and felt connectedness to the natural world. The significance of this catalogue for educational theory is considerable, because essentially every item on it is malleable—altered by experience, teaching, and community membership. Stewardship orientation is not a fixed trait; it is cultivated through specific types of formative engagement. That conclusion is the empirical bedrock on which this chapter's argument rests.

Althnayan et al. (2022) provide an additional dimension worth noting. Their quantitative study demonstrates that environmental transformational leadership—operationalised as leadership behaviour that models environmental values and inspires environmental citizenship in followers—is significantly associated with organisational sustainability performance. The implication is not simply that stewardship leaders do more environmental good themselves; they amplify the environmental behaviour of those around them. This multiplier logic strengthens the case for treating stewardship leadership as a high-leverage educational investment, since the returns extend considerably beyond the individual who receives the education.

2.2 Education for Sustainable Development: Promise and Limitations

Education for Sustainable Development (ESD) has operated as the dominant international framework for embedding sustainability into formal education since at least the United Nations Decade of ESD (2005–2014), and its influence on national curriculum policy—if not always on classroom practice—has been substantial. Vasconcelos and Orion (2021) note that ESD represents a significant departure from earlier environmental education traditions in its explicit attention to futures thinking, social justice dimensions, and the transformation of learning processes themselves, not only of learning content. Rather than treating environmental knowledge as the primary educational objective, ESD positions ecological understanding as one component of a broader orientation that includes ethical reasoning, civic agency, and the capacity for action under uncertainty.

Gajparia et al. (2022) offer a taxonomy that clarifies what, precisely, is at stake in different modes of sustainability education. Education about sustainability delivers information about ecological systems and environmental problems—necessary, but by itself insufficient for leadership development. Education for sustainability aims to cultivate the agency, values, and competencies needed to act on environmental problems—closer to what stewardship leadership requires. Education as sustainability goes further still, treating the learning process itself as a site of sustainability practice, embedding participatory, democratic, and place-responsive values in how teaching and learning are organised, not merely in what is taught. The distinction matters in practice: curricula that address sustainability as content without attending to the pedagogical relationship through which that content is transmitted rarely produce the dispositional shifts that stewardship leadership requires.

VanWynsberghe (2022) identifies a tension within ESD that is particularly relevant here. Individual transformation—changing what a person knows, values, and does—and collective social transformation—changing what communities, institutions, and systems do—are both objectives of sustainability education, but they operate on different timescales and through different mechanisms. Stewardship leadership development sits at this intersection: it is simultaneously a project of personal formation and a form of social reproduction, requiring educational designs that can honour both dimensions without collapsing one into the other. The conceptual framework developed for this chapter (Figure 1) attempts to map these considerations onto a developmental model that connects educational inputs to stewardship leadership outputs.

Figure 1. Conceptual framework: Education as the bridge to environmental stewardship leadership

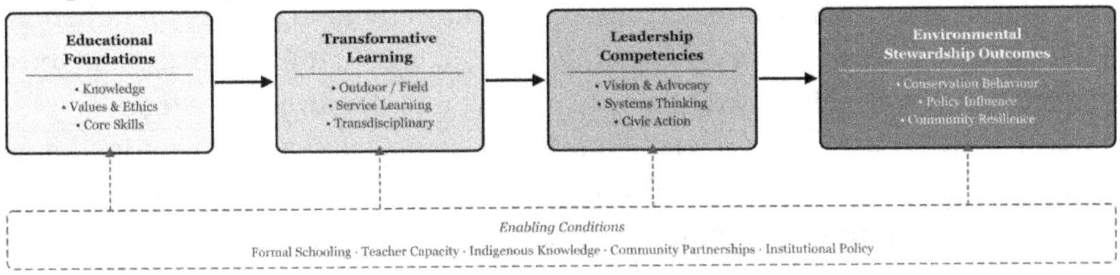

Note. Arrows indicate progressive developmental pathways. Dashed lines represent enabling contextual conditions.
Sources: Boeske (2023); McLeod et al. (2024); Redman & Wiek (2021); Barth et al. (2023).

Figure 1 illustrates the progressive developmental pathways; dashed lines denote enabling contextual conditions (teacher capacity, institutional policy, community partnership

The framework in Figure 1 maps a progression from foundational educational provision through transformative learning experiences to the leadership competency development that generates environmental stewardship outcomes. What is important about the framework is not its linearity—which is schematic rather than descriptive of actual developmental trajectories—but its insistence that foundational education (knowledge, values, skills) is necessary but insufficient. Without activation through genuine transformative experience—outdoor encounter, service to community, critical dialogue about real dilemmas—foundational provision rarely crosses the threshold into the identity commitment that stewardship leadership requires. The enabling conditions running beneath this progression—teacher preparation, institutional culture, indigenous knowledge systems, policy environment—are not incidental; they determine whether the progression can occur at all for most learners.

3. FORMAL EDUCATION AND SUSTAINABILITY LEADERSHIP: CONTRIBUTIONS AND CONSTRAINTS

3.1 Schools as Sites of Ecological Identity Formation

The school is, for most children and adolescents, the primary institutional space in which they encounter organised environmental content—and, more importantly, in which ecological attitudes and relational orientations toward the natural world are formed in the company of peers. Imran et al. (2024), working with teachers across multiple institutional contexts, found broad professional recognition of environmental education's developmental potential alongside a persistent gap between that recognition and actual classroom practice. What constrained teachers was not motivation but capacity: inadequate pre-service preparation in environmental pedagogy, curriculum frameworks that treated environmental content as peripheral rather than cross-curricular, limited access to outdoor spaces, and examination accountability structures that systematically de-prioritised anything not directly tested.

These constraints are not trivial, and they deserve acknowledgment before any idealisation of what schools can accomplish in stewardship leadership development. When environmental content appears only in isolated science units, when field experience is a once-yearly event rather than a regular pedagogical practice, and when teachers themselves have never been trained in facilitating outdoor or reflective learning, the contribution of school-based environmental education to stewardship leadership development is necessarily limited. Dastgir et al. (2024), studying Pakistani schools, found that this limitation is not inevitable: schools that actively engaged students in green land practices—composting programmes, tree nurseries, water harvesting projects embedded in the school grounds—produced markedly stronger environmental attitudes and behavioural intentions than comparison schools using conventional didactic methods. The lesson is not simply that practical engagement outperforms passive reception; it is that embodied, place-specific ecological experience generates a felt sense of responsibility that abstract content delivery rarely achieves.

There is also something to be said about developmental timing. Environmental psychologists have consistently found that formative encounters with the natural world during childhood and early adolescence are disproportionately influential in shaping long-term ecological identity—the sense of self as embedded in and responsible for natural systems (Mann et al., 2022). Schools that provide regular, well-designed contact with natural environments during these developmental windows are not merely improving environmental knowledge scores; they are, whether they recognise it or not, shaping the motivational architecture through which students will later decide whether environmental leadership feels like a vocation or an inconvenience. The stakes of getting this right during the school years are therefore higher than curriculum policy discussions typically acknowledge.

3.2 Higher Education: Where Competency Meets Commitment

If schools carry primary responsibility for the affective foundations of stewardship—wonder, care, sense of place, ecological empathy—higher education institutions bear particular responsibility for the strategic and competency layers: systems analysis, policy literacy, professional capacity, and the ability to translate environmental conviction into institutional change. This division is, of course, schematic; good higher education also attends to values and emotions, and good schools also develop analytical skills. But the distinction captures something real about the different developmental tasks that the two levels of education typically address.

Ghasemy et al. (2024), in an international comparative study of sustainability leadership education across the Asia Pacific region, document considerable variation in how universities are approaching this task. The most effective institutions in their sample—measured by the sustainability leadership orientations of graduates—treat sustainability leadership as an explicit, assessed educational objective, with competency frameworks that students are expected to achieve and that faculty are expected to design toward. Less effective institutions treat sustainability as a curricular add-on: a module available for interested students, a research theme for willing faculty, a value statement in the institutional charter. The structural differences between these two approaches—in curriculum design, faculty preparation, resource allocation, and assessment architecture—are substantial, and their effects on student outcomes are correspondingly significant (Gupta et al., 2026).

Bryant et al. (2023) provide perhaps the most granular account currently available of what actually changes in learners who participate in well-designed sustainability leadership programmes at postgraduate level. Their qualitative study of a dedicated sustainability leadership master's programme tracked participants through a series of transformative learning episodes—encounters with real-world environmental dilemmas, facilitated critical dialogue about competing values, structured reflection on their own assumptions and leadership orientations. The changes that emerged were not primarily in knowledge—participants often already knew a great deal about environmental problems—but in identity: in how they understood themselves as actors in environmental systems, and in the kind of leadership they felt called to exercise. The shift from sustainability manager (technical problem-solver) to stewardship leader (values-driven civic actor) is precisely what this kind of educational intervention, when well-designed, appears to catalyse.

Aslam et al. (2022), examining student perspectives on higher education in the immediate post-pandemic period, find considerable generational appetite for exactly this kind of formation. Respondents across multiple national contexts expressed dissatisfaction with educational experiences that addressed sustainability as abstraction—as content to be assessed rather than as a challenge to be addressed. What they wanted, and what many felt they were not getting, was learning that made genuine demands on them: that gave them real agency, real accountability, and real stakes. HEIs that respond to this aspiration are not merely improving student satisfaction; they are engaging with a demand that is, in substantive terms, a demand for stewardship leadership education (Agrawal et al., 2026).

One often-overlooked leverage point in this picture is the educator. Persson et al. (2023), studying faculty development in higher education sustainability, argue that investing in the sustainability competencies of academic staff yields larger systemic returns than investing in new courses or programmes in isolation. When faculty are equipped and given permission to redesign their teaching around transformative rather than transmissive models—when a chemist can reframe a course around environmental chemistry as a site of ethical reasoning, or a law professor can build land rights and ecological governance into foundational modules—the sustainability learning environment shifts for entire cohorts of students, not just for those who elect specialist sustainability programmes.

4. TRANSFORMATIVE AND EXPERIENTIAL PEDAGOGIES

4.1 The Transformative Learning Premise

The concept of transformative learning—associated principally with Mezirow's work in adult education, though significantly extended by subsequent scholars—rests on a claim that is both familiar and, in its implications, quite radical: that meaningful change in how people act in the world does not follow reliably from providing them with better information. People are quite capable of knowing a great deal about environmental degradation and continuing to behave as though they know nothing. What actually changes behaviour—durably, at the level of values and identity rather than surface compliance—is a particular kind of educational experience: one that disrupts existing frames of reference, creates conditions for critical reflection on the assumptions embedded in those frames, and provides the relational support necessary to construct new orientations in their place. Singer-Brodowski et al. (2022), O'Grady (2023), and Brooks and Brooks (2024) have each explored what this means in sustainability and environmental leadership contexts specifically.

O'Grady (2023), writing from the perspective of a higher education faculty member engaged in sustainability education, identifies the pedagogical conditions under which this kind of reconstruction is most likely: genuinely safe spaces for critical dialogue—not comfortable spaces, but ones where discomfort can be expressed without professional consequence; exposure to real-world environmental situations that resist simple resolution; mentorship from practitioners who model the integration of values and professional action; and sustained, structured opportunity for reflection. This last condition—the sustained, structured nature of the reflection—is important and often underappreciated. Transformative learning is not a single disorienting event; it is a process of reconstruction that typically requires repeated cycles of experience, dialogue, and reflection over extended time.

The emotional dimension of transformative learning in sustainability contexts deserves particular attention. Singer-Brodowski et al. (2022) note, drawing on accounts from university-level sustainability educators, that the affective weight of ecological crisis creates pedagogical challenges that earlier transformative learning theorists—working primarily with adult literacy and vocational retraining—did not anticipate. Students who are drawn into genuine awareness of climate trajectories, species extinctions, or the environmental injustices borne by marginalised communities frequently encounter grief, anger, or something close to despair before they encounter motivation. Containing and working with these emotional responses—what Singer-Brodowski et al. describe as creating 'safe enough' spaces—is a sophisticated pedagogical skill. When educators get this right, the emotional intensity of ecological crisis becomes a catalyst for committed leadership. When they handle it badly, or ignore it entirely, it produces what researchers in adjacent literatures have called eco-anxiety and disengagement.

4.2 Outdoor and Nature-Based Learning: What the Evidence Says

Among all the pedagogical modalities reviewed in the sustainability education literature, outdoor and nature-based learning has perhaps the most consistent track record across diverse age groups and national contexts. Mann et al. (2022) systematically reviewed nature-specific outdoor learning programmes, drawing on studies from Europe, North America, Australasia, and East Asia, and found robust evidence of positive effects across a range of outcomes: ecological knowledge, pro-environmental attitudes, nature connectedness, intrinsic motivation to protect the environment, and, notably, collaborative and social capacities that are directly relevant to leadership development. These effects appear across a wide range of programme types—forest schools, environmental field trips, place-based immersion programmes, wilderness experiences—suggesting that the active ingredient is not a particular programme format but the quality of engagement with natural systems that programmes facilitate.

This last point is worth dwelling on. Powell et al. (2023), in a national-scale study of environmental education field trip programmes across the United States, found that programme quality predicted learning outcomes far more strongly than either programme length or natural setting type. What distinguished effective from ineffective programmes was the coherence among pre-visit preparation (which allowed students to arrive with questions rather than blank curiosity), structured in-field inquiry (which channelled engagement toward specific learning objectives), and post-visit reflection (which converted experience into transferable meaning). Field trips that were isolated events—disconnected from prior classroom learning and followed by no sustained reflection—produced weak and transient effects. Those embedded within coherent pedagogical units produced effects that remained measurable months later. The implication is important: the challenge for schools is not simply to provide students with more time in nature (though that matters) but to design the surrounding pedagogical architecture that makes time in nature educationally transformative.

Iriarte Flores et al. (2023) add a contextual complication that is valuable precisely because it resists the temptation to treat nature contact as universally and automatically beneficial. Their research with youth in a high-biodiversity Colombian region found that stewardship orientation was strongly mediated by community and cultural factors—the presence of adults who modelled stewardship as a community value, the connection of ecological care to local identity and livelihood, the embeddedness of ecological knowledge in relationships rather than in decontextualised instruction. Youth who learned about their local ecosystems in the company of community elders who demonstrably cared about those ecosystems developed stronger stewardship orientations than those who received ecologically equivalent information from outside experts with no local roots. The finding cautions against treating outdoor education as an educational technology that can be imported into any context and will produce predictable results; its effects depend significantly on the human relationships and cultural meanings through which ecological experience is mediated.

4.3 Service Learning as Civic Stewardship Formation

Service learning occupies a distinctive position in the stewardship leadership literature partly because it explicitly combines two dimensions of educational experience—structured academic learning and community engagement—that more conventional educational models keep separate. Unlike outdoor education, which primarily develops the ecological and affective dimensions of stewardship, service learning develops the civic and relational dimensions: the capacity to work with communities rather than for them, to understand environmental problems as socially embedded, and to exercise leadership through facilitation and partnership rather than through technical expertise and individual agency.

Rodríguez-Zurita et al. (2024) provide the most comprehensive recent synthesis of service learning evidence in higher education sustainability contexts. Their systematic review identifies three features that consistently distinguish effective from ineffective programmes. First, authentic partnership: communities must be genuine co-investigators in the educational project, with their own goals, knowledges, and timescales respected, rather than being treated as passive sites in which students accumulate experiential credit. Second, learning integration: the service experience must be formally connected to academic learning through structured reflection, assessment, and conceptual discussion; without this integration, service learning degenerates into volunteering that produces goodwill but limited transformative development. Third, reciprocity: both students and community partners must benefit substantively from the collaboration; when benefit flows only toward students, the relational dynamics that generate the most powerful stewardship learning—working alongside community members as mutual actors—are undermined.

Brooks and Brooks (2024), focusing specifically on sustainability leadership professional development contexts, describe how service learning in adult programmes often triggers a reconceptualisation of leadership itself. Participants in their study described arriving at programmes with leadership understood primarily as direction-setting and decision-making—the executive model—and leaving with understandings far more consistent with facilitative, place-responsive, and adaptive orientations. This reconceptualisation is arguably more fundamental than any specific skill or knowledge acquisition, because it determines the kind of leadership that participants will subsequently seek to exercise. A programme that changes what people know about sustainability but leaves their model of leadership intact has achieved less than it might appear.

5. INDIGENOUS KNOWLEDGE AND CULTURALLY GROUNDED STEWARDSHIP EDUCATION

5.1 The Case Beyond Cultural Inclusion

Advocates for integrating indigenous and local knowledge (ILK) into sustainability curricula have sometimes framed that case primarily in terms of representational equity: ensuring that non-Western knowledge traditions are included in educational spaces that have historically privileged European scientific epistemologies. That argument is valid as far as it goes. But it understates what ILK integration contributes to stewardship leadership education, because the value of traditional ecological knowledge (TEK) is not merely representational—it is substantive. Jessen et al. (2022) document, in a synthesis of indigenous knowledge contributions to ecological and evolutionary science, multiple domains in which TEK provides information that formal scientific monitoring has not generated: long-term documentation of species population trajectories, ecological change across generational timescales, the functional significance of relationships among organisms in specific landscapes, and the behavioural responses of species to environmental conditions. These are not curiosities; they are data that conservation managers need and that TEK holders possess.

Beyond its informational value, indigenous ecological knowledge introduces an ontological dimension that sustainability education rarely accesses through other means. Western ecological science, whatever its epistemological achievements, is largely structured by a subject-object model: the scientist studies nature as an external entity to be understood, managed, and occasionally protected. Indigenous ontologies—internally diverse, but characteristically relational—tend to position humans as members of ecological communities rather than their observers or managers, bound by reciprocal obligations to other species and to the ecological systems they share. This relational ontology aligns more closely with what contemporary ecological theory increasingly recognises as appropriate to complex adaptive systems, where top-down management by external experts routinely fails and adaptive co-management by embedded community actors tends to succeed. Educators who bring indigenous frameworks into sustainability curricula are, therefore, not merely diversifying their content; they are potentially reorganising the conceptual framework through which learners understand stewardship itself (Buendia-Martinez & Carrasco, 2022).

5.2 Integration in Practice

Buendia-Martinez and Carrasco (2022) provide the most systematic recent account of how ILK integration is being attempted across different national and institutional contexts. Their review identifies several approaches that recur across high-quality programmes: place-based pedagogies that anchor learning in locally significant ecosystems rather than in universal ecological principles; collaborative curriculum co-design that positions indigenous knowledge holders as authors rather than informants; the incorporation of indigenous languages, oral traditions, and narrative forms as vehicles for ecological knowledge—not translations into academic prose; and land-based educational practices (harvesting, prescribed burning, planting, fishing) that simultaneously transmit ecological knowledge and enact stewardship as a relational practice.

Effective integration, these authors argue, requires a disposition they call epistemic humility—not the performative acknowledgment of indigenous knowledge as valuable, but a genuine willingness to let that knowledge challenge and reorganise existing conceptual frameworks rather than simply supplement them. This is harder than it sounds, particularly for institutions whose authority rests substantially on the credibility of scientific methodology and whose faculty may have spent careers operating within epistemological assumptions that TEK integration implicitly questions. Developing epistemic humility as an institutional disposition, rather than merely an individual one, is itself a significant governance challenge. But it is worth noting that the capacity for epistemic humility—for learning from knowledge traditions different from one's own—is precisely a core stewardship leadership competency in pluralistic conservation contexts. Educational experiences that develop it in learners are therefore doing double pedagogical work.

Dearborn and Tansey (2024) connect ILK integration to a broader framework of global citizenship education for stewardship. Their argument is that effective stewardship leaders in a globalised world need to be simultaneously locally grounded and globally aware—capable of acting from a place of deep knowledge of and care for specific ecosystems while understanding those places within larger patterns of environmental change and international environmental governance. Indigenous knowledge integration is one of the primary educational mechanisms through which this dual grounding can be achieved: it roots learners in the ecological and cultural specificity of particular places while revealing how those places have been shaped by and now participate in global processes. The glocal competency that results—thinking systemically across scales while acting from locally grounded knowledge—is among the most consistently identified attributes of effective environmental stewardship leadership.

6. EMERGING APPROACHES: DIGITAL LEARNING, MARINE EDUCATION, AND BIODIVERSITY CURRICULA

6.1 Digital and Mobile Platforms: Potential and Cautions

The integration of digital technologies into environmental education has accelerated substantially over the past decade, driven by the proliferation of mobile sensing technologies, open ecological datasets, and educational applications designed to connect classroom learning with field observation. The theoretical appeal of these tools is genuine: they can extend environmental monitoring across spatial and temporal scales that classroom or field trip education cannot reach; they can make real ecological data available to learners who are not geographically proximate to the ecosystems in question; and they can provide interactive simulation environments that allow learners to explore ecological dynamics in ways that physical field experiences cannot.

Narboshova et al. (2026) offer an empirically grounded account of mobile experiential learning applied to environmental pollution science in secondary schools. Their study found that students who used mobile platforms to collect real measurements from local environmental sites—integrating field data collection with classroom analysis and environmental decision-making simulation—demonstrated significantly stronger pro-environmental behavioural intentions than those receiving conventional instruction about equivalent content. The key mechanism, according to their analysis, was not the technology itself but the authenticity of the data: students who recognised that their measurements were real, that their findings would be shared with environmental monitoring bodies, and that their analyses could potentially influence local environmental decisions responded with substantially higher engagement and stronger identity investment in the environmental monitoring role. The platform was a vehicle for authentic participation in environmental stewardship, not a substitute for it.

Onyeaka and Akinsemolu (2024), examining green education in the MENA region, introduce an equity dimension that optimistic accounts of digital environmental education frequently overlook. Digital tools are unevenly distributed: hardware costs, connectivity infrastructure, teacher preparation, and the cultural fit of technology-mediated learning vary enormously across and within countries. If digital environmental education is primarily accessible to students in well-resourced urban schools, it may reinforce rather than challenge the socio-economic gradients that already structure access to environmental leadership roles. More fundamentally, digital mediation of ecological experience carries a substitution risk: a student who is thoroughly competent in augmented-reality species identification but has never spent an afternoon sitting quietly in a functioning ecosystem has a relationship to the natural world that is, in important respects, impoverished—and it is from relationships of embodied ecological familiarity, not informational competence, that the deepest stewardship commitments tend to grow.

6.2 Marine Biodiversity Education and the Ocean Literacy Gap

Ocean ecosystems are, by any ecological measure, among the most important natural systems on Earth. They cover the majority of the planet's surface, support an extraordinary proportion of global biodiversity, regulate atmospheric chemistry and global climate, and provide nutrition and economic livelihood to hundreds of millions of people. Yet they feature remarkably rarely in school curricula outside coastal regions, and ocean literacy—understood as both factual knowledge of marine systems and a sense of personal connection to and responsibility for them—is demonstrably low among populations that will need to make consequential decisions about marine environments in the coming decades. Xamrakulova et al. (2026) address this gap through a study of project-based marine biodiversity education with secondary students, finding substantial gains in both ocean knowledge and conservation-oriented attitudes when students engaged in sustained, collaborative, inquiry-led marine biodiversity assessment.

The programme design details in that study repay attention, because they illustrate several of the principles identified elsewhere in this review. Students worked in small groups on real biodiversity assessment tasks, using data collection protocols shared with regional marine monitoring programmes. Results were shared publicly. Reflection sessions connected field findings to broader marine conservation policy. The programme was sustained over a semester rather than compressed into a day visit. Each of these features—authenticity, publicness, policy connection, duration—appears in the broader effectiveness literature as a predictor of stronger stewardship outcomes. The study suggests that marine environments, despite their historical marginalisation in environmental education, are no less effective than terrestrial ones as sites for stewardship leadership development when educational design principles are properly applied.

6.3 Conservation Genetics and Competency-Based Biodiversity Literacy

Among the most intellectually ambitious recent contributions to stewardship leadership curriculum design is the competency-based biodiversity literacy framework proposed by Urolov et al. (2026) for undergraduate-level education. Rather than treating conservation genetics as a specialist biological subdiscipline, their framework embeds molecular ecology within a transdisciplinary curriculum that integrates conservation ethics, environmental law and policy, socio-economic analysis of biodiversity loss drivers, and sustained field practicum experience. The result is not a conventional course but a competency architecture—a designed environment in which the knowledge domains and practical skills required for conservation leadership are systematically developed and assessed in integrated rather than isolated ways.

This approach maps well onto the five-competency framework that Redman and Wiek (2021) identified through their systematic review of sustainability transformation research: systems thinking, anticipatory capacity, normative reasoning, strategic competency, and interpersonal competency. The Urolov et al. (2026) curriculum is notably ambitious in addressing all five rather than privileging the first two (which conventional science education tends to develop) at the expense of the latter three (which require pedagogical contexts that science education rarely provides). The result, if the framework is implemented as designed, should be graduates who are capable not only of analysing biodiversity problems with technical sophistication but of reasoning about the values at stake in conservation decisions, devising strategies that can be realistically pursued in specific governance contexts, and building the coalitions and relationships that effective conservation leadership requires. That is a comprehensive stewardship leadership profile, and the fact that it can be framed within a university biodiversity programme illustrates the degree to which curriculum design ambition can determine educational outcome.

7. SYNTHESIS: DEVELOPMENTAL MODEL, COMPARATIVE TABLE, AND EFFECTIVENESS EVIDENCE

7.1 A Five-Stage Developmental Model of Stewardship Leadership

The educational contributions reviewed across the preceding sections can be organised into a developmental model that maps the progression from initial ecological awareness through to sustained public environmental leadership. The model proposed here is explicitly non-deterministic: not every individual moves through all five stages in sequence, stages frequently co-develop, and entry points vary by biography and context. Nevertheless, the model offers a useful heuristic for understanding how different educational interventions relate to one another in the broader developmental arc of stewardship leaders (see Figure 2).

Figure 2. The five-stage stewardship leadership development model

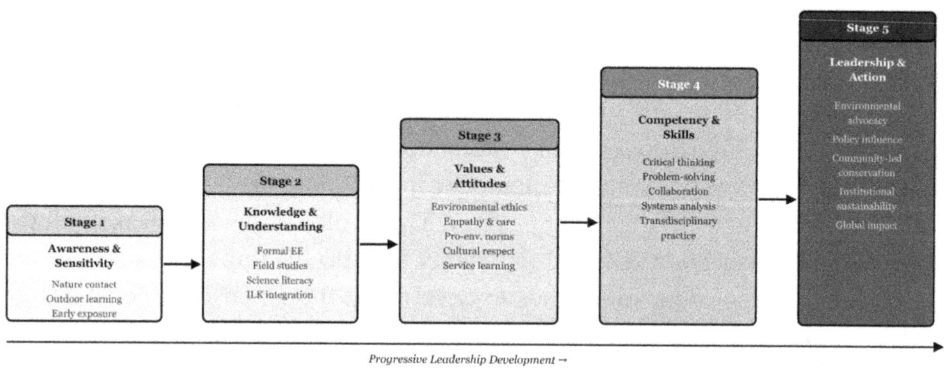

Note. Stage heights denote increasing depth of competency. Sources: Bryant et al. (2023); Redman & Wiek (2021); McLeod et al. (2024); Brooks & Brooks (2024).

Figure 2 illustrate Stage heights schematically represent increasing depth of competency development; the model is broadly progressive but not strictly sequential.

Stage 1 (Awareness and Sensitivity) is primarily a product of early and repeated contact with natural environments, mediated through outdoor learning, unstructured nature experience, and place-based educational provision. The developmental task at this stage is less cognitive than affective: cultivating a sense of wonder, emotional attachment to particular places, and an incipient sense of ecological belonging. Research on environmental identity formation suggests that these affective foundations, laid during childhood and early adolescence, exert lasting influence on how individuals subsequently relate to environmental issues and whether they experience environmental leadership as a calling or an imposition (Mann et al., 2022; Iriarte Flores et al., 2023).

Stage 2 (Knowledge and Understanding) is developed through formal environmental education, field studies, and structured engagement with ecological science—including, where available, exposure to indigenous knowledge frameworks. The knowledge built at this stage provides the analytical infrastructure through which stewardship commitment can be channelled strategically: understanding nutrient cycling, species interdependency, climate feedbacks, and the socio-economic drivers of ecological change allows intuitive care for the environment to become informed, targeted action. Without this knowledge layer, stewardship motivation remains emotionally genuine but strategically constrained.

Stage 3 (Values and Attitudes) represents the affective and normative deepening that distinguishes environmentally knowledgeable individuals from committed stewards. This transformation is most reliably catalysed by experiences that require learners to take ethical positions in real-world environmental dilemmas, engage with communities whose lives are materially bound up with ecological health, or confront evidence of environmental injustice in ways that cannot be intellectually resolved without personal normative commitment. Service learning, indigenous knowledge integration, and transformative pedagogy are the primary educational vehicles for Stage 3 development.

Stage 4 (Competency and Skills) involves the explicit development of leadership-relevant capabilities through transdisciplinary programmes, professional learning contexts, and sustained community engagement. Systems thinking, evidence-based strategic communication, collaborative problem-solving across difference, adaptive management of complex socio-ecological challenges, and cross-cultural facilitation are the competency targets at this stage. The competency frameworks documented by Redman and Wiek (2021), Ghasemy et al. (2024), and Urolov et al. (2026) are directly relevant resources for institutions seeking to design Stage 4 educational experiences.

Stage 5 (Leadership and Action) emerges when the preceding developmental layers are brought to bear in authentic contexts of environmental leadership—community conservation management, environmental policy advocacy, institutional sustainability governance, or professional roles in conservation organisations. At this stage, what the educational system has invested across earlier stages is converted into the sustained, values-driven, strategically competent environmental leadership that stewardship requires. Institutions can support this transition through internship placements, alumni mentorship networks, community partnership structures, and continuing professional development provisions that maintain connections between educational identity and leadership practice.

7.2 Comparing Educational Approaches

Table 1, below, compares the major educational approaches examined in this chapter against four criteria: key features, effectiveness for stewardship leadership development, and principal scholarly sources. The comparison is not intended to produce a simple ranking—the evidence base does not support that degree of precision—but rather to clarify the distinctive contributions of different approaches and indicate the kinds of outcomes most reliably associated with each.

Table 1. Comparative overview of educational approaches for cultivating environmental stewardship leaders

Educational Approach	Key Features	Effectiveness for Stewardship Leadership	Primary Sources
Formal Classroom EE	Structured curricula; textbook-led ESD; lecture-based delivery	Moderate — adequate for foundational ecological knowledge; on its own, rarely sufficient to shift leadership identity or produce durable behavioural change	Boeske (2023); Imran et al. (2024); Dastgir et al. (2024)
Outdoor and Nature-Based Learning	Forest schools; environmental field trips; place-based immersion; wilderness programmes	High — consistent association with pro-environmental attitudes, nature connectedness, and intrinsic stewardship motivation across age groups	Mann et al. (2022); Powell et al. (2023); Iriarte Flores et al. (2023)
Transformative Learning Programmes	Critical reflection pedagogy; sustainability master's programmes; disorienting-dilemma curriculum design	Very high — well-documented identity shifts, deepened leadership commitment, and sustained stewardship orientation in participants	Bryant et al. (2023); Brooks & Brooks (2024); O'Grady (2023); Singer-Brodowski et al. (2022)
Service Learning & Community Engagement	Reciprocal university–community sustainability partnerships; civic action-learning; collaborative field projects	High — cultivates civic stewardship identity, stakeholder engagement capacity, and practical conservation skills in context	Rodríguez-Zurita et al. (2024); Dearborn & Tansey (2024)
Indigenous & Local Knowledge Integration	TEK transmission; land-based practices; co-designed curricula with knowledge holders; storytelling	High — deepens ecological understanding, introduces relational stewardship ontology, develops epistemic humility	Buendia-Martinez & Carrasco (2022); Jessen et al. (2022)

continued on following page

Table 1. Continued

Educational Approach	Key Features	Effectiveness for Stewardship Leadership	Primary Sources
Digital and Mobile Experiential Learning	Field data collection apps; AR/VR ecological simulations; mobile pollution monitoring	Moderate-emerging — engagement gains documented; robust evidence on sustained leadership outcomes still accumulating	Narboshova et al. (2026); Onyeaka & Akinsemolu (2024)
Transdisciplinary Learning	Cross-faculty problem-based design; systems-thinking frameworks; real-world sustainability challenges as curriculum spine	Very high — generates integrated competency bundles (systems reasoning, normative capacity, collaboration) central to stewardship leadership	Barth et al. (2023); Redman & Wiek (2021); Ghasemy et al. (2024)

Several patterns emerge from this comparison and merit emphasis. The first is the consistent superiority of experiential, reflective, and community-embedded approaches over content-transmission models, across all educational levels and cultural contexts. This pattern holds regardless of whether the outcome variable is ecological knowledge, pro-environmental attitudes, leadership identity, or stewardship behaviour—though the magnitude of the advantage varies across outcomes. The second pattern is the complementarity rather than substitutability of different approaches: outdoor learning and service learning are not two ways of achieving the same educational outcome; they develop different but mutually reinforcing components of stewardship leadership. A curriculum that offers one without the other is more limited than one that combines them thoughtfully. The third pattern is the relatively weak performance of formal classroom environmental education when used as the primary modality—not because knowledge transmission is unimportant, but because it is insufficient on its own to drive the identity-level shifts that stewardship leadership development requires.

7.3 Synthesised Effectiveness Evidence

Graph 1 presents a visual synthesis of the comparative effectiveness evidence. Composite effectiveness scores were derived by weighing the direction, consistency, and magnitude of effects documented in the reviewed literature for each educational approach, giving greater weight to systematic reviews and multi-site empirical studies than to single-context case studies. These scores are explicitly not the product of a formal meta-analysis—the methodological diversity of the relevant literature precludes that—but they represent a considered and consistent application of the same evaluative criteria across all seven approaches.

Figure 3. Relative effectiveness of educational approaches in developing environmental stewardship leadership competencies (composite score, %).

(Synthesised from systematic reviews and empirical studies in the reference pool; scores reflect aggregate evidence ratings)

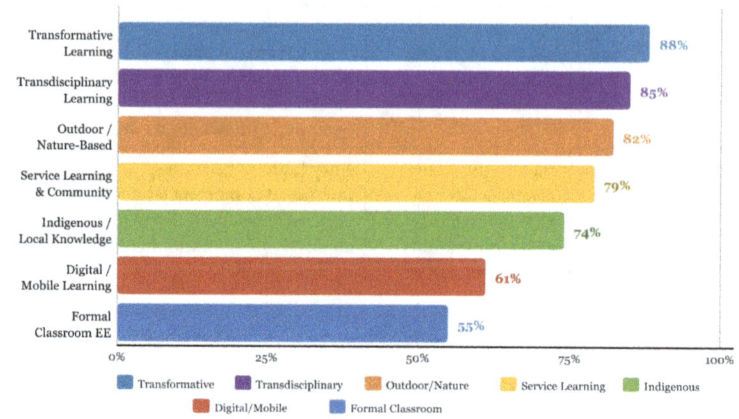

Figure 3 illustrate Higher scores indicate stronger aggregate evidence for stewardship leadership competency development.

Transformative learning programmes (88%) and transdisciplinary approaches (85%) score highest, reflecting consistent evidence across multiple national and institutional contexts for their capacity to generate the identity shifts and integrated competency development most closely associated with stewardship leadership. Outdoor and nature-based learning (82%) scores third, supported by a substantial body of systematic review evidence. Service learning (79%) and indigenous knowledge integration (74%) follow, with the latter's lower score reflecting less empirical documentation rather than weaker effects in practice—a methodological limitation of the available literature rather than a substantive one. Digital and mobile learning (61%) reflects a promising but still-emerging evidence base; formal classroom environmental education (55%) scores lowest among all modalities, underscoring the limitations of content-delivery approaches when not combined with experiential and reflective elements.

8. IMPLICATIONS FOR POLICY AND PRACTICE

8.1 What Institutions Need to Change

The synthesis presented in this chapter does not generate comfortable institutional conclusions. The approaches most reliably associated with stewardship leadership development—transformative learning, transdisciplinary education, service learning, sustained outdoor experience—all require more of institutions than adding a module to an existing curriculum. They require redesigning the learning environment: building community partnerships that can sustain authentic service learning over years rather than semesters; establishing outdoor learning as a regular curricular practice rather than an occasional enrichment activity; creating assessment structures that can evaluate systems thinking, collaborative capacity, and normative reasoning rather than just content recall; and developing faculty who are equipped and supported to facilitate transformative learning rather than to deliver information.

HEIs should, as a concrete institutional priority, develop explicit sustainability leadership competency frameworks and use them to design, sequence, and assess sustainability-related programmes. Ghasemy et al. (2024) demonstrate that this structural commitment produces significantly better stewardship leadership outcomes than treating sustainability as a value rather than an educational objective. The specific competency domains identified by Redman and Wiek (2021)—systems thinking, anticipatory capacity, normative reasoning, strategic competency, interpersonal facilitation—provide a well-validated starting point for such frameworks, though adaptation to institutional context and cultural specificity is necessary.

For schools, the most actionable priority is treating outdoor and nature-based learning as a structural element of the curriculum rather than a discretionary supplement. Mann et al. (2022) and Powell et al. (2023) collectively provide sufficient evidence to support policy mandates for regular, coherently designed outdoor learning provision at primary and secondary levels. The pedagogical architecture matters as much as the outdoor time itself; investment in teacher preparation for outdoor learning design is as important as investment in outdoor provision.

8.2 Policy Architecture for Stewardship Leadership Education

At policy level, the most consequential change available to national governments and international bodies is the reframing of ESD from content requirement to pedagogical mandate. Environmental education policy is, in most jurisdictions, primarily about content—specifying what students should know about ecological systems and environmental issues. The evidence reviewed in this chapter consistently suggests that this framing is insufficient. What matters for stewardship leadership development is not what is taught but how: whether the learning environment creates opportunities for authentic experience, critical reflection, ethical reasoning, and collaborative action. Policy that specifies only content leaves the pedagogical model—the critical variable—unaddressed (Onyeaka & Akinsemolu, 2024; Imran et al., 2024; Dastgir et al., 2024).

Teacher preparation deserves sustained policy attention. The gap between teachers' recognition of environmental education's importance and their practical capacity to deliver transformative sustainability learning—documented consistently across very different national contexts—reflects a structural underinvestment in educator preparation that content-focused curriculum reform cannot address. Pre-service teacher education programmes need to develop not merely environmental content knowledge but the facilitative, reflective, and community-partnership competencies that transformative ESD requires. This is a more demanding and more expensive form of teacher preparation than current norms in most jurisdictions; the evidence suggests it is nonetheless a more productive educational investment (Persson et al., 2023; Imran et al., 2024).

Funding architectures within HEIs need to be revised to support transdisciplinary, community-engaged approaches to sustainability research and teaching. Barth et al. (2023) document how existing academic structures—departmental organisation, research assessment metrics that prioritise disciplinary publication, short-term project funding cycles—systematically disadvantage the collaborative, problem-oriented, practice-embedded approaches that the sustainability leadership evidence base consistently identifies as most effective. Policy change at the level of research funding criteria, academic reward systems, and institutional governance is necessary to make these approaches financially and professionally viable at scale.

8.3 Research Priorities

Several important gaps in the evidence base warrant emphasis. Longitudinal studies tracking stewardship leadership development across educational and professional careers are rare; most existing research captures learning outcomes at a single time point, making it impossible to determine whether educational gains in stewardship orientation are sustained across the life course or whether they depend on ongoing environmental community membership to be maintained. This is a significant limitation for investment decisions, since the case for expensive transformative educational interventions rests partly on the assumption that their effects are durable.

Cross-cultural comparative research on stewardship leadership education remains limited in geographical scope. The strongest evidence comes from North American, European, and Australasian contexts; the generalisability of these findings to South and Southeast Asian, sub-Saharan African, and Latin American contexts—where biodiversity stakes are often highest and educational development challenges most acute—has not been established. The studies of Iriarte Flores et al. (2023), Dastgir et al. (2024), and Onyeaka and Akinsemolu (2024) begin to address this gap; more is needed.

The digital and mobile learning evidence base, while promising, requires more rigorous investigation before it can support confident policy prescription. Experimental or quasi-experimental designs, longitudinal follow-up of participants, and replication across diverse contextual conditions would substantially strengthen the evidence in this domain. The growing integration of digital tools into environmental education makes this a pressing research priority, since educators and policymakers need evidence-based guidance on how to leverage digital technologies to support—rather than substitute for—the embodied ecological engagement that appears to be foundational to stewardship development.

9. CONCLUSION

There is something both hopeful and demanding in the conclusion to which this review points. Hopeful, because it insists that the characteristics of stewardship leadership—ecological literacy, values-driven commitment, civic capacity, adaptive competence—are not distributed by luck or temperament but are cultivatable through intentional educational experience. The research base reviewed here, for all its limitations, is consistent enough to say that: schools that provide regular, well-designed outdoor learning; universities that embed transformative pedagogy in sustainability programmes; institutions that pursue genuine community partnerships for service learning; curricula that integrate indigenous knowledge as epistemological resource rather than decorative supplement; assessment frameworks that hold students accountable for systems thinking and normative reasoning—each of these contributes, in measurable and reproducible ways, to the development of people who lead environments toward health rather than away from it.

Demanding, because none of this is easy or cheap. Transformative learning cannot be produced by a revised lecture series. Service learning cannot be generated by adding a community volunteering requirement. Indigenous knowledge integration cannot be accomplished by inviting an elder to speak once a semester. Outdoor education cannot be reduced to an annual field trip. Each of these approaches, done well, requires sustained institutional commitment, significant faculty development, genuine community partnership, and the willingness to evaluate educational success by criteria—identity development, leadership orientation, civic stewardship—that conventional academic assessment machinery is not designed to measure.

The evidence reviewed here does not underestimate these difficulties. But it does insist that they are not reasons for inaction. The ecological challenges that future stewardship leaders will face—climate disruption, biodiversity collapse, land degradation, freshwater scarcity—will not accommodate the timescale of gradual institutional reform. Educational institutions, policymakers, and the scholarly communities that advise them have both the evidence and the professional obligation to move with more urgency than has characterised most sustainability education reform to date. The question is not whether we know enough to act differently. We do. The question is whether the will to translate that knowledge into institutional practice can be cultivated at the speed the situation requires.

REFERENCES

Agrawal, A., Tiwari, V., Banerjee, A. R. S., Jhariya, M. K., Hashmi, I., Garg, N., Soni, R., & Gupta, D. (2026). Factors Influencing Student Withdrawal and Predictive Approaches for Retention in Higher Education. In Sam, R., Serey, M., & Em, S. (Eds.), *Student Belonging and Well-Being in Higher Education Retention* (pp. 1–38). IGI Global Scientific Publishing., DOI: 10.4018/979-8-2600-2168-2.ch001

Al-khamaiseh, R. A. M., Bailey, R. P., & Jarvis, A. (2025). Definitions of sustainable leadership in education: A systematic review and analysis. *Journal of Research on Leadership Education*, *20*(3), 280–297. Advance online publication. DOI: 10.1177/19427751241283029

Althnayan, S., Alarifi, A., Bajaba, S., & Alsabban, A. (2022). Linking environmental transformational leadership, environmental organizational citizenship behaviour, and organizational sustainability performance. *Sustainability*, *14*(14), 8779. DOI: 10.3390/su14148779

Aslam, S., Parveen, K., Alghamdi, A. A., Abbas, S., Shah, A. H., & Elumalai, K. V. (2022). Hopes for the future: Capturing the perspectives of students on higher education and sustainable development in the post-pandemic era. *Sustainability (Basel)*, *14*(19), 12531. DOI: 10.3390/su141912531

Barth, M., Jiménez-Aceituno, A., Lam, D. P., Bürgener, L., & Lang, D. J. (2023). Transdisciplinary learning as a key leverage for sustainability transformations. *Current Opinion in Environmental Sustainability*, *64*, 101361. DOI: 10.1016/j.cosust.2023.101361

Boeske, J. (2023). Leadership towards sustainability: A review of sustainable, sustainability, and environmental leadership. *Sustainability (Basel)*, *15*(16), 12626. DOI: 10.3390/su151612626

Brooks, R. E., & Brooks, A. K. (2024). An emerging transformative learning journey to foster sustainability leadership in professional development programs. *Journal of Transformative Education*, *23*(3), 1–20. DOI: 10.1177/15413446241255909

Bryant, J., Ayers, J., & Missimer, M. (2023). What transforms? – Transformative learning in a sustainability leadership master's program. *International Journal of Sustainability in Higher Education*, *24*(9), 231–251. DOI: 10.1108/IJSHE-03-2022-0086

Buendia-Martinez, I., & Carrasco, I. (2022). Integration of indigenous and local knowledge into sustainability education: A systematic literature review. *Environmental Education Research*, *28*(8), 1209–1236. DOI: 10.1080/13504622.2022.2083081

Chaudhry, M. J., & Noureen, S. (2023). Exploring sustainable leadership. *International Journal of Multidisciplinary Research and Analysis*, *6*(10), 1–13. DOI: 10.47191/ijmra/v6-i10-13

Dastgir, G., Zaman, F. U., & Qayyum, A. (2024). Empowering environmental stewardship: Pedagogical approaches for integrating green and clean land practices in educational institutions of Pakistan. *Pakistan Social Sciences Review*, *8*(1), 478–487. DOI: 10.35484/pssr.2024(8-I)42

Dearborn, H., & Tansey, R. (2024). 21st-century stewardship: Infusing environmental stewardship education with global citizenship. *Environmental Education Research*, *31*(2), 364–389. DOI: 10.1080/13504622.2024.2335614

Gajparia, J., Strachan, G., & Leverton, K. (2022). Transformation through learning: Education about, for, and as sustainability. *Frontiers in Sustainability*, *3*, 982718. DOI: 10.3389/frsus.2022.982718

Ghasemy, M., Elwood, J. A., & Scott, G. (2024). Approaches to education for sustainability leadership development in higher education: An international comparative study in the Asia Pacific region. *International Journal of Sustainability in Higher Education*, *25*(8), 1870–1894. DOI: 10.1108/IJSHE-11-2023-0558

Gupta, D., Ashirova, A., Ilhom, J., Sanzhar, D., Giyosiddinov, M., Urozboev, K., & Nodira, E. (2026). Digital Leadership and Educational Transformation in ASEAN: Competencies, Contexts, and Strategic Pathways. In Em, S. (Ed.), *Digital Leadership and Innovation in ASEAN Education* (pp. 97–124). IGI Global Scientific Publishing., DOI: 10.4018/979-8-2600-1142-3.ch004

Imran, M., Almusharraf, N., & Abdellatif, M. S. (2024). Shaping a sustainable future: A study of teachers' perspectives on environmental education awareness. *Cogent Education*, *11*(1), 2422223. DOI: 10.1080/2331186X.2024.2422223

Iriarte Flores, A., Vergara Romero, A., Forero-Medina, G., & Biggs, D. (2023). Exploring environmental stewardship among youth from a high-biodiverse region in Colombia. *The Journal of Environmental Education*, *55*(1), 1–18. DOI: 10.1080/00958964.2023.2238649

Jessen, T. D., Ban, N. C., Claxton, N. X., & Darimont, C. T. (2022). Contributions of Indigenous knowledge to ecological and evolutionary understanding. *Frontiers in Ecology and the Environment*, *20*(2), 93–101. DOI: 10.1002/fee.2435

Mann, J., Gray, T., Truong, S., Brymer, E., Passy, R., Ho, S., Sahlberg, P., Ward, K., Bentsen, P., Curry, C., & Cowper, R. (2022). Getting out of the classroom and into nature: A systematic review of nature-specific outdoor learning on school children's learning and development. *Frontiers in Public Health, 10*, 877058. DOI: 10.3389/fpubh.2022.877058 PMID: 35651851

McLeod, L. J., Kitson, J. C., Dorner, Z., Tassell-Matamua, N. A., Stahlmann-Brown, P., Milfont, T. L., & Hine, D. W. (2024). Environmental stewardship: A systematic scoping review. *PLoS One*, *19*(5), e0284255. DOI: 10.1371/journal.pone.0284255 PMID: 38713707

Narboshova, M., Gafurova, S., Abdukadirova, M., Alimardonov, Z., Nishanov, S., Musinova, N., & Muxtarova, L. (2026). Mobile experiential learning for environmental pollution science with linking authentic field data to pro-environmental behaviour in secondary schools. *Natural and Engineering Sciences*, *11*(1), 271–282. DOI: 10.28978/nesciences.261020

O'Grady, M. (2023). Transformative education for sustainable development: A faculty perspective. *Environment, Development and Sustainability*, *27*(9), 20745–20761. DOI: 10.1007/s10668-023-03609-y

Onyeaka, H., & Akinsemolu, A. A. (2024). Advancing green education in MENA region: Challenges, opportunities and best practices. *Sustainable Development (Bradford)*, *33*(1), 1–15. DOI: 10.1002/sd.3182

Persson, C., Einarson, D., & Melén, M. (2023). Educating the educators to be a driving force in higher education towards sustainable development. *International Journal of Sustainability in Higher Education*, *24*(1), 197–212. DOI: 10.1108/IJSHE-10-2022-0332

Powell, R. B., Stern, M. J., & Fresley, B. T. (2023). Which approaches are associated with better outcomes? Evidence from a national study of environmental education field trip programs for adolescent youth in the United States. *Environmental Education Research*, *29*(3), 331–356. DOI: 10.1080/13504622.2022.2145270

Redman, A., & Wiek, A. (2021). Competencies for advancing transformations towards sustainability. *Frontiers in Education*, *6*, 785163. DOI: 10.3389/feduc.2021.785163

Rodríguez-Zurita, D., Jaya-Montalvo, M., Moreira-Arboleda, J., Raya-Diez, E., & Carrión-Mero, P. (2024). Sustainable development through service learning and community engagement in higher education: A systematic literature review. International Journal of Sustainability in Higher Education, ahead-of-print. DOI: 10.1108/IJSHE-10-2023-0461

Singer-Brodowski, M., Förster, R., Eschenbacher, S., Biberhofer, P., & Getzin, S. (2022). Facing crises of unsustainability: Creating and holding safe enough spaces for transformative learning in higher education for sustainable development. *Frontiers in Education*, *7*, 787490. DOI: 10.3389/feduc.2022.787490

Urolov, B., Abdunazarova, Z., Saidov, J., Nasirova, N., Salomov, G., Masharipov, A., & Shayeva, R. (2026). Embedding conservation genetics and invasion biology into undergraduate curricula with competency based framework for biodiversity literacy. *Natural and Engineering Sciences*, *11*(1), 249–259. DOI: 10.28978/nesciences.261018

VanWynsberghe, R. (2022). Education for sustainability, transformational learning time and the individual ↔ collective dialectic. *Frontiers in Education*, *7*, 838388. DOI: 10.3389/feduc.2022.838388

Vasconcelos, C., & Orion, N. (2021). Earth science education as a key component of education for sustainability. *Sustainability (Basel)*, *13*(3), 1316. DOI: 10.3390/su13031316

Xamrakulova, X., Mirzaeva, Z., Shabbazova, D., Sabirov, A., Rasulova, Z., Ismoilov, B., & Mukhriddin, A. (2026). Project based marine biodiversity education to improve ocean literacy and conservation attitudes among secondary students. *Natural and Engineering Sciences*, *11*(1), 211–220. DOI: 10.28978/nesciences.261015

KEY TERMS AND DEFINITIONS

Education for Sustainable Development (ESD): An approach to learning that equips individuals with knowledge, skills, values, and attitudes needed to build a sustainable future. It integrates environmental, social, and economic issues, encouraging critical thinking, responsible behavior, and informed decision-making in everyday life.

Environmental Stewardship: Responsible use and protection of the natural environment through conservation, sustainable practices, and ethical decision-making. It emphasizes caring for ecosystems, reducing harm, and ensuring resources remain available for future generations while balancing human needs with ecological health.

Outdoor Education: A form of experiential learning conducted in natural environments, focusing on personal development, environmental awareness, and practical skills. It fosters connection with nature, teamwork, problem-solving, and appreciation for ecological systems through activities like hiking, camping, and field studies.

Sustainability Leadership: The ability to guide individuals, organizations, or communities toward sustainable practices by promoting environmental responsibility, social equity, and economic viability. It involves vision, innovation, and influencing change to address global challenges like climate change and resource depletion.

Transformative Learning: A deep learning process that changes an individual's perspectives, beliefs, and assumptions through critical reflection and experience. It leads to more inclusive, open, and adaptive ways of thinking, often inspiring personal growth and meaningful action toward social or environmental change.

Chapter 14

The Role of Education in Sustainability Through Project-Based Learning

J. Ho Siew Ching
http://orcid.org/0000-0001-6756-9937
Politeknik Tuanku Sultanah Bahiyah, Malaysia

Krishnavehni A. P. Gopal
http://orcid.org/0000-0001-7403-3524
Kolej Komuniti Sungai Petani, Malaysia

ABSTRACT

In traditional learning environments, there is less interaction between learners and the surrounding environment. Considering employers' needs and graduates' adaptive profile that meets the labor market, transformation of the learning environment should be done. The shift in pedagogy from the traditional learning approach to project-based learning (PrBL) is more suitable for new human capital demands of industries and workplaces, as well as sustainable development specialists. By participation in PrBL, both educator and student can gain information, skills, and attitudes for a sustainable environment in the community; beside developing other soft skills such as critical thinking, problem-solving skills, teamwork, adaptability, promoting opportunities to practice and solve real contextual problems. The aim of this chapter is to provide the leader with a better understanding on PrBL, integration of PrBL on sustainability and its impacts.

DOI: 10.4018/979-8-3373-7877-0.ch014

INTRODUCTION

Education is increasingly recognized as a transformative force in advancing sustainable development. UNESCO (2025) has highlighted the role of education in achieving the Agenda for Sustainable Development Goals (SDGs) and its 17 SDGs in cultivating social and environmental responsibility as global awareness. SDGs address ecological issues which are essential to be exposed from an early age, especially through educational institutions as transformative agents in making SDGs principles a success practice (UNESCO, 2025). As global industries undergo rapid technological, environmental, and economic transitions, the demand for graduates equipped with sustainability competencies, adaptive thinking, and interdisciplinary problem-solving skills has intensified. Traditional teacher-centered learning approaches, which emphasize content delivery, memorization, and standardized assessment, often limit learners' engagement with authentic sustainability challenges. Such approaches may develop theoretical knowledge but frequently fail to cultivate the applied competencies, ethical awareness, and collaborative skills required in contemporary green economies.

UNESCO's 2030 Education for Sustainable Development (ESD) framework calls for a comprehensive education across three critical dimensions of (i) what is learned as curriculum content, (ii) the way it is learned as pedagogical approaches, and (iii) the place of learning as learning environments, which empower learners at any age to become active agents in undertake sustainability challenges confronting our world (UNESCO, 2024). As expressed by UNESCO (2024), ESD functions as a strategic educational approach that systematically develops individuals' capacity for critical thinking, systems analysis, values clarification, and sophisticated problem-solving competencies that collectively enable meaningful contributions to sustainable development pathways. Advanced by UNESCO, ESD is an educational approach and guidelines that equips learners with the knowledge, skills, values, and competencies needed to help realize those goals in practice (Sari et al., 2025; Choi, 2025), and positively link with environmental responsiveness in achieving SDGs by year of 2030 (Sari et al., 2025). In this theoretical and practical integration, ESD establishes a critical nexus between individual agency and collective sustainability imperatives, positioning education as an indispensable mechanism through which societies can cultivate the intellectual acuity, moral foundation, and transformative capacity requisite for addressing the multifaceted sustainability challenges confronting contemporary civilization (Choi, 2025). By incorporating case studies and real-world contexts, this approach shifts the classroom from a teacher-centered model to a student-centered environment that empowers learners to apply knowledge critically and effectively, emphasizing the capacity to act as well as to understand (AlAfnan & Dishari, 2024).

ESD has emerged as a transformative educational paradigm designed to address complex sustainability challenges through systemic and institution-wide reform, and promotes environmental, social, and economic considerations (AlAfnan & Dishari, 2024; Ukamaka, 2024). ESD is crucial in enabling people to alter their perspectives on the world, interact with it differently, and actively contribute to society's transformation. Three principal strategies operationalize ESD in practice, (1) Cross-curricular integration weaves sustainability themes across subject boundaries, linking environmental concerns with social justice, cultural contexts, and global issues while enabling discipline-specific adaptation; (2) Project-based learning provides experiential and inquiry-driven opportunities for students to engage directly and live practice with sustainability problems such as food security, waste management, water conservation, and climate adaptation; (3) Community partnerships extend learning beyond institutional boundaries, fostering collaboration with local stakeholders to research and implement context-responsive solutions. Such partnerships strengthen civic engagement, enrich academic learning, and sustainability education within real-world application (Ukamaka, 2024). Thus, understanding environmental phenomena such as climate change, the greenhouse effect, and water pollution is not merely scientific literacy but a foundation for sustainability awareness. By fostering critical awareness of these global challenges, ESD stimulates sustained intellectual engagement across multi disciplines and educational levels (Ukamaka, 2024).

To map the role of PrBL in sustainability education, this chapter draws on an integrative narrative review that brings together recent peer-reviewed scholarship, international policy frameworks, and illustrative case studies. The literature was drawn primarily from 2020 to 2026 and selected where it spoke directly to the intersection of PrBL and ESD, as well as competency development across primary, secondary, and higher education contexts. The intent is not to test a single hypothesis but to weave a fragmented and fast-moving body of evidence into a clear conceptual account, and identifies gaps that future empirical work will need to address.

BACKGROUND

Learning theories change across time. The implementation of traditional learning activities started with the educator according to learning goals or objectives; while the learning process was carried out in the class with effective communication based on text books. Behaviourism Learning Theory by Skinner, Cognitivism Learning Theory by Piaget, and Constructivism Learning Theory by Vygotsky are categories as traditional learning theories; while Connectivism Learning Theory by Siemens and Downes is recognized as theory of digital age (Sahin, 2024).

Further explanation by Sahin (2024), behaviourism is based on objective where human behaviour is observed; cognitivists focused on mental processes; while constructivism emphasized the social and cultural factors in learning activities (Table 1). Digitalization has changed education and become more effective with the help of the Internet, latest technology devices, and services; where texts, images, audio, and video material are applied in learning and teaching processes. As a contemporary theoretical framework, connectivism emerges in response to the limitations of traditional learning theories amidst rapid technological change, emphasizing the location and accessibility of functional knowledge rather than solely its content or methods of acquisition (Sahin, 2024).

Table 1. Summary of learning theories

Learning Theory	Details
Behaviorism (B. F. Skinner, 1904-1990)	Focuses on observable behaviours shaped by reinforcement and punishment (stimulus-response behaviour)
Cognitivism (Jean Piaget, 1986-1980)	Emphasizes internal mental processes, memory, and how information is structured (knowledge constructs behaviour)
Constructivism (Lev Vygotsky, 1896-1934)	Views learning as actively building knowledge through culture and social interaction (socialization).
Connectivism (G. Siemens & S. Downes, 2009)	Explains learning in the digital age, focusing on making connections across networks (diversity of network / experience learning).

Within the constructivism framework, teachers act as facilitators who provide flexible guidance, encourage dialogue, collaboration, and problem-based inquiry, and integrate appropriate resources and media to support meaningful engagement grounded in students' prior knowledge and experiences. At the same time, students are motivated to analyse, predict, summarize, question, experiment, and conduct independent inquiry through interactive, hands-on learning experiences that actively engage them in constructing knowledge, thereby shifting the classroom from a teacher-centered to a student-centered environment that prioritizes the capacity to apply understanding and fosters the development of autonomous, confident learners. Research indicates that such approaches are effective even in large classes, enhancing student participation, collaboration, independent learning, and overall positive academic behavior (AlAfnan & Dishari, 2024).

Meanwhile, in sustainability education, constructivism principle emphasizes active knowledge formation through experience and social interaction, encouraging learners to relate new insights to prior understanding while engaging with external stakeholders such as companies, communities, and end-users in applied projects. The constructivism principle asserts that learners actively build knowledge by connecting new information to prior understanding through experience and social interaction, fostering the development of metacognitive capacities that support complex sustainability concerns (Sundman et al., 2025). Constructivism grounds PrBL's active, social nature, while connectivism reframes it for the digital age, where knowledge is distributed across networks, and navigating that information becomes a core competency particularly vital to sustainability learning (Sahin, 2024). Implementation of PrBL allows learners to situate a local project for instance, better prepare students for future challenge, transforming an isolated classroom activity into participation in a wider community environment practice, beside proven PrBL more effective than traditional teacher-centered pedagogy in improving students' understanding and commitment to SDGs especially (Espino-Diaz, et al., 2025).

As the concepts are frequently mixed, it is worth establishing at the outset how they are nested within one another. At the broadest level sits constructivism, the philosophy that knowledge is actively built by the learner through experience and social interaction rather than passively received. Within this philosophy lies experiential learning, which operationalised most influentially through Kolb's four-stage cycle that comprised structured process of acting, reflecting, conceptualising, and experimenting. PrBL in turn, is a specific method within the experiential family project-driven mode of instruction that enacts the experiential cycle around an authentic problem. Constructivism thus explains *why* active learning works, experiential learning describes *how* it is structured, and PrBL specifies *what* learners actually do.

THE ROLE OF EDUCATION IN SUSTAINABILITY

What Is an Experiential Learning Approach?

In line with the passage of time, the terms of constructivist learning theory were replaced with the word "experiential learning", which have emerged across diverse disciplines in higher education. According to Kolb & Kolb (2017), experiential learning is learning through experience, where knowledge is created through hands-on experience and reflection. This learning approach has been widely used as a guide for more than 30 fields in academic disciplines from the U.S., Canada, Brazil, China, India, Australia, Japan, Thailand, Netherland, Sweden, Norway, U.K., and Finland. It has been proven to strengthen and deepen student engagement especially in higher education through experiential activities which are shown in Table 2, and the key elements of experiential learning are:

- Concrete Experience: Actively participating in an experience.
- Reflective Observation: Reviewing or reflecting on the experience.
- Abstract Conceptualization: Drawing conclusions based on reflection.
- Active Experimentation: Planning and trying out what have been learned.

Table 2. Examples of experiential learning and outcome

Examples	Outcome
Internships / Field Work	Working directly in a professional environment able to apply and solve real-world problems.
Simulations / Gaming	Virtual simulations are able to practice decision-making in a safe and controlled environment.
Laboratory Experiments	Conducting experiments able to test hypotheses and solve real-world problems.
Case study	Analyzing real-life scenarios is able to develop critical thinking and problem-solving skills.

Experiential learning acknowledged benefits such as improving student engagement and motivation, increased learning effectiveness, and long-term knowledge retention, strengthening of both technical and interpersonal competencies which are necessary for professional contexts (Sundman et al., 2025; Kolb & Kolb, 2017). Case studies from India, Pakistan, and Kenya reveal that ESD projects strengthen problem-solving abilities, critical thinking, teamwork, and community engagement, adaptability of individuals to respond effectively to complex challenges, proving the capacity of such pedagogy to cultivate active citizenship and a sustainability-oriented disposition (Choi, 2025). Although experiential approaches demand educators confident in their pedagogy, yet many reports show inadequately of sustainability principles in their teaching; bridging this gap calls for targeted professional development centred on systems thinking, collaborative learning, and action-oriented pedagogy (Choi, 2025). Thus, sustainable development can be taught and implemented through experiential-based learning as a non-conventional method in education which engages active participation (Leal Filho, 2021).

A study done by Astuti et al. (2026) undertakes a comparative analysis of nature-based schools and conventional schools, focusing on learning approaches, classroom climate, character and soft-skill formation, student motivation, and academic performance. The findings reinforced through triangulation of data sources and methods, reveal a clear philosophical and pedagogical distinction between the two models. Nature-based schools foreground experiential-based learning which focuses on student-centered learning, utilizing natural environments as active learning spaces that cultivate curiosity, creativity, emotional engagement, intrinsic motivation, and improvement of soft-skills. By contrast, conventional schools operate within a structured, teacher-directed framework closely aligned with national curriculum standards which place more emphasis in systematic knowledge transmission and academic performance outcomes. Overall, the study demonstrates that nature-based schools as experiential-based learning, while conventional schools as traditional learning (Table 3). Although conventional schools demonstrate stronger academic achievement compared to nature-based schools, the needs of industry or working environments nowadays require workers with higher creativity and soft-skills.

Table 3. Experiential-based learning vs traditional learning

Aspects	Experiential-based learning	Traditional learning
Terms	Nature-based schools	Conventional schools
Space / Venue	Open, close to nature	Close, classroom-based
Learning activities	Exploration, projects, hands-on	Lectures, curriculum attainment
Learning methods	Experiential, outdoor, projects	Lectures, assignments, textbook
Soft-skills	Leadership, collaboration, creativity	Discipline, collaboration, responsibility
Character building	Integrated into daily learning activities	Separate from core learning
Opportunities	Real-life situations in nature	Dependent on non-academic activities
Academic strengths	Contextual concepts, field practices	Memorizations, strong in theoretical
Evaluation model	Project and observation assessments	Written tests, quizzes, examinations

What Is Project-based learning (PrBL)?

Project-based learning (PrBL) is a student-centered environment that emphasizes a dynamic classroom approach, where students gain deeper knowledge by the involvement in the structured project that is related to real world difficulties or challenges (Al-Bahadli et al., 2023). This learning approach enables students to apply knowledge to address complex problems or produce tangible outcomes, participate in inquiry-driven investigations, enabling them to construct understanding by asking questions as autonomous learning, exchanging ideas, developing plans, decision-making, collaborating and communicate with others to pursue solutions (Al-Bahadli et al., 2023; Dacumos, 2023; Lozano et al., 2022).

As the climate emergency demands urgent action, educational institutions are increasingly recognizing their vital responsibility to cultivate environmental stewardship and sustainability leadership among students. In response to this critical need, PrBL has emerged as a powerful pedagogical approach that transforms schools into spaces where students engage meaningfully with sustainability challenges (Lozano et al., 2022). A conceptual framework for sustainability-focused PrBL includes input, process, output, and outcome components. A conceptual framework for sustainability-focused PrBL includes input, process, output, and outcome components. Inputs include project design, instructional strategies, and sustainability topics. The learning process involves experiential learning, collaboration, and reflection. Outputs include knowledge acquisition and competency development, while outcomes include workforce readiness and sustainability leadership. By integrating knowledge acquisition with authentic engagement in the communities, involvement such as green space stewardship and sustainable lifestyle initiatives offers a holistic model that prepares students not merely with facts, but with the mindsets and capabilities required to address the pressing environmental challenges in ecological responsibility and social cooperation (Lozano et al., 2022).

A successful implementation of PrBL requires careful planning and structured instructional design. Educators must first identify relevant sustainability challenges aligned with curriculum objectives. Clear learning objectives and expected outcomes should be established to guide student learning and assessment. During the project implementation phase, students engage in research, data collection, analysis, and solution development. Educators serve as facilitators, providing guidance, feedback, and support while encouraging student autonomy. This learner-centered approach promotes critical thinking and independent learning. Effective assessment strategies include formative and summative assessment methods, such as reflective journals, project presentations, peer evaluations, and portfolio assessment. These assessment methods provide comprehensive evaluation of student learning and competency development.

PrBL is strongly grounded in constructivist learning theory, which emphasizes that knowledge is actively constructed by learners rather than passively received from instructors. According to constructivist perspectives, meaningful learning occurs when learners engage in authentic tasks, reflect on their experiences, and connect new knowledge to prior understanding. This approach is particularly relevant to sustainability education, as sustainability challenges are complex, interdisciplinary, and require learners to synthesize knowledge from multiple domains. By engaging in sustainability-focused projects, students actively participate in knowledge construction, enabling deeper cognitive processing and long-term retention of sustainability concepts.

PrBL and experiential learning approaches are the same terms that address learning processes that are suitable for the engagement of complex sustainability problems, managing cognitive demands while fostering deeper understanding, and problem-solving skills, alongside affective and socio-emotional development (Sundman et al., 2025). Experiential learning is a broad philosophy that covers various methods, such as simulations, internships, or field research; whereas PrBL is a pedagogical method using projects to achieve specific academic goals over time. At the same time, experiential and PrBL often described as "learning by doing", demonstrated value in engaging students directly with authentic sustainability problems, strengthening critical thinking, collaboration, and practical competence (Alhassani et al., 2025; Avsec, S., & Jagiełło-Kowalczyk, 2021).

In short, all PrBL is considered as experiential learning, but not all experiential learning is necessarily PrBL. Constructivist approaches are learning theories asserting that learners actively construct their own understanding by building on prior knowledge rather than passively absorbing information. PrBL is a specific, hands-on instructional method derived from constructivist theory, where students create projects to solve problems. Through hands-on sustainability projects, students develop an understanding of ecological systems, resource conservation, and environmental protection strategies. These experiences foster environmental stewardship and encourage learners to adopt sustainable practices in their personal and professional lives. Some of the important points can be summarized as below: -

- Knowledge Construction: Learners actively create meaning from experience.
- Active Learning: Students engage in dialogue, inquiry, and critical thinking.
- Student-Centered: Lessons start with big concepts and prioritize student questions, rather than following a rigid, linear curriculum.
- Application: Specific method, project, or model.

Role of Education Institution and PrBL

Education institutions are responsible to improve institution quality, increase competitiveness and efficiency for better achievement. In response to United Nations SDGs, higher education institutions have increasingly assumed responsibility for cultivating environmental awareness and preparing future leaders committed to sustainability through structured educational initiatives. Different learning curricula and approaches are found across the world in fulfilling sustainability needs. Variety of hands-on activities such as participating in ecological projects, awareness campaigns, recycling, and gardening are important initiatives of PrBL which indirectly increase sustainability conscious behaviors (Friman et al., 2024). Application of sustainability into education may involve many subjects or curriculum. Curriculum addressed key topics such as ecology, renewable energy, waste management, and energy efficiency, employing interactive pedagogical strategies and collaborative experiential learning between students and elementary school pupils (Friman et al., 2024; Chavula et al., 2024).

Integrating environmental education into both formal and informal learning systems is essential for developing environmental literacy, fostering sustainable behaviours, and supporting the achievement of the United Nations SDGs. Education provides a critical platform for equipping learners with the knowledge, values, and competencies needed to understand environmental challenges and make responsible decisions that contribute to long-term ecological and societal well-being. Sustainability itself encompasses interconnected environmental, social, and economic dimensions, requiring a balanced approach that protects ecosystems and natural resources, promotes social equity and quality of life, and supports responsible economic development. Rather than focusing solely on theoretical knowledge, effective environmental education emphasizes critical thinking, problem-solving, and real-world application, enabling students to understand complex sustainability issues and develop practical solutions (Chavula et al., 2024)

Global collaborations, such as UNESCO's ESD initiative, support knowledge sharing, curriculum development, and capacity building across countries. Few case studies demonstrate that integrating environmental education at multiple educational levels enhances environmental awareness, encourages community engagement, and empowers learners to become responsible global citizens capable of contributing to a more sustainable and resilient future (Chavula et al., 2024).

- Case Study 1: Kenya – Integrating Environmental Education Primary Schools
- Case Study 2: South Africa – Green Schools Program
- Case Study 3: Uganda – Climate Change Education in Secondary Schools
- Case Study 4: Ghana – Community Based Environmental Education

Taken together, these four initiatives show how environmental education can take root at different stages of schooling and through very different methods across the African continent. In Kenya, it has been built into primary education, where young children meet ecological ideas not through textbooks but through their hands. For example, tending a school garden, planting trees, sorting waste activities which had become something they have actually touched. South Africa's Green Schools programme works at the level of the whole institution, with students audit how much water and electricity their own school uses, form eco-committees to act on what they find, and connect classroom lessons to real reductions in waste. Meanwhile, Uganda's climate-change initiatives focus on the secondary years, giving teenagers both the scientific understanding to grasp what drives climate change and the confidence to design responses for their own communities, such as planning a local tree-planting scheme to slow soil erosion. Ghana's community-based model carries learning beyond the school gate altogether, drawing in families, traditional leaders, and local associations as fellow teachers in conservation and sanitation work (Chavula et al., 2024).

Read side by side, the four cases reveal a shared logic running beneath their differences. Each moves the learner from simply knowing about the environment toward actively caring for it, and each succeeds to the degree that it is rooted in real local concerns rather than delivered as content disconnected from daily life. The Kenyan and Ugandan cases show why matching the learning to the learner's age matters, where hands-on activity-based learning suits the primary years, while inquiry and adaptation planning fit the secondary stage, when students are ready to grapple with causes and solutions. The South African and Ghanaian cases, by contrast, show how lasting structures as an eco-committee, a network of community elders, keep engagement alive long after an initial burst of enthusiasm fades. Together, they point to a clear conclusion that in the Global South, the success of environmental education depends far less on sophisticated technology than on the strength of the partnerships, leadership, and curricular integration that hold it together (Chavula et al., 2024).

Meanwhile, a study has been done by Van Den Beemt et al. (2023) in response to evolving economic expectations, to find pedagogical innovations in higher education that can enhance learning outcomes across four different regional countries of Mexico, Netherlands, Ireland, and China. Findings show that Dublin, Mexico, and China have served as a reference point for a broader innovative teaching and learning approach, while in the Netherlands, embedded curriculum practices and faculty-led challenge-based learning experiments have influenced student learning behaviors and outcomes. Collectively, these efforts highlight the establishment of innovation spaces as collaborative hubs that connect academia with industry, where stakeholders actively mentor students, support decision-making processes, and provide essential resources.

Embedding sustainable development principles and frameworks into Science, Technology, Engineering and Mathematics (STEM) teaching through the implementation of PrBL had been done for a group of undergraduates in Romania. According to Bărbulescu et al. (2025), PrBL modules were designed addressing authentic challenges including renewable energy development, waste management systems, and climate adaptation strategies for situating learning within real-world contexts. Pre- and post-intervention assessments indicate measurable gains in sustainability knowledge, critical and creative problem-solving capacities, and student engagement, alongside a notable shift in career aspirations toward environmentally responsible professions. The findings affirm that PrBL serves as an effective pedagogical method for sustainability education in STEM (Bărbulescu et al., 2025; Dacumos, 2023). Meanwhile, PrBL is also found as one of the good approaches in teaching English language. According to study done by Al-Bahadli et al. (2023), the experimental group that participated in a two-month weekly lesson using the PrBL method underlined English language skill improvement.

Integration of PrBL on Sustainability

PrBL plays a critical role in developing environmental competencies by increasing students' awareness of environmental challenges and promoting responsible environmental behavior. Sustainability challenges require collaboration among diverse stakeholders, including communities, industries, and policymakers. Through collaborative project work, students develop communication skills, teamwork abilities, and ethical awareness. These competencies enable learners to engage effectively with stakeholders and contribute to collective sustainability efforts. According to Leal Filho et al. (2022), the term sustainability is a social one and not just a technical challenge. Professional capital now demands transdisciplinary competencies, the capability to converse across the limitations of science, community, and policy. Hence, PrBL students often play the role to bridge between academic research and community needs.

In Europe, study done by Ariza and Olatunde-Aiyedun (2023) on PrBL for renewable and sustainable energies. The findings demonstrate that the implemented methodology effectively engaged students by situating learning within real-world challenges in renewable energy, thereby promoting meaningful learning, motivation, peer collaboration, and problem-solving skills. Consistent with the vision of the European Commission, higher education institutions are called to cultivate critical thinkers, problem solvers, and responsible agents of change through engagement with complex societal realities and participatory decision-making processes. Despite the endorsement of SDG since 1990, the implementation practices remain necessary for most higher education institutions in Europe, especially in organizing sustainability project-based activities to enhance active learning initiatives besides fostering dynamic collaboration between students and educators.

In Portugal, PrBL was designed by Filipe et al. (2023) for approximately 300 students from the field of business and marketing, University of Aveiro. PrBL aligned with SDGs through engagement with companies and the integration of technology, with the aim of enhancing learning quality, motivation, and academic achievement. Developed under the title *Higher Education for Good*, the project embedded SDG-related content across four curricular units, combining theoretical exploration, practical in-class activities, group work, examinations, and visual communication outputs such as posters, thereby fostering sustainable co-creation and awareness of the societal implications of marketing, strategy, and competitiveness decisions. Although the SDG framework was introduced in a prototypical manner and requires further consolidation, the initiative achieved its primary objectives on SDGs awareness and knowledge, as well as demonstrated complementary soft skills within higher education curricula. Business students have also engaged in sustainability-focused projects, such as developing sustainable business models and conducting environmental impact assessments. These projects enhance students' understanding of sustainable business practices and prepare them for careers in green industries. Thus, PrBL is proven as pedagogical innovation projects which effectively improve students' marketing strategy and competitiveness decisions in contemporary society.

A study reviews a collection of research within the discipline of Architectural Engineering, found that sustainability competence is no longer peripheral but foundational, given engineers' central role in shaping resilient infrastructure and environmentally responsible (Alhassani et al., 2025). The curricula encourage engineers and architects to be environmentally conscious and ethically responsive to complex global challenges, align with the three pillars of sustainability global agenda of United Nations' 17 SDGs, economic, social, and environmental. Thus, experiential and PrBL as active learning strategies, hands-on assessments method can enhance knowledge retention, teamwork, and applied skills in sustainability education for diverse learners with varied educational backgrounds and learning preferences.

Beside engineering as a field of study, a study was done by Marques (2025) in Portuguese for historical and heritage context by using HISTOESE as PrBL framework. The longitudinal study concluded three main findings: (1) grounded pedagogies enhanced children's historical awareness and heritage literacy; (2) low-cost and upcycling-based strategies broadened local community participation; (3) corporations with museums, tourism offices, and cultural associations produced more cultural initiatives. Positioned within the "Knowledge of the World" curricular domain, the approach encourages students to investigate their natural and social environments through observation, inquiry, and experiential learning. Consistent with UNESCO's agenda on SDGs, the model bridging formal, non-formal, and informal learning contexts, and by connecting education with tourism and community development, PrBL implementation produces tangible cultural outcomes aligned with sustainable development. Effective pedagogy is important by linking educator, community engagement, and cultural tourism sustainability.

PrBL Towards Green and Economy

The term 'Green Economy "is less likely to be linked by any technical literacy but it solely accommodates the ability to navigate complexity. According to Valls-Val and Bovea (2021), PrBL plays an important role in bridging the gap between industry needs and theoretical knowledge by a few roles. The first one is Authentic Assessment; this will function as a factor to move from 'rote memorization' to "doing." This mirrors the modern workplace. The next role is about Cross-Disciplinary Literacy, which caters on sustainability issues such as carbon sequestration and circular economics. These undeniably require the knowledge of both science and policy. The third role is about Value Creation in which students will be able to learn to point out market driven solutions for environmental issues, which will directly make them highly employable in environmental, social, and governance sectors. Hence, it is a known fact that modern labour markets are now undergoing a "Green Transformation" and human capital is no longer defined solely by technical proficiency but also the ability to tackle the transition to a low carbon economy.

The labour market now, 2026 does no longer perceive "sustainability" as a specialized domain but as interdisciplinary competency that is required across sectors from engineering to finance. Valls-Val and Bovea (2021) argue that the traditional learning system produces more "linear thinkers" whereas the modern economy requires 'system thinkers". This is where PrBL addresses the issues by incorporating the Competency-Based Learning (CBL). Whenever the students involved in a project, for example in redesigning a local supply chain to attain "Net Zero," they are indirectly obliged to embed disparate knowledge streams as such

- Technical Literacy: Understanding carbon sequestration or life-cycle assessment (LCA).
- Economic Literacy: Navigating the cost-benefit analysis of sustainable transitions.
- Regulatory Literacy: Understanding the European Green Deal or Asian Carbon Tax frameworks.

As per the report by UNESCO, 'Green Skills Gap' is a significant bottleneck for industries striving to achieve Net Zero. PrBL mitigates this issue by embedding Authentic Task Exposure. Students will indirectly develop a Professional Agency by dealing with real world core issues. This will add up to their confidence to act as change agents within an institution. Compared with peers from conventional, transmission-based programmes, graduates of immersive PrBL initiatives emerge better prepared and more inclined to take on leadership roles in emerging green sectors such as renewable energy and sustainable agriculture. This positions PrBL as a powerful means of educating the leaders that sustainability transitions demand. Meanwhile, the "Green Ambassadors in the Community" program integrated hands-on activities, ecological projects, and systematic monitoring of outcomes to enable participants to observe the tangible effects of their actions on energy use and environmental practices is one of the successful applications of experiential learning for sustainability (Friman et al., 2024).

Economies continue to grow due to technological advancement, globalization, and the rise of innovation-driven industries, universities are expected not only to impart theoretical knowledge but also to cultivate entrepreneurial mindsets and practical competencies that enable students to create opportunities, adapt to change, and contribute meaningfully to economic development. PrBL enables learners to actively apply theoretical concepts to practical situations, thereby strengthening their ability to think critically, solve problems creatively, and collaborate effectively with others. These experiential learning opportunities encourage students to move beyond passive knowledge acquisition and instead participate in meaningful learning processes that mirror real entrepreneurial environments (Chitamba et al., 2025). Importantly, PrBL plays a crucial role in preparing students for workforce transition and entrepreneurial, with the competencies, confidence, and practical experience necessary to become innovative entrepreneurs and proactive contributors to sustainable economic growth career pathways.

Impact of PrBL towards Soft Skills

A study conducted by Akhimien & Adekunle (2023) stated that AI automates routine cognitive tasks, the human capital that remains valuable is that which involves creative synthesis. Sustainability challenges are "wicked problems" with no single answer. PrBL trains students in the iterative process of prototyping and failure. This builds a high level of adaptability (Akhimien & Adekunle, 2023; Leal Filho et al., 2022; Almulla, 2020), a soft skill that is now considered "Green Currency" in the 2026 economy. As environmental data shifts, students must pivot their project goals, building resilience. Students who engage in PrBL develop a "strategic flexibility" that allows them to pivot their career paths toward emerging green sectors with higher efficiency. When a student's certain project fails to produce the expected outcome, they must quickly adapt their design which builds resilience, a much-needed soft skill for the 2020s. Moreover, PrBL motivated the students to foster opportunities by tackling the market gaps. Thus, through instilling an entrepreneurial mindset within a sustainability framework, students will be able to see environmental waste as a resource rather than waste solely and this will undeniably create the Green Startups.

Integrated curriculum alignment for sustainability education enhances sustainability competency of creative problem-solving abilities (Alhassani et al., 2025; Bărbulescu et al., 2025) as students demonstrated an increased experience when facing sustainability challenges in PrBL activities. At the same time, emerging sustainability demands to tackle real-world problems, engage in hands-on activities and project management also enhance problem-solving skills (Sundman et al., 2025). Studies also show that PrBL enhances students' communication and problem-solving abilities in programming education (Ruslan et al., 2024); food industries (Brundiers et al., 2021); sustainability assessments (Leal Filho et al., 2022).

Soft skills are always hard to teach to the students in isolation. However, PrBL had instilled these skills in the students by getting them the tools needed to complete the whole project. Critical thinking is the core ability to deconstruct complex systems in sustainability. Moreover, PrBL creates an opportunity to do so by instilling students with unstructured problems. The initiative to solve the problem demonstrated the value of cultivating critical thinking improving students' theoretical knowledge with real-life business application (Filipe et al., 2023; Tadjer et al., 2022; Kolb & Kolb, 2017), as well as communication skills and decision-making (Alhassani et al., 2025; Pons et al., 2024; Al-Bahadli et al., 2023).

PrBL also embeds collaborative leadership by insisting students to negotiate roles between the team. This is opposite to the traditional learning activities whereby the tasks are frequently split and reassembled. Sustainability PrBL requires trans disciplinary cooperation. Students ought to lead by influence rather than authority especially when the project involves persuading community members to adopt new habits or technologies. Soft skills of leadership, teamwork, collaboration, and empathy are implanted through direct interaction with community outreach, project management, nature and real-life challenges (Astuti et al., 2026; Alhassani et al., 2025; Sundman et al., 2025; Filipe et al., 2023; Akhimien & Adekunle, 2023; Leal Filho et al., 2022).

Students, for their part, practise leadership through the transdisciplinary stakeholder management that authentic sustainability projects demand. Because sustainability challenges cut across science, community, and policy, learners cannot fall back on positional authority. They must lead by influence, in negotiating roles within diverse teams, communicating across professional and cultural boundaries, and reconciling the competing interests of residents, regulators, and industry partners. This is leadership of that often harder than the command-and-control model that students may expect. A team might, for example, redesign a proposal not because the engineering was wrong but because residents distrusted the metering data as an intensely instructive lesson in the politics of implementation. Through such experiences, learners develop the preventive and interpersonal competencies that sustainable leadership requires, rehearsing in a supported setting the stakeholder-management capabilities they will later exercise across the green workforce (Leal Filho et al., 2022; Brundiers et al., 2021).

PrBL transforms students from a passive observer to an active stakeholder. Recent research stated that this transformation fosters a specific suite of "sustainability competencies". The findings of the study indicate that participation of sustainability community programs led to notable positive shifts in attitudes toward sustainability, strengthened commitment to environmentally responsible behavior (Alhassani et al., 2025; Friman et al., 2024), and extended influence beyond individual participants to their wider communities. PrBL approaches have indirectly linked to increased community engagement with local climate-action initiatives, as students' participation outspreads the reach of sustainability efforts beyond the classroom. The students' enthusiasm and evidence-based solutions serve to lower the community's resistance to sustainable transitions.

Sustainability Competencies Developed through PrBL

Five competencies from the Brundiers et al. (2021) framework, each with its term bolded and tied directly to PrBL mechanics:

- **Systems-thinking** — tracing interconnections across domains (water-quality example)
- **Anticipatory** — scenario-building, forecasting, reasoning under uncertainty
- **Strategic** — planning, prototyping, and acting (the action-oriented core of PrBL)
- **Normative** — weighing values, justice, and ethical trade-offs
- **Interpersonal** — collaboration, stakeholder facilitation, leading by influence

Rather than just collecting facts or leaning on vague soft skills, sustainability education is pivoting toward what students can actually do. It is a shift from memorizing data to building the practical, specialized capacity needed to tackle messy environmental and social crises. While soft skills are useful anywhere, sustainability competencies are highly specific tools designed for real-world problem-solving. Looking back at twenty years of research, Brundiers et al. (2021) mapped out five core, interconnected competencies that capture this mindset: systems-thinking, anticipatory, strategic, normative, and interpersonal. Because PrBL throws students headfirst into unpredictable, real-world challenges, it provides the perfect training ground to develop all five of these skills naturally and simultaneously, rather than tackling them in isolation (Brundiers et al., 2021; Sundman et al., 2025).

Systems-thinking is essentially the ability to untangle messy, interconnected problems without stripping away their context. It means looking at how a problem behaves across different fields, scales, and timelines, while watching out for hidden feedback loops, trade-offs, and accidental side effects. As real-world sustainability issues don't stop at the borders of a single academic department, PrBL forces students to face this chaos head-on. Take a project on local water quality: students can't just isolate a single chemical variable in a lab. Instead, they are forced to track how local land use, industrial practices, public health concerns, and political policies all collide. By pushing learners to map out these messy relationships and think about how the entire system behaves as a whole, PrBL naturally builds the kind of big-picture, integrated thinking that these crises demand (Alhassani et al., 2025).

Anticipatory competence is all about the ability to look ahead, and to imagine and weigh different futures through long-term consequences when nothing is certain.

Since sustainability is inherently focused on the future, project work like scenario-building, forecasting, or climate adaptation planning gives students a safe space to practice this exact foresight. When students design a project, they can't just think about the immediate present. They are forced to consider how their solution will hold up over time, map out alternative paths for a community, and weigh risks that haven't even happened yet. This is exactly the kind of forward-thinking and precautionary judgment needed for responsible environmental choices. In PrBL, this mindset gets reinforced every time a project forces students to adapt their plans based on what might happen down the road, rather than just reacting to what is happening right now.

Strategic competence is what bridges the gap between understanding a problem and actually doing something about it. It's the ability to design and carry out real-world interventions, smooth transitions, and management plans that push a system toward sustainability. What makes PrBL so effective here is that it always builds toward action. Students don't just sit back and analyze a crisis; they are expected to plan, build prototypes, and test tangible solutions. Along the way, they have to manage resources, map out timelines, and tweak their approach whenever they run into feedback or failure. This constant loop of planning, testing, and adjusting is exactly how students build a practical, action-oriented mindset. It moves them past purely theoretical knowledge and gives them the tools to turn good intentions into real, measurable change (Leal Filho et al., 2022).

Normative competence is essentially about handling the ethical side of sustainability. It is the ability to spot, debate, and apply core principles like justice, equity, and our responsibility to future generations. Sustainability choices are never just neutral, technical decisions; they force us to weigh competing interests and figure out what should be done, rather than just what is technically possible. PrBL brings these ethical dilemmas to life by anchoring projects in real communities. When students have to balance something like local economic growth against environmental protection, they are forced to confront these hidden tensions head-on. They can't just hide behind data, but they have to clearly state and defend the values driving their choices. This kind of debate helps students realize that sustainability isn't just a technical puzzle to solve, but a deeply social and ethical commitment (Sundman et al., 2025).

Interpersonal competence is all about the human element of sustainability, which included the ability to collaborate, communicate, and lead across completely different groups, fields, and cultures. It is also about facilitating real, community-driven problem-solving. This is probably the most natural skill developed through PrBL. Projects are built around teamwork, students are instantly thrown into a mix where they have to negotiate roles, resolve conflicts, and coordinate with real external partners. Learners have to learn how to lead by influence and communicate persuasively across professional and cultural divides. This hands-on teamwork builds the genuine relationships and empathy that collective sustainability action absolutely depends on (Brundiers et al., 2021; Sundman et al., 2025).

Crucially, these five competencies don't exist in separate silos. A single, well-designed sustainability project forces student to use all of them at the exact same time. They need systems-thinking to map out the problem, anticipatory thinking to see where it's heading, strategic competence to act, normative competence to make ethical judgments, and interpersonal competence to work together. This is exactly why PrBL fits so perfectly with the core goals of sustainability education. By tying skill development directly to real-world action, PrBL helps students build true "action competence." Ultimately, that ability to step up and make a difference is both the ultimate goal of sustainability education and the truest measure of its success.

Educator, Digital Tools, AI and Connectivism in PrBL

Beyond the institutional and methodological hurdles already discussed, sustainability-focused PrBL raises ethical questions of its own that deserve direct attention. These projects so often involve real communities and outside partners, educators must make sure the relationship gives back as much as it takes. That means seeking people's informed consent and avoiding what are sometimes called 'parachute' projects, where students drop into a community, collect their data or trial a solution, and leave without anything of lasting value behind them. Inside the classroom, fairness needs protecting too, because the team-based nature of PrBL can quietly hide unequal effort, one student carrying the group while a shy or less confident peer fade into the background. Where learners gather environmental or social data, questions of who owns that data, how privacy is protected, and how the information is used must be handled openly. Educators also have a duty to present sustainability issues in a balanced, evidence-based way, one that builds students' sense of agency rather than leaving them anxious and overwhelmed, or pushing a cause without teaching them to question it. Finally, the honesty of the work itself matters, artificial intelligence should be used responsibly and openly declared, and no evidence should be fabricated or left unverifiable. Building these principles into the design of a project is, in itself, part of the sustainable-leadership stance this chapter calls for.

If educators are to become the sustainable leaders that effective PrBL depends on, then steady investment in their professional growth is not optional but it is essential. Many were themselves taught in traditional, lecture-led classrooms, and may feel unsure about guiding open-ended, cross-subject projects or building the outside partnerships that sustainability work calls for (Vasileva et al., 2025). A science teacher comfortable delivering a fixed syllabus, for example, may find it genuinely discouraging to hand students a messy, real-world problem with no single right answer. Ongoing communities where teachers learn from one another, mentoring from more experienced colleagues, and the chance for teachers to take on real sustainability projects themselves because it is hard to guide an experience you have never had. Institutional backing matters just as much. Timetables need to allow for project work that runs over several weeks rather than single lessons; interdisciplinary teaching should count in workload and promotion decisions rather than going unrewarded; and schools need partnerships with industry, government, and community groups that supply genuine problems and resources. Without this supporting structure, PrBL risks staying a fringe activity that survives only on the energy of a few committed individuals, rather than becoming a lasting, school-wide way of teaching.

New technologies are changing the conditions under which sustainability-focused PrBL takes place. Digital learning platforms, mapping tools such as geographic information systems, satellite and remote-sensing data, and online collaboration software now stretch a project far beyond its immediate surroundings. A class studying deforestation, for example, can pull up live satellite images of a shrinking forest rather than relying on a years-old textbook map, or compare notes with students in another country facing the same problem in learning, in line with connectivism ideas, that knowledge often lives in the connections between people and sources rather than in any single classroom. Generative artificial intelligence brings both promise and danger. Used wisely, it can help at the early stages of a project, where student might use it to quickly scan possible solutions to urban flooding or to sketch out different scenarios before testing them. Used carelessly, it can do the opposite, handing over a ready-made answer and skipping the very inquiry, struggle, and reflection that give PrBL its value, much as a calculator handed to a child too early can stunt their feel for numbers. The key teaching challenge, then, is to treat these tools as supports for students' own thinking rather than replacements for it, the field has only begun to explore, and one the closing section flags as a priority for future research.

Global Comparison on Sustainability Education Across Asia and Europe

Cross-level comparisons indicate notable pedagogical differences. Systematic review by Zhang et al. (2025) shows that sustainability activities in primary education are more prominent in creative tasks, role-playing, and whole-school initiatives. By contrast, secondary education on sustainability more often relies on textbook-centered teaching, though it places relatively greater emphasis on multi-perspective inquiry, critical thinking, and project-based or student-centered learning. Most study shows sustainability education integrated to higher institutions as one of the subjects in environmental studies.

Regional disparities are also evident. In wealthier regions of the Global North, Policy-driven via the European Green Deal. Strong emphasis on ESD as a right, which commonly incorporates collaborative strategies, experimental activities, project-based learning reflecting interdisciplinary orientations. Conversely, in many parts of the Global South, instruction remains more textbook-driven and teacher-centered education. This imbalance is mirrored in the research landscape itself, where studies disproportionately focus on the Global North, underscoring the need for expanded scholarship in underrepresented regions such as the Global South and East Asia, where pressing environmental and socio-ecological challenges demand contextually responsive approaches (Zhang et al., 2025).

As in Asia, economic necessity and climate vulnerability led most of the projects to focus on tech-heavy solutions, for example Smart Cities and water management to meet immediate urban challenges. Within the broader Global South, Asia presents a distinctive and internally diverse landscape for sustainability education. Across much of the region, the motivation for embedding sustainability springs less from rights-based policy mandates than from immediate economic necessity and acute climate vulnerability. Rapid urbanisation, exposure to flooding and water stress, and the pressures of industrial growth give projects on smart cities, water management, waste, and renewable energy a pronounced practical orientation (Zhang et al., 2025). This grounding in lived resolve is one of the region's quiet strengths, which relevance is rarely something the educator must manufacture. Yet the region's research base remains comparatively thin, with East Asia in particular underrepresented in the international literature despite the scale of its environmental challenges, with a gap that leaves effective local practice invisible to the wider field and constrains the cross-national learning from which all parties stand to gain.

Concrete initiatives nonetheless reveal the breadth of the region's pedagogical imagination. In Indonesia, comparative research on nature-based and conventional schools shows how experiential, outdoor learning can nurture curiosity, intrinsic motivation, and soft-skill development, rather than conventional schooling still outperforms on standardised academic measures (Astuti et al., 2026). Complementary work shows that flipped, project-based designs delivered through learning-management systems lift students' communication and problem-solving in technical subjects (Ruslan et al., 2024). In China, the rise of campus innovation spaces and faculty-led challenge-based learning has bridged universities and industry, letting students tackle authentic engineering and sustainability problems alongside external mentors, for example designing, a waste-to-energy prototype with a manufacturing partner rather than a textbook (Van Den Beemt et al., 2023). Even in language education, evidence from Iraq shows that PrBL raise both engagement and achievement, indicating how readily the approach travels across diverse Asian curricula (Al-Bahadli et al., 2023). Taken together, these cases suggest that, while system-level integration remains uneven, pockets of innovative, experientially grounded practice are now firmly established across the continent.

Setting the Asian experience alongside the European illuminates how context shapes commitments that, on the surface, look shared. In Europe, sustainability education is strongly policy-driven, secured in multinational frameworks such as the European Green Deal and an ESD agenda often framed as a learner's right. This top-down scaffolding is visible in embedded, interdisciplinary curricula and a mature research culture in Spain (Ariza & Olatunde-Aiyedun, 2023), SDG-aligned business education in Portugal (Filipe et al., 2023), STEM project modules in Romania (Bărbulescu et al., 2025), and cross-institutional challenge-based learning spanning the Netherlands and Ireland (Van Den Beemt et al., 2023). In Asia, by contrast, the driver is more often economic transformation and climate adaptation than rights-based policy; infrastructural solutions, secondary schooling frequently remains more textbook-centred, and the supporting research literature is still maturing (Zhang et al., 2025). These differences are summarised in Table 4.

Table 4. Comparison of sustainability education in Asia and Europe

Dimension	Europe	Asia
Primary driver	Policy mandate and rights-based ESD	Economic transformation and acute climate vulnerability
Policy framework	Universal and embedded across curricula	Varied and largely national; less unified
Dominant pedagogy	Interdisciplinary, collaborative, project-based, challenge-based	Mixed, textbook-centered at secondary level, with growing experiential and PrBL pockets
Project focus	SDG, civic and systemic (Energy, business, heritage)	Technological and infrastructural (Smart cities, water, waste, renewable energy)
Research maturity	Extensive and well-represented in the literature	Emerging, East Asia notably underrepresented
Representative studies	Ariza & Olatunde-Aiyedun (2023); Filipe et al. (2023); Bărbulescu et al. (20265)	Astuti et al. (2026); Van Den Beemt et al. (2023)

CONCLUSION

PrBL represents a transformative approach to sustainability education. A careful synthesis of recent empirical research affirms that PrBL constitutes a powerful, effective, and evidence-based approach to sustainability education, particularly when it is interdisciplinary and firmly connected to community contexts. A recurring theme in the work of Rekalde-Rodríguez (2026) is the "transformative impact" on both the students and educators. Their study suggests that teachers who facilitate sustainability-focused PrBL undergo a shift in pedagogical identity, moving from "content deliverers" to "co-researchers" alongside their students. Furthermore, global findings reported by UNESCO demonstrate that the effectiveness of sustainability-focused PrBL is best gauged through the development of "action competence."

The theoretical grounding of PrBL in constructivist thought, experiential learning cycles, and social learning perspectives reinforces its capacity to situate knowledge within lived experience, collective problem-solving, active learning, emphasizes learning within social context, and is crucial for community-based sustainability projects. Nevertheless, the statistically significant relationship between sustainability exposure and career interest underscores the transformative potential of a well-designed, interdisciplinary curriculum grounded in sustainability principles. At the educational level, recalibrating project complexity, enhancing systematic monitoring and reporting of activities, reinforcing the sense of community are essential measures for improving the effectiveness of the methodology (Ariza & Olatunde-Aiyedun, 2023). PrBL represents a transformative approach to sustainability education. Educational institutions play a critical role in implementing PrBL and fostering sustainability competencies. The integration of PrBL into sustainability education supports workforce readiness and contributes to the development of sustainable societies. Summary of PrBL, education and sustainability are shown in Table 5.

Table 5. PrBL, education, and sustainability

Economic Demand	PrBL Activity	Skill Developed	Long-term Impact
ESD Compliance	Life-Cycle Assessment Projects	Technical Green Literacy	Reduced corporate environmental risk.
Innovation	Prototyping Circular Products	Entrepreneurial Mindset	Growth of the Circular Economy.
Stakeholder Management	Community Outreach Tasks	Transdisciplinary Communication	Enhanced social cohesion/ cooperation.
Strategic Planning	Scenario-based Climate Modeling	Critical Thinking/ Forecasting	Robust long-term climate adaptation.

Challenge and Future Research Direction

Statistical evidence supports the view that PrBL has effective pedagogies in fostering sustainability-related competencies. Meaningful progress depends on a holistic modernization of environmental education, encompassing content, pedagogy, interdisciplinary collaboration, practical engagement, regional relevance, and a coherent conceptual foundation (Vasileva et al., 2025). Community-based sustainability projects provide additional opportunities for experiential learning. Students collaborate with local communities to address sustainability challenges, such as waste management and water conservation. These projects foster civic engagement and social responsibility while contributing to community sustainability.

Despite its benefits, the implementation of PrBL presents several challenges. Persistent challenges include the limited integration of sustainable development principles into curricula, inadequate resources, and a gap between theoretical instruction and practical application within local contexts (Vasileva et al., 2025). Besides, weak professional preparation in environmental education and sustainable development remains a widespread issue, underscoring the importance of systematic and specialized training for educators (Vasileva et al., 2025). Educators may require additional training to effectively facilitate PrBL, and institutions may face resource constraints. Addressing these challenges requires institutional support, professional development programs, and investment in educational resources.

Methodologically, existing research relies heavily on self-reported data and interviews, which may not fully capture actual classroom practices. Future investigations would benefit from triangulated and mixed-methods designs, incorporating classroom observations and the perspectives of diverse stakeholders, including students, parents, and community partners, to develop a more understanding of sustainability implementation. Besides, future research should explore the long-term impact of PrBL on sustainability competencies and career outcomes. Emerging technologies, such as artificial intelligence and digital learning platforms, provide new opportunities to enhance PrBL and sustainability education. Other suggestion may include:

- **AI Integration:** How generative AI can assist in the "ideation" phase of sustainability PrBL without bypassing the critical thinking process.
- **Longitudinal Impact:** Tracking students 5–10 years post-PrBL to see if "green attitudes" translate into long-term lifestyle and career choices.
- **Scalability in Low-Resource Settings:** Adapting PrBL for schools with limited access to technology or funding.

REFERENCES

Akhimien, O. G., & Adekunle, S. A. (2023). Technological environment and sustainable performance of oil and gas firms: A structural equation modelling approach. *Future Business Journal*, *9*(1), 24. DOI: 10.1186/s43093-023-00204-5

Al-Bahadli, K. H., Al-Obaydi, L. H., & Pikhart, M. (2023). The Impact of the Online Project-Based Learning on Students' Communication, Engagement, Motivation, and Academic Achievement. *Psycholinguistics*, *33*(2), 217–237. DOI: 10.31470/2309-1797-2023-33-2-217-237

AlAfnan, M. A., & Dishari, S. (2024). ESD Goals and Soft Skills Competencies through Constructivist Approaches to Teaching: An Integrative Review. [EduLearn]. *Journal of Education and Learning*, *18*(3), 708–718.

Alhassani, F., Saleem, M. R., & Messner, J. (2025). Integrating sustainability in engineering: A global review. *Sustainability (Basel)*, *17*(15), 6930. DOI: 10.3390/su17156930

Almulla, M. A. (2020). The effectiveness of the project-based learning (PBL) approach as a way to engage students in learning. *SAGE Open*, *10*(3), 2158244020938702. DOI: 10.1177/2158244020938702

Ariza, J. Á., & Olatunde-Aiyedun, T. G. (2023). Bringing project-based learning into renewable and sustainable energy education: A case study on the development of the electric vehicle EOLO. *Sustainability (Basel)*, *15*(13), 10275. DOI: 10.3390/su151310275

Astuti, C. T., Leovanny, S., Aulia, L., Caesaria, N. Z., & Siswanto, D. H. (2026). Learning Approaches, Climate, Motivation, Soft Skills, and Achievement: A Comparative Study of Nature-Based and Conventional Educational Systems. *Jurnal Penelitian Ilmiah Multidisipliner, 2*(04).

Avsec, S., & Jagiełło-Kowalczyk, M. (2021). Investigating possibilities of developing self-directed learning in architecture students using design thinking. *Sustainability (Basel)*, *13*(8), 4369. DOI: 10.3390/su13084369

Bărbulescu, R., Enachescu, V. A., & Baltaretu, C. (2025). Integrating sustainability into stem education through project-based learning: empowering undergraduate students for eco-conscious innovation. *EDULEARN25 Proceedings*, 10428-10438.

Brundiers, K., Barth, M., Cebrián, G., Cohen, M., Diaz, L., Doucette-Remington, S., Dripps, W., Habron, G., Harré, N., Jarchow, M., Losch, K., Michel, J., Mochizuki, Y., Rieckmann, M., Parnell, R., Walker, P., & Zint, M. (2021). Key competencies in sustainability in higher education toward an agreed-upon reference framework. *Sustainability Science*, *16*(1), 13–29. DOI: 10.1007/s11625-020-00838-2

Chavula, P., Yusuf, U., Elias, A., Agnes, U., Chebelo, H., George, B. M., & Grace, M., Ntezimana, Lydia, A., Gilbert, L., & Fredrick, K. (2024). Bridging environmental education and sustainable development: An integrated approach for a greener future. *Asian Basic and Applied Research Journal*, *6*(01), 161–171.

Chitamba, A., Yearwood, V., Swanepoel, M., & Myeza, G. (2025). Project-Based Learning and The Development of Entrepreneurial Skills in Higher Education: An Integrative Approach. *International Journal of Business & Management Studies*, *6*(04), 123–133. DOI: 10.56734/ijbms.v6n4a12

Choi, S. J. (2025). Education for Sustainable Development: Challenges and Opportunities. *Journal of Global Social Transformation*, *1*(2), 14–24.

Dacumos, L. P. (2023). STEM education and the project-based learning: A review article. *STEM Education Review, 1*.

Espino-Diaz, L., Luque-González, R., Fernández-Caminero, G., & Álvarez-Castillo, J. L. (2025). Exploring the impact of project-based learning on sustainable development goals awareness and university students' growth. *European Journal of Educational Research*, *14*(1), 283–296. DOI: 10.12973/eu-jer.14.1.283

Filipe, S., Aragonez, T., & Saur-Amaral, I. (2023). Innovative Approaches to Teaching and Learning: Project Based Learning (PBL) Practices for Good. *International Journal of Marketing. Innovation and Strategy*, *1*(2), 131–138.

Friman, H., Banner, I., Sitbon, Y., Sahar-Inbar, L., & Shaked, N. (2024). Experiential learning for sustainability: A catalyst for global change. *Educ. Adm. Theory Pract*, *30*, 8508–8514.

Kolb, A. Y., & Kolb, D. A. (2017). Experiential learning theory as a guide for experiential educators in higher education. *Experiential Learning & Teaching in Higher Education*, *1*(1), 7–44.

Leal Filho, W. (2021). Non-conventional learning on sustainable development: Achieving the SDGs. *Environmental Sciences Europe*, *33*(1), 97. DOI: 10.1186/s12302-021-00525-8 PMID: 34603904

Leal Filho, W., Levesque, V., Sivapalan, S., Salvia, A. L., Fritzen, B., Deckert, R., Kozlova, V., LeVasseur, T. J., Emblen-Perry, K., Azeiteiro, U. M., Paço, A., Borsari, B., & Shiel, C. (2022). Social values and sustainable development: Community experiences. *Environmental Sciences Europe, 34*(1), 67. DOI: 10.1186/s12302-022-00641-z PMID: 35967983

Lozano, A., López, R., Pereira, F. J., & Blanco Fontao, C. (2022). Impact of cooperative learning and project-based learning through emotional intelligence: A comparison of methodologies for implementing SDGs. *International Journal of Environmental Research and Public Health, 19*(24), 16977. DOI: 10.3390/ijerph192416977 PMID: 36554857

Marques, G. M. (2025). Heritage education, sustainability and community resilience: The HISTOESE project-based learning model. *Sustainability (Basel), 17*(21), 9891. DOI: 10.3390/su17219891

Pons, S., Khalilzadeh, J., Weber, M. R., & Smith, R. A. (2024). *Cultivating sustainability savvy: the role of soft skills in shaping sustainable practices*. International Hospitality Review.

Rekalde-Rodríguez, I. (2026). The transformative power of PBL in teacher training for sustainability. *Journal of Project-Based Research, 14*(2).

Ruslan, R., Lu'mu, L. M., Fakhri, M. M., Ahmar, A. S., & Fadhilatunisa, D. (2024). Effectiveness of the flipped project-based learning model based on moodle LMS to improve student communication and problem-solving skills in learning programming. *Education Sciences, 14*(9), 1021. DOI: 10.3390/educsci14091021

Sahin, M. (2024). Approaches to learning: From traditional theories to connectivism. *Education & Science 2024-iv*, 7-34.

Sari, W. A., Artika, W., Oktari, R. S., Rahmatan, H., & Sasaki, D. (2025). Assessing Education for Sustainable Development (ESD) Competencies and Environmental Empathy in Disaster and Environment Knowledge to Support SDGs 2030. *Jurnal Pendidikan IPA Indonesia, 14*(1).

Sundman, J., Feng, X., Shrestha, A., Johri, A., Varis, O., & Taka, M. (2025). Experiential learning for sustainability: A systematic review and research agenda for engineering education. *European Journal of Engineering Education*, 1–31. DOI: 10.1080/03043797.2025.2532591

Tadjer, H., Lafifi, Y., Seridi-Bouchelaghem, H., & Gulsecen, S. (2022). Improving soft skills based on students' traces in problem-based learning environments. *Interactive Learning Environments, 33*(4), 1879–1896. DOI: 10.1080/10494820.2020.1753215

Ukamaka, E. (2024). Education for Sustainable Development: Integrating Sustainability into Curricula. *Research Output Journal of Arts and Management, 3*(3), 53–57.

UNESCO. (2024). *Education for Sustainable Development for 2030: Framework for implementation.* UNESCO.

UNESCO. (2025). *SDG 4 – Education 2030.* https://www.unesco.org/sdg4education2030/en

Valls-Val, K., & Bovea, M. D. (2021). Carbon footprint in Higher Education Institutions: A literature review and prospects for future research. *Clean Technologies and Environmental Policy, 23*(9), 2523–2542. DOI: 10.1007/s10098-021-02180-2 PMID: 34456663

Van Den Beemt, A., Vázquez-Villegas, P., Gómez Puente, S., O'Riordan, F., Gormley, C., Chiang, F. K., Leng, C., Caratozzolo, P., Zavala, G., & Membrillo-Hernández, J. (2023). Taking the challenge: An exploratory study of the challenge-based learning context in higher education institutions across three different continents. *Education Sciences, 13*(3), 234. DOI: 10.3390/educsci13030234

Vasileva, A., Troitskaya, T., Dronova, S., Muravyov, K., Matkarimov, A., & Tatlyev, R. (2025). Environmental education as a key factor in achieving sustainable development goals: Challenges and prospects. *International Journal of Ecosystems and Ecology Science, 15*(3), 301–308. DOI: 10.31407/ijees15.334

Zhang, X., Jung, W., & Asari, M. (2025). Systematic review of environmental education teaching practices in schools: Trends and gaps (2015–2024). *Sustainability (Basel), 17*(19), 8561. DOI: 10.3390/su17198561

KEY TERMS AND DEFINITIONS

Curriculum: Structured framework, content, objectives, instructional strategies, assessment methods, educational outcomes for a course of study.

ESD: Education for Sustainable Development (ESD) is an educational approach that equips learners with the knowledge, skills, values, and competencies to achieve SDGs.

Experiential Learning: Experiential learning is an umbrella term covering various methods, such as simulations, internships, or field research.

HEI: Higher Educational Institution: Institution provides education beyond the secondary (high school) level.

PrBL: Project-based Learning (PrBL): Pedagogical method using projects way to achieve specific academic goals over time.

SDGs: Sustainable Development Goals (SDGs) are a global policy framework established by the United Nations outlining 17 measurable goals.

Soft Skills: Personal qualities and social abilities on how individuals interact, communicate, and work with others.

Sustainability: Encourages balanced decision-making that protects the environment, people and the planet.

Traditional Learning: Conventional method of education, teacher-directed classroom setting.

Compilation of References

Abbas, J., Mahmood, S., Ali, H., Ali Raza, M., Ali, G., Aman, J., Bano, S., & Nurunnabi, M. (2019). The effects of corporate social responsibility practices and environmental factors through a moderating role of social media marketing on sustainable performance of business firms. *Sustainability (Basel)*, *11*(12), 3434. DOI: 10.3390/su11123434

Abdelraouf, M., & Nagasawa, A. (2023). Saudi green initiatives and their role in achieving environmental goals in the Middle East. Gulf Research Center, 19.

Abeng, T. (1997). Business Ethics in Islamic Context: Perspectives of a Muslim. *Business Ethics Quarterly*, *7*(3), 47–54. DOI: 10.2307/3857312

Abiddin, N. Z., Ibrahim, I., & Abdul Aziz, S. A. (2022). Non-governmental organisations (NGOs) and their part towards sustainable community development. *Sustainability (Basel)*, *14*(8), 4386. DOI: 10.3390/su14084386

Abouarab, R., Mishra, T., & Wolfe, S. (2025). Spotting Portfolio Greenwashing in Environmental Funds. *Journal of Business Ethics*, *197*(4), 811–839. DOI: 10.1007/s10551-024-05783-z

Adadi, A., & Berrada, M. (2018). Peeking inside the black-box: A survey on explainable artificial intelligence (XAI). *IEEE Access : Practical Innovations, Open Solutions*, *6*, 52138–52160. DOI: 10.1109/ACCESS.2018.2870052

Adams, C. A., Heijltjes, M. G., Jack, G., Marjoribanks, T., & Powell, M. (2011). The development of leaders able to respond to climate change and sustainability challenges: The role of business schools. *Sustainability Accounting. Management and Policy Journal*, *2*(1), 165–171. DOI: 10.1108/20408021111162191

Adanma, U. M., & Ogunbiyi, E. O.Uwaga Monica AdanmaEmmanuel Olurotimi Ogunbiyi. (2024). A comparative review of global environmental policies for promoting sustainable development and economic growth. *International Journal of Applied Research in Social Sciences*, *6*(5), 954–977. DOI: 10.51594/ijarss.v6i5.1147

Adeniran, I. A., Efunniyi, C. P., Osundare, O. S., & Abhulimen, A. O.Ibrahim Adedeji AdeniranChristianah Pelumi EfunniyiOlajide Soji OsundareAngela Omozele Abhulimen. (2024). Transforming marketing strategies with data analytics: A study on customer behavior and personalization. *International Journal of Management & Entrepreneurship Research*, *6*(8), 41–51. DOI: 10.56781/ijsret.2024.4.1.0022

Afshar Jahanshahi, A., & Kalaycioglu, O. (2025). The Dual Faces of Leadership: CEOs' Dark and Light Traits in Shaping Business Ethical Climate and Sustainability. *Journal of Business Ethics*, •••, 1–21.

Agrawal, A., Tiwari, V., Banerjee, A. R. S., Jhariya, M. K., Hashmi, I., Garg, N., Soni, R., & Gupta, D. (2026). Factors Influencing Student Withdrawal and Predictive Approaches for Retention in Higher Education. In Sam, R., Serey, M., & Em, S. (Eds.), *Student Belonging and Well-Being in Higher Education Retention* (pp. 1–38). IGI Global Scientific Publishing., DOI: 10.4018/979-8-2600-2168-2.ch001

Aguilera, R. V., Filatotchev, I., Gospel, H., & Jackson, G. (2008). An organizational approach to comparative corporate governance: Costs, contingencies, and complementarities. *Organization Science*, *19*(3), 475–492. DOI: 10.1287/orsc.1070.0322

Aguilera, R. V., Rupp, D. E., Williams, C. A., & Ganapathi, J. (2007). Putting the S back in corporate social responsibility: A multilevel theory of social change in organizations. *Academy of Management Review*, *32*(3), 836–863. DOI: 10.5465/amr.2007.25275678

Ahmed, A. (2024). Sustainable leadership: Integrating environmental responsibility into corporate governance. *Journal of Reproducible Research*, *2*(2), 45–54. https://journalrrsite.com/index.php/Myjrr/article/view/69

Ahmed, S. K. (2024). The pillars of trustworthiness in qualitative research. *Journal of Medicine, Surgery, and Public Health*, *2*, 100051. DOI: 10.1016/j.glmedi.2024.100051

Ahsan, M. J. (2025). Green leadership and innovation: Catalysts for environmental performance in Italian manufacturing. *The International Journal of Organizational Analysis*, *33*(7), 1646–1669. DOI: 10.1108/IJOA-04-2024-4450

Ahuja, J., Yadav, M., & Sergio, R. P. (2023). Green leadership and pro-environmental behaviour: A moderated mediation model with rewards, self-efficacy and training. *International Journal of Ethics and Systems*, *39*(2), 481–501. DOI: 10.1108/IJOES-02-2022-0041

Aisha, R., Channa, N. A., Casali, L., Bhutto, M. H., & Qureshi, N. A. (2026). Servant leadership for a sustainable future: transforming organisations through environmental stewardship. Journal of Knowledge, Risk, and Sustainable Management, 1-21.

Akbar, M. A., Khan, A. A., Mahmood, S., Rafi, S., & Demi, S. (2024). Trustworthy artificial intelligence: A decision-making taxonomy of potential challenges. *Software, Practice & Experience*, *54*(9), 1621–1650. DOI: 10.1002/spe.3216

Akhimien, O. G., & Adekunle, S. A. (2023). Technological environment and sustainable performance of oil and gas firms: A structural equation modelling approach. *Future Business Journal*, *9*(1), 24. DOI: 10.1186/s43093-023-00204-5

Akpan, J. U., & Oluwagbade, O. (2023). Social and environmental responsibility in accounting: Beyond financial metrics. *International Journal of social sciences and management research, 9*(9), 163-188. https://doi.org/DOI: 10.56201/ijssmr.v9.no9.2023.pg148.173

Aksoy, C. (2023). Digital business ecosystems: An environment of collaboration, innovation, and value creation in the digital age. *Journal of business and trade, 4*(2), 156-180. DOI: 10.58767/joinbat.1358560

AlAfnan, M. A., & Dishari, S. (2024). ESD Goals and Soft Skills Competencies through Constructivist Approaches to Teaching: An Integrative Review. [EduLearn]. *Journal of Education and Learning*, *18*(3), 708–718.

Alavi, S. B. (2024). The Making of an Authentic Leader's Internalized Moral Perspective: The Role of Internalized Ethical Philosophies in the Development of Authentic Leaders' Moral Identity. *Journal of Business Ethics*, *190*(1), 77–92. DOI: 10.1007/s10551-023-05442-9

Al-Bahadli, K. H., Al-Obaydi, L. H., & Pikhart, M. (2023). The Impact of the Online Project-Based Learning on Students' Communication, Engagement, Motivation, and Academic Achievement. *Psycholinguistics*, *33*(2), 217–237. DOI: 10.31470/2309-1797-2023-33-2-217-237

Albuquerque Pai, A., Anand, A., Pazhoothundathil, N., & Ashok, L. (2024). Leadership perspectives on resilience capabilities for navigating disruption. *Journal of Asia Business Studies*, *18*(1), 103–123. DOI: 10.1108/JABS-03-2023-0081

Alderman, H. (2024). *Leadership for environmental sustainability in education abroad* [Doctoral dissertation, University of North Georgia]. Nighthawks Open Institutional Repository. https://ir.ung.edu/work/sc/569236aa-6121-4d0f-a8ff-2ba42eba8861

Alemu, B. A. (2025). Cultivating a Culture of Sustainability: The Role of Organizational Values and Leadership in Driving Sustainable Practices. *Business Ethics and Leadership*, *9*(1), 79–94. DOI: 10.61093/bel.9(1).79-94.2025

Alexander, C. (2024). *Relationship Between Performance Feedback and Job Satisfaction Environment of Accountants at Accounting Firms* (Doctoral dissertation, Capella University).

Alhassani, F., Saleem, M. R., & Messner, J. (2025). Integrating sustainability in engineering: A global review. *Sustainability (Basel)*, *17*(15), 6930. DOI: 10.3390/su17156930

Ali, N., & Shabn, O. S. (2024). Customer lifetime value (CLV) insights for strategic marketing success and its impact on organizational financial performance. *Cogent Business & Management*, *11*(1), 2361321. DOI: 10.1080/23311975.2024.2361321

Al-khamaiseh, R. A. M., Bailey, R. P., & Jarvis, A. (2025). Definitions of sustainable leadership in education: A systematic review and analysis. *Journal of Research on Leadership Education*, *20*(3), 280–297. Advance online publication. DOI: 10.1177/19427751241283029

Alkhodary, D. (2023). Integrating sustainability into strategic management: A path towards long-term business success. *International Journal of Professional Business Review: Int. J. Prof. Bus. Rev.*, *8*(4), 39.

Allen, I., Brown, K., & Pelican, K. (2024). Connecting higher education to the SDGs: Strategies from the University of Minnesota SDG Initiative. In Shoemaker, J., & Regulska, J. (Eds.), *Global Goals, Global Education: Advancing the United Nations Sustainable Development Goals* (pp. 55–62). NAFSA Association of International Educators.

Almadadha, R. (2024). Blockchain Technology in Financial Accounting: Enhancing Transparency, Security, and ESG Reporting. *Blockchains*, *2*(3), 312–333. DOI: 10.3390/blockchains2030015

Almalchi, H. N. H., Abedin, B., Safari, M., Sani, M. A., & Mat, A. (2024). Identifying Factors Affecting Ethical Consumption among Customers: Evidence from Iraq. *Pakistan Journal of Life and Social Sciences*, *22*(2). Advance online publication. DOI: 10.57239/PJLSS-2024-22.2.00190

Almatar, K. M. (2025). Sustainable Urban Futures: Transportation and Development in Riyadh, Jeddah, and Neom. *Sustainability (Basel)*, *17*(22), 10133. DOI: 10.3390/su172210133

Almulla, M. A. (2020). The effectiveness of the project-based learning (PBL) approach as a way to engage students in learning. *SAGE Open*, *10*(3), 2158244020938702. DOI: 10.1177/2158244020938702

Al-Qutb, H. R., & Ghaith, N. A. (2018). The impact of job localization on organizational commitment: A field study of employees at the Saudi Geological Survey Authority in Jeddah. Journal of Economic. *Administrative and Legal Sciences*, *2*(6), 81–96.

Alshuwaikhat, H. M., & Mohammed, I. (2017). Sustainability matters in national development visions—Evidence from Saudi Arabia's Vision for 2030. *Sustainability (Basel)*, *9*(3), 408. DOI: 10.3390/su9030408

Althnayan, S., Alarifi, A., Bajaba, S., & Alsabban, A. (2022). Linking Environmental Transformational Leadership, Environmental Organizational Citizenship Behavior, and Organizational Sustainability Performance: A Moderated Mediation Model. *Sustainability (Basel)*, *14*(14), 8779. DOI: 10.3390/su14148779

Alvarez, S. A., & Sachs, S. (2023). Where do stakeholders come from? *Academy of Management Review*, *48*(2), 187–202. DOI: 10.5465/amr.2019.0077

Amalia, R., Fadhilah, N., & Putri, A. (2025). Strategic sustainability and CSR transformation in Indonesian energy companies. *Journal of Asian Business and Sustainability*, *14*(1), 22–38.

American Council on Education. (2023, February 18). *About us*. https://www.acenet.edu/About/Pages/default.aspx

Amin, A., Wiyarni, W., & Kadarusman, K. (2025). Corporate governance principles and their influence on sustainability reporting in Indonesian public companies: The moderating role of associations and regulators. *Edelweiss Applied Science and Technology*, *9*(10), 1459–1471. DOI: 10.55214/2576-8484.v9i10.10680

Amjad, M., Abbas, M., & Saeed, R. (2023). Linking green transformational leadership and green creativity: The role of green HRM practices. *Environmental Science and Pollution Research International*, *30*(12), 14572–14589.

Amran, Y. A., Amran, Y. M., Alyousef, R., & Alabduljabbar, H. (2020). Renewable and sustainable energy production in Saudi Arabia according to Saudi Vision 2030; Current status and future prospects. *Journal of Cleaner Production*, *247*, 119602. DOI: 10.1016/j.jclepro.2019.119602

Ananda, A. R., Putra, M. N. T., & Azhar, M. I. Y. (2025). Virtual Influencers, Real Impact: A Narrative Review on Credibility, Generational Trust, and Purchase Intention in Digital Marketing. *Sinergi International Journal of Communication Sciences*, *3*(2), 99–112. DOI: 10.61194/ijcs.v3i2.690

Anos, C. (2025). Ensuring leadership continuity: An integrative review of succession planning, leadership development and organisational memory. *Edelweiss Applied Science and Technology*, *9*(5), 2869–2877. DOI: 10.55214/25768484.v9i5.7607

Antonaras, A., Iacovidou, M., & Memtsa, C. (2011). Measuring social return on investment using the EBEN GR Business Ethics Excellence Model. *Current Issues of Business and Law*, *6*(1), 69–89. DOI: 10.5200/1822-9530.2011.04

Apostol, O., Dey, C., & Thomson, I. (2025). Introducing a Themed Issue on Sustainability Reporting Regulation in Europe. *Social and Environmental Accountability Journal*, *45*(2), 89–94. DOI: 10.1080/0969160X.2025.2537994

Aragón-Correa, J. A., Hurtado-Torres, N., Sharma, S., & García-Morales, V. J. (2008). Environmental strategy and performance in small firms: A resource-based perspective. *Journal of Environmental Management*, *86*(1), 88–103. DOI: 10.1016/j.jenvman.2006.11.022 PMID: 17239519

Aragón-Correa, J. A., & Sharma, S. (2003). A contingent resource-based view of proactive corporate environmental strategy. *Academy of Management Review*, *28*(1), 71–88. DOI: 10.5465/amr.2003.8925233

Ariffin, M. H., Abdul Rahman, A. R., Arshad, R. A., & Zahari, R. (2019b). Bourdieu Capitals And Leadership: The Case Of Design Consultant Firms' Managers. In *The Malaysian Construction Industry*. EDP Sciences., DOI: 10.1051/matecconf/201926603005

Ariffin, M. H., Arshad, R. A., Rahim, A. R. A., & Zahari, R. (2019a). Leadership-Relevant Bourdieu Capitals Of Design Consultant Firms' Managers in The Malaysian Construction Industry. [MCRJ]. *Malaysian Construction Research Journal*, *8*(3), 60–76.

Ariza, J. Á., & Olatunde-Aiyedun, T. G. (2023). Bringing project-based learning into renewable and sustainable energy education: A case study on the development of the electric vehicle EOLO. *Sustainability (Basel)*, *15*(13), 10275. DOI: 10.3390/su151310275

Arora, M., & Sharma, R. L. (2023). Artificial intelligence and big data: Ontological and communicative perspectives in multi-sectoral scenarios of modern businesses. *foresight*, *25*(1), 126-143. DOI: 10.1108/FS-10-2021-0216

Arrieta, A. B., Díaz-Rodríguez, N., Del Ser, J., Bennetot, A., Tabik, S., Barbado, A., & Herrera, F. (2020). Explainable Artificial Intelligence (XAI): Concepts, taxonomies, opportunities and challenges toward responsible AI. *Information Fusion, 58*, 82–115. DOI: 10.1016/j.inffus.2019.12.012

Aru, S., & Srivastava, A. (2025). The role of ai in shaping influencer marketing strategies for genz and millennials on instagram. *Learnovate-International*, 1-7. https://doi.org/DOI: 10.70604/learnint.v2i1.34

Asad, M. (2024). Impact of Environmental Management on Sustainable Performance of Pakistani Entrepreneurial Firms: The Mediating Role of Green Product Innovation and the Moderating Effect of Transformational Leadership. *Sustainability (Basel), 16*(24), 10935. DOI: 10.3390/su162410935

Askland, H. H., Gajendran, T., & Brewer, G. (2013). Project organisations as organisational fields: Expanding the level of analysis through Pierre Bourdieu's Theory of Practice. *Engineering Project Organisation Journal, 3*(2), 116–126. DOI: 10.1080/21573727.2013.768986

Aslam, S., Parveen, K., Alghamdi, A. A., Abbas, S., Shah, A. H., & Elumalai, K. V. (2022). Hopes for the future: Capturing the perspectives of students on higher education and sustainable development in the post-pandemic era. *Sustainability (Basel), 14*(19), 12531. DOI: 10.3390/su141912531

Association of American Colleges and Universities. (2014). *Global learning VALUE rubric*. https://www.aacu.org/initiatives/value-initiative/value-rubrics/value-rubrics-global-learning

Association of International Education Administrators. (2016, November). *Standards of professional practice for international education leaders and senior international officers*. https://www.aieaworld.org/standards-of-professional-practice.html#gsc.tab=0

Association of International Education Administrators. (2022, February 1). *What is an SIO?* https://www.aieaworld.org/sio

Association of International Education Administrators. (2023). *2022-23 AIEA annual report*. https://aieaworld.org/wp-content/uploads/2026/03/2022-23_aiea_annual_report-LR.pdf

Astuti, C. T., Leovanny, S., Aulia, L., Caesaria, N. Z., & Siswanto, D. H. (2026). Learning Approaches, Climate, Motivation, Soft Skills, and Achievement: A Comparative Study of Nature-Based and Conventional Educational Systems. *Jurnal Penelitian Ilmiah Multidisipliner, 2*(04).

Avery, G. C. (2005). *Leadership for sustainable futures: Achieving success in a competitive world.* Edward Elgar Publishing. DOI: 10.4337/9781845425494

Avery, G. C., & Bergsteiner, H. (2011). Sustainable leadership practices for enhancing business resilience and performance. *Strategy and Leadership, 39*(3), 5–15. DOI: 10.1108/10878571111128766

Avsec, S., & Jagiełło-Kowalczyk, M. (2021). Investigating possibilities of developing self-directed learning in architecture students using design thinking. *Sustainability (Basel), 13*(8), 4369. DOI: 10.3390/su13084369

Awan, A. B. (2019). Performance analysis and optimization of a hybrid renewable energy system for sustainable NEOM city in Saudi Arabia. *Journal of Renewable and Sustainable Energy, 11*(2), 025905. DOI: 10.1063/1.5071449

Awashreh, R. (2025). Leadership in the pursuit of net-zero emissions: Overcoming barriers and embracing strategic innovation. *Arab Gulf Journal of Scientific Research, 43*(1), 18–36. DOI: 10.1108/AGJSR-02-2025-0030

Azizi, Z. M., & Abidin, N. Z. (2025). Green Construction Project Design Decision-making: The Role of Design Leaders. In Developing a Body of Knowledge for Green Construction Project Management (pp. 275-298).

Badshah, S. (2012). Historical study of leadership theories. *Journal of Strategic Human Resource Management, 1*(1), 49–59. DOI: 10.18533/journal.v8i2.1582

Bai, A., Vahedian, M., Ghahreman, R., & Piri, H. (2023). Elevating women in the workplace: The dual influence of spiritual intelligence and ethical environments on job satisfaction. *arXiv preprint arXiv:2310.16341.*

Bajani, D. K., & Rotimi, J. O. B. (2024, November). Construction Briefs and the Sustainability Agenda: A Scoping Study to Enhance Brief Quality. In *The International Conference of Sustainable Development and Smart Built Environments* (pp. 534-545). Singapore: Springer Nature Singapore.

Bajra, U., & Čadež, S. (2020). Alternative regulatory policies, compliance, and corporate governance quality. *Baltic Journal of Management, 15*(1), 42–60. DOI: 10.1108/BJM-11-2018-0373

Banerjee, S. B. (2002). Corporate environmentalism: The construct and its measurement. *Journal of Business Research, 55*(3), 177–191. DOI: 10.1016/S0148-2963(00)00135-1

Bansal, P. (2005). Evolving sustainably: A longitudinal study of corporate sustainable development. *Strategic Management Journal, 26*(3), 197–218. DOI: 10.1002/smj.441

Bansal, P., & Clelland, I. (2004). Talking trash: Legitimacy, impression management, and unsystematic risk in the context of the natural environment. *Academy of Management Journal*, *47*(1), 93–103. DOI: 10.2307/20159562

Bansal, P., & Roth, K. (2000). Why companies go green: A model of ecological responsiveness. *Academy of Management Journal*, *43*(4), 717–736. DOI: 10.2307/1556363

Bappenas. (2024). *Indonesia Low Carbon Development Update 2024*. Ministry of National Development Planning.

Barber, N., Taylor, C., & Strick, S. (2009). Wine consumers' environmental knowledge and attitudes: Influence on willingness to purchase. *International Journal of Wine Research*, 59-72. https://doi.org/DOI: 10.2147/IJWR.S4649

Bărbulescu, R., Enachescu, V. A., & Baltaretu, C. (2025). Integrating sustainability into stem education through project-based learning: empowering undergraduate students for eco-conscious innovation. *EDULEARN25 Proceedings*, 10428-10438.

Barkhuizen, E. N., Masakane, G., & van der Sluis, L. (2022). In search of factors that hinder the career advancement of women to senior leadership positions. *SA Journal of Industrial Psychology*, *48*, 1986. DOI: 10.4102/sajip.v48i0.1986

Barney, J., Wright, M., & Ketchen, D. J.Jr. (2001). The resource-based view of the firm: Ten years after 1991. *Journal of Management*, *27*(6), 625–641. DOI: 10.1177/014920630102700601

Barron, A.R., Domeshek M., Metz, L.E., Draucker, L.C, & Strong, A.L. (2021). Carbon neutrality should not be the end goal: Lessons for institutional climate action from U.S. higher education. *One Earth, 4*, 1248-1258. Elsevier Inc. https://doi.org/ DOI: 10.1016/j.oneear.2021.08.014

Barth, M., Jiménez-Aceituno, A., Lam, D. P., Bürgener, L., & Lang, D. J. (2023). Transdisciplinary learning as a key leverage for sustainability transformations. *Current Opinion in Environmental Sustainability*, *64*, 101361. DOI: 10.1016/j.cosust.2023.101361

Baruah, A., Shaikh, M., Kumar, R. M., Shaikh, I. A. K., & Thomas, S. N. (2024, April). Artificial intelligence influence on leadership styles in human resource management for employee engagement. In *2024 Ninth International Conference on Science Technology Engineering and Mathematics (ICONSTEM)* (pp. 1-5). IEEE. DOI: 10.1109/ICONSTEM60960.2024.10568819

Basel Committee on Banking Supervision (BCBS). (2024). The role of climate scenario analysis in strengthening the management and supervision of climate-related financial risks. Bank for International Settlements. https://www.bis.org/bcbs/publ/d572.pdf

Bass, B. M. (1985). *Leadership and performance beyond expectations*. Free Press.

Bass, B. M., & Avolio, B. J. (1994). Transformational leadership and organizational culture. *International Journal of Public Administration*, *17*(3-4), 541–554. DOI: 10.1080/01900699408524907

Bass, B. M., & Riggio, R. E. (2006). *Transformational leadership*. Psychology press., DOI: 10.4324/9781410617095

Batten, S., Sowerbutts, R., & Tanaka, M. (2020). Climate change: Macroeconomic impact and implications for monetary policy. *Ecological, societal, and technological risks and the financial sector*, 13-38.

Bell, K. (2010, December 4). *Where travel abroad meets sustainability: The green passport*. https://melibeeglobal.com/blog/2010/12/where-travel-abroad-meets-sustainability-the-green-passport/

Ben Mahjoub, L. (2025). Greenwashing practices and ESG reporting: An international review. *The International Journal of Sociology and Social Policy*, *45*(1/2), 173–188. DOI: 10.1108/IJSSP-08-2024-0365

Bennett, N. J., Whitty, T. S., Finkbeiner, E., Pittman, J., Bassett, H., Gelcich, S., & Allison, E. H. (2018). Environmental stewardship: A conceptual review and analytical framework. *Environmental Management*, *61*(4), 597–614. DOI: 10.1007/s00267-017-0993-2 PMID: 29387947

Benveniste, J., & Mizrahi, S. (2023). Accountability, governance, and institutional change in intergovernmental relations: A theoretical framework. *Politics & Policy*, *51*(2), 242–255. DOI: 10.1111/polp.12520

Berkes, F. (2009). Evolution of co-management: Role of knowledge generation, bridging organizations and social learning. *Journal of Environmental Management*, *90*(5), 1692–1702. DOI: 10.1016/j.jenvman.2008.12.001 PMID: 19110363

Berniak-Woźny, J., & Rataj, M. (2023). Towards green and sustainable healthcare: A literature review and research agenda for green leadership in the healthcare sector. *International Journal of Environmental Research and Public Health*, *20*(2), 908. DOI: 10.3390/ijerph20020908 PMID: 36673663

Bernini, F., & La Rosa, F. (2024). Research in the greenwashing field: Concepts, theories, and potential impacts on economic and social value. *The Journal of Management and Governance*, *28*(2), 405–444. DOI: 10.1007/s10997-023-09686-5

Bertels, S., Papania, L., & Papania, D. (2010). Embedding sustainability in organizational culture. *A systematic review of the body of knowledge*, 25.

Besharat, A., Nardini, G., & Mesler, R. M. (2024). Bringing ethical consumption to the forefront in emerging markets: The role of product categorization. *Journal of Business Ethics*, *191*(4), 777–792. DOI: 10.1007/s10551-024-05667-2

Binu Raj, A., Subramani, A. K., & Jan, N. A. (2024). Faculty engagement, quality of work-life, organizational commitment and spiritual leadership–examining the mediation and moderation effects. *Industrial and Commercial Training*, *56*(1), 41–52. DOI: 10.1108/ICT-03-2023-0018

Bitra, S. K.Sai Kumar Bitra. (2025). Ethical AI and privacy in digital personalization: Balancing personalization and user trust. *World Journal of Advanced Engineering Technology and Sciences*, *15*(2), 774–779. DOI: 10.30574/wjaets.2025.15.2.0596

Bîzoi, A. C., & Bîzoi, C. G. (2025). Essence and Appearance: A Critical Exploration of Corporate Greenwashing Through Ethical Dilemmas in Global Supply Chains. *Green Wealth: Navigating towards a Sustainable Future*, 211-237.

Boeske, J. (2023). Leadership towards sustainability: A review of sustainable, sustainability, and environmental leadership. *Sustainability (Basel)*, *15*(16), 12626. DOI: 10.3390/su151612626

Bolton, P., Despres, M., Da Silva, L. A. P., Samama, F., & Svartzman, R. (2020). *The green swan*. BIS Books.

Boström, M., Uggla, Y., & Hansson, V. (2018). Environmental representatives: Whom, what, and how are they representing? *Journal of Environmental Policy and Planning*, *20*(1), 114–127. DOI: 10.1080/1523908X.2017.1332522

Boudhane, Y. (2025). Event Communication as a Catalyst for Corporate Identity Promotion: A Case Study of Ooredoo Qatar. *Modern Management Review*, *30*(3), 7–21. DOI: 10.7862/rz.2025.mmr.13

Bourdieu, P. (1977). Outline of a theory of practice (R. Nice, Trans.). Cambridge University Press. (Original work published in 1972)

Bourdieu, P. (1984). Distinction: A social critique of the judgement of taste (R. Nice, Trans.). Harvard University Press. (Original work published 1979)

Bourdieu, P. (1986). The forms of capital. In J. Richardson (Ed.), Handbook of theory and research for the sociology of education (pp. 241–258). Greenwood Press.

Bourdieu, P. (1990). The logic of practice (R. Nice, Trans.). Stanford University Press. (Original work published 1980)

Bourdieu, P. (1992). The logic of practice (R. Nice, Trans.). Stanford University Press. (Original work published 1975)

Bourdieu, P., & Passeron, J.-C. (1970). *La reproduction: Éléments pour une théorie du système d'enseignement*. Paris: Éditions de Minuit. (Translated as: Bourdieu, P., & Passeron, J.-C. (1990). Reproduction in education, society and culture (2nd ed., R. Nice, Trans.). Sage.)

Bourdieu, P. (1988). *Homo academicus* (Collier, P., Trans.). Stanford University Press. (Original work published 1984)

Bowdin, G. A., Allen, J., Harris, R., Jago, L., O'Toole, W., & McDonnell, I. (2023). *Events management*. Routledge. DOI: 10.4324/9781003044963

Bowen, F., & Aragon-Correa, J. A. (2014). Greenwashing in Corporate Environmentalism Research and Practice: The Importance of What We Say and Do. In *Organization and Environment* (Vol. 27, Number 2, pp. 107–112). SAGE Publications Inc. DOI: 10.1177/1086026614537078

Bowman, K. D. (2022, July 12). Future-proofing: How international education leaders foster resilience and fuel innovation. *International Educator*. https://www.nafsa.org/ie-magazine/2022/7/12/future-proofing-international-education

Bramston, P., Pretty, G., & Zammit, C. (2011). Assessing environmental stewardship motivation. *Environment and Behavior*, *43*(6), 776–788. DOI: 10.1177/0013916510382875

Brand, B. M., Rausch, T. M., & Brandel, J. (2022). The importance of sustainability aspects when purchasing online: Comparing generation X and generation Z. *Sustainability (Basel)*, *14*(9), 5689. DOI: 10.3390/su14095689

Braun, V., & Clarke, V. (2006). Using thematic analysis in psychology. *Qualitative Research in Psychology*, *3*(2), 77–101. DOI: 10.1191/1478088706qp063oa

Bridges, T. S. (2009). Gender capital and male bodybuilders. *Body & Society*, *15*(1), 83–107. DOI: 10.1177/1357034X08100148

Brooks, R. E., & Brooks, A. K. (2024). An emerging transformative learning journey to foster sustainability leadership in professional development programs. *Journal of Transformative Education*, *23*(3), 1–20. DOI: 10.1177/15413446241255909

Brown, K. M. (2004). Leadership for social justice and equity: Weaving a transformative framework and pedagogy. *Educational Administration Quarterly*, *40*(1), 77–108. DOI: 10.1177/0013161X03259147

Brown, M. E., Treviño, L. K., & Harrison, D. A. (2005). Ethical leadership: A social learning perspective for construct development and testing. *Organizational Behavior and Human Decision Processes*, *97*(2), 117–134. DOI: 10.1016/j.obhdp.2005.03.002

Brundiers, K., Barth, M., Cebrián, G., Cohen, M., Diaz, L., Doucette-Remington, S., Dripps, W., Habron, G., Harré, N., Jarchow, M., Losch, K., Michel, J., Mochizuki, Y., Rieckmann, M., Parnell, R., Walker, P., & Zint, M. (2021). Key competencies in sustainability in higher education toward an agreed-upon reference framework. *Sustainability Science*, *16*(1), 13–29. DOI: 10.1007/s11625-020-00838-2

Brundtland Commission. (1987). *Our common future*. Oxford University Press.

Bryant, J., Ayers, J., & Missimer, M. (2023). What transforms? – Transformative learning in a sustainability leadership master's program. *International Journal of Sustainability in Higher Education*, *24*(9), 231–251. DOI: 10.1108/IJSHE-03-2022-0086

Bucic, T., Harris, J., & Arli, D. (2012). Ethical consumers among the millennials: A cross-national study. *Journal of Business Ethics*, *110*(1), 113–131. DOI: 10.1007/s10551-011-1151-z

Buendia-Martinez, I., & Carrasco, I. (2022). Integration of indigenous and local knowledge into sustainability education: A systematic literature review. *Environmental Education Research*, *28*(8), 1209–1236. DOI: 10.1080/13504622.2022.2083081

Buhr, M., & Weissbrod, I. (2026). The Role of Individual Beliefs for Corporate Sustainability Transformations. *Business Ethics, the Environment & Responsibility*, *35*(1), 340–366. DOI: 10.1111/beer.12783

Bui, T. M. C., & Fifka, M. (2025). *From adversaries to allies: Cross-sector partnerships for sustainability between businesses and civil society organisations—a systematic literature review and future research avenues*. Management Review Quarterly., DOI: 10.1007/s11301-025-00573-x

Bu, M., Liu, X., Zhang, B., Hazaea, S. A., Fan, R., & Wang, Z. (2024). Governance of Corporate Greenwashing through ESG Assurance. *Systems*, *12*(9), 365. DOI: 10.3390/systems12090365

Burmann, C., Garcia, F., Guijarro, F., & Oliver, J. (2021). Ranking the performance of universities: The role of sustainability. *Sustainability (Basel)*, *13*(23), 13286. Advance online publication. DOI: 10.3390/su132313286

Burns, J. M. (1978). *Leadership*. Harper & Row.

Buttigieg, C. P., & Pulis, S. (2024). Strategic initiatives to address greenwashing. *ERA Forum, 25*, 327–337. https://doi.org/DOI: 10.1007/s12027-024-00815-7

Campbell, A. C., Nguyen, T., & Steward, M. (2022). *Promoting international student mobility for sustainability? Navigating conflicting realities and emotions of international educators*. Association for Studies in International Education., 6.DOI: 10.1177/1028315322112138

Caraiani, C., Lungu, C. I., Dascalu, C., Cimpoeru, M. V., & Dinu, M. (2012). Social and environmental performance indicators: Dimensions of integrated reporting and benefits for responsible management and sustainability. *African Journal of Business Management*, *6*(14), 4990.

Cavallin Toscani, A., Vendraminelli, L., & Vinelli, A. (2024). Environmental sustainability in the event industry: A systematic review and a research agenda. *Journal of Sustainable Tourism*, *32*(12), 2663–2697. DOI: 10.1080/09669582.2024.2309544

Cebrián, G., Segalàs, J., & Hernández, À. (2019). Assessment of sustainability competencies: A literature review and pathways for future research and practice. *Central European Review of Economics and Management*, *3*(3), 19–44.

Chaaben, N., Elleuch, Z., Hamdi, B., & Kahouli B. (2024). Green economy performance and sustainable development achievement: Empirical evidence from Saudi Arabia. *Environment, Development and Sustainability*, *26*(1), 549–564. DOI: 10.1007/s10668-022-02722-8 PMID: 36320556

Chang, R. D., Zuo, J., Soebarto, V., Zhao, Z. Y., & Zillante, G. (2017). Dynamic interactions between sustainability and competitiveness in construction firms: A transition perspective. *Engineering, Construction, and Architectural Management*, *24*(5), 842–859. DOI: 10.1108/ECAM-01-2016-0025

Charles, J., & Duffett, R. G. (2025). *The role of brand awareness and trust on purchase intent in Google shopping ads and demographic factor influences among Millennials and Generation Z*. BP International., DOI: 10.9734/bpi/mono/978-93-49238-37-4/CH9

Chatzopoulou, E., & de Kiewiet, A. (2021). Millennials' evaluation of corporate social responsibility: The wants and needs of the largest and most ethical generation. *Journal of Consumer Behaviour*, *20*(3), 521–534. DOI: 10.1002/cb.1882

Chaudhry, M. J., & Noureen, S. (2023). Exploring sustainable leadership. *International Journal of Multidisciplinary Research and Analysis*, *6*(10), 1–13. DOI: 10.47191/ijmra/v6-i10-13

Chaudhuri, S., Yoon, S. W., Hagen, M., & Legrand, A. (2022). Inter-generational developmental network and the impact of technology on knowledge creation and sharing. In *HRD Perspectives on Developmental Relationships: Connecting and Relating at Work* (pp. 123–150). Springer International Publishing., DOI: 10.1007/978-3-030-85033-3_6

Chauhan, R. S., Connelly, S., Howe, D. C., Soderberg, A. T., & Crisostomo, M. (2022). The danger of "fake news": How using social media for information dissemination can inhibit the ethical decision making process. *Ethics & Behavior, 32*(4), 287–306. DOI: 10.1080/10508422.2021.1890598

Chavula, P., Yusuf, U., Elias, A., Agnes, U., Chebelo, H., George, B. M., & Grace, M., Ntezimana, Lydia, A., Gilbert, L., & Fredrick, K. (2024). Bridging environmental education and sustainable development: An integrated approach for a greener future. *Asian Basic and Applied Research Journal, 6*(01), 161–171.

Chen, C. W. (2024). Utilizing a hybrid approach to identify the importance of factors that influence consumer decision-making behavior in purchasing sustainable products. *Sustainability (Basel), 16*(11), 4432. DOI: 10.3390/su16114432

Chen, Y., Tang, G., Jin, J., Li, J., & Paillé, P. (2020). Green human resource management and employee green behavior: The role of green self-efficacy and individual green values. *Journal of Business Ethics, 164*(2), 1–16. DOI: 10.1007/s10551-019-04300-1

Chhabra, D. (2010). Back to the past: A sub-segment of Generation Y's perceptions of authenticity. *Journal of Sustainable Tourism, 18*(6), 793–809. DOI: 10.1080/09669582.2010.483280

Chitamba, A., Yearwood, V., Swanepoel, M., & Myeza, G. (2025). Project-Based Learning and The Development of Entrepreneurial Skills in Higher Education: An Integrative Approach. *International Journal of Business & Management Studies, 6*(04), 123–133. DOI: 10.56734/ijbms.v6n4a12

Choi, S. J. (2025). Education for Sustainable Development: Challenges and Opportunities. *Journal of Global Social Transformation, 1*(2), 14–24.

Chopdar, P. K., & Paul, J. (2024). The impact of brand transparency of food delivery apps in interactive brand communication. *Journal of Research in Interactive Marketing, 18*(2), 238–256. DOI: 10.1108/JRIM-12-2022-0368

Christensen, D. M., Serafeim, G., & Sikochi, A. (2022). Why is Corporate Virtue in the Eye of The Beholder? The Case of ESG Ratings. *The Accounting Review, 97*(1), 147–175. DOI: 10.2308/TAR-2019-0506

Christensen, L. T., & Cornelissen, J. (2015). Organizational transparency as myth and metaphor. *European Journal of Social Theory*, *18*(2), 132–149. DOI: 10.1177/1368431014555256

CIDB (Construction Industry Development Board). (2024, December 12)Part 1: Malaysia's Construction Industry Roars Back to Life: A Boom Fueled by Investment and Infrastructure https://www.cidb.gov.my/eng/part-1-malaysias-construction-industry-roars-back-to-life-a-boom-fueled-by-investment-and-infrastructure/

Clugston, R., & Calder, W. (2014). Higher education for strong sustainability. In Corcoran, P. B., & Hollingshead, B. P. (Eds.), *Intergenerational learning and transformative leadership for sustainable futures*. Wageningen Academic Publishers. DOI: 10.3920/978-90-8686-802-5_6

Cohen, S., Kadach, I., Ormazabal, G., & Reichelstein, S. (2023). Executive Compensation Tied to ESG Performance: International Evidence. *Journal of Accounting Research*, *61*(3), 805–853. DOI: 10.1111/1475-679X.12481

Cole, R., Johan, S., & Schweizer, D. (2021). Corporate failures: Declines, collapses, and scandals. *Journal of Corporate Finance*, *67*, 101872. Advance online publication. DOI: 10.1016/j.jcorpfin.2020.101872

Collevecchio, F., Temperini, V., Barba-Sanchez, V., & Meseguer-Martinez, A. (2025). Sustainable Governance: Board Sustainability Experience and the Interplay with Board Age for Firm Sustainability. *Journal of Business Ethics*, *197*(2), 371–389. DOI: 10.1007/s10551-024-05739-3

Coman, I. A., Yuan, S., & Tsai, J. Y. (2022). Toward an audience-centric framework of corporate social advocacy strategy: An exploratory study of young consumers from Generation Z. *Sustainability (Basel)*, *14*(7), 4099. DOI: 10.3390/su14074099

Confetto, M. G., Covucci, C., Addeo, F., & Normando, M. (2023). Sustainability advocacy antecedents: How social media content influences sustainable behaviours among Generation Z. *Journal of Consumer Marketing*, *40*(6), 758–774. DOI: 10.1108/JCM-11-2021-5038

Connerley, M. L., & Pedersen, P. B. (2005). *Leadership in a diverse and multicultural environment: Developing awareness, knowledge, and skills*. Sage Publications.

Cooper, K. J., & Gibson, R. B. (2022). A novel framework for inner-outer sustainability assessment. *Challenges*, *13*(64), 64. Advance online publication. DOI: 10.3390/challe13020064

Çop, S., Olorunsola, V. O., & Alola, U. V. (2021). Achieving environmental sustainability through green transformational leadership policy: Can green team resilience help? *Business Strategy and the Environment*, *30*(1), 671–682. DOI: 10.1002/bse.2646

Cortese, A. (2010, January 28). Higher education's purpose: a healthy, just, and sustainable society. *Fast Company*. https://www.fastcompany.com/1527689/higher-educations-purpose-healthy-just-and-sustainable-society

Costa, C., & Azevedo, A. (2024). Antecedents and consequences of the "cancel culture" firestorm journey for brands: Is there a possibility for forgiveness? *Journal of Marketing Management*, *40*(3-4), 289–312. DOI: 10.1080/0267257X.2023.2266465

Cotton, D. R. E., Morrison, D., Magne, P., Payne, S., & Heffernan, T. (2019). Global citizenship and cross-cultural competency: Student and expert understandings of internationalization terminology. *Journal of Studies in International Education*, *23*(3), 346–364. DOI: 10.1177/1028315318789337

Crilly, D., Hansen, M., & Zollo, M. (2016). The grammar of decoupling: A cognitive-linguistic perspective on firms' sustainability claims and stakeholders' interpretation. *Academy of Management Journal*, *59*(2), 705–729. DOI: 10.5465/amj.2015.0171

Cubukcu Cerasi, C., & Balcioglu, Y. S. (2024). Bridging Generations and Values: Understanding Generation Z's Organizational Preferences and the Mediating Role of Sustainability and Innovation Attitudes in Turkey. *Administrative Sciences*, *14*(9), 229. DOI: 10.3390/admsci14090229

Cui, G. Q., Zeng, J. Y., & Jin, C. H. (2022). The impact of vertical/horizontal individualism and collectivism on ethical consumption. *Sustainability (Basel)*, *14*(21), 14254. DOI: 10.3390/su142114254

Curtin, J., McInerney, C., Gallachóir, B. Ó., Hickey, C., Deane, P., & Deeney, P. (2019). Quantifying stranding risk for fossil fuel assets and implications for renewable energy investment: A review of the literature. *Renewable & Sustainable Energy Reviews*, *116*, 109402. DOI: 10.1016/j.rser.2019.109402

Dacumos, L. P. (2023). STEM education and the project-based learning: A review article. *STEM Education Review, 1*.

Dadabaeva, R. A., & Jamoliddinov, F. S. (2024). Digital marketing as an element of sustainable development: Trends, challenges and opportunities. *Digital Models and Solutions*, *3*(1), 65–79. DOI: 10.29141/2949-477X-2024-3-1-6

Daikeler, J., Bosnjak, M., & Manfeda, K. L. (2020). Web versus other survey modes: An updated and extended meta-analysis comparing response rates. *Journal of Survey Statistics and Methodology*, *9*(3), 513–539. DOI: 10.1093/jssam/smz008

Daily, B. F., Bishop, J. W., & Govindarajulu, N. (2009). A conceptual model for organizational citizenship behavior directed toward the environment. *Business & Society*, *48*(2), 243–256. DOI: 10.1177/0007650308315439

Daily, B. F., Bishop, J. W., & Massoud, J. A. (2012). The role of training and empowerment in environmental performance: A study of the Mexican maquiladora industry. *International Journal of Operations & Production Management*, *32*(5), 631–647. DOI: 10.1108/01443571211226524

Daily, B. F., & Huang, S. C. (2001). Achieving sustainability through attention to human resource factors in environmental management. *International Journal of Operations & Production Management*, *21*(12), 1539–1552. DOI: 10.1108/01443570110410892

Dar, A. A., Arora, V., & Yas, H. (2026). Sustainable Intelligence: How AI Powers a Greener, Smarter Planet. In *Predicting and Monitoring Climate Risks Through Green AI: Weather Forecasting, Disaster Prediction, and Biodiversity Monitoring* (pp. 147-178). IGI Global Scientific Publishing.

Dar, A. A., & Lone, W. Z. (Eds.). (2026). *Integrative Approaches to Quality, Data Analysis, and Interdisciplinary Research*. CRC Press. DOI: 10.1201/9781779641014

Dar, A. A., & Shaik, A. J. (2025). Spatiotemporal clustering and multivariate forecasting of air quality index across Indian cities using machine learning and deep learning models. *Franklin Open*, *13*, 100435. DOI: 10.1016/j.fraope.2025.100435

Dastgir, G., Zaman, F. U., & Qayyum, A. (2024). Empowering environmental stewardship: Pedagogical approaches for integrating green and clean land practices in educational institutions of Pakistan. *Pakistan Social Sciences Review*, *8*(1), 478–487. DOI: 10.35484/pssr.2024(8-I)42

Davidescu, A. A., Manta, E. M., Bîrlan, I., Miler, A. M., & Ni ă, S. C. (2026). Detecting greenwashing in ESG disclosure: An NLP-based analysis of Central and Eastern European firms. *Sustainability (Basel)*, *18*(3), 1486. DOI: 10.3390/su18031486

Davies, J. C. (2016). *Exploring the inherent conflict between the internationalization and carbon management agendas in the UK higher education sector*. [Doctoral thesis, Manchester Metropolitan University]. e-space Manchester Metropolitan University's Research Repository. https://e-space.mmu.ac.uk/id/eprint/619558/

Davis, J. H., Schoorman, F. D., & Donaldson, L. (1997). Toward a stewardship theory of management. *Academy of Management Review*, *22*(1), 20–47. https://www.jstor.org/stable/259223. DOI: 10.2307/259223

Dawley, D. D., Andrews, M. C., & Bucklew, N. S. (2008). Mentoring, supervisor support, and perceived organizational support: What matters most? *Leadership and Organization Development Journal, 29*(3), 235–247. DOI: 10.1108/01437730810861290

Dearborn, H., & Tansey, R. (2024). 21st-century stewardship: Infusing environmental stewardship education with global citizenship. *Environmental Education Research, 31*(2), 364–389. DOI: 10.1080/13504622.2024.2335614

Deardorff, D. K., Diabate, D. B., & Rosenbaum, K. L. (2018). Foundations in leading internationalization: Lessons from Mestenhouser. In Deardorff, D. K., & Charles, H. (Eds.), *Leading internationalization: A handbook for international education leaders* (pp. 16–22). Stylus Publishing.

Deghaim, S., Albanaqi, M., Albathali, M. A., Saad, N., Alotaibi, T., Ghazai, B., Alotaibi, G., Ali, W., Otayn, M., Awad, F., Alsharari, M., Shtewi, S., Alsharari, Q., Awad, A., Alsharari, S., Tabia, F., Alsharari, Q., Alenazi, J. H. A., Hathal, M., & Alotaibi, H. (2024). Strategic data stewardship in modern healthcare: An integrated governance framework for ensuring quality, privacy, and ethical disclosure across clinical, operational, and analytical domains. *Saudi Journal of Medicine and Public Health, 1*(2), 1619–1626. DOI: 10.64483/202412450

Dehdasht, G., Ferwati, M. S., Abidin, N. Z., & Oyedeji, M. O. (2022). Trends of the construction industry in Malaysia and its emerging challenges. *Journal of Financial Management of Property and Construction, 27*(2), 161–178. DOI: 10.1108/JFMPC-08-2020-0054

Delmas, M. A., & Toffel, M. W. (2008). Organizational responses to environmental demands: Opening the black box. *Strategic Management Journal, 29*(10), 1027–1055. DOI: 10.1002/smj.701

Dempere, J., Alamash, E., & Mattos, P. (2024). Unveiling the truth: greenwashing in sustainable finance. In *Frontiers in Sustainability* (Vol. 5). Frontiers Media SA. DOI: 10.3389/frsus.2024.1362051

Den Hartog, D. N., & De Hoogh, A. H. (2024). Cross-cultural leadership: What we know, what we need to know, and where we need to go. *Annual Review of Organizational Psychology and Organizational Behavior, 11*(1), 535–566. DOI: 10.1146/annurev-orgpsych-110721-033711

DePew, R., & Gonzales, G. (2020). Differences in health outcomes between millennials and generation X in the USA: Evidence from the National Health Interview Survey. *Population Research and Policy Review, 39*(3), 605–616. DOI: 10.1007/s11113-019-09563-w

DeWeese, G. J. (2023). A Theocentric Environmental Ethic. *Religions*, *14*(7), 913. Advance online publication. DOI: 10.3390/rel14070913

deWit, H., & Merkx, G. (2021). The history of the internationalization of higher education. In Leask, et al. (Eds.), pp. 23-52. *The Handbook of International Higher Education*. Stylus Publishing. https://ebookcentral.proquest.com/lib/northgeorgia/detail.action?docID=6822677

Dhanapal, S., Vashu, D., & Subramaniam, T. (2015). Perceptions on the challenges of online purchasing: A study from "baby boomers", generation "X" and generation "Y" point of views. *Contaduría y Administración*, *60*, 107–132. DOI: 10.1016/j.cya.2015.08.003

Dhiman, S. (2017). Leadership and Sustainability. In Marques, J., & Dhiman, S. (Eds.), *Leadership Today: Practices for Personal and Professional Performance* (pp. 161–177). Springer International Publishing., DOI: 10.1007/978-3-319-31036-7_9

Di Chiacchio, L., Martínez-Caro, E., Cegarra-Navarro, J. G., & Garcia-Perez, A. (2024). Eco-innovation in the textile industry: The strategic importance of data privacy and environmental knowledge management. *Kybernetes*, *53*(10), 3444–3461. DOI: 10.1108/K-02-2023-0222

Dickson, R. K. (2023, August). Analysis of the traditional leadership theories: A review of contemporary leadership approaches and management effectiveness. In Information and Knowledge Management (Vol. 13, No. 5, pp. 9-23). DOI: DOI: 10.7176/IKM/13-5-02

Djafarova, E., & Foots, S. (2022). Exploring ethical consumption of generation Z: Theory of planned behaviour. *Young Consumers*, *23*(3), 413–431. DOI: 10.1108/YC-10-2021-1405

Donaldson, T., & Preston, L. E. (1995). The stakeholder theory of the corporation: Concepts, evidence, and implications. *Academy of Management Review*, *20*(1), 65–91. DOI: 10.2307/258887

Doppelt, B. (2009). *Leading change toward sustainability: A change-management guide for business, government and civil society*. Taylor & Francis Group.

Dorfleitner, G., & Utz, S. (2024). Green, green, it's green they say: A conceptual framework for measuring greenwashing on firm level. *Review of Managerial Science*, *18*(12), 3463–3486. DOI: 10.1007/s11846-023-00718-w

Dounavi, L. E., Dermitzakis, E., Chatzistelios, G., & Kirytopoulos, K. (2022). Project management for corporate events: A set of tools to manage risk and increase quality outcomes. *Sustainability (Basel)*, *14*(4), 2009. DOI: 10.3390/su14042009

Du Toit, E. (2024). Thirty Years of Sustainability Reporting: Insights, Gaps and an Agenda for Future Research Through a Systematic Literature Review. *Sustainability (Basel), 16*(23), 10750. DOI: 10.3390/su162310750

Dumitru, D. (2021). Critical thinking, moral courage, and sustainable development goals. Universities as a driving force for societal change. In *Universities, Sustainability and Society: Supporting the Implementation of the Sustainable Development Goals* (pp. 171–179). Springer International Publishing. DOI: 10.1007/978-3-030-63399-8_12

Dvorak, A. M. W., Christiansen, L. D., Fischer, N. L., & Underhill, J. B. (2011). A necessary partnership: Study abroad and sustainability in higher education. *Frontiers: The Interdisciplinary Journal of Study Abroad, 21*(1), 143–166. DOI: 10.36366/frontiers.v21i1.307

Eccles, R. G., Ioannou, I., & Serafeim, G. (2014). The impact of corporate sustainability on organizational processes and performance. *Management Science, 60*(11), 2835–2857. DOI: 10.1287/mnsc.2014.1984

Eckstein, D., Künzel, V., Schäfer, L., & Winges, M. (2019). Global climate risk index 2020. *Bonn. Germanwatch, 1*, 1–50.

Edgell, D. L.Sr. (2019). *Managing sustainable tourism: A legacy for the future.* Routledge.

Egri, C. P., & Herman, S. (2000). Leadership in the North American environmental sector: Values, leadership styles, and contexts of environmental leaders and their organizations. *Academy of Management Journal, 43*(4), 571–604. DOI: 10.2307/1556356

Elkington, J. (1997). *Cannibals with forks: The triple bottom line of 21st century business.* Capstone Publishing.

Elshaer, I. A., Azazz, A. M., Fayyad, S., Kooli, C., Fouad, A. M., Hamdy, A., & Fathy, E. A. (2025). Consumer boycotts and fast-food chains: Economic consequences and reputational damage. *Societies (Basel, Switzerland), 15*(5), 114. DOI: 10.3390/soc15050114

Emini, A., & Zeqiri, J. (2021). Social media marketing and purchase intention: Evidence from Kosovo. *Ekonomska Misao i Praksa, 30*(2), 475–492. DOI: 10.17818/EMIP/2021/2.8

Engelbrecht, A. S., Heine, G., & Mahembe, B. (2014). The influence of ethical leadership on trust and work engagement: An exploratory study. *SA Journal of Industrial Psychology, 40*(1), 1–9. DOI: 10.4102/sajip.v40i1.1210

Eom, K., & Ng, S. T. (2023). The Potential of Religion for Promoting Sustainability: The Role of Stewardship. *Topics in Cognitive Science, 15*(3), 480–499. DOI: 10.1111/tops.12641 PMID: 36780334

Epstein, M. J. (2018). *Making sustainability work: Best practices in managing and measuring corporate social, environmental and economic impacts.* Routledge., DOI: 10.4324/9781351280129

Eriyawan, A. F., & Utama, C. A. (2026). The Effect of Greenwashing on Firm Performance with Multiple Directorships as Moderating Variable. [ATT]. *Aptisi Transactions on Technopreneurship, 8*(1), 74–84. DOI: 10.34306/att.v8i1.606

Espino-Diaz, L., Luque-González, R., Fernández-Caminero, G., & Álvarez-Castillo, J. L. (2025). Exploring the impact of project-based learning on sustainable development goals awareness and university students' growth. *European Journal of Educational Research, 14*(1), 283–296. DOI: 10.12973/eu-jer.14.1.283

European Commission. (2021). *Screening of websites for "greenwashing": Analysis of environmental claims in the EU consumer market.* European Commission. https://environment.ec.europa.eu/topics/circular-economy/green-claims_en

European Union. (2024). Regulation (EU) 2024/1689 of the European Parliament and of the Council laying down harmonised rules on artificial intelligence (Artificial Intelligence Act). Official Journal of the European Union.

Eyo-Udo, N. L., Mokogwu, C., Olufemi-Phillips, A. Q., & Adewale, T. T. (2024). Developing ethical frameworks for sustainable food pricing through supply chain transparency. *International Journal of Innovation and Scientific Research, 11*(12), 919–947. DOI: 10.51244/IJRSI.2024.11120081

Fang, E., Kotlar, J., Memili, E., Chrisman, J. J., & De Massis, A. (2016). Tapping the power of local knowledge: A local-global interactive perspective. *Journal of World Business, 51*(4), 641–653. DOI: 10.1016/j.jwb.2016.04.001

Farag, M. I. H. (2026). Sustainability as a Management Strategy: Integrating Environmental, Social, and Governance Practices into Business Administration. *Management Science Advances, 3*(1), 20–44. DOI: 10.31181/msa31202630

Fasasi, S. T., Adebowale, O. J., Abdulsalam, A., & Nwokediegwu, Z. Q. S. (2021). Framework for integrating methane monitoring data into carbon credit verification systems. *International Journal of Multidisciplinary Research and Growth Evaluation, 2*(2), 538–543. DOI: 10.54660/.IJMRGE.2021.2.2.538-543

Feeney, B. C., & Lemay, E. P.Jr. (2012). Surviving relationship threats: The role of emotional capital. *Personality and Social Psychology Bulletin*, *38*(8), 1004–1017. DOI: 10.1177/0146167212442971 PMID: 22535924

Ferdig, M. A. (2007). Sustainability leadership: Co-creating a sustainable future. *Journal of Change Management*, *7*(1), 25–35. DOI: 10.1080/14697010701233809

Filipe, S., Aragonez, T., & Saur-Amaral, I. (2023). Innovative Approaches to Teaching and Learning: Project Based Learning (PBL) Practices for Good. *International Journal of Marketing. Innovation and Strategy*, *1*(2), 131–138.

Fischer, K. (2022, April 27). An accord aims to make international education more sustainable. *Latitudes, a newsletter of The Chronicle of Higher Education.*

Fisher, D. R., Campbell, L. K., & Svendsen, E. S. (2012). The organisational structure of urban environmental stewardship. *Environmental Politics*, *21*(1), 26–48. DOI: 10.1080/09644016.2011.643367

Flammer, C. (2021). Corporate green bonds. *Journal of Financial Economics*, *142*(2), 499–516. DOI: 10.1016/j.jfineco.2021.01.010

Flammer, C. (2021). Short-termism and executive compensation. [München: ifo Institut-Leibniz-Institut für Wirtschaftsforschung an der Universität München.]. *CESifo Forum*, *22*(3), 17–19.

Folke, C., Österblom, H., Schultz, L., Olsson, P., Galaz, V., Hentati-Sundberg, J., & de Zeeuw, A. (2021). Governing complex commons: Social–ecological innovations and transformations. *Annual Review of Environment and Resources*, *46*, 319–349. DOI: 10.1146/annurev-environ-012220-011104

Fombrun, C. J. (2005). Corporate reputations as economic assets. The Blackwell handbook of strategic management, 285-308.DOI: 10.1111/b.9780631218616.2006.00011.x

Forliano, C., Battisti, E., de Bernardi, P., & Kliestik, T. (2025). Mapping the greenwashing research landscape: A theoretical and field analysis. *Review of Managerial Science*, *19*(11), 3407–3456. DOI: 10.1007/s11846-025-00856-3

Forum on Education Abroad. (2020). *Standards of good practice for education abroad.* https://www.forumea.org/standards-of-good-practice.html

Forum on Education Abroad. (2021). *Advancing the UN sustainable development goals through education abroad.* https://forumea.org/resources/guidelines/advancing-the-un-sdgs/

Franklin, R. S. (2010). *Exploring the moral development and moral outcomes of authentic leaders*. Regent University.

Frantz, P., Rego, F., & Barbas, S. (2025). Ecocentrism vs. Anthropocentrism: To the Core of the Dilemma to Overcome It. *The Linacre Quarterly*, *92*(4), 449–459. DOI: 10.1177/00243639251339844 PMID: 40510561

Fraser, J. R., Chung, M., & Cheon, H. J. (2023). Ethical consumption in the digital age: Analyzing benefit types, temporal distance, and normative factors for Gen Z. [GBFR]. *Global Business & Finance Review*, *28*(3), 50–67. DOI: 10.17549/gbfr.2023.28.3.50

Free, C., Jones, S., & Tremblay, M. S. (2025). Greenwashing and sustainability assurance: A review and call for future research. *Journal of Accounting Literature*, *47*(4), 739–764. DOI: 10.1108/JAL-11-2023-0201

Freeman, J., & Boeker, W. (1984). The ecological analysis of business strategy. *California Management Review*, *26*(3), 73–86. DOI: 10.2307/41165081

Freeman, R. E. (1984). *Strategic management: A stakeholder approach*. Pitman Publishing.

Freidheim, S. J. (2012). *Valuing study abroad: the global mandate for higher education* [Policy brief]. Institute of International Education, Center for Academic Mobility Research. https://www.iie.org/publications/valuing-study-abroad/

French, R. D. (2011). Political capital. *Representation (McDougall Trust)*, *47*(2), 215–230. DOI: 10.1080/00344893.2011.581086

Friede, G., Busch, T., & Bassen, A. (2015). ESG and financial performance: Aggregated evidence from more than 2000 empirical studies. *Journal of Sustainable Finance & Investment*, *5*(4), 210–233. DOI: 10.1080/20430795.2015.1118917

Friman, H., Banner, I., Sitbon, Y., Sahar-Inbar, L., & Shaked, N. (2024). Experiential learning for sustainability: A catalyst for global change. *Educ. Adm. Theory Pract*, *30*, 8508–8514.

Fry, L. W., & Egel, E. (2021). Global leadership for sustainability. *Sustainability (Basel)*, *13*(11), 6360. Advance online publication. DOI: 10.3390/su13116360

Fusarelli, B. C., Fusarelli, L. D., & Riddick, F. (2018). Planning for the future: Leadership development and succession planning in education. *Journal of Research on Leadership Education*, *13*(3), 286–313. DOI: 10.1177/1942775118771671

Gabrielova, K., & Buchko, A. A. (2021). Here comes Generation Z: Millennials as managers. *Business Horizons*, *64*(4), 489–499. DOI: 10.1016/j.bushor.2021.02.013

Gadinis, S., & Fregoni, S. (2026). Beyond the Brussels Effect: The surprising rise of the International Sustainability Standards Board. *Journal of International Economic Law*, •••, jgaf049. Advance online publication. DOI: 10.1093/jiel/jgaf049

Gajparia, J., Strachan, G., & Leverton, K. (2022). Transformation through learning: Education about, for, and as sustainability. *Frontiers in Sustainability*, *3*, 982718. DOI: 10.3389/frsus.2022.982718

Galpin, T., & Lee Whittington, J. (2012). Sustainability leadership: From strategy to results. *The Journal of Business Strategy*, *33*(4), 40–48. DOI: 10.1108/02756661211242690

Garcia-Sanchez, I. M., Hussain, N., Martínez-Ferrero, J., & Ruiz-Barbadillo, E. (2019). Impact of disclosure and assurance quality of corporate sustainability reports on access to finance. *Corporate Social Responsibility and Environmental Management*, *26*(4), 832–848. DOI: 10.1002/csr.1724

General Authority for Statistics. (2023). *Labor market statistics: Unemployment rate among Saudi nationals*. General Authority for Statistics. Retrieved from the official website of the General Authority for Statistics, Saudi Arabia.

Ghanem, A. M., & Alamri, Y. A. (2023). The impact of the green Middle East initiative on sustainable development in the Kingdom of Saudi Arabia. *Journal of the Saudi Society of Agricultural Sciences*, *22*(1), 35–46. DOI: 10.1016/j.jssas.2022.06.001

Ghasemy, M., Elwood, J. A., & Scott, G. (2024). Approaches to education for sustainability leadership development in higher education: An international comparative study in the Asia Pacific region. *International Journal of Sustainability in Higher Education*, *25*(8), 1870–1894. DOI: 10.1108/IJSHE-11-2023-0558

Ghitti, M., Gianfrate, G., & Palma, L. (2024). The agency of greenwashing. *The Journal of Management and Governance*, *28*(3), 905–941. DOI: 10.1007/s10997-023-09683-8

Gipper, B., Ross, S., & Shi, S. X. (2025). ESG assurance in the United States. *Review of Accounting Studies*, *30*(2), 1753–1803. DOI: 10.1007/s11142-024-09856-2

Gizyatova, A., Dlusskaya, V., & Bisultanova, A. (2024). Behavioural economics and sustainability: Aligning human incentives with green growth strategies. *Reliability: Theory & Applications, 19*(SI 6 (81)), 1145-1150.

Glaser, I. (1998). Roles and relationships: Reflections on the khalifah and the image of God. *Transformation (Durban)*, *15*(1), 18–23. DOI: 10.1177/026537889801500108

Goerzen, A., Van Assche, A., Zhan, J. X., & Zhang, L. (2025). From the editors: Global sustainability reporting standards and the future of international business. *Journal of International Business Policy, 8*(2), 125-136. https://doi.org/DOI: 10.1057/s42214-025-00213-x

Goldin, C. (2024). Human capital. In *Handbook of cliometrics* (pp. 353–383). Springer International Publishing. DOI: 10.1007/978-3-031-35583-7_23

Gomes, S., Lopes, J. M., & Nogueira, S. (2023). Willingness to pay more for green products: A critical challenge for Gen Z. *Journal of Cleaner Production, 390*, 136092. DOI: 10.1016/j.jclepro.2023.136092

Gopal, S., & Pitts, J. (2025). ESG Integration: Unveiling Risk and Driving Innovation in Sustainable Finance. In *The FinTech Revolution: Bridging Geospatial Data Science, AI, and Sustainability* (pp. 35-81). Cham: Springer Nature Switzerland. DOI: 10.1007/978-3-031-74418-1_2

Gopikrishna, M., Burri, R. D., Kumar, N. M., Mahesh, D., Sankar, G. S., & Rao, N. K. K. (2023, October). Green Marketing Analysis for significant Investment Utilizing NLP Process. In *2023 International Conference on New Frontiers in Communication, Automation, Management and Security (ICCAMS)* (Vol. 1, pp. 1-8). IEEE. DOI: 10.1109/ICCAMS60113.2023.10525932

Greenberg, D. (2011, February 17). *5 ways to design sustainable study abroad.* https://my3.my.umbc.edu/gropus/sustainability/posts/5279

Gregersen-Hermans, J. (2021). Toward a curriculum for the future: Synthesizing education for sustainable development and internationalization of the curriculum. *Journal of Studies in International Education, 4*(25), 461–481. DOI: 10.1177/10283153211031033

Grenfell, M. (2014). *Pierre Bourdieu: Key Concepts* (2nd ed.). Routledge. DOI: 10.4324/9781315729923

Grocutt, A., Gulseren, D., Weatherhead, J. G., & Turner, N. (2022). Can mentoring programmes develop leadership? *Human Resource Development International, 25*(4), 404–414. DOI: 10.1080/13678868.2020.1850090

Grooms, L. D., & Reid-Martinez, K. (2011). Sustainable leadership development: A conceptual model of a cross-cultural blended learning program. *International Journal of Leadership Studies, 6*(3), 412–429.

Groves, K. S. (2007). Integrating leadership development and succession planning best practices. *Journal of Management Development, 26*(3), 239–260. DOI: 10.1108/02621710710732146

Gullahorn, J. T., & Gullahorn, J. E. (1958). American objectives in study abroad. *The Journal of Higher Education*, *7*(29), 369–374. https://www.jstor.org/stable/1978752. DOI: 10.1080/00221546.1958.11779560

Gunduzyeli, B. (2024). Artificial Intelligence in Digital Marketing Within the Framework of Sustainable Management. *Sustainability (Basel)*, *16*(23), 10511. DOI: 10.3390/su162310511

Gunningham, N. (2009). Environment Law, Regulation and Governance: Shifting Architectures. *Journal of Environmental Law*, *21*(2), 179–212. DOI: 10.1093/jel/eqp011

Gupta, D., Ashirova, A., Ilhom, J., Sanzhar, D., Giyosiddinov, M., Urozboev, K., & Nodira, E. (2026). Digital Leadership and Educational Transformation in ASEAN: Competencies, Contexts, and Strategic Pathways. In Em, S. (Ed.), *Digital Leadership and Innovation in ASEAN Education* (pp. 97–124). IGI Global Scientific Publishing., DOI: 10.4018/979-8-2600-1142-3.ch004

Haile, E. A. (2025). The nexus between green transformational leadership and financial performance of hotels: The mediating effect of environmental management practices. *Journal of Hospitality and Tourism Insights*, *8*(4), 1416–1435. DOI: 10.1108/JHTI-05-2024-0524

Hale, B. W. (2019). Understanding potential impacts from university-led educational travel. *International Journal of Sustainability in Higher Education*, *20*(2), 245–263. DOI: 10.1108/IJSHE-09-2018-0163

Hale, B. W., Vogelaar, A., & Long, J. (2013). A-broad spectrum: Sustainability in educational travel. *International Journal of Sustainability in Higher Education*, *4*(14), 349–366. DOI: 10.1108/IJSHE-07-2011-0049

Halibas, A., Akram, U., Hoang, A. P., & Thi Hoang, M. D. (2025). Unveiling the future of responsible, sustainable, and ethical consumption: A bibliometric study on Gen Z and young consumers. *Young Consumers*, *26*(7), 142–171. DOI: 10.1108/YC-11-2024-2327

Hallegatte, S. (2016). *Shock waves: managing the impacts of climate change on poverty*. World Bank Publications.

Hanif, S., Ahmed, A., & Younas, N. (2023). Examining the impact of environmental management accounting practices and green transformational leadership on corporate environmental performance: The mediating role of green process innovation. *Journal of Cleaner Production*, *414*, 137584. DOI: 10.1016/j.jclepro.2023.137584

Hao, X., Tian, T., Dong, L., Wong, C. W. Y., & Lai, K. (2025). Unmasking greenwashing in ESG disclosure: Insights from evolutionary game analysis. *Annals of Operations Research, 356*(1), 667–695. DOI: 10.1007/s10479-025-06538-3

Hardman, J. (2012). *Leading for regeneration: Going beyond sustainability in business, education, and community.* Routledge; Earthscan.

Hardman, J. (2017). Educational leadership and environmental justice in a climate-challenged world. In White, D., & Bogotch, I. (Eds.), *The Wiley international handbook of educational leadership* (1st ed., pp. 139–156). John Wiley & Sons, Inc. DOI: 10.1002/9781118956717.ch8

Hargreaves, A., & Fink, D. (2006). Redistributed leadership for sustainable professional learning communities. *Journal of School Leadership, 16*(5), 550–565. DOI: 10.1177/105268460601600507

Harrison, J. S., Bosse, D. A., & Phillips, R. A. (2010). Managing for stakeholders, stakeholder utility functions, and competitive advantage. *Strategic Management Journal, 31*(1), 58–74. DOI: 10.1002/smj.801

Hart, S. L. (1995). A natural-resource-based view of the firm. *Academy of Management Review, 20*(4), 986–1014. DOI: 10.2307/258963

Hart, S. L., & Dowell, G. (2011). Invited editorial: A natural-resource-based view of the firm: Fifteen years after. *Journal of Management, 37*(5), 1464–1479. DOI: 10.1177/0149206310390219

Hasan, M. M., Ali, M., Alam, M. M., & Jan, A. (2024). Sustainability leadership and strategic planning for SDGs. *Journal of Cleaner Production, 412*, 137405.

Hasan, S., Wooliscroft, B., & Ganglmair-Wooliscroft, A. (2023). Drivers of ethical consumption: Insights from a developing country. *Journal of Macromarketing, 43*(2), 175–189. DOI: 10.1177/02761467231168045

Heifetz, R., Grashow, A., & Linsky, M. (2009). *The practice of adaptive leadership.* Harvard Business Press.

Herijanto, H. (2022). Al amanah in al qur'an vs trust: A comparative study. *International Journal of Ethics and Systems, 38*(4), 549–575. DOI: 10.1108/IJOES-03-2021-0064

Heslin, P. A., & Keating, L. A. (2017). In learning mode? The role of mindsets in derailing and enabling experiential leadership development. *The Leadership Quarterly, 28*(3), 367–384. DOI: 10.1016/j.leaqua.2016.10.010

Heyl, J. D., & Hunter, F. J. H. (2019). *The senior international officer as change agent* (2nd ed.). Association of International Education Administrators.

Heyl, J. D., & Tullbane, J. (2012). Leadership in international higher education. In Deardorff, D. K. (Eds.), *The sage handbook of international higher education* (pp. 113–130). SAGE Publications., DOI: 10.4135/9781452218397.n7

Hidayat, D., Tjandra, E. U., Herawati, N., & Simajuntak, M. R. A. (2023). Gen Z digital leadership through social media. *Widyakala Journal: Journal of Pembangunan Jaya University*, *10*(2), 62–71. DOI: 10.36262/widyakala.v10i2.779

Hidayati, N. U., Effendi, M. I., & Kristanto, R. H. (2024). Effect of Green Transformational Leadership and Green Knowledge on Environmental Performance with Green Creativity as Intervening Variable in Tourism Villages in DIY. *West Science Journal Economic and Entreprenurship*, *2*(4), 581–593. Advance online publication. DOI: 10.58812/wsjee.v2i04.1425

Higher Education Opportunity Act, 110 U.S.C.§4137 (2008). https://www.congress.gov/bill/110th-congress/house-bill/4137

Hill, R. (1998). What sample size is "enough" in internet survey research? *Interpersonal Computing and Technology, 6*(3-4), 1-10. https://cadcommunity.pbworks.com/f/what%20sample%20size.pdf

Hoang, T. (2018). The role of the integrated reporting in raising awareness of environmental, social and corporate governance (ESG) performance. In *Stakeholders, governance and responsibility* (pp. 47-69). Emerald Publishing Limited.

Hoffman, A. J., & Sandelands, L. E. (2005). Getting right with nature: Anthropocentrism, ecocentrism, and theocentrism. *Organization & Environment*, *18*(2), 141–162. DOI: 10.1177/1086026605276197

Hofstede, G. (2011). Dimensionalizing cultures: The Hofstede model in context. *Online Readings in Psychology and Culture*, *2*(1), 8. DOI: 10.9707/2307-0919.1014

Hossain, M. Z., Rashed, J. R., & Hredoy, A. I. (2025). Greenwashing in corporate sustainability: Implications for financial reporting and accounting practices. *European Journal of Innovative Studies and Sustainability*, *1*(6), 45–57. DOI: 10.59324/ejiss.2025.1(6).05

Huang, Z., Shi, Y., & Jia, M. (2025). Greenwashing: A systematic literature review. *Accounting and Finance*, *65*(1), 857–882. DOI: 10.1111/acfi.13352

Hua, S., Jin, S., & Jiang, S. (2024). The limitations and ethical considerations of chatgpt. *Data Intelligence*, *6*(1), 201–239. DOI: 10.1162/dint_a_00243

Huckle, J., & Wals, A. E. J. (2015). The UN decade of education for sustainable development: Business as usual in the end. *Environmental Education Research, 21*(3), 491–505. DOI: 10.1080/13504622.2015.1011084

Hudzik, J. K. (2018). *Comprehensive and strategic internationalization: lessons learned and prospects.* NAFSA Association of International Educators.

Hughes, H., Hanneman, K., & Patlas, M. N. (2025). First year in a new leadership role: Lessons learned. *Canadian Association of Radiologists Journal, 76*(4), 659–663. DOI: 10.1177/08465371251332504 PMID: 40245170

Hunt, S. D. (2019). The ethics of branding, customer-brand relationships, brand-equity strategy, and branding as a societal institution. *Journal of Business Research, 95*, 408–416. DOI: 10.1016/j.jbusres.2018.07.044

Huppatz, K., & Goodwin, S. (2013). Masculinised jobs, feminised jobs and men's 'gender capital' experiences: Understanding occupational segregation in Australia. *Journal of Sociology (Melbourne, Vic.), 49*(2-3), 291–308. DOI: 10.1177/1440783313481743

Huron Consulting Group. American Council on Education, & Georgia Institution of Technology. (2019). *The transformation-ready higher education institution: How leaders can prepare for and promote change.* https://www.huronconsultinggroup.com/insights/transformation-ready-higher-education-institution

Husamah, R. A., Permana, T. I., & Lestari, N. (2025). Islam and sustainability issues, how far has the relationship progressed? A bibliometric analysis. In *Social Sciences and Humanities Open* (Vol. 12). Elsevier Ltd. DOI: 10.1016/j.ssaho.2025.101703

Hu, X., Li, R. Y. M., Kumari, K., Ben Belgacem, S., Fu, Q., Khan, M. A., & Alkhuraydili, A. A. (2022). Relationship between green leaders' emotional intelligence and employees' green behavior: A PLS-SEM approach. *Behavioral Sciences (Basel, Switzerland), 13*(1), 25. DOI: 10.3390/bs13010025 PMID: 36661597

Huyler, D., Gomez, L., Rocco, T. S., & Plakhotnik, M. S. (2025). Leading different generational cohorts in the workplace: Focus on situational and inclusive leadership. *New Horizons in Adult Education and Human Resource Development, 37*(1), 6–19. DOI: 10.1177/19394225241297230

Imran, M., Almusharraf, N., & Abdellatif, M. S. (2024). Shaping a sustainable future: A study of teachers' perspectives on environmental education awareness. *Cogent Education, 11*(1), 2422223. DOI: 10.1080/2331186X.2024.2422223

Ims, K. J. (2019). Caring Entrepreneurship and Ecological Conscience—The Case of Patagonia Inc. In *Caring management in the new economy: socially responsible behaviour through spirituality* (pp. 197–220). Springer International Publishing. DOI: 10.1007/978-3-030-14199-8_11

Indrawati, N., Sensuse, D., Sumirat, D., & Purwaningsih, E. H. (2025). Blockchain governance assessment tools: A conceptual model from a business sustainability perspective. *Journal of Information Systems Engineering & Management, 10*(25), 828–850. DOI: 10.52783/jisem.v10i25s.4173

Ines, A., & Moreira, A. C. (2025). Exploring the Intersection of Environmental Sustainability and Anti-Consumption: A Review and Research Agenda. *Journal of Consumer Behaviour, 24*(4), 2121–2142. Advance online publication. DOI: 10.1002/cb.2513

Innovation Hub, B. I. S. (2024). The use of artificial intelligence for policy purposes. Bank for International Settlements. https://www.bis.org/publ/othp100.pdf

Institute of International Education. (2021). *Open doors report: 2021 fast facts.* https://opendoorsdata.org/wp-content/uploads/2021/11/OD21_Fast-Facts-2021.pdf

Institute of International Education. (2023). *U.S. study abroad for academic credit trends, 1989/90-2021/22.* https://opendoorsdata.org/data/us-study-abroad/u-s-study-abroad-for-academic-credit-trends/

Institute of International Education. (n.d.). *A brief history* of *senior international officers on U.S. higher education campuses.* https://www.iie.org/publications/a-brief-history-of-senior-international-officers-on-u-s-higher-education-campuses/

Intergovernmental Panel on Climate Change. (2014). *Climate change 2014 synthesis report summary for policymakers.* United Nations Framework Convention on Climate Change. https://www.ipcc.ch/site/assets/uploads/2018/02/AR5_SYR_FINAL_SPM.pdf

International Monetary Fund (IMF). (2024). *Global Financial Stability Report, October 2024: Steadying the Course: Uncertainty, Artificial Intelligence, and Financial Stability.* IMF.

Internationalization Lab. (2022). *Model for comprehensive internationalization.* American Council on Education. https://www.acenet.edu/Research-Insights/Pages/Internationalization/CIGE-Model-for-Comprehensive-Internationalization.aspx

Ioannou, I., & Serafeim, G. (2015). The impact of corporate social responsibility on investment recommendations: Analysts' perceptions and shifting institutional logics. *Strategic Management Journal, 36*(7), 1053–1081. DOI: 10.1002/smj.2268

IPCC. (2023). [Synthesis Report. Contribution of Working Groups I, II and III to the Sixth Assessment Report of the Intergovernmental Panel on Climate Change. IPCC.]. *Climatic Change*, 2023.

Iqbal, Q., Ahmad, N. H., Nasim, A., & Khan, S. A. R. (2020). A moderated-mediation analysis of psychological empowerment: Sustainable leadership and sustainable performance. *Journal of Cleaner Production*, *262*, 121429. Advance online publication. DOI: 10.1016/j.jclepro.2020.121429

Iqbal, Q., & Piwowar-Sulej, K. (2022). Sustainable leadership in higher education institutions: Social innovation as a mechanism. *International Journal of Sustainability in Higher Education*, *23*(8), 1–20. DOI: 10.1108/IJSHE-04-2021-0162

Iqbal, Q., Piwowar-Sulej, K., & Andersén, J. (2026). From Eco-Policy to Green Practice: Government Eco-Initiatives, Managerial Attention and Frugal Green Innovation in Saudi Arabia. *Business Strategy and the Environment*, bse.70821. Advance online publication. DOI: 10.1002/bse.70821

Iriarte Flores, A., Vergara Romero, A., Forero-Medina, G., & Biggs, D. (2023). Exploring environmental stewardship among youth from a high-biodiverse region in Colombia. *The Journal of Environmental Education*, *55*(1), 1–18. DOI: 10.1080/00958964.2023.2238649

Islam, F. A. S. (2025). Artificial intelligence's hidden footprint in environmental engineering: A life-cycle risk assessment and the AI-ERAF governance framework. *American Journal of Data Science and Artificial Intelligence*, *1*(2), 40–57. DOI: 10.54536/ajdsai.v1i2.6293

Islam, M. S., & Samsudin, S. (2018). Interpretations of Al-amanah Among Muslim Scholars and Its Role in Establishing Peace in Society. *Social Change*, *48*(3), 437–450. DOI: 10.1177/0049085718781689

Ismail, H. W., & Berghout, A. (2024). *AL-SHAJARAH ISTAC Journal of Islamic Thought and Civilization*.

Ismail, I. J. (2025). Transforming small business ventures for a greener tomorrow: The interplay between green transformational leadership, organizational green culture, and environmental sustainability. *Cogent Social Sciences*, *11*(1), 2441400. Advance online publication. DOI: 10.1080/23311886.2024.2441400

Ivanova, E., Rimanoczy, I., & Singhal, D. (Eds.). (2026). *The Power of Arts-Based Pedagogies: Transforming Mindsets to Shape Our Future*. Taylor & Francis. DOI: 10.4324/9781003666325

Ivasciuc, I. S., & Ispas, A. (2025). Communicating Sustainability in Hospitality: A Multi-Layer Analysis of Transparency, Green Claims, and Corporate Value Construction. *Sustainability (Basel)*, *18*(1), 172. DOI: 10.3390/su18010172

Jaas, A. (2022). E-marketing and its strategies: Digital opportunities and challenges. *Open Journal of Business and Management*, *10*(2), 822–845. DOI: 10.4236/ojbm.2022.102046

Jabbour, C. J. C., Santos, F. C. A., & Nagano, M. S. (2010). Contributions of HRM throughout the stages of environmental management: Methodological triangulation applied to companies in Brazil. *International Journal of Human Resource Management*, *21*(7), 1049–1089. DOI: 10.1080/09585191003783512

Jackson, P. (2007). From Stockholm to Kyoto: A brief history of climate change. *UN Chronicle*, *44*(2), 1. https://www.un.org/en/chronicle/article/stockholm-kyoto-brief-history-climate-change

Jain, V., & Mitra, A. (2025). *Navigating Environmental Regulations in Digital Marketing: Ensuring Compliance While Promoting Green Claims*. Regulatory Frameworks and Digital Compliance in Green Marketing, 125-164. IGI Global. DOI: 10.4018/979-8-3693-9831-9.ch005

Jamali, D., & Karam, C. (2018). Corporate social responsibility in developing countries as an emerging field of study. *International Journal of Management Reviews*, *20*(1), 32–61. DOI: 10.1111/ijmr.12112

Jamil, N. N., & Wahyuni, E. T. (2024). Unveiling greenwashing: Risks in sustainability and ESG reporting. *International Journal of Accounting, Finance and Business*, *9*(57), 30-42. https://academicinspired.com/ijafb/article/view/2608

Janik, A., & Ryszko, A. (2026). Greenwashing in Sustainability Reporting: A Systematic Literature Review of Strategic Typologies and Content-Analysis-Based Measurement Approaches. *Sustainability (Basel)*, *18*(1), 17. DOI: 10.3390/su18010017

Jasrotia, S. S., Darda, P., & Pandey, S. (2023). Changing values of millennials and centennials towards responsible consumption and sustainable society. *Society and Business Review*, *18*(2), 244–263. DOI: 10.1108/SBR-01-2022-0013

Javidan, M., Dorfman, P. W., De Luque, M. S., & House, R. J. (2006). In the eye of the beholder: Cross cultural lessons in leadership from project GLOBE. *The Academy of Management Perspectives*, *20*(1), 67–90. DOI: 10.5465/amp.2006.19873410

Jayashree, P., El Barachi, M., & Hamza, F. (2022). Practice of sustainability leadership: A multi-stakeholder inclusive framework. *Sustainability (Basel)*, *14*(10), 6346. DOI: 10.3390/su14106346

Jennings, P. D., & Zandbergen, P. A. (1995). Ecologically sustainable organizations: An institutional approach. *Academy of Management Review, 20*(4), 1015–1052. DOI: 10.2307/258964

Jensen, M. C., & Meckling, W. H. (1976). Theory of the firm: Managerial behavior, agency costs and ownership structure. *Journal of Financial Economics, 3*(4), 305–360. DOI: 10.1016/0304-405X(76)90026-X

Jessen, T. D., Ban, N. C., Claxton, N. X., & Darimont, C. T. (2022). Contributions of Indigenous knowledge to ecological and evolutionary understanding. *Frontiers in Ecology and the Environment, 20*(2), 93–101. DOI: 10.1002/fee.2435

Jhunior, R. D. O. S., de Araujo Góes, H. A., & Valentinov, V. (2026). Stakeholder washing: Theorizing symbolic stakeholder engagement. *Journal of Business Research, 215*(11631), 8.

Jia, K., Zhu, T., Zhang, W., Rasool, S. F., Asghar, A., & Chin, T. (2022). The linkage between ethical leadership, well-being, work engagement, and innovative work behavior: The empirical evidence from the higher education sector of China. *International Journal of Environmental Research and Public Health, 19*(9), 5414. DOI: 10.3390/ijerph19095414 PMID: 35564809

Jiang, X., Mastromartino, B., Li, X., & Zhang, J. J. (2025). Value co-creation in participatory sports event tourism: A mixed methods study. *Behavioral Sciences (Basel, Switzerland), 15*(3), 368. DOI: 10.3390/bs15030368 PMID: 40150263

Jimenez, D., Franco, I. B., & Smith, T. (2021). A review of corporate purpose: An approach to actioning the sustainable development goals (SDGs). *Sustainability (Basel), 13*(7), 3899. DOI: 10.3390/su13073899

JINI. T. (2024). *A STUDY ON AWARENESS OF GREEN WASHING IN ERNAKULAM* (Doctoral dissertation, St. Teresa's College (autonomous) Ernakulam).

Job, J. S., & Khanna, S. (2024). Building a resilient future: Collaborative sustainability regulation. *Clean Energy, 8*(3), 223–237. DOI: 10.1093/ce/zkae026

Jones, G. G., Lopes, T. D. S., Pananond, P., van Tulder, R., Sinkovics, N., & Sinkovics, R. R. (2025). Transitioning from responsible and reactive to deeply responsible and proactive international business. *Critical Perspectives on International Business, 21*(2), 196–225. DOI: 10.1108/cpoib-08-2024-0092

Jones, L., Bradbury, T., & Boutillier, S. (2011). Understanding doxa: The common sense of field-specific knowledge. *Journal of Organisational Studies, 18*(3), 45–59.

Joo, B. K., Lim, D. H., & Kim, S. (2016). Enhancing work engagement: The roles of psychological capital, authentic leadership, and work empowerment. *Leadership and Organization Development Journal*, *37*(8), 1117–1134. DOI: 10.1108/LODJ-01-2015-0005

Jorgensen, B. (2003). Baby Boomers, Generation X and Generation Y? Policy implications for defence forces in the modern era. *Foresight*, *5*(4), 41–49. DOI: 10.1108/14636680310494753

Joseph, S., & Rao, C. B. N. (2022). Transformation leadership and work performance: Mediating role of faculty engagement in higher educational institutions during COVID-19 pandemic. *International Journal of Management Education*, *16*(6), 664–680. DOI: 10.1504/IJMIE.2022.126588

Jung, S., & Lee, J. (2022). Current and future influences of COVID-19 on the knowledge management function of conventions and exhibitions. *Service Business*, *16*(2), 391–410. DOI: 10.1007/s11628-022-00488-7

Kang, J., Faria, A. A., Lee, J., & Choi, W. J. (2023). Will consumers give us another chance to bounce back? Effects of precrisis commitments to social and product responsibility on brand resilience. *Journal of Product and Brand Management*, *32*(6), 927–941. DOI: 10.1108/JPBM-03-2022-3899

Kao, F. C. (2023). How do ESG activities affect corporate performance? *MDE. Managerial and Decision Economics*, *44*(7), 4099–4116. DOI: 10.1002/mde.3944

Karamidehkordi, E., Karimi, V., Hallaj, Z., Karimi, M., & Naderi, L. (2024). Adaptable leadership for arid/semi-arid wetlands conservation under climate change: Using analytical hierarchy process (AHP) approach. *Journal of Environmental Management*, *351*, 119860. DOI: 10.1016/j.jenvman.2023.119860 PMID: 38128210

Karuppiah, K., Virmani, N., & Sindhwani, R. (2024). Toward a sustainable future: Integrating circular economy in the digitally advanced supply chain. *Journal of Business and Industrial Marketing*, *39*(12), 2605–2619. DOI: 10.1108/JBIM-12-2023-0742

Kebe, I. A., Liu, Y., & Kahl, C. (2025). Greener solutions: Green transformational leadership and sustainable environmental outcomes within Sierra Leone's telecom sector. *Journal of Cleaner Production*, *517*, 145811. DOI: 10.1016/j.jclepro.2025.145811

Kejora, M. T. B., Komariah, A., Herawan, E., & Sudarsyah, A. (2025). Environmental Leadership of Ecopesantren Supports Sustainable Development Goals. *Didaktika: Jurnal Kependidikan, 14*(2 Mei), 2065-2076.

Kelvin, K. (2024). Global Leadership and Sustainability. [IJSRM]. *Int. J. Sci. Res. Manag., 12*, 7015–7026.

Ke, Y., Wen, Y., & Teng, L. (2025). Chain Leader Policy and Corporate Environmental Sustainability: A Multi-Level Analysis of Greenwashing Mitigation Mechanisms. *Sustainability (Basel), 17*(19), 8871. DOI: 10.3390/su17198871

Kezar, A., & Eckel, P. D. (2002). The effect of institutional culture on change strategies in higher education: Universal principles or culturally responsive concepts? *The Journal of Higher Education, 73*(4), 435–460. https://www.jstor.org/stable/1558422. DOI: 10.1080/00221546.2002.11777159

Khalili, N.R. (2011). *Practical sustainability: From grounded theory to emerging strategies.* Palgrave Macmillan; ProQuest Ebook Central.

Khan, N., Jabeen, S., Qureshi, M. I., & Falahat, M. (2026). Greenwashing Communication Deficiencies and Consumer Response: Implications for Sustainable Technology Entrepreneurship. *Sustainable Technology and Entrepreneurship, 100132*(2), 100132. Advance online publication. DOI: 10.1016/j.stae.2026.100132

Knihova, L. (2024). AI-Powered Immersive Learning: Transforming Tertiary Education. DOI: 10.24132/jbt.2024.14.2.14_24

Kocaman, R. (2025). A practical guideline for addressing data trustworthiness in qualitative research. In *Qualitative Research Methods for Dissertation Research* (pp. 317-346). IGI Global Scientific Publishing.

Koch, T., & Denner, N. (2025). Different shades of green deception. Greenwashing's adverse effects on corporate image and credibility. *Public Relations Review, 51*(1), 102521. Advance online publication. DOI: 10.1016/j.pubrev.2024.102521

Koch, T., & Denner, N. (2025). What is greenwashing – a scoping review of greenwashing definitions and development of the Need-for-Balance Model. *Journal of Sustainable Business, 10*(1), 17. DOI: 10.1186/s40991-025-00124-3

Koehn, P. H., Deardorff, D. K., & Bolognese, K. D. (2011). Enhancing international research and development project activity on university campuses: Insights from U.S. senior international officers. *Journal of Studies in International Education, 15*(4), 332–350. DOI: 10.1177/1028315310365540

Kohl, K., Hopkins, C., Barth, M., Michelsen, G., Dlouha, J., Razak, D. A., Bin Sanusi, Z. A., & Toman, I. (2022). A whole-institution approach towards sustainability: A crucial aspect of higher education's individual and collective engagement with the SDGs and beyond. *International Journal of Sustainability in Higher Education, 23*(2), 218–236. DOI: 10.1108/IJSHE-10-2020-0398

Kolb, A. Y., & Kolb, D. A. (2017). Experiential learning theory as a guide for experiential educators in higher education. *Experiential Learning & Teaching in Higher Education*, *1*(1), 7–44.

Kopnina, H., Washington, H., Taylor, B., & Piccolo, J. J. (2018). Anthropocentrism: More than Just a Misunderstood Problem. In *Journal of Agricultural and Environmental Ethics* (Vol. 31, Number 1, pp. 109–127). Springer Netherlands. DOI: 10.1007/s10806-018-9711-1

Kouhizadeh, M., & Sarkis, J. (2018). Blockchain practices, potentials, and perspectives in greening supply chains. *Sustainability (Basel)*, *10*(10), 3652. DOI: 10.3390/su10103652

Krasodomska, J., Zarzycka, E., Street, D. L., & Grabowski, W. (2025). The impact of companies' trust-building efforts on sustainability reporting assurance quality: Insights from Europe. *Meditari Accountancy Research*, *33*(7), 246–279. DOI: 10.1108/MEDAR-05-2024-2496

Kuh, G. D. (2008). *High impact educational practices: What they are, who has access to them, and why they matter.* Association of American Colleges and Universities. https://dgmg81phhvh63.cloudfront.net/content/user-photos/Publications/PDFs/E-HIGHIMP.pdf

Kuokkanen, H., & Sun, W. (2020). Companies, meet ethical consumers: Strategic CSR management to impact consumer choice. *Journal of Business Ethics*, *166*(2), 403–423. DOI: 10.1007/s10551-019-04145-4

Kwai, C. K. (2020). *The SIO profile: A preliminary analysis of the survey on senior international education officers, their institutions and offices*. Association of International Education Administrators. https://www.aieaworld.org/uploads/1/4/5/2/145269424/2020_aiea_sio_profile_survey_executive_summary.pdf

Kwon, K., Lee, J., Wang, C., & Diwanji, V. S. (2024). From green advertising to greenwashing: Content analysis of global corporations' green advertising on social media. *International Journal of Advertising*, *43*(1), 97–124. DOI: 10.1080/02650487.2023.2208489

Lahmar, F. (2024). Redefining Leadership: The Role of Spirituality and Motherhood in Muslim Women's Educational Leadership. *Religions*, *15*(12), 1565. Advance online publication. DOI: 10.3390/rel15121565

Lamm, E., Tosti-Kharas, J., & King, C. E. (2015). Empowering employee sustainability: Perceived organizational support toward the environment. *Journal of Business Ethics*, *128*(1), 207–220. DOI: 10.1007/s10551-014-2093-z

Lang, Q., Ma, F., Mirza, N., & Umar, M. (2023). The interaction of climate risk and bank liquidity: An emerging market perspective for transitions to low carbon energy. *Technological Forecasting and Social Change*, *191*, 122480. DOI: 10.1016/j.techfore.2023.122480

Leal Filho, W. (2021). Non-conventional learning on sustainable development: Achieving the SDGs. *Environmental Sciences Europe*, *33*(1), 97. DOI: 10.1186/s12302-021-00525-8 PMID: 34603904

Leal Filho, W., Eustachio, J. H. P. P., Caldana, A. C. F., Will, M., Lange Salvia, A., Rampasso, I. S., Anholon, R., Platje, J., & Kovaleva, M. (2020). Sustainability leadership in higher education institutions: An overview of challenges. *Sustainability (Basel)*, *12*(9), 3761. DOI: 10.3390/su12093761

Leal Filho, W., Levesque, V., Sivapalan, S., Salvia, A. L., Fritzen, B., Deckert, R., Kozlova, V., LeVasseur, T. J., Emblen-Perry, K., Azeiteiro, U. M., Paço, A., Borsari, B., & Shiel, C. (2022). Social values and sustainable development: Community experiences. *Environmental Sciences Europe*, *34*(1), 67. DOI: 10.1186/s12302-022-00641-z PMID: 35967983

Leal Filho, W., Salvia, A. L., Ulluwishewa, R., Abubakar, I. R., Mifsud, M., LeVasseur, T. J., Correia, V., Consorte-McCrea, A., Paço, A., Fritzen, B., Ray, S., Gordon, N., Luetz, J. M., Borsari, B., Venkatesan, M., Mukul, S. A., Carp, R. M., Begum, H., Nunoo, E. K., & Farrugia, E. (2022). Linking sustainability and spirituality: A preliminary assessment in pursuit of a sustainable and ethically correct world. *Journal of Cleaner Production*, *380*, 135091. Advance online publication. DOI: 10.1016/j.jclepro.2022.135091

Lee, J. L., & Lundemo, O. A. (2021, June 5). *Why sustainability is not enough in international education.* University World News. https://www.universityworldnews.com/post.php?story=20210531123436636

Lee, G. K. S. (2025). Building Brands for a Sustainable Future: Integrating Strategy, Consumer Response and Ethical Communication. *Asian Journal of Economics. Business and Accounting*, *25*(5), 10–9734. DOI: 10.9734/ajeba/2025/v25i51800

Lee, S., Isa, M., & Shaiban, M. (2025). Corporate governance and greenwashing: The moderating role of political connection and sensitive industries. *SAGE Open*, *15*(3), 1–15. DOI: 10.1177/21582440251369200

Leggett, K. (2012). Leaving light footprints. [NAFSA Association of International Educators.]. *International Educator*, (May-June), 41–52.

Le, T. H., Novais, M. A., Arcodia, C., Berchtenbreiter, R., Humpe, A., & Nguyen, N. (2024). How authenticity in events fosters social sustainability: Towards an authenticity ecosystem and implications for destination management. *Tourism Management Perspectives*, *51*, 101222. DOI: 10.1016/j.tmp.2024.101222

Liao, Y. (2022). Sustainable leadership: A literature review and prospects for future research. *Frontiers in Psychology*, *13*, 1045570. DOI: 10.3389/fpsyg.2022.1045570 PMID: 36420377

Lincoln, Y. S., & Guba, E. G. (1986). But is it rigorous? Trustworthiness and authenticity in naturalistic evaluation. *New Directions for Program Evaluation*, *1986*(30), 73–84. DOI: 10.1002/ev.1427

Linnenluecke, M. K., & Griffiths, A. (2010). Corporate sustainability and organizational culture. *Journal of World Business*, *45*(4), 357–366. DOI: 10.1016/j.jwb.2009.08.006

Li, S., & Liu, C. (2025). Corporate strategic greenwashing under ESG disclosure uncertainty: Financing incentives and nonlinear effects. *Journal of Environmental Management*, *394*, 127473. DOI: 10.1016/j.jenvman.2025.127473 PMID: 41038102

Lissitsa, S., & Kol, O. (2016). Generation X vs. Generation Y–A decade of online shopping. *Journal of Retailing and Consumer Services*, *31*, 304–312. DOI: 10.1016/j.jretconser.2016.04.015

Lissitsa, S., & Kol, O. (2021). Four generational cohorts and hedonic m-shopping: Association between personality traits and purchase intention. *Electronic Commerce Research*, *21*(2), 545–570. DOI: 10.1007/s10660-019-09381-4

Liu, C. W., & Cheng, J. S. (2022). Low-carbon literacy of exhibitors in the exhibition industry in China. *Sustainability (Basel)*, *14*(4), 2262. DOI: 10.3390/su14042262

Liu, M., Zhao, J., & Liu, H. (2025). Digital transformation, employee and executive compensation, and sustained green innovation. *International Review of Financial Analysis*, *97*, 103873. DOI: 10.1016/j.irfa.2024.103873

Liu, Y., Wang, H., Chen, J., Zhang, X., Yue, X., Ke, J., Wang, B., & Peng, C. (2020). Emergency management of nursing human resources and supplies to respond to coronavirus disease 2019 epidemic. *International Journal of Nursing Sciences*, *7*(2), 135–138. DOI: 10.1016/j.ijnss.2020.03.011 PMID: 32292632

Lizardo, O. (2004). The cognitive origins of Bourdieu's habitus. *Journal for the Theory of Social Behaviour*, *34*(4), 375–401. DOI: 10.1111/j.1468-5914.2004.00255.x

Long, J., Vogelaar, A., & Hale, B. (2014). Toward sustainable educational travel. *Journal of Sustainable Tourism*, *22*(3), 421–439. DOI: 10.1080/09669582.2013.819877

López, S. M. C., Guerra, I. R., & Pernas, J. B. (2025). Sustainable Leadership as a Governance Mechanism in the ESG Era: A Systematic Review of Organizational Transformation in the Hospitality Sector. *Merits*, *5*(4), 22. DOI: 10.3390/merits5040022

Lovelace, K. J., Manz, C. C., & Alves, J. C. (2007). Work stress and leadership development: The role of self-leadership, shared leadership, physical fitness and flow in managing demands and increasing job control. *Human Resource Management Review*, *17*(4), 374–387. DOI: 10.1016/j.hrmr.2007.08.001

Lozano, A., López, R., Pereira, F. J., & Blanco Fontao, C. (2022). Impact of cooperative learning and project-based learning through emotional intelligence: A comparison of methodologies for implementing SDGs. *International Journal of Environmental Research and Public Health*, *19*(24), 16977. DOI: 10.3390/ijerph192416977 PMID: 36554857

Maak, T., & Pless, N. M. (2006). Responsible leadership in a stakeholder society–a relational perspective. *Journal of Business Ethics*, *66*(1), 99–115. DOI: 10.1007/s10551-006-9047-z

Maak, T., Pless, N. M., & Voegtlin, C. (2016). Business statesman or shareholder advocate? CEO responsible leadership styles and the micro-foundations of political CSR. *Journal of Management Studies*, *53*(3), 463–493. DOI: 10.1111/joms.12195

Mageto, J. (2021). Big data analytics in sustainable supply chain management: A focus on manufacturing supply chains. *Sustainability (Basel)*, *13*(13), 7101. DOI: 10.3390/su13137101

Mahran, K., Albarrak, H., Ibrahim, B. A., & Elamer, A. A. (2025). Leadership and sustainability in tourism and hospitality: A systematic review and future research agenda. *International Journal of Contemporary Hospitality Management*, *37*(7), 2219–2242. DOI: 10.1108/IJCHM-07-2024-0985

Mair, J., & Smith, A. (2021). Events and sustainability: Why making events more sustainable is not enough. *Journal of Sustainable Tourism*, *29*(11-12), 1739–1755. DOI: 10.1080/09669582.2021.1942480

Malik, H., & Mehmood, M. (2022). Impact of authentic leadership, sustainable leadership on sustainable growth with mediating role of organizational learning. [AJSS]. *Academic Journal of Social Sciences*, *6*(1), 48–69. DOI: 10.54692/ajss.2022.06011681

Mallow, S., Toman, I., & van't Land, H. (2020). *Higher education and the 2030 Agenda: Moving into the 'Decade of Action and Delivery for the SDGs'*. International Association of Universities. https://static1.squarespace.com/static/64ca66e5f086975ce8783d53/t/653bcfbcaf643c254532ba19/1698418623774/iau_hesd_survey_report_final_jan2020.pdf

Manning, B., Braam, G., & Reimsbach, D. (2019). Corporate governance and sustainable business conduct—E ffects of board monitoring effectiveness and stakeholder engagement on corporate sustainability performance and disclosure choices. *Corporate Social Responsibility and Environmental Management*, *26*(2), 351–366. DOI: 10.1002/csr.1687

Manning, S., & Reinecke, J. (2016). A modular governance architecture in-the-making: How transnational standard-setters govern sustainability transitions. *Research Policy*, *45*(3), 618–633. DOI: 10.1016/j.respol.2015.11.007

Mann, J., Gray, T., Truong, S., Brymer, E., Passy, R., Ho, S., Sahlberg, P., Ward, K., Bentsen, P., Curry, C., & Cowper, R. (2022). Getting out of the classroom and into nature: A systematic review of nature-specific outdoor learning on school children's learning and development. *Frontiers in Public Health*, *10*, 877058. DOI: 10.3389/fpubh.2022.877058 PMID: 35651851

Maragakis, A., & Van den Dobbelsteen, A. (2013). Higher education: Features, trends, and needs in relation to sustainability. *Journal of Sustainability Education*, *4*, 1–20. https://repository.tudelft.nl/islandora/object/uuid%3A4b307380-2960-48e1-a6cc-bb186d518095

Marinakou, E. L., & Mathew, A. (2025). Hybrid Festival Visitors Experience Using the Event Experience Theory in the Post Covid Era. *Event Management*, *29*(6), 817–835. DOI: 10.3727/152599525X17367484906426

Marques, G. M. (2025). Heritage education, sustainability and community resilience: The HISTOESE project-based learning model. *Sustainability (Basel)*, *17*(21), 9891. DOI: 10.3390/su17219891

Marquis, C., Toffel, M. W., & Zhou, Y. (2016). Scrutiny, norms, and selective disclosure: A global study of greenwashing. *Organization Science*, *27*(2), 483–504. DOI: 10.1287/orsc.2015.1039

Marrucci, L., Iovino, R., & Iraldo, F. (2025). Environmental sustainability award winners: Do they communicate their environmental performance without potential greenwashing? *Corporate Social Responsibility and Environmental Management*, *32*(2), 2783–2794. DOI: 10.1002/csr.3088

Marshall, A., & Sandberg, J. (2011). Sensemaking in'real'versus virtual environments: A comparison and challenge. In *3rd International Symposium on Process Organization Studies.*

Marta, D. J., & Rake, P. (2025). The success of the whistleblowing system in preventing corruption in village financial management in Indonesia. *Jurnal Inspektorat, 1*(2), 72–88. DOI: 10.64527/inspektorat.v1i2.39

Masyhuri, M., Muttaqien, F., Rizki, V. L., Fitria, N., Sabrina, C., & Jawo, M. (2025). Empowering leadership and employee agility: The mediating role of knowledge sharing and perceived organizational support. *International Journal of Business Ecosystem & Strategy (2687-2293), 7*(6), 160-170.

Matasick, C. (2017). Open government: How transparency and inclusiveness can reshape public trust. *Trust and Public Policy*, 105.

Ma, X., Ma, X., Fei, W., Jiang, Q., & Zhang, Y. (2026). Stop Where it Should Stop: Board Faultlines, Stakeholder Supervision and Greenwashing Behaviors of Firms. *Journal of Business Ethics, 203*(2), 313–340. DOI: 10.1007/s10551-025-06046-1

McBride, K. (2022, July 25). *Scope 3 emissions: The big exclusion.* Bound International. https://bound-intl.com/scope-3-emissions-the-big-exclusion/

McBride, K., & Greenberg, D. (2021). *Survey of sustainability in education abroad.* Bound International. https://bound-intl.com/wp-content/uploads/2021/12/Survey-of-Sustainability-in-Education-Abroad-2.pdf

McBride, K., & Ponce-Taylor, D. (2022). *Environmental sustainability and internationalization: challenges and opportunities* (Trends & Insights, May). NAFSA Association of International Educators. https://www.nafsa.org/professional-resources/research-and-trends/trends-insights/environmental-sustainability-and-internationalization

McCann, J. T., & Holt, R. A. (2010). Defining sustainable leadership. *International Journal of Sustainable Strategic Management, 2*(2), 204–210. DOI: 10.1504/IJSSM.2010.032561

McClelland, K. (2015). *Cultivating social responsibility in study abroad: A phenomenological exploration of study abroad program design* [Doctoral dissertation, Plymouth State University]. ProQuest Dissertations and Theses Global.

McIntyre, O. (2018). Transnational environmental regulation and the normativisation of global environmental governance standards: The promise of order from chaos? *Journal of Property. Planning and Environmental Law, 10*(2), 92–112. DOI: 10.1108/JPPEL-01-2018-0003

McLeod, L. J., Kitson, J. C., Dorner, Z., Tassell-Matamua, N. A., Stahlmann-Brown, P., Milfont, T. L., & Hine, D. W. (2024). Environmental stewardship: A systematic scoping review. *PLoS One*, *19*(5), e0284255. DOI: 10.1371/journal.pone.0284255 PMID: 38713707

Mebratu, D. (1998). Sustainability and sustainable development: Historical and conceptual review. *Environmental Impact Assessment Review*, *18*(6), 493–520. DOI: 10.1016/S0195-9255(98)00019-5

Mehra, V., Bharany, S., Singh, P., Sawhney, R. S., Kaur, U., Rehman, A. U., & Hussen, S. (2025). Impacts of digital technologies and social media platforms on advocating environmental sustainability in sports sector. *Discover Sustainability*, *6*(1), 1–25. DOI: 10.1007/s43621-025-00932-4

Mehryar, S., Yazdanpanah, V., & Tong, J. (2024). AI and climate resilience governance. *iScience*, *27*(5), 109812. DOI: 10.1016/j.isci.2024.109812 PMID: 38784017

Mei, Y., & Broyde, M. J. (2025). Reclaiming constitutional authority of algorithmic power. *arXiv*. https://arxiv.org/abs/2508.11699 DOI: 10.2139/ssrn.5389563

Metcalf, L., & Benn, S. (2013). Leadership for sustainability: An evolution of leadership ability. *Journal of Business Ethics*, *112*(3), 369–384. DOI: 10.1007/s10551-012-1278-6

Metz, D., Ilies, L., & Metz, M. (2020). The impact of organisational culture on leadership effectiveness and performance. In Proceedings of the 14th International Management Conference" Managing Sustainable Organizations (pp. 776-788).DOI: DOI: 10.24818/IMC/2020/04.09

Meyer, J. W., & Rowan, B. (1977). Institutionalized organizations: Formal structure as myth and ceremony. *American Journal of Sociology*, *83*(2), 340–363. DOI: 10.1086/226550

Mhatre, K. H., & Conger, J. A. (2011). Bridging the gap between gen x and gen y: Lessons from authentic leadership. *Journal of Leadership Studies*, *5*(3), 72–76. DOI: 10.1002/jls.20235

Miao, Q., & Nduneseokwu, C. (2025). Applications of leadership theories and concepts in environmental leadership. In *Environmental leadership in a VUCA era: An interdisciplinary handbook* (pp. 237–275). Springer Nature Singapore.

Miao, Q., & Nduneseokwu, C. (2025). *Environmental leadership in Private organizations. Environmental leadership in a VUCA Era: An interdisciplinary handbook.* Springer Nature.

Mikhailova, K. (2025). Exhibitions as a marketing tool: strategies for booth design, visitor engagement.

Ministry of Environment and Forestry, Republic of Indonesia. (2023). *Indonesia's climate actions towards 2030: FOLU Net Sink.*

Mishchuk, H., Bilan, Y., Androniceanu, A., & Krol, V. (2023). Social capital: Evaluating its roles in competitiveness and ensuring human development. *Journal of Competitiveness, 15*(2).

Mitchell, V. W., Makri, K., Schlegelmilch, B., & Blaha, P. (2023). Exhibition event evaluation revisited: An individual value approach. *Event Management, 27*(1), 107–125. DOI: 10.3727/152599522X16419948695044

Mol, A. P. (2015). Transparency and value chain sustainability. *Journal of Cleaner Production, 107*, 154–161. DOI: 10.1016/j.jclepro.2013.11.012

Molnár, P., Lukács, B., & Tóth, Á. (2025). Toward Social Disclosure Alignment: Evaluating the Interoperability of ISSB S2 with ESRS and GRI Standards. *Societies (Basel, Switzerland), 15*(10), 273. DOI: 10.3390/soc15100273

Monasterolo, I. (2020). Climate change and the financial system. *Annual Review of Resource Economics, 12*(1), 299–320. DOI: 10.1146/annurev-resource-110119-031134

Montgomery, A. W., Lyon, T. P., & Barg, J. (2024). No end in sight? A greenwash review and research agenda. *Organization & Environment, 37*(2), 221–256. DOI: 10.1177/10860266231168905 PMID: 39091586

Morais, D. B., & Ogden, A. C. (2011). Initial development and validation of the Global Citizenship Scale. *Journal of Studies in International Education, 5*(15), 445–466. DOI: 10.1177/1028315310375308

Moşteanu, N. R. (2024). Adapting to the unpredictable: Building resilience for business continuity in an ever-changing landscape. *European Journal of Theoretical and Applied Sciences, 2*(1), 444–457. DOI: 10.59324/ejtas.2024.2(1).37

Mousavi, K., Kowsari, E., Ramakrishna, S., Chinnappan, A., & Gheibi, M. (2024). A comprehensive review of greenwashing in the textile industry (life cycle assessment, life cycle cost, and eco-labeling). *Environment, Development and Sustainability, 27*(9), 1–41. DOI: 10.1007/s10668-024-04508-6

Muff, K., Liechti, A., & Dyllick, T. (2020). How to apply responsible leadership theory in practice: A competency tool to collaborate on the sustainable development goals. *Corporate Social Responsibility and Environmental Management*, *27*(5), 2254–2274. DOI: 10.1002/csr.1962

Mullins, M., Holland, C. P., & Cunneen, M. (2021). Creating ethics guidelines for artificial intelligence and big data analytics customers: The case of the consumer European insurance market. *Patterns (New York, N.Y.)*, *2*(10), 100362. Advance online publication. DOI: 10.1016/j.patter.2021.100362 PMID: 34693379

Muneer, S., Singh, A., Choudhary, M. H., Alshammari, A. S., & Butt, N. A. (2025). Does Environmental Disclosure and Corporate Governance Ensure the Financial Sustainability of Islamic Banks? *Administrative Sciences*, *15*(2), 54. DOI: 10.3390/admsci15020054

Murillo, E., & Atristain-Suárez, C. (2023). Leveraging Strategic Planning to Navigate Volatile Environments. In *Creating Economic Stability Amid Global Uncertainty: Post-Pandemic Recovery in Mexico's Emerging Economy* (pp. 189-217). Cham: Springer Nature Switzerland. DOI: 10.1007/978-3-031-41386-5_9

Murphy, C. (2022). *Sustainable leadership: Lessons of vision, courage, and grit from the CEOs who dared to build a better world*. John Wiley & Sons.

Myles, J. (2004). From Doxa to Experience: Issues in Bourdieu's Adoption of Husserlian Phenomenology. *Theory, Culture & Society*, *21*(2), 91–107. DOI: 10.1177/0263276404042136

NAFSA Association of International Educators. (2019, March). *NAFSA's statement of ethical principles*. https://www.nafsa.org/about-us/about-nafsa/nafsas-statement-ethical-principles

NAFSA Strategic Task Force on Education Abroad. (2003). *Securing America's future: global education for a global age*. NAFSA Association of International Educators. https://www.nafsa.org/policy-and-advocacy/policy-resources/securing-americas-future

Nakamura, Y. T., Williams, E. N., Pyle, L., & Grande, S. E. (2023). Cultivating authentic leaders: Toward conceptual coherence and sustainable practice. *The Journal of Values Based Leadership*, *16*(1), 19.

Narboshova, M., Gafurova, S., Abdukadirova, M., Alimardonov, Z., Nishanov, S., Musinova, N., & Muxtarova, L. (2026). Mobile experiential learning for environmental pollution science with linking authentic field data to pro-environmental behaviour in secondary schools. *Natural and Engineering Sciences*, *11*(1), 271–282. DOI: 10.28978/nesciences.261020

Nashrullah, H., Maemunah, M., & Purnamasari, P. (2025). Building integrity from within: The role of organizational culture and commitment in preventing fraud in student organizations at Universitas Islam Bandung. *Journal of Economics and Business UBS*, *14*(5), 1313–1323. DOI: 10.52644/n2ydyt42

National Center for Education Statistics. (2023). *Fast facts: enrollment*. U.S. Department of Education. https://nces.ed.gov/fastfacts/display.asp?id=98

Nawaz Khan, A. (2023). Is green leadership associated with employees' green behaviour? Role of green human resource management. *Journal of Environmental Planning and Management*, *66*(9), 1962–1982. DOI: 10.1080/09640568.2022.2049595

Nazlim, O. F., & Ozgen, N. (2024). Perceived information and disinformation of generation Z in the context of the changing consumer society: A research on news consumption behavior on social media. *Turkish Online Journal of Design Art and Communication*, *15*(1), 309–332. DOI: 10.7456/tojdac.1569297

Nduneseokwu, C. K., & Harder, M. K. (2023). Developing environmental transformational leadership with training: Leaders and subordinates' environmental behaviour outcomes. *Journal of Cleaner Production*, *403*, 136790. DOI: 10.1016/j.jclepro.2023.136790

Nee, V., & Opper, S. (2010). Political capital in a market economy. *Social Forces*, *88*(5), 2105–2132. DOI: 10.1353/sof.2010.0039

Newstead, T., Dawkins, S., Macklin, R., & Martin, A. (2020). The virtues project: An approach to developing good leaders. *Journal of Business Ethics*, *167*(4), 605–622. DOI: 10.1007/s10551-019-04163-2

Nicolo, G., Zanellato, G., Tiron-Tudor, A., & Tartaglia Polcini, P. (2023). Revealing the corporate contribution to sustainable development goals through integrated reporting: A worldwide perspective. *Social Responsibility Journal*, *19*(5), 829–857. DOI: 10.1108/SRJ-09-2021-0373

Nielsen, K. (2022). Anxiety, trust, leadership, communication and stress. *Nordic Psychology*, *74*(3), 150–151. DOI: 10.1080/19012276.2022.2124079

Nolan, K. (2012). Dispositions in the field: Viewing mathematics teacher education through the lens of Bourdieu's social field theory. *Educational Studies in Mathematics*, *80*(1-2), 201–215. DOI: 10.1007/s10649-011-9355-9

Noor, F. A. M., Said, R., Arsad, S., & Shafie, N. A. (2025). Bridging governance and transparency: The role of board age diversity in shaping integrity disclosure. *International Journal of Research and Innovation in Social Science*, *9*(8), 4682–4690. DOI: 10.47772/IJRISS.2025.908000375

Northouse, P. G. (2010). *Introduction. Leadership theory and practice* (5th ed.). SAGE Publications.

Norton, T. A., Parker, S. L., Zacher, H., & Ashkanasy, N. M. (2015). Employee green behavior: A theoretical framework, multilevel review, and future research agenda. *Organization & Environment*, *28*(1), 103–125. DOI: 10.1177/1086026615575773

Nurunnabi, M. (2017). Transformation from an oil-based economy to a knowledge-based economy in Saudi Arabia: The direction of Saudi vision 2030. *Journal of the Knowledge Economy*, *8*(2), 536–564. DOI: 10.1007/s13132-017-0479-8

O'Donoghue, D., & Van Der Werff, L. (2022). Empowering leadership: Balancing self-determination and accountability for motivation. *Personnel Review*, *51*(4), 1205–1220. DOI: 10.1108/PR-11-2019-0619

O'Grady, M. (2023). Transformative education for sustainable development: A faculty perspective. *Environment, Development and Sustainability*, *27*(9), 20745–20761. DOI: 10.1007/s10668-023-03609-y

OECD. (2024). OECD Principles on Artificial Intelligence (updated). Organisation for Economic Co-operation and Development. https://oecd.ai/en/ai-principles

Oklevik, O., Kwiatkowski, G., Preuss, H., & Kurdyś-Kujawska, A. (2022). Contextual engagement in event visitors' experience and satisfaction. *Scandinavian Journal of Hospitality and Tourism*, *22*(1), 58–79. DOI: 10.1080/15022250.2021.1927829

Oktay, D., & Acaralp, M. C. (2025). Leadership approach for generation alpha and a greener future: The health professionals of tomorrow. In *Generation Alpha and Next Generation Leadership for Greener Futures* (pp. 63–86). IGI Global Scientific Publishing. DOI: 10.4018/979-8-3693-9281-2.ch004

Olu-Owolabi, F. E., Oladoyin, A. M., Odukoya, O., & Osimen, G. U. (2024). Harnessing traditional institutions for sustainable leadership and good governance in Nigeria. *Journal of Infrastructure. Policy and Development*, *8*(11), 6951. DOI: 10.24294/jipd.v8i11.6951

Omisola, J. O., Bihani, D., Daraojimba, A. I., Osho, G. O., Ubamadu, B. C., & Etukudoh, E. A. (2023). Blockchain in supply chain transparency: A conceptual framework for real-time data tracking and reporting using blockchain and AI. *International Journal of Multidisciplinary Research and Growth Evaluation*, *4*(1), 1238–1253. Advance online publication. DOI: 10.54660/.IJMRGE.2023.4.1.1238-1253

Ones, D. S., & Dilchert, S. (2012). Environmental sustainability at work: A call to action. *Industrial and Organizational Psychology: Perspectives on Science and Practice*, *5*(4), 444–466. DOI: 10.1111/j.1754-9434.2012.01478.x

Ong, B., & Toh, D. J. (2023). Digital Dominance and Social Media Platforms: Are Competition Authorities Up to the Task? *IIC; International Review of Industrial Property and Copyright Law*, *54*(4), 527–572. DOI: 10.1007/s40319-023-01302-1 PMID: 37073397

Onyeaka, H., & Akinsemolu, A. A. (2024). Advancing green education in MENA region: Challenges, opportunities and best practices. *Sustainable Development (Bradford)*, *33*(1), 1–15. DOI: 10.1002/sd.3182

Ooi, R. (2025). The Regenerative Wealth Economy: From ESG to EEoM in Transforming Capitalism via Multi-Capital Reinvestment & AI Governance. https://doi: DOI: 10.20944/preprints202503.1365.v2

P., R., Gupta, A. K., Satyavathi, M., Chaitanya, M. S. K., Parameswari, M., & Kalpana, R. (2025). Sustainable IT governance models using blockchain for transparent decision-making. *Proceedings of the 4th International Conference on Innovative Mechanisms for Industry Applications (ICIMIA 2025)*, 1948–1954. https://doi.org/ DOI: 10.1109/ICIMIA67127.2025.11200918

Pacelli, V., Foglia, M., & Mariano, D. (2025). The link between climate and systemic risk: A bibliometric and systematic literature review. *Research in International Business and Finance*, *79*, 103072. DOI: 10.1016/j.ribaf.2025.103072

Parment, A. (2013). Generation Y vs. Baby Boomers: Shopping behavior, buyer involvement and implications for retailing. *Journal of Retailing and Consumer Services*, *20*(2), 189–199. DOI: 10.1016/j.jretconser.2012.12.001

Parvez, M. O., Kim, W. G., Patwary, A. K., & Zainol, N. A. (2026). Active participation matters: Impacts of green organizational identity on employees' voluntary green behaviors. *International Journal of Hospitality Management*, *133*, 104447. DOI: 10.1016/j.ijhm.2025.104447

Pasban, M., & Nojedeh, S. H. (2016). A Review of the Role of Human Capital in the Organisation.Procedia-social and behaviou*ral sciences,230*, 249-253.

Patil, P. A., Jadhav, R. S., Jagtap, P., Thorat, K. D., & Patil, H. K. (2025). Citizen-led AI audit platform for transparency and accountability in automated decision-making. *International Journal on Advanced Computer Engineering and Communication Technology*, *14*(1), 676–681. DOI: 10.65521/ijacect.v14i1.751

Pavlakova Docekalova, M., & Kocmanova, A. (2018). Comparison of sustainable environmental, social, and corporate governance value added models for investors decision making. *Sustainability (Basel)*, *10*(3), 649. DOI: 10.3390/su10030649

Pedersen, E. R. G., Lüdeke-Freund, F., Henriques, I., & Seitanidi, M. M. (2021). Toward collaborative cross-sector business models for sustainability. *Business & Society*, *60*(5), 1039–1058. DOI: 10.1177/0007650320959027

Perez-Aranda, J., Tolkach, D., & Panchal, J. H. (2024). Reputation and eWOM in accommodation decision-making: Insights from Generation Z users. *Tourism Review*. Advance online publication. DOI: 10.1108/TR-03-2024-0185

Pérez-Durán, I. (2024). Twenty-five years of accountability research in public administration: Authorship, themes, methods, and future trends. *International Review of Administrative Sciences*, *90*(3), 546–562. DOI: 10.1177/00208523231211751

Permatasari, Y., Rosyidah, M. A., Sabaruddin, L., Embuningtiyas, S. S., & Yuniawati, R. A. (2025). Sustainability governance and greenwashing: The role of sustainability committees in ASEAN-5 countries. *Asian Review of Accounting*, •••, 1–26. DOI: 10.1108/ARA-03-2025-0067

Persakis, A., Nikolopoulos, T., Negkakis, I. C., & Pavlopoulos, A. (2025). Greenwashing in marketing: A systematic literature review and bibliometric analysis. *International Review on Public and Nonprofit Marketing*, *22*(4), 957–992. DOI: 10.1007/s12208-025-00452-x

Persson, C., Einarson, D., & Melén, M. (2023). Educating the educators to be a driving force in higher education towards sustainable development. *International Journal of Sustainability in Higher Education*, *24*(1), 197–212. DOI: 10.1108/IJSHE-10-2022-0332

Pfister, J. A., Otley, D., Ahrens, T., Dambrin, C., Darwin, S., Granlund, M., Jack, S. L., Lassila, E. M., Millo, Y., Peda, P., Sherman, Z., & Wilson, D. S. (2024). Performance management in the prosocial market economy: A new paradigm for economic performance and sustainability. *Qualitative Research in Accounting & Management*, *21*(5), 397–443. DOI: 10.1108/QRAM-02-2024-0031

Pham, H. T., Pham, T., Truong Quang, H., & Dang, C. N. (2023). Impact of transformational leadership on green learning and green innovation in construction supply chains. *Engineering, Construction, and Architectural Management, 30*(5), 1883–1901. DOI: 10.1108/ECAM-05-2021-0379

Phan, T. A., & Ninh, T. T. D. (2026). The impact of company field trips and representative image on students' CSR knowledge sharing intentions. *Business and Professional Communication Quarterly, 89*(2), 400–418. DOI: 10.1177/23294906241237721

Piercy, N. F., Cravens, D. W., & Lane, N. (2010). Thinking strategically about pricing decisions. *The Journal of Business Strategy, 31*(5), 38–48. DOI: 10.1108/02756661011076309

Pillay, S. (2021). The influence of electronic word-of-mouth adoption on brand love amongst Generation Z consumers. *Acta Commercii, 21*(1), 1–11. DOI: 10.4102/ac.v21i1.928

Piwowar-Sulej, K., & Iqbal, Q. (2023). Leadership styles and sustainable performance: A systematic literature review. *Journal of Cleaner Production, 382*, 134600. Advance online publication. DOI: 10.1016/j.jclepro.2022.134600

Piwowar-Sulej, K., & Iqbal, Q. (2025). *Sustainability-focused leadership and pro-environmental behavior in SMEs: through the lens of conservation of resources theory.* Central European Management Journal., DOI: 10.1108/CEMJ-11-2024-0375

Pless, N. M., & Maak, T. (2011). Responsible leadership: Pathways to the future. *Journal of Business Ethics, 98*(1), 3–13. DOI: 10.1007/s10551-011-1114-4

Pless, N. M., Maak, T., & Waldman, D. A. (2012). Different approaches toward doing the right thing: Mapping the responsibility orientations of leaders. *The Academy of Management Perspectives, 26*(4), 51–65. DOI: 10.5465/amp.2012.0028

PLN (Persero). (2024). *Climate-related disclosure report 2024.*

Podsakoff, N. P., Podsakoff, P. M., MacKenzie, S. B., Maynes, T. D., & Spoelma, T. M. (2014). Consequences of unit-level organizational citizenship behaviors: A review and recommendations for future research. *Journal of Organizational Behavior, 35*(S1), S87–S119. DOI: 10.1002/job.1911

Podsakoff, P. M., MacKenzie, S. B., Podsakoff, N. P., & Lee, J. Y. (2003). The mismeasure of man (agement) and its implications for leadership research. *The Leadership Quarterly, 14*(6), 615–656. DOI: 10.1016/j.leaqua.2003.08.002

Pokharel, J., & Maharjan, I. (2024). Financial behavior of generation Z and Millennials. *Journal of Emerging Management Studies*, *1*(2), 148–170. DOI: 10.3126/jems.v1i2.71522

Pons, S., Khalilzadeh, J., Weber, M. R., & Smith, R. A. (2024). *Cultivating sustainability savvy: the role of soft skills in shaping sustainable practices*. International Hospitality Review.

Porter, M. E., & Linde, C. V. D. (1995). Toward a new conception of the environment-competitiveness relationship. *The Journal of Economic Perspectives*, *9*(4), 97–118. DOI: 10.1257/jep.9.4.97

Powell, R. B., Stern, M. J., & Fresley, B. T. (2023). Which approaches are associated with better outcomes? Evidence from a national study of environmental education field trip programs for adolescent youth in the United States. *Environmental Education Research*, *29*(3), 331–356. DOI: 10.1080/13504622.2022.2145270

Powell, W. W., & DiMaggio, P. J. (1983). The iron cage revisited: Institutional isomorphism and collective rationality in organizational fields. *American Sociological Review*, *48*(2), 147–160. DOI: 10.2307/2095101

Prabhu, J. (2017). Frugal innovation: Doing more with less for more. *Philosophical Transactions. Series A, Mathematical, Physical, and Engineering Sciences*, *375*(2095), 20160372. DOI: 10.1098/rsta.2016.0372 PMID: 28461436

Prabowo, T., & Kim, S. (2024). Environmental justice and community vulnerability in Southeast Asia: A case study of extractive regions in Indonesia. *Environmental Policy and Governance*, *34*(2), 101–118.

Prashad, A., Rogers, H., & Srivastava, M. (2025, March). Pay-per-use business models as a driver for additive manufacturing adoption: supply chain implications. In *Supply Chain Forum: An International Journal* (pp. 1-22). Taylor & Francis. DOI: 10.1080/16258312.2025.2476383

Pratama, A., Yusoff, H., Yadiati, W., & Jaenudin, E. (2025). Sustainability-related corporate governance and greenwashing practices: Preliminary evidence from Southeast Asian companies. *Meditari Accountancy Research*, *33*(3), 853–885. DOI: 10.1108/MEDAR-07-2024-2579

Qin, X., Luo, Q., Wang, X., & Muskat, B. (2024). Green innovation in events: The role of institutional pressures, future orientation, and past experience. *Journal of Sustainable Tourism*, *32*(4), 753–772. DOI: 10.1080/09669582.2023.2168008

Rachbini, W., Anggraeni, D., Burhan, A. B., Sovianti, R., & Rahmawati, E. (2025). Marketing Communications Across Generations: Insight from Indonesia's Online Marketplaces. *Jurnal The Messenger, 17*(1), 37–53. DOI: 10.26623/themessenger.v17i1.11832

Rahman, R., & Qattan, A. (2021). Vision 2030 and sustainable development: state capacity to revitalize the healthcare system in Saudi Arabia. INQUIRY: The Journal of Health Care Organization, Provision, and Financing, 58.

Rakhmat, A. (2022). Islamic Ecotheology: Understanding the Concept of Khalifah and the Ethical Responsibility of The Environment. *Academic Journal of Islamic Principles and Philosophy, 3*(1), 1–24. DOI: 10.22515/ajipp.v3i1.5104

Ramaswamy, M., Marciniuk, D. D., Csonka, V., Colo, L., & Saso, L. (2021). Reimagining internationalization in higher education through the United Nations Sustainable Development Goals for the betterment of society. *Journal of Studies in International Education, 4*(25), 388–406. DOI: 10.1177/10283153211031046

Rami, A. M., Aziz, F., Zaremohzzabieh, Z., & Ibrahim, A. (2021). Assessing the Challenges of Local Leaders in Rural Community Development: A Qualitative Study in Malaysia. *Pertanika Journal of Social Science & Humanities*, •••, 29.

Ramirez, A. G., Monsalve, J., González-Ruiz, J. D., Almonacid, P., & Peña, A. (2022). Relationship between the cost of capital and environmental, social, and governance scores: Evidence from Latin America. *Sustainability (Basel), 14*(9), 5012. DOI: 10.3390/su14095012

Redman, A., & Wiek, A. (2021). Competencies for advancing transformations towards sustainability. *Frontiers in Education, 6*, 785163. DOI: 10.3389/feduc.2021.785163

Redman, A., Wiek, A., & Barth, M. (2021). Current practice of assessing students' sustainability competencies: A review of tools. *Sustainability Science, 16*(1), 117–135. DOI: 10.1007/s11625-020-00855-1

Rekalde-Rodríguez, I. (2026). The transformative power of PBL in teacher training for sustainability. *Journal of Project-Based Research, 14*(2).

Ren, S., Cooke, F. L., Stahl, G. K., Fan, D., & Timming, A. R. (2023). Advancing the sustainability agenda through strategic human resource management: Insights and suggestions for future research. *Human Resource Management, 62*(3), 251–265. DOI: 10.1002/hrm.22169

Renwick, D. W., Redman, T., & Maguire, S. (2013). Green human resource management: A review and research agenda. *International Journal of Management Reviews, 15*(1), 1–14. DOI: 10.1111/j.1468-2370.2011.00328.x

Ren, X., Liu, X., Li, L., & Lin, F. (2024). How perceived environmental leadership influences employee green behaviour in the construction industry: The mediating role of environmental awareness and the moderating role of perceived pro-environmental climate. *Current Psychology (New Brunswick, N.J.)*, *43*(37), 29419–29432. DOI: 10.1007/s12144-024-06565-6

Reyes, J. P., Malabanan, M., & Monterroyo, C. B. (2026). A Contemporary Evidence Synthesis of Leadership Styles and Their Impact on Organizational Performance: A Systematic Review. *Journal of Interdisciplinary Perspectives*, *4*(1), 38–45. DOI: 10.69569/jip.2025.707

Rickly, J. M. (2022). A review of authenticity research in tourism: Launching the Annals of Tourism Research Curated Collection on authenticity. *Annals of Tourism Research*, *92*, 103349. DOI: 10.1016/j.annals.2021.103349

Rihal, J., Alzubi, A., Aljuhmani, H. Y., & Berberoğlu, A. (2025). From ethics to ecology: How ethical leadership drives environmental performance through green organizational identity and culture. *PLoS One*, *20*(11), e0336608. Advance online publication. DOI: 10.1371/journal.pone.0336608 PMID: 41231815

Robertson, J. L., & Barling, J. (2013). Greening organizations through leaders' influence on employees' pro-environmental behaviors. *Journal of Organizational Behavior*, *34*(2), 176–194. DOI: 10.1002/job.1820

Rockström, J., Falkenmark, M., Karlberg, L., Hoff, H., Rost, S., & Gerten, D. (2009). Future water availability for global food production: The potential of green water for increasing resilience to global change. *Water Resources Research*, *45*(7), 2007WR006767. Advance online publication. DOI: 10.1029/2007WR006767

Rodrigues, P., Sousa, A., & Pinto Borges, A. (2024). Generation Z's attitudes toward virtual tourism: The impact of implicit and explicit attitudes. *EuroMed Journal of Business*, *19*(1), 21–45. DOI: 10.1108/EMJB-05-2023-0135

Rodriguez-Anton, J. M., Rubio-Andrada, L., Celemín-Pedroche, M. S., & Alonso-Almeida, M. D. M. (2019). Analysis of the relations between circular economy and sustainable development goals. *International Journal of Sustainable Development and World Ecology*, *26*(8), 708–720. DOI: 10.1080/13504509.2019.1666754

Rodríguez-Zurita, D., Jaya-Montalvo, M., Moreira-Arboleda, J., Raya-Diez, E., & Carrión-Mero, P. (2024). Sustainable development through service learning and community engagement in higher education: A systematic literature review. International Journal of Sustainability in Higher Education, ahead-of-print. DOI: 10.1108/IJSHE-10-2023-0461

Roldan, C. J. L., Ong, A. K. S., & Tomas, D. Q. (2024). Cancel culture in a developing country: A belief in a just world behavioral analysis among generation Z. *Acta Psychologica*, *248*, 104378. DOI: 10.1016/j.actpsy.2024.104378 PMID: 38941914

Rolfe, P. (2011). Transformational leadership theory: What every leader needs to know. *Nurse Leader*, *9*(2), 54–57. DOI: 10.1016/j.mnl.2011.01.014

Roosen, N. (2022). *Regenerative leadership: Understanding the possibilities of a new leadership paradigm for the corporate sector* [Master's thesis, Swedish University of Agricultural Sciences]. SLU Epsilon Archive for Student Projects. https://stud.epsilon.slu.se/18030/

Rosa, A., Romeo, E., & Durst, S. (2026). Aligning Perceptions and Action: The Role of Organisational Coherence in Driving Employee Engagement for SDG 13 Implementation. *Business Strategy and the Environment*.

Rosario, A. T., & Boechat, A. C. (2025). How Sustainable Leadership Can Leverage Sustainable Development. *Sustainability (Basel)*, *17*(8), 3499. DOI: 10.3390/su17083499

Ross-Smith, A., & Huppatz, K. (2010). Management, women and gender capital. *Gender, Work and Organization*, *17*(5), 547–566. DOI: 10.1111/j.1468-0432.2010.00523.x

Rumbley, L. E. (2020, Winter). Internationalization of higher education and the future of the planet. *International Higher Education*, (100), 32–34. https://ihe.bc.edu/pub/usg1w08q

Ruslan, R., Lu'mu, L. M., Fakhri, M. M., Ahmar, A. S., & Fadhilatunisa, D. (2024). Effectiveness of the flipped project-based learning model based on moodle LMS to improve student communication and problem-solving skills in learning programming. *Education Sciences*, *14*(9), 1021. DOI: 10.3390/educsci14091021

Ruther, N. L. (2002). *Barely there, powerfully present: Years of U.S. policy on international higher education*. Taylor & Francis Group. https://ebookcentral.proquest.com/lib/northgeorgia/detail.action?docID=1602099

Ryan, R. L., Kaplan, R., & Grese, R. E. (2001). Predicting volunteer commitment in environmental stewardship programmes. *Journal of Environmental Planning and Management*, *44*(5), 629–648. DOI: 10.1080/09640560120079948

Saberi, S., Kouhizadeh, M., Sarkis, J., & Shen, L. (2019). Blockchain technology and its relationships to sustainable supply chain management. *International Journal of Production Research*, *57*(7), 2117–2135. DOI: 10.1080/00207543.2018.1533261

Sadiq, M., Leong, M. K., Jiayao, Z., & Siddiqui, K. (2026). The effect of perceived firm diversity, equity,y and inclusion (DEI) on employee wellbeing and green performance: A moderated mediated model. *Corporate Social Responsibility and Environmental Management*, *33*(1), 500–518. DOI: 10.1002/csr.70186

Safari, M., & Parker, L. D. (2024). Understanding multiple accountability logics within corporate governance policy discourse: Resistance, compromise, or selective coupling? *European Accounting Review*, *33*(4), 1467–1496. DOI: 10.1080/09638180.2023.2194028

Sahin, M. (2024). Approaches to learning: From traditional theories to connectivism. *Education & Science 2024-iv*, 7-34.

Sahu, M., Mishra, A., Alahdal, W. M., & Sami, M. (2025). ESG performance and audit committee expertise: Advancing sustainable development goals in leading nations. *International Review of Economics & Finance*, *103*, 104445. DOI: 10.1016/j.iref.2025.104445

Sajid, M., Zakkariya, K. A., Suki, N. M., & Islam, J. U. (2024). When going green goes wrong: The effects of greenwashing on brand avoidance and negative word-of-mouth. *Journal of Retailing and Consumer Services*, *78*, 103773. DOI: 10.1016/j.jretconser.2024.103773

Sajjad, A., Eweje, G., & Raziq, M. M. (2024). Sustainability leadership: An integrative review and conceptual synthesis. *Business Strategy and the Environment*, *33*(4), 2849–2867. DOI: 10.1002/bse.3631

Saleem, F., & Mateou, S. (2025). Servant leadership for sustainability: A serial mediation model. *Administrative Sciences*, *15*(10), 380. DOI: 10.3390/admsci15100380

Saleem, F., Sundarasen, S., & Malik, M. I. (2025). Green leadership and environmental performance in hospitals: A multi-mediator study. *Sustainability (Basel)*, *17*(12), 5376. DOI: 10.3390/su17125376

Saleem, M., & Mateou, A. (2025). Servant leadership for sustainability and proactive environmental behavior: A multilevel perspective. *Sustainability*, *17*(2), 509.

Sanchez-Cartas, J. M., & León, G. (2021). Multisided platforms and markets: A survey of the theoretical literature. *Journal of Economic Surveys*, *35*(2), 452–487. DOI: 10.1111/joes.12409

Sánchez, V. P. D., Zorio-Grima, A., & Merello, P. (2026). Sustainability reporting assurance: Knowledge, lags, levels, and providers in leading global companies. *Journal of Innovation & Knowledge*, *12*, 100899. DOI: 10.1016/j.jik.2025.100899

Sang, K., & Powell, A. (2012). Gender inequality in the construction industry: Lessons from Pierre Bourdieu. *Management, 237*, 247.

Santos de Souza, F. D. O., & Chimenti, P. (2024). Emotions in organizational change: An integrative review. *Journal of Change Management, 24*(2), 137–176. DOI: 10.1080/14697017.2024.2345078

Santoso, A., & Jaya, E. (2022). Policy coherence challenges in Indonesia's land-use governance. *Journal of Environmental Management, 318*, 115–128.

Sari, W. A., Artika, W., Oktari, R. S., Rahmatan, H., & Sasaki, D. (2025). Assessing Education for Sustainable Development (ESD) Competencies and Environmental Empathy in Disaster and Environment Knowledge to Support SDGs 2030. *Jurnal Pendidikan IPA Indonesia, 14*(1).

Sasaki, N. (2025). Addressing scandals and greenwashing in carbon offset markets: A framework for reform. *Global Transitions, 7*, 375–382. DOI: 10.1016/j.glt.2025.06.003

Saxena, A., Ramaswamy, M., Beale, J., Marciniuk, D., & Smith, P. (2021). Striving for the United Nations (UN) sustainable development goals (SDGs): What will it take? *Discover Sustainability, 2*(1), 20. DOI: 10.1007/s43621-021-00029-8

Schmidt, M. J. (2010). Is there a place for emotions within leadership preparation programmes? *Journal of Educational Administration, 48*(5), 626–641. DOI: 10.1108/09578231011067776

Schroeder, K., Wood, C., Galiardi, S., & Koehn, J. (2009). First, do no harm: Ideas for mitigating negative community impacts of short-term study abroad. *The Journal of Geography, 108*(3), 141–147. DOI: 10.1080/00221340903120866

Schroeder, P., Anggraeni, K., & Weber, U. (2019). The relevance of circular economy practices to the sustainable development goals. *Journal of Industrial Ecology, 23*(1), 77–95. DOI: 10.1111/jiec.12732

Searcy, C., & Buslovich, R. (2014). Corporate perspectives on the development and use of sustainability reports. *Journal of Business Ethics, 121*(2), 149–169. DOI: 10.1007/s10551-013-1701-7

Seele, P., Dierksmeier, C., Hofstetter, R., & Schultz, M. D. (2021). Mapping the ethicality of algorithmic pricing: A review of dynamic and personalized pricing. *Journal of Business Ethics, 170*(4), 697–719. DOI: 10.1007/s10551-019-04371-w

Seema, S., Saini, G., Mittal, A., & Dhankar, N. (2025). Green transformational leadership and green management practices-based environmental stewardship: A SEM model with mediation and parallel moderation effects. *Global Knowledge, Memory and Communication*. Advance online publication. DOI: 10.1108/GKMC-07-2024-0445

Selim, M. M., & Alshareef, N. (2025). Trends and opportunities in renewable energy investment in Saudi Arabia: Insights for achieving vision 2030 and enhancing environmental sustainability. *Alexandria Engineering Journal*, *112*, 224–234. DOI: 10.1016/j.aej.2024.10.107

Senge, P. M. (2006). *The fifth discipline: The art and practice of the learning organization*. Broadway Business.

Sergi, A. (2024). 'Familiness' diversification and recognition: Intergenerational changes in mobile 'ndrangheta families in Australia. *Trends in Organized Crime*, •••, 1–22. DOI: 10.1007/s12117-024-09537-1

Seyfi, S., Michael Hall, C., & Strzelecka, M. (2025). Gen Z–pioneers or paradoxes in sustainable tourism? *Journal of Sustainable Tourism*, *33*(6), 987–1015. DOI: 10.1080/09669582.2025.2491702

Shaheen Hosany, A. R., & Serdiuk, K. (2025). Understanding Gen Z Consumers: A Typology of (Un) sustainable Purchases. *Psychology and Marketing*, *42*(11), 2820–2832. Advance online publication. DOI: 10.1002/mar.70013

Shahid, A. U., Nagarajan, C. D., Tufail, H. S., & Imran, M. (2025). Investors' pro-environmental attitude, spirituality and socially responsible investing: A mediating role of greenwashing practices. *Journal of Accounting & Organizational Change*, *21*(4), 711–735. DOI: 10.1108/JAOC-03-2024-0102

Shao, M., Liu, Y., Zhao, G., Sun, H., & Zhao, P. (2025). Mitigating Involutionary Competition Through Corporate ESG Adoption: Evidence from the Consumer Electronics Manufacturing Industry. *Sustainability (Basel)*, *17*(20), 8998. DOI: 10.3390/su17208998

Sherif, K., Mohamed, S. S. A., & Amanulla, R. (2025). Managing the environmental sustainability performance of mega-sport events: A blockchain-based smart control system. *Sport, Business and Management*, *15*(3), 245–263. DOI: 10.1108/SBM-04-2024-0044

Shields, C. M. (2010). Transformative leadership: Working for equity in diverse contexts. *Educational Administration Quarterly*, *46*(4), 558–589. https://DOI.org/10.1177/0013161X10375609. DOI: 10.1177/0013161X10375609

Shields, R. (2019). The sustainability of international higher education: Student mobility and global climate change. *Journal of Cleaner Production, 217*, 594–602. DOI: 10.1016/j.jclepro.2019.01.291

Shriberg, M., & MacDonald, L. (2013). Sustainability leadership programs: Emerging goals, methods & best practices. The Journal of Sustainability Education.

Shriberg, M., & MacDonald, L. (2013). Sustainability leadership programs in higher education: Alumni outcomes and impacts. *Journal of Sustainability Education, 5*, 1–21.

Shrivastava, P. (1995). The role of corporations in achieving ecological sustainability. *Academy of Management Review, 20*(4), 936–960. DOI: 10.2307/258961

Shukla, T. (2025). Leveraging blockchain for ESG disclosure: Enhancing transparency and accountability in sustainability reporting. *International Journal of Engineering Technologies and Management Research, 12*(3). Advance online publication. DOI: 10.29121/ijetmr.v12.i3.2025.1546

Singer-Brodowski, M., Förster, R., Eschenbacher, S., Biberhofer, P., & Getzin, S. (2022). Facing crises of unsustainability: Creating and holding safe enough spaces for transformative learning in higher education for sustainable development. *Frontiers in Education, 7*, 787490. DOI: 10.3389/feduc.2022.787490

Singh, V., & Sangal, S. (2025). *Evolving CSR: Ethical Leadership in Environmental Challenges and Sustainable Consumption*. Enhancing Social Sustainability in Manufacturing Supply Chains, 1-32. IGI Global Scientific Publishing.

Singh, A. S., & Segatto, A. P. (2020). When relational capabilities walk in education for sustainability scenario. *Journal of Cleaner Production, 263*, 1–10. DOI: 10.1016/j.jclepro.2020.121478

Singh, S., Subramani, A. K., David, R., & Jan, N. A. (2024). Workplace ostracism influencing turnover intentions: Moderating roles of perceptions of organizational virtuousness and authentic leadership. *Acta Psychologica, 243*, 104136. DOI: 10.1016/j.actpsy.2024.104136 PMID: 38244371

Sklavos, G., Zournatzidou, G., & Sariannidis, N. (2026). Unveiling ESG controversy risks: A multi-criteria evaluation of whistleblowing performance in European financial institutions. *Risks, 14*(1), 10. DOI: 10.3390/risks14010010

Sneideriene, A., & Legenzova, R. (2025). Greenwashing prevention in environmental, social, and governance (ESG) disclosures: A bibliometric analysis. *Research in International Business and Finance, 74*, 102720. Advance online publication. DOI: 10.1016/j.ribaf.2024.102720

Sneideriene, A., & Legenzova, R. (2026). A framework for mitigating greenwashing in sustainability reporting. *Sustainability (Basel), 18*(1), 524. DOI: 10.3390/su18010524

Snowden, A. W., Hancock, A. S., Buhusi, C. V., & Warren, C. M. (2022). Event-related correlates of evolving trust evaluations. *Social Neuroscience, 17*(2), 154–169. DOI: 10.1080/17470919.2022.2043935 PMID: 35184692

Soler, M. C., Kim, J. H., & Cecil, B. G. (2022). *Mapping internationalization on US campuses: 2022 edition.* American Council on Education. https://www.acenet.edu/Documents/Mapping-Internationalization-2022.pdf

Soni, V., Gnekpe, C., Roux, M., Anand, R., Yaroson, E. V., & Banwet, D. K. (2023). Adaptive distributed leadership and circular economy adoption by emerging SMEs. *Journal of Business Research, 156*, 113488. DOI: 10.1016/j.jbusres.2022.113488

Sotelo-Duarte, M., & Gonzalez-Cavazos, B. (2023). Young consumers influence by older generations: Developing the construct and scale to measure intergenerational brand influence. *Young Consumers, 24*(3), 253–271. DOI: 10.1108/YC-06-2022-1539

Spaniol, M. J., Danilova-Jensen, E., Nielsen, M., Rosdahl, C. G., & Schmidt, C. J. (2024). Defining greenwashing: A concept analysis. *Sustainability (Basel), 16*(20), 9055. DOI: 10.3390/su16209055

Sposato, M., & Carlos Dittmar, E. (2025). Leadership Strategies for Implementing Environmental Management Systems. In Ordóñez de Pablos, P., Lytras, M., Alkhaldi, A., & Malik, S. (Eds.), *Insights Into Digital Business, Human Resource Management, and Competitiveness* (pp. 277–300). IGI Global Scientific Publishing., DOI: 10.4018/979-8-3693-9440-3.ch012

Steffen, W., Richardson, K., Rockström, J., Cornell, S. E., Fetzer, I., Bennett, E. M., ... & Sörlin, S. (2015). Planetary boundaries: Guiding human development on a changing planet. science, 347(6223), 1259855.DOI: DOI: 10.1126/science.1259855

Stewart, J. S., Oliver, E. G., Cravens, K. S., & Oishi, S. (2017). Managing millennials: Embracing generational differences. *Business Horizons, 60*(1), 45–54. DOI: 10.1016/j.bushor.2016.08.011

Stober, H., Gaebel, M., & Morrisroe, A. (2021). *Greening in European higher education institutions.* European University Association. https://www.eua.eu/publications/reports/greening-in-european-higher-education-institutions2.html

Stouten, J., Rousseau, D. M., & De Cremer, D. (2018). Successful organizational change: Integrating the management practice and scholarly literatures. *The Academy of Management Annals, 12*(2), 752–788. DOI: 10.5465/annals.2016.0095

Strömgren, M., Eriksson, A., Bergman, D., & Dellve, L. (2016). Social capital among healthcare professionals: A prospective study of its importance for job satisfaction, work engagement and engagement in clinical improvements. *International Journal of Nursing Studies*, *53*, 116–125. DOI: 10.1016/j.ijnurstu.2015.07.012 PMID: 26315780

Stromquist, N. P. (2007). Internationalization as a response to globalization: Radical shifts in university environments. *Higher Education*, *53*(1), 81–105. https://link.springer.com/article/10.1007/s10734-005-1975-5. DOI: 10.1007/s10734-005-1975-5

Subaşı, E., Karatepe, O. M., Rezapouraghdam, H., & Kim, T. T. (2026). Green transformational leadership and green recovery performance: Green adaptability and green creativity as mediators and emotional intelligence as a moderator. *International Journal of Hospitality Management*, *133*, 104508. DOI: 10.1016/j.ijhm.2025.104508

Suchman, M. C. (1995). Managing legitimacy: Strategic and institutional approaches. *Academy of Management Review*, *20*(3), 571–610. DOI: 10.2307/258788

Sundarasen, S., Zyznarska-Dworczak, B., & Goel, S. (2024). Sustainability reporting and greenwashing: A bibliometrics assessment in G7 and non-G7 nations. *Cogent Business & Management*, *11*(1), 2320812. DOI: 10.1080/23311975.2024.2320812

Sundman, J., Feng, X., Shrestha, A., Johri, A., Varis, O., & Taka, M. (2025). Experiential learning for sustainability: A systematic review and research agenda for engineering education. *European Journal of Engineering Education*, 1–31. DOI: 10.1080/03043797.2025.2532591

Sun, Y., & Xing, J. (2022). The impact of social media information sharing on the green purchase intention among generation Z. *Sustainability (Basel)*, *14*(11), 6879. DOI: 10.3390/su14116879

Suriyankietkaew, S., & Avery, G. C. (2014). Leadership practices influencing stakeholder satisfaction in Thai SMEs. *Asia-Pacific Journal of Business Administration*, *6*(3), 247–261. DOI: 10.1108/APJBA-01-2014-0010

Su, X., Xu, A., Lin, W., Chen, Y., Liu, S., & Xu, W. (2020). Environmental leadership, green innovation practices, environmental knowledge learning, and firm performance. *SAGE Open*, *10*(2), 2158244020922909. DOI: 10.1177/2158244020922909

Szabo, S., & Webster, J. (2021). Perceived greenwashing: The effects of green marketing on environmental and product perceptions. *Journal of Business Ethics*, *171*(4), 719–739. DOI: 10.1007/s10551-020-04461-0

Tadjer, H., Lafifi, Y., Seridi-Bouchelaghem, H., & Gulsecen, S. (2022). Improving soft skills based on students' traces in problem-based learning environments. *Interactive Learning Environments, 33*(4), 1879–1896. DOI: 10.1080/10494820.2020.1753215

Tan, J., & Sim, J. (2023). Making Employee-Driven Innovation Achievable: Paving the Way. In *Making Employee-Driven Innovation Achievable* (pp. 1-10). Routledge.

Tapsir, S. H., Yusof, R., & Hashmi, A. (2026). Transformational leadership, Green HRM and top management commitment: Pathways to sustainability in Malaysian higher education. *Journal of Applied Research in Higher Education, 18*(2), 478–494. DOI: 10.1108/JARHE-09-2024-0493

Taylor, R. M. (2024). Methodological reflections on normative case studies: What they are and why we need better quality criteria to inform their use. *Educational Theory, 74*(3), 301–311. DOI: 10.1111/edth.12555

Thapa, R. (2024). An application of Pierre Bourdieu's key concepts to the domain of climate change education. *Journal of Teacher Education for Sustainability, 26*(2), 105–120. DOI: 10.2478/jtes-2024-0018

Thawani, B., Bhatia, M., Panigrahi, T., & Mall, S. (2025). Research trends on use of technology in sustainability reporting: A keyword-based bibliometric analysis. *International Journal of Disclosure and Governance, 22*(4), 910–940. DOI: 10.1057/s41310-025-00315-1

Theocharis, D., & Tsekouropoulos, G. (2025). Sustainable consumption and branding for Gen Z: How brand dimensions influence consumer behavior and adoption of newly launched technological products. *Sustainability (Basel), 17*(9), 4124. DOI: 10.3390/su17094124

Theocharis, D., Tsekouropoulos, G., & Athanasios, G. (2025). The Digital Generation: Branding and Consumer Behavior in Tech Adoption. *Rural and Regional Development, 3*(3), 10009. DOI: 10.70322/rrd.2025.10009

Theodorakopoulos, L., & Theodoropoulou, A. (2024). Leveraging big data analytics for understanding consumer behavior in digital marketing: A systematic review. *Human Behavior and Emerging Technologies, 2024*(1), 3641502. DOI: 10.1155/2024/3641502

Thomas, J. (2021). Inner Development Goals: Background, method, and the IDG framework. https://drive.google.com/file/d/13fcf9xmYrX9wrsh3PC3aeRDs0rWsWCpA/

Thoradeniya, P., Lee, J., Tan, R., & Ferreira, A. (2015). Sustainability reporting and the theory of planned behaviour. *Accounting, Auditing & Accountability Journal, 28*(7), 1099–1137. DOI: 10.1108/AAAJ-08-2013-1449

Tian, Y., Wen, H., & Guo, K. (2025). Machine learning applications in climate finance: An overview. *Research in International Business and Finance, 79*, 103063. DOI: 10.1016/j.ribaf.2025.103063

Tin, S. T., & Yong, J. (2025). From promise to proof: GreenBrew Malaysia's sustainability challenge. *Semarak International Journal of Entrepreneurship, Economics, and Business Development, 4*(1), 25–48. DOI: 10.37934/sijeebd.4.1.2548

Tkalac Vercic, A., & Vercic, D. (2025). Attitudes toward sustainable development and employer brands: Comparing generations X, Y and Z in two countries. *Corporate Communications, 30*(2), 355–371. DOI: 10.1108/CCIJ-11-2023-0161

Tolegenov, B., Rafif, K., Rafi, M., Herawati, R., Abdullah, A., & Sari, L. K. (2024). Digital Transparency and Consumer Awareness: Digital Platforms for Educating Consumers about Product Origins and Sustainability. *AL-MUZARA'AH, 12*(2), 277–287. DOI: 10.29244/jam.12.2.277-287

Tran, K., Nguyen, T., Tran, Y., Nguyen, A., Luu, K., & Nguyen, Y. (2022). Eco-friendly fashion among generation Z: Mixed-methods study on price value image, customer fulfillment, and pro-environmental behavior. *PLoS One, 17*(8), e0272789. DOI: 10.1371/journal.pone.0272789 PMID: 35972928

Tremblay, C. (Host). (2021, November 03). *Sustainability is not enough, enter regenerativity* (No. 13) [Audio podcast episode]. In CANIE Climate Dialogues. https://canie.org/learn/podcast/13-sustainability-is-not-enough-enter-regenerativity

Tumpa, R. J., & Naeni, L. (2025). Improving decision-making and stakeholder engagement at project governance using digital technology for sustainable infrastructure projects. *Smart and Sustainable Built Environment, 14*(4), 1292–1329. DOI: 10.1108/SASBE-10-2024-0451

Turunen, L. L. M., & Halme, M. (2021). Communicating actionable sustainability information to consumers: The Shades of Green instrument for fashion. *Journal of Cleaner Production, 297*, 126605. DOI: 10.1016/j.jclepro.2021.126605

Twesigye, M., Obedgiu, V., Nkurunziza, G., & Oceng, P. (2026). The mediating role of stewardship behaviour on leadership competencies and sustainability of government projects in developing economies. *The International Journal of Organizational Analysis, 34*(1), 143–164. DOI: 10.1108/IJOA-08-2023-3935

U.S. Environmental Protection Agency. (2021, November 22). *Sustainability Primer*. https://www.epa.gov/sites/default/files/2015-05/documents/sustainability_primer_v9.pdf

U.S. Securities and Exchange Commission (SEC). (2023). *SEC charges DWS Investment Management Americas Inc. with misstatements regarding ESG investments and anti-money laundering violations.* https://www.sec.gov/newsroom/press-releases/2023-194

Udin, U., Chanthes, S., & Dananjoyo, R. (2025). Green transformational leadership and environmental performance: Insights from bibliometric analysis for future research agenda. *International Journal of Sustainable Development and Planning*, *20*(6), 2331. DOI: 10.18280/ijsdp.200606

Udokwu, C., Voicu-Doroban u, R., Ogunyemi, A. A., Norta, A., Sturua, N., & Craß, S. (2025). Leveraging Blockchain for Ethical AI: Mitigating Digital Threats and Strengthening Societal Resilience. *Future Internet*, *17*(7), 309. DOI: 10.3390/fi17070309

Ukamaka, E. (2024). Education for Sustainable Development: Integrating Sustainability into Curricula. *Research Output Journal of Arts and Management*, *3*(3), 53–57.

Umar, S. B., Ahmad, J., Mohd Bukhori, M. A. B., Ali, K. A. M., & Hussain, W. M. H. W. (2025). Transforming Higher-Education-Intitutes: Impact of Change Management on Sustainable Performance Through Transformational Leadership and Knowledge Management. *Sustainability (Basel)*, *17*(6), 2445. DOI: 10.3390/su17062445

UNESCO. (2024). *Education for Sustainable Development for 2030: Framework for implementation*. UNESCO.

UNESCO. (2025). *SDG 4 – Education 2030*. https://www.unesco.org/sdg4education2030/en

Urolov, B., Abdunazarova, Z., Saidov, J., Nasirova, N., Salomov, G., Masharipov, A., & Shayeva, R. (2026). Embedding conservation genetics and invasion biology into undergraduate curricula with competency based framework for biodiversity literacy. *Natural and Engineering Sciences*, *11*(1), 249–259. DOI: 10.28978/nesciences.261018

Valls-Val, K., & Bovea, M. D. (2021). Carbon footprint in Higher Education Institutions: A literature review and prospects for future research. *Clean Technologies and Environmental Policy*, *23*(9), 2523–2542. DOI: 10.1007/s10098-021-02180-2 PMID: 34456663

Van Den Beemt, A., Vázquez-Villegas, P., Gómez Puente, S., O'Riordan, F., Gormley, C., Chiang, F. K., Leng, C., Caratozzolo, P., Zavala, G., & Membrillo-Hernández, J. (2023). Taking the challenge: An exploratory study of the challenge-based learning context in higher education institutions across three different continents. *Education Sciences, 13*(3), 234. DOI: 10.3390/educsci13030234

Vandini, V. A. D., & Ala'aldin, J. (2025). Strategic Initiatives of Multinational Corporations in Environmental Sustainability in Indonesia: Case Study of Allianz and PT Azbil Corporation. *PROIROFONIC, 1*(1), 627–633.

VanWynsberghe, R. (2022). Education for sustainability, transformational learning time and the individual ↔ collective dialectic. *Frontiers in Education, 7*, 838388. DOI: 10.3389/feduc.2022.838388

Vargas-Hernández, J. G. (2022). Implications Between Environmental Management and Green Transformational Leadership in Greening Organizations. In Ortiz-Rodriguez, F., Tiwari, S., Iyer, S., & Medina-Quintero, J. (Eds.), *Futuristic Trends for Sustainable Development and Sustainable Ecosystems* (pp. 165–190). IGI Global Scientific Publishing., DOI: 10.4018/978-1-6684-4225-8.ch011

Varsha, M., Dhanaraj, D. V. T., & Info, A. (2024). Analysis of the influence of accounting information on stock price with the extent of corporate social responsibility disclosures as a moderation. *African Journal of Commercial Studies, 5*(3), 157–163. DOI: 10.59413/ajocs/v5.i.3.2

Vasanicova, P., & Melnyk, K. (2025). Generational Differences in Motivational Drivers and Travel Preferences: An Exploration of International Travel Behavior of Slovak Women. *Tourism and Hospitality, 6*(2), 117. DOI: 10.3390/tourhosp6020117

Vasconcelos, C., & Orion, N. (2021). Earth science education as a key component of education for sustainability. *Sustainability (Basel), 13*(3), 1316. DOI: 10.3390/su13031316

Vasileva, A., Troitskaya, T., Dronova, S., Muravyov, K., Matkarimov, A., & Tatlyev, R. (2025). Environmental education as a key factor in achieving sustainable development goals: Challenges and prospects. *International Journal of Ecosystems and Ecology Science, 15*(3), 301–308. DOI: 10.31407/ijees15.334

Veerasanai, R., & Alias, A. (2022). A review of foreign construction firm presence in the Malaysian construction industry. *International Journal of Construction Management, 22*(10), 1830–1843. DOI: 10.1080/15623599.2020.1742635

Virapongse, A., Brooks, S., Metcalf, E. C., Zedalis, M., Gosz, J., Kliskey, A., & Alessa, L. (2016). A social-ecological systems approach for environmental management. *Journal of Environmental Management, 178*, 83–91. DOI: 10.1016/j.jenvman.2016.02.028 PMID: 27131638

Visser, W., & Courtice, P. (2011). Sustainability Leadership: Linking Theory and Practice, SSRN *Working Paper Series*, 2011. https://www.cisl.cam.ac.uk/resources/sustainability-leadership/sustainability-leadership-linking-theory-and-practice?utm_source=chatgpt.com DOI: 10.2139/ssrn.1947221

Visser, W., & Courtice, P. (2011). *Sustainability leadership: Linking theory and practice. Cambridge Programme for Sustainability Leadership*. University of Cambridge.

Volz, U., Beirne, J., Ambrosio Preudhomme, N., Fenton, A., Mazzacurati, E., Renzhi, N., & Stampe, J. (2020). Climate change and sovereign risk.

Voulvoulis, N., Giakoumis, T., Hunt, C., Kioupi, V., Petrou, N., Souliotis, I., Vaghela, C. J. G. E. C., & binti Wan Rosely, W. (2022). Systems thinking as a paradigm shift for sustainability transformation. *Global Environmental Change, 75*, 102544. DOI: 10.1016/j.gloenvcha.2022.102544

Waldman, D. A., & Siegel, D. (2008). Defining the socially responsible leader. *The Leadership Quarterly, 19*(1), 117–131. DOI: 10.1016/j.leaqua.2007.12.008

Waldman, D. A., Siegel, D. S., & Javidan, M. (2006). Components of CEO transformational leadership and corporate social responsibility. *Journal of Management Studies, 43*(8), 1703–1725. DOI: 10.1111/j.1467-6486.2006.00642.x

Walkiewicz, J., Lay-Kumar, J., & Herzig, C. (2021). The integration of sustainability and externalities into the "corporate DNA": A practice-oriented approach. *Corporate Governance (Bradford), 21*(3), 479–496. DOI: 10.1108/CG-06-2020-0244

Wals, A. E. (2010). Mirroring, Gestaltswitching and transformative social learning: Stepping stones for developing sustainability competence. *International Journal of Sustainability in Higher Education, 11*(4), 380–390. DOI: 10.1108/14676371011077595

Wals, A. E. (2011). Learning our way to sustainability. *Journal of Education for Sustainable Development, 5*(2), 177–186. DOI: 10.1177/097340821100500208

Walters, P. (2021). Are generation Z ethical consumers? In *Generation Z marketing and management in tourism and hospitality: The future of the industry* (pp. 303–325). Springer International Publishing., DOI: 10.1007/978-3-030-70695-1_12

Walumbwa, F. O., & Schaubroeck, J. (2009). Leader personality traits and employee voice behavior: Mediating roles of ethical leadership and work group psychological safety. *The Journal of Applied Psychology, 94*(5), 1275–1286. DOI: 10.1037/a0015848 PMID: 19702370

Wamsler, C., Osberg, G., Janss, J., & Stephan, L. (2024). Revolutionising sustainability leadership and education: Addressing the human dimension to support flourishing, culture and system transformation. *Climatic Change, 177*(1), 4. DOI: 10.1007/s10584-023-03636-8

Wang, M., Guo, T., Ni, Y., Shang, S., & Tang, Z. (2019). The effect of spiritual leadership on employee effectiveness: An intrinsic motivation perspective. *Frontiers in Psychology, 9*(JAN), 2627. Advance online publication. DOI: 10.3389/fpsyg.2018.02627 PMID: 30662419

Wang, W., Sun, Z., Zhu, W., Ma, L., Dong, Y., Sun, X., & Wu, F. (2023). How does multi-agent govern corporate greenwashing? A stakeholder engagement perspective from "common" to "collaborative" governance. *Corporate Social Responsibility and Environmental Management, 30*(1), 291–307. DOI: 10.1002/csr.2355

Wang, Z., & Tian, X. (2025). From Greenwashing to Sustainability: The Mediating Effect of Green Innovation in the Agribusiness Sector on Financial Performance. *Agriculture, 15*(12), 1316. DOI: 10.3390/agriculture15121316

Warhurst, A. (2002). Sustainability indicators and sustainability performance management. *Mining, Minerals and Sustainable Development [MMSD] project report, 43*, 129.

Washington, H., & Maloney, M. (2020). The need for ecological ethics in a new ecological economics. In *Ecological Economics* (Vol. 169). Elsevier B.V., DOI: 10.1016/j.ecolecon.2019.106478

Waters, A. (2022). Activating Cosmic Capital: Evaluating the Role of Social Capital and Sense of Place in Environmental Stewardship and Adaptive Governance.

Weber, J., & Urick, M. J. (2023). Searching for a stronger generational understanding than just age: A multi-country analysis of millennials' personal values and moral reasoning. *Journal of Intergenerational Relationships, 21*(1), 110–135. DOI: 10.1080/15350770.2021.1951920

Wee, H., Kamarulzaman, N. D., & Anas, M. S. (2022). MICE industry survival: A systematic literature review. *International Journal of Academic Research in Business & Social Sciences, 12*(11). Advance online publication. DOI: 10.6007/IJARBSS/v12-i11/15622

Wee, H., Mustapha, N. A., & Anas, M. S. (2021). Characteristic of green event practices in mice tourism: A systematic literature review. *International Journal of Academic Research in Business & Social Sciences, 11*(16), 271–291. DOI: 10.6007/IJARBSS/v11-i16/11234

Wegner, R. da S. Estivalete, V. de F. B., Matheis, T. K., & Vieira, K. M. (2025). The influence of transformational leadership on environmental sustainability and organizational citizenship behaviors (Version 1.0) [Data set]. Harvard Dataverse. https://doi.org/DOI: 10.7910/DVN/BTSE7R

Wernerfelt, B. (1984). A resource-based view of the firm. *Strategic Management Journal, 5*(2), 171–180. DOI: 10.1002/smj.4250050207

West, C. (2014). Puzzle masters. *International Educator,* (January-February), 26-31. https://www.nafsa.org/sites/default/files/ektron/files/underscore/ie_janfeb14_puzzlemasters.pdf

Wijaya, A., & Ernst, N. (2023). Corporate climate transition strategies in emerging economies: Lessons from Indonesia. *Asia Pacific Business Review, 29*(5), 765–784.

Williams, P., Escalas, J. E., & Morningstar, A. (2022). Conceptualizing brand purpose and considering its implications for consumer eudaimonic well-being. *Journal of Consumer Psychology, 32*(4), 699–723. DOI: 10.1002/jcpy.1324

Woodfield, S. (2013). Key trends and emerging issues in international student mobility (ISM). In Maringe, F., & Foskett, N. (Eds.), *Globalization and internationalization in higher education: Theoretical, strategic, and management perspectives.* Bloomsbury Publishing.

Xamrakulova, X., Mirzaeva, Z., Shabbazova, D., Sabirov, A., Rasulova, Z., Ismoilov, B., & Mukhriddin, A. (2026). Project based marine biodiversity education to improve ocean literacy and conservation attitudes among secondary students. *Natural and Engineering Sciences, 11*(1), 211–220. DOI: 10.28978/nesciences.261015

Xie, X., Huo, J., & Zou, H. (2019). Green process innovation, green product innovation, and corporate financial performance: A content analysis method. *Journal of Business Research, 101*, 697–706. DOI: 10.1016/j.jbusres.2019.01.010

Xu, M., Tse, Y. K., Geng, R., Liu, Z., & Potter, A. (2025). Greenwashing and market value of firms: An empirical study. *International Journal of Production Economics, 284*, 109606. DOI: 10.1016/j.ijpe.2025.109606

Yang, Z., Nguyen, T. T. H., Nguyen, H. N., Nguyen, T. T. N., & Cao, T. T. (2020). Greenwashing behaviours: Causes, taxonomy and consequences based on a systematic literature review. In *Journal of Business Economics and Management* (Vol. 21, Number 5, pp. 1486–1507). VGTU. DOI: 10.3846/jbem.2020.13225

Yang, C., Ye, C., Ding, Y., Mi, J., Liu, Y., & Zhou, L. (2026). Life Cycle Participation in Urban Regeneration: A Policy Design–Implementation–Evaluation Assessment of Guangzhou. *Land (Basel)*, *15*(3), 402. DOI: 10.3390/land15030402

Ye, M., & Guo, R. (2025). Regulatory overload or incentive deficit? Forty years of structural imbalances in China's marine environmental policy instruments. *Frontiers in Marine Science*, *12*, 1674863. DOI: 10.3389/fmars.2025.1674863

Yong, C. K., Alfan, E., & Mustapha, M. Z. (2024). Sustainability Committee and Environmental, Social and Governance (ESG). *International Journal of Business and Society*, *25*(2), 484–508. DOI: 10.33736/ijbs.7603.2024

Yukl, G., Gordon, A., & Taber, T. (2022). A hierarchical taxonomy of leadership behaviour: Integrating a half century of behaviour research. *Journal of Leadership & Organizational Studies*, *9*(1), 15–32. DOI: 10.1177/107179190200900102

Yukl, G., Mahsud, R., Hassan, S., & Prussia, G. E. (2013). An improved measure of ethical leadership. *Journal of Leadership & Organizational Studies*, *20*(1), 38–48. DOI: 10.1177/1548051811429352

Yusuf, M., Saliha, R., & Ikbal, M. (2026). Symbolic Participation in Village Budgeting: A Legal Analysis of Community Involvement in Tangkulowi Village. *JIHK*, *7*(2), 1480–1492. DOI: 10.46924/jihk.v7i2.413

Zafari, Z., Golzary, A., Rouhi, K., & Mansourihanis, O. (2025). From conventional approaches to circular systems: Evolution of waste management in mega-sporting events. *Journal of the Air & Waste Management Association*, *75*(5), 368–386. DOI: 10.1080/10962247.2025.2462005 PMID: 39907531

Zahari, R., Ariffin, M. H., & Othman, N. (2018, February). Influential aspects of the leader's Bourdieu capital on subordinates' creativity in Malaysian landscape architecture. In IOP *Conference Series: Earth and Environmental Science*(Vol. 117, No. 1, p. 012008). IOP Publishing.

Zahari, R., Ariffin, M. H., & Othman, N. (2019a). Leaders' Bourdieu Capitals and Subordinates' Creative Behaviours: Empirical Study In Malaysian Landscape Architecture Firms. [MRCJ]. *Malaysian Construction Research Journal*, *8*(3), 133–148.

Zahari, R., Ariffin, M. H., & Othman, N. (2019b). Measuring Bourdieu Capitals of Landscape Architects. In *Malaysian Landscape Architecture Firms*. EDP Sciences.

Zanella, M., Ashraf, N., & Mahmood, Z. (2026). Carbon accounting for sustainability management: Case study of MICE events. *Journal of Accounting & Organizational Change*, *22*(1), 89–103. DOI: 10.1108/JAOC-08-2024-0259

Zehir, C., Müceldili, B., Altindağ, E., Şehitoğlu, Y., & Zehir, S. (2014). Charismatic leadership and organizational citizenship behavior: The mediating role of ethical climate. *Social Behavior and Personality*, *42*(8), 1365–1375. DOI: 10.2224/sbp.2014.42.8.1365

Zervoudi, E. K., Moschos, N., & Christopoulos, A. G. (2025). From the Corporate Social Responsibility (CSR) and the Environmental, Social and Governance (ESG) Criteria to the Greenwashing Phenomenon: A Comprehensive Literature Review About the Causes, Consequences and Solutions of the Phenomenon with Specific Case Studies. In *Sustainability (Switzerland)* (Vol. 17, Number 5). Multidisciplinary Digital Publishing Institute (MDPI). DOI: 10.3390/su17052222

Zhang, X., Jung, W., & Asari, M. (2025). Systematic review of environmental education teaching practices in schools: Trends and gaps (2015–2024). *Sustainability (Basel)*, *17*(19), 8561. DOI: 10.3390/su17198561

Zhang, X., Singh, S., Li, J., & Shao, X. (2024). Exploring the effects of value co-creation strategies in event services on attendees' citizenship behaviors: The roles of customer empowerment and psychological ownership. *Journal of Retailing and Consumer Services*, *76*, 103619. DOI: 10.1016/j.jretconser.2023.103619

Zhao, H., Xu, Y., & Wang, X. (2024). Greenwashing in workplace: The influencing mechanism of psychological contract fulfillment and the moderating role of ethical work climate. *Journal of Hospitality & Tourism Research (Washington, D.C.)*, *48*(3), 420–433. DOI: 10.1177/10963480221109551

Zhu, Q., Sarkis, J., & Lai, K. H. (2013). Institutional-based antecedents and performance outcomes of internal and external green supply chain management practices. *Journal of Purchasing and Supply Management*, *19*(2), 106–117. DOI: 10.1016/j.pursup.2012.12.001

Zimand-Sheiner, D., & Lissitsa, S. (2024). Generation Z-factors predicting decline in purchase intentions after receiving negative environmental information: Fast fashion brand SHEIN as a case study. *Journal of Retailing and Consumer Services*, *81*, 103999. DOI: 10.1016/j.jretconser.2024.103999

Zioło, M., Bąk, I., & Spoz, A. (2024). Literature review of greenwashing research: State of the art. *Corporate Social Responsibility and Environmental Management*, *31*(6), 5343–5356. DOI: 10.1002/csr.2842

Zouari-Hadiji, R., & Chouaibi, Y. (2021). Corporate ethical behavior and the cost of equity capital: Evidence from the world's most ethical companies. *Journal of Financial Reporting and Accounting*, *19*(5), 939–964. DOI: 10.1108/JFRA-08-2020-0223

Zurba, M., Dhyani, S., Mwaura, G., Sivadas, D., Elegbede, I., & Williamson, D. F. (2023). Pathway to mainstream youth engagement and intergenerational partnership in nature conservation. *Frontiers in Ecology and the Environment*, *21*(4), 175–181. DOI: 10.1002/fee.2612

About the Contributors

Qaisar Iqbal is currently serving as Assistant Professor of Management at Al Yamamah University, Riyadh, Saudi Arabia, bringing with him extensive international academic experience from esteemed institutions such as King Fahd University of Petroleum & Minerals (KFUPM), Saudi Arabia; Sichuan University of Science & Engineering (SUSE), China; and Universiti Sains Malaysia (USM). Renowned for the depth and impact of his scholarship, Dr. Iqbal was ranked among Stanford University's World's Top 2% Scientists (2024) for his notable contributions to Business and Economics. His prolific research portfolio encompasses publications in leading high-impact journals, including Human Resource Management (FT-50), Technological Forecasting and Social Change, German Journal of Human Resource Management, International Journal of Manpower, Journal of Cleaner Production, and Business Strategy and the Environment, and. His work frequently investigates the nexus of sustainable leadership, green HRM, frugal innovation, and organizational behavior, offering valuable theoretical and practical insights to both academia and industry. His academic excellence has been recognized through numerous accolades, including the Top Impact Researcher Award from KFUPM, the USM's Sanggar Sanjung Award, multiple Wiley Top Cited Article Awards, and the Emerald Literati Outstanding Paper Award. Beyond research, Dr. Iqbal is a sought-after speaker and has been invited to deliver keynote lectures, workshops, and training sessions on advanced research methodologies and contemporary management issues across Asia.

***

Sharmila A is an Assistant Professor with a background in dentistry, hospital administration, and academic research. She earned her Bachelor of Dental Surgery (BDS) from Saveetha University and subsequently completed a Fellowship in Medical Cosmetology (FMC) from ILAMED. She also holds a Postgraduate Diploma in Medico Legal Systems from Symbiosis University and an MBA in

Hospital Management from Saveetha School of Management. Currently, she is pursuing her PhD in Hospital Management, further deepening her expertise in healthcare systems and operations. Her career spans over a decade with 4 years of clinical experience, 3 years in hospital administration, 1 year in human resources, and 3 years in academia. Dr. Sharmila brings a well-rounded perspective to healthcare, combining medical knowledge, legal understanding, and managerial skills to drive quality care and organizational efficiency.

Holly Alderman is currently serving as Associate Director of International Business Programs at the University of Georgia, USA, where she develops international education strategy, curriculum, programs and partnerships for the Terry College of Business. Dr. Alderman's accomplishments and service have been recognized by numerous higher education awards, and she has chaired and presented sessions and panels to international educators, administrators, and others to help advance the field.

Emmanuel Amanquandoh is a research assistant at the department of management studies, Pentecost University.

Bismark Amponsah is a research assistant at the department of management studies, Pentecost University.

Andika Andika is a research fellow at Universitas Brawijaya (UB), Indonesia, whose scholarly expertise lies at the intersection of Human Resource Management, Entrepreneurship, and Organizational Sustainability. His research focuses on leadership, talent management, organizational behavior, entrepreneurial development, and the strategic role of human capital in fostering sustainable organizational performance. He has contributed to academic scholarship through publications in Scopus-indexed and other international journals, as well as the authorship of higher education textbooks, including Organizational Behavior.

G. N. P. V. Babu is an Associate Professor at Christ University, Bengaluru, with more than 20 years of academic experience and a strong research orientation in Marketing and Business Analytics. He holds an MBA in Marketing Management and IT Management, and a Ph.D. in Management from JNTU Hyderabad. His research interests include technology in marketing, customer experience, branding, continuous usage, and university branding. Dr. Babu has published extensively in UGC–Care, Scopus-indexed, and ABDC journals, and has authored a book on digital marketing. He has presented research papers at prestigious national and international conferences, contributing to scholarly debates on consumer behaviour, analytics, and digital transformation in marketing. He has also completed certifications in

Business Analytics from IIM Kozhikode and other institutions, further strengthening his research expertise. Through his academic contributions, Dr. Babu continues to advance knowledge at the intersection of marketing, analytics, and technology.

Mohd Ruzaini Bin Che Zahari is a senior lecturer from Landscape Architecture Department, Faculty of Built Environment and Surveying, Universiti Teknologi Malaysia. His research focused on Bourdieu's theory integration into creativity, leadership, technology adoption in landscape architecture, and cultural and heritage preservation. In his spare time, he enjoys sketching, playing guitar, and thinking to solve world's problems.

Amir Dar is the Faculty in the Department of Mathematics at Lovely Professional University Punjab. The area of interest: Financial mathematics, financial economics, statistics and business mathematics.

Ibrahim Eriqat holds a PhD in Accounting from Universiti Sains Malaysia (USM). He is currently a faculty member at Smart University College for Modern Education in Hebron, Palestine. His academic and research interests focus on sustainability reporting, corporate governance, corporate social responsibility (CSR), and environmental, social, and governance (ESG) practices. His work also examines financial inclusion, digital finance, and FinTech, with particular attention to their implications for financial sector development and organisational performance.

Gigi G.S., Associate Professor at Saveetha School of Management, SIMATS, Chennai, brings 16 years of academic experience with a strong focus on research in human resource management, particularly employee engagement in the IT sector. She is currently guiding three Ph.D. research scholars and serves as a Research Advisory Committee (RAC) member for ten others across various universities. Her scholarly work includes numerous presentations and publications at national and international conferences and in reputed journals. Her research is driven by a commitment to addressing real-world HR challenges, blending academic rigor with practical insight to advance the field meaningfully.

Krishnavehni Gopal is currently a Lecturer in Kolej Komuniti Sungai Petani, Malaysia.

Deepak Gupta is a distinguished academician with over 20 years of teaching experience and expertise in Computer Science and Engineering. He holds a Ph.D. in Computer Engineering and has an extensive research portfolio, including many international journal papers, multiple conference proceedings, and several books. His research interests span machine learning, artificial intelligence, and data mining.

Currently, Dr. Gupta serves as an Associate Professor and Head of Department of CSE (Emerging Branches) at ITM Gwalior, MP, India. He has also contributed significantly to the field through patents and copyrights and actively engages as a reviewer, conference chair and editor.

Raja Shree K. C is a Biomedical Engineer with a Master's in Business Administration from Saveetha University, specializing in Hospital Management and Marketing, and holds a Ph.D. in Healthcare Management. An energetic and enthusiastic professor, she excels in academics and as a multitasker, organizing seminars, workshops, cultural programs, and other extra-curricular activities. She serves as the Coordinator of NSS Unit of Saveetha School of Management, leading them to achieve multiple accolades including the Best Club award. A prominent researcher with publications in Web of Science, Scopus, and UGC-listed journals, she actively enhances her expertise through conferences, workshops, MOOCs, and certifications, including Quality Coordination in Hospitals. Awarded the "Young Academician Award" and "Young Teacher Award," she is passionate about innovation, student skill development, and inspiring young minds through exemplary teaching and mentorship.

Hong Jin Liew holds a Masters in Finance from Universiti Sains Malaysia, Penang, under the School of Management and is also a Lecturer in HELP University.

Lovleen Marwaha, Professor Department of Zoology, School of Bioengineering and Biosciences, Lovely Professional University, Phagwara, Punjab-144411, India Email: lovleenmarwaha@gmail.com ORCID ID-0000-0003-4250-5317

Masyhuri Masyhuri is a faculty member at the Institute of Technology and Business Widya Gama Lumajang, specializing in human resource management, organizational behavior, and sustainable leadership. His research interests focus on green transformational leadership, employee environmental behavior, and sustainability-oriented organizational practices. He has authored and co-authored academic works related to leadership, HRM, and environmental governance, and has been actively involved in developing learning modules, community empowerment programs, and organizational capacity-building initiatives in Indonesia. Beyond academic research, he collaborates with local institutions, cooperatives, and community groups to promote sustainable practices, green entrepreneurship, and environmentally responsible organizational strategies. His scholarly and practical contributions reflect a commitment to advancing leadership models that support climate resilience, environmental stewardship, and sustainable development.

Nitish Minz is a multidisciplinary researcher and academic editor currently pursuing his MBA at Vellore Institute of Technology, Vellore (AACSB & NAAC A++ accredited). He holds a background in commerce and has authored several research papers and book chapters published in Scopus and Web of Science-indexed journals. His areas of expertise include human resource management, artificial intelligence, sustainability, gender equity, digital transformation, and cybersecurity. His current research focuses on AI-driven inclusive hiring, digitalization in higher education HRM, and entrepreneurship for women's economic resilience in the Global South. Nitish serves as the Managing Editor of the International Journal of Convergent Research (IJCR), overseeing editorial operations and indexing initiatives. Nitish's scholarly contributions have earned him accolades such as the Chancellor's Excellence Award and Best Paper Presenter recognition at multiple conferences. Deeply committed to bridging research, policy, and practice, he strives to contribute meaningful insights to academia and society through innovation-driven, inclusive, and sustainability-focused scholarship.

Praveen Kumar P. holds an MCom, MBA, and PhD in Finance from Pondicherry University. He secured the University 1st Rank in his Under Graduation. His teaching interests include Financial Accounting, Financial Management, Mergers & Acquisitions, Security Analysis & Portfolio Management, and other finance-related courses. His research interests extend to interdisciplinary areas.

Thanuja Rathakrishnan is a Senior Lecturer and the Program Coordinator for the Bachelor of Business Franchise Program (IIMS, Nepal) at Taylor's University, Malaysia. She holds a Ph.D. in Business Economics from Universiti Putra Malaysia (UPM) and an MBA from Putra Business School. Thanuja is actively involved in Entrepreneurship modules, particularly within the Team Entrepreneurship program. Alongside her academic work, she engages in consultancy projects and research grants from both government and non-government organizations. Her primary research interests include entrepreneurship, entrepreneurship studies, SMEs, and family businesses.

J-Ho Siew Ching, currently Lecturer in Politeknik Tuanku Sultanah Bahiyah, Malaysia.

KM Richa Singh, Research Scholar, Department of Zoology, School of Bioengineering and Biosciences, Lovely Professional University, Phagwara-144411, India

Xueying Yan is a PhD candidate at the School of Management, Universiti Sains Malaysia. Her research focuses on change leadership and digital leadership, with interests in innovation and organizational transformation. Prior to her doctoral studies, she worked in management consulting for two years and participated in more than ten corporate consulting projects, including work supporting a major power utility in China in the green energy domain. This professional experience informs her academic interest in how leadership enables sustainability oriented change under organizational constraints. Methodologically, her research draws on both quantitative and qualitative approaches, including statistical analysis and interviews, to examine leadership and change in practice.

Index

A

B

C

D

E

F

G

H

R

S

T

V

W

www.ingramcontent.com/pod-product-compliance
Lightning Source LLC
LaVergne TN
LVHW081317300926
849161LV00008B/1794